AF650934

Soil
A Vulnerable Natural Resource

NIPA® GENX ELECTRONIC RESOURCES & SOLUTIONS P. LTD.
New Delhi-110 034

About Author

Dr. Venkatesh Bharadwaj is retired as Professor, Soil Science and Head, Agrometeorology departments from the G.B. Pant University of Agriculture & Technology, Pant Nagar, 263145, Uttarakhand after serving for more than 38 years at various capacities. He taught almost all UG and PG courses and guided M.Sc. (Ag.) and Ph.D. students of Soil Science and Agrometeorology. Numerous books, chapters in books, research papers, technical and popular articles and radio talks are in his credit besides conducting refresher courses and completing national and international research projects on climate change, greenhouse gas emission and mitigation, heavy metals, long term use of agrochemicals etc. Dr. Bharadwaj is also active as member, expert and advisor in various commissions, boards and universities and presently enjoying his hobbies of Technical writing, editing, gardening and travelling.

Soil
A Vulnerable Natural Resource

Venkatesh Bharadwaj
Former Professor
Soil Science Department and
Head, Agrometeorology Department
G.B. Pant University of Agriculture and Technology
Pant Nagar, 263145, U.S. Nagar (Uttarakhand), India

NIPA® GENX ELECTRONIC RESOURCES & SOLUTIONS P. LTD.
New Delhi-110 034

NIPA® GENX ELECTRONIC
RESOURCES & SOLUTIONS P. LTD.

101,103, Vikas Surya Plaza, CU Block
L.S.C.Market, Pitam Pura, New Delhi-110 034
Ph : +91 11 4386 0225, 9717133558, 9540816132
E-mail: newindiapublishingagency@gmail.com
Website: www.niparesources.com

© 2025, Author

Print ISBN: 978-81-19235-76-6

ebook ISBN: 978-81-19235-78-0

All rights reserved. No part of this publication may be reproduced, stored in a retrieval system or transmitted in any form or by any means, including electronic, mechanical, photocopying recording or otherwise without the prior written permission of the publisher or the copyright holder.

This book contains information obtained from authentic and highly reliable sources. Reasonable efforts have been made to publish reliable data and information, but the author/s, editor/s and publisher cannot assume responsibility for the validity, accuracy or completeness of all materials or information published herein or the consequences of their use. The work is published with the understanding that the publisher and author/s are not attempting to render any professional services. The author/s, editor/s and publisher have attempted to trace and acknowledge the copyright holders of all material reproduced in this publication and apologize to copyright holders if permission and/or acknowledgements to publish in this form have not been taken. If any copyrighted material has not been acknowledged, please write to us and let us know so that we may rectify the error, in subsequent reprints.

Trademark Notice: NIPA®, the NIPA® logos and their presentations (the way they are written/presented) in this book are the trademarks of the publisher and hence may not be used without written permission, if copied or used without authorization, the infringer will be prosecuted as per law.

NIPA® also publishes books in a variety of electronic formats. Some content that appears in print may not be available in electronic books, and vice versa

Composed and Designed by NIPA®.

Preface

The soil is the soul of infinite life and Sadguru says that our body is essentially soil and water. The quality of our soil and water determines the quality of our food, our body, and our life. These two sayings show the importance of soils in our life, nature and environment.

This book, **"Soil: A Vulnerable Natural Resource"** is the sequel of my more than 38 years of teaching, research and extension of knowledge of soil science. The course content of this book is based on the syllabi of courses, Fundamentals of soil science of B.Sc. (Agriculture) and B.Sc. (Horticulture) and course, Geology & Soil Science of B.Sc. (Forestry) according to the Dean's Committee of ICAR, New Delhi. This book is also useful for the course, Principles of soil science of B.Tech. (Agriculture Engineering) besides PG students, NET, JRF, SRF, ARS etc. aspirants' and extension personnel.

Soils are used in buildings, roads, brick kilns, naturopathy etc. Soils are a basic requirement of agriculture and provide nutrients, water, air, heat and mechanical support to arable plants to produce the food, fodder, fibre, fuel wood and other plant products for humans and animals. Soils also naturally treat depression by stimulating serotonin production, which makes you relaxed and happier. The earthy fragrance, Petrichor and Geosmin, coming out from soils just after the first summer rain fascinates us. Now Soil Scent, Mitti attar, is also available in the market. Even use of Multani Matti (Fuller's Earth) as soap, shampoo, and face pack as well as mud bath are common in rural folk to high end spa.

The subject matter of this book is divided into five chapters. First chapter is Introduction. This chapter gives a general idea about soil and soil as a natural body of earth and soil as a medium for plant growth. Second chapter is Soils in nature and environment and deals with Genesis, survey, morphology, classification, geography etc. of soils. Third chapter describes Physical, Chemical, Biological and biochemical, Geological and Mineralogical properties of soil, Soil fertility and integrated plant nutrition and various natural cycles of molecules and elements in soils.

The terms vulnerable and vulnerability are now bites of every one and used for everything. Vulnerable means to the possibility of attack or harm by any means. Human beings are vulnerable to Covid-19, computers are vulnerable to viruses, softwares are vulnerable to hackers and attackers to control systems, and the environment is vulnerable to pollution. Vulnerabilities refer to the lack of ability to resist the effects of the aggressive environment.

Soils are vulnerable to numerous threats like water and wind erosion, chemical deterioration, loss of nutrients and organic matter, salinity, alkalinity, acidity, pollution, contamination, change in biological activities, compaction, sealing, crusting, water-logging, aridity, lowering of water table, overgrazing, deforestation, landslide etc. However, some of the most pertinent vulnerabilities for soils are described in Chapter four.

Prevention, control, reduction, conservation, and management of soils from vulnerabilities are the matter of increasing awareness and concern for food security of an ever-increasing population which is depending upon this ever shrinking and non-renewable natural resource. All these can be achieved first by identifying the vulnerability and later by adopting various mechanical, physical, chemical and biological methods as and when required to remove that vulnerability. In this regard remote sensing and geographical information systems in relation to soil is essential as these provide spatial and temporal information of soil surface reflectance. Restoration of soil from the vulnerabilities and management of water eroded and wind eroded soils, Soils with loss of nutrients and organic matter, Acidic, sodic, and salt affected soils, Aridic soils, Polluted and chemically contaminated soils, Water logged soils, Soils with compaction, sealing, crusting, Soils of abandoned mines and landslide and Watershed management are dealt in chapter five.

I acknowledge to all the authors, scientists, scholars, artists, publishers etc. whose works I used in this book. I am highly thankful to them.

I am highly indebted to my most honourable ancestors including my Late parents Shri Mathura Prasad Sharma 'Mathuresh ji' and Smt. Shanti Devi Sharma ji and my father-in-law (Late) Pt. Narendra Mohan Sharma ji for their unlimited showering of blessings on me to complete this task. Therefore, I am dedicating this book in the memory of them.

I am thankful to the NIPA, GENX Electronic Resources & Solutions P Ltd., New Delhi for very nicely preparing and presenting the book. I am also grateful to my brothers, Dr. Ramanuj, Dr. Yatindra and Dr. Gopal and sons-in-law, Er.

Rahul and Er. Ravi for their best wishes during the publication of this book.

Finally, I am highly obliged to my wife Smt. Neelam, my daughters Smt. Sumedha and Smt. Akanksha and sister-in-law Smt. Dr. Kumkum for their untiring help from writing to publication of this creation.

Lastly, I am deeply indebted to my all the time Guru and Ph.D. Advisor (Late) Dr. P. K. Omanwar, Former ADG, ICAR, New Delhi for his helps in my all academic presentations.

Venkatesh Bharadwaj

Contents

Preface .... *ix*

1. Introduction **1**

Soil as a Natural Body 1

Earth 2

Land 3

Landscape 3

Soil 4

Components and phases of soil 5

Mineral Matter 6

Organic Matter 6

Soil Water 7

Soil Air 7

Soil Organism 8

Phases of Soil 8

Soil as a Natural Medium for Plant Growth 8

Soil – plant Relationship 10

Definition of Soil and Branches of Soil Science 13

Definition of Soils 13

Branches of Soil Science 14

Soil vulnerability 15

2. Soil in Nature and Environment **17**

Genesis of Soil 17

Organic Parent Material 18

Sedentary Parent Material 18

Transported Parent Material 18

Parent Material Transported by Water 19

Alluvial 19

Lacustrine 19

Marine 19

Parent Material Transported by Wind 19

Parent Material Transported by Snow 20
Parent Material Transported by Gravity 20
Factors of Soil Formation 20
Climate 22
Organisms 23
Relief And Topography 25
Parent Material 27
Time 28
Process of Soil Formation 29
Types of Soil Forming Processes 30
Survey of Soil 33
Soil Survey 34
Utility of Soil Survey 34
Objectives of Soil Survey 35
Types of Soil Survey 35
Equipments and Instruments used in Soil Survey 37
Soil Survey Techniques 37
Soil Survey Methodology 38
Soil Survey Interpretation 40
Soil Map Preparation 41
Soil Survey Report Presentation 42
Morphology of Soil 42
Soil Profile, Soil Horizons and Layer 43
Main Characteristics of Master Horizons 44
Subordinate Distinction of Master Horizons 45
Vertical Sub Divisions and Discontinuities 46
Soil Micromorphology 47
Soil Monolith 48
Classification of Soil 49
Purpose of Soil Classification 49
System of Soil Classification 49
Soil Taxonomy 51
Specialties snd Drawback of Soil Taxonomy 51
Structure of Soil Taxonomy 52
Order 53
Sub Order 58
Great Group 59

Sub Group 61
Family 61
Soil Series 61
Soil Phase 62
World Reference Base For Soil Resources (WRB) 62
History of WRB 62
Importance and Specialties of WRB 63
Element of World Reference Base (WRB) 63
Classification of Soil According to WRB 65
Numerical Classification of Soil 66
Universal Soil Classification System and Unified Soil 67
Classification System 67
Land Capability Classification 68
Limitations Of Affecting Land Capability 68
Land Suitability Classification 69
Geography and Cartography of Soil 71
Soil Mapping 71
Preparation of Soil Maps 72
Types of Soils of World, Asia and India 73
Types of Soils of World 73
Types of Soils of Asia 75
Types of Soils of India, Their Distribution, Nature, Properties and Uses 75
Agroclimatic Zones or Regions of India 76
Agroecological Zones or Regions of India 78
Physiographical Zones or Regions of India 79
Physiographical Regions or Divisions of India 79
Crustal Segments of India 80
Soils of India 80
Alluvial Soils 81
Subgroups of Alluvial Soils 82
Subgroups of Red Soils 84
Black Soils 85
Subgroups of Black Soils 86
Desert and arid Soils 87
Laterite and Lateritic Soils 88
Tarai and Sub-Montane Soils 89

Forest and Hill/Mountain Soils ... 90
Hill / Mountain Soil ... 92
Saline and Alkali/ Salt Affected Soils ... 93
Peaty, Marshy and Skeletal Soils ... 94
Soils of Indian States ... 95
3. Properties and Processes of Soil ... 99
Physical Properties of Soil ... 99
Soil Texture ... 99
Importance of Soil Texture ... 101
Determination of Soil Texture ... 102
Fingers Feel Method to Determine Soil Texture ... 103
Soil Texture and Plant Growth ... 104
Soil Structure ... 104
Types of Soil Structure ... 105
Class of Soil Structure ... 107
Grade of Soil Structure ... 107
Formation and Development of Soil Structure ... 108
Soil Aeration and Plant Growth ... 110
Management of Soil Structure ... 111
Composition of Soil Air ... 112
Soil Aeration and Plant Growth ... 113
Renewal of Air in Soil ... 114
Air Exchange Between Soil And Atmosphere ... 114
Oxygen Diffusion On Rate (ODR) ... 115
Oxidation – Reduction Potential (Redox Potential) ... 116
Soil Density ... 117
Calculation of Soil Weight ... 118
Factors Affecting Bulk Density of Soil ... 118
Estimation of Bulk Density ... 119
Particle Density ... 120
Soil Pore Space And Porosity ... 121
Calculation of Percent Pore Space ... 121
Soil Temperature ... 123
Thermal Capacity ... 123
Volumetric Heat Capacity of Soil ... 123
Thermal Transmission in Soil ... 125
Factors Affecting Thermal Conductivity of Soil ... 125

Thermal Diffusivity 126
Heat Balance of Earth and Soil 127
Soil Heat Flux 128
Soil Temperature Fluctuations 129
Control and Management of Soil Temperature 130
Soil Water 132
Energy Concept of Soil Water 133
Soil Water Potential 134
Movement or Flow of Water in Soil 136
Water Movement Under Saturated Condition 136
Horizontal Saturated Movement or Flow of Water 138
Water Movement or Flow Under Unsaturated Condition 138
Water Movement or Flow Under Stratified Soils 139
Movement or Flow of Water Vapour in Soil 139
Retention of Water in Soil 140
Gravitational Water 141
Ground Water 141
Constants And Units To Denote Energy State Of Water 142
Soil Colour 143
Soil Consistence 145
Significance of Atterburg Limit 148
Chemical Properties of Soil 150
Chemical Composition of Soils 150
Chemical Potentials in Soils 152
Oxidation – Reduction Potential 153
Donan or Membrane Potential 155
Zeta Potential 155
Lime Potential 157
Phosphate Potential 157
Potassium Potential 157
Adsorption by Soil 158
Types of Adsorption 158
Adsorption Isotherms 159
Adsorption of Adsorbates in Soils 161
Soil Reaction 164
Normal Ranges of Soil pH 165
Numerical on pH 167

Determination of pH 172
Methods of pH Determination 172
Buffering in Soils 173
Buffering Agents 174
Mechanism of Buffering 174
Buffer Curves 175
Ion Exchange in Soil 176
Ion Exchange Equations 176
Types of Ion Exchange 178
Importance of Cation Exchange 179
Cation Exchange Capacity of Soils 180
Factors Affecting 180
CEC Determination 181
Units of Cation Exchange Capacity 181
Molarity 182
Anion Exchange in Soils 184
Factors Affecting Anion Exchange Capacity 185
Base Saturation of Soil 185
Biological And Biochemical Properties Of Soil 186
Organisms in Soil 187
Classification of Soil Organisms 188
Soil Macrofauna 199
Soil Magna Fauna 200
Biologically Active Chemicals In Soil 201
Factors Affecting Soil Organisms 201
Nitrogen Fixation in Soil 203
Biological Nitrogen Fixation 204
Necessities or Requirements of Biological Nitrogen Fixation 204
Organisms Involved in Nitrogen Fixation 207
Types of Biological Nitrogen Fixation 207
Asymbiotic Nitrogen Fixation 207
Nitrogen Fixation by Free Living Bacteria 207
Autotrophic Cyanobacteria 208
Factors Affecting Asymbiotic Nitrogen Fixation 209
Biochemical Process of Asymbiotic Nitrogen Fixation 211
Symbiotic Nitrogen Fixation 211
Symbiotic Nitrogen Fixation by Legume Crops 211

Cross Inoculation Crops 212
Process of Nitrogen Fixation by Symbiotic Bacteria in Legume 213
Factors Affecting Symbiotic Nitrogen Fixation 217
Biochemistry of Symbiotic Nitrogen Fixation 220
Nodulation on Stems of Legume Crops 221
Symbiotic Nitrogen Fixation In Non-Legume Plants 222
Biochemistry of Nitrogen Fixation by Non- Legume Crops 223
Plant And Algae Symbiosis 223
Azolla 223
Azolla Pinnata 224
Anabaena Azollae 225
Biochemistry of Nitrogen Fixation by Azolla- Anabaena Symbiosis 225
Factors Affecting Growth And Nitrogen Fixation by Azolla 226
Geological and Mineralogical Properties of Soil 227
Rock 229
Igneous Rocks 230
Sedimentary Rocks 231
Fragmental Rocks 232
Chemical Rocks 232
Organogenic Rocks 232
Metamorphic Rocks 234
Foliated Rocks 234
Unfoliated Rocks 234
Minerals 235
Minerals Forming Elements 236
Identification of Minerals 236
Crystal System 236
Physical Properties of Minerals 237
Optical Properties of Minerals 239
Chemical Properties of Minerals 239
Occurrence of Minerals 241
Minerals Present in Soils 242
Minerals Present in Fine Fraction of Soils 249
Weathering of Rocks and Minerals 250
Types of Weathering 251
Physical Weathering 251

Chemical Weathering ... 252
Biological Weathering ... 254
Factors Affecting the Rate of Weathering ... 255
Weathering Sequence and Products ... 256
Weathering Sequence ... 257
Weathering Products ... 257
Soil Colloids ... 257
Basic Properties of Colloids ... 258
Importance of Soil Colloids ... 259
Types of Soil Colloids ... 260
Inorganic Colloids of Soil ... 260
Silicate Clays ... 260
Silicon Tetrahedron ... 260
Aluminium Octahedron ... 262
Formation Layer of Silicate Tetrahedron and Alumina Octahedron Sheet ... 263
Groups of Silicate Clays ... 265
Mixed And Interstratified Minerals ... 270
Sesquioxide Or Hydrous Oxide Clays ... 270
Amorphous or Non-Crystalline Clays/Allophane ... 271
Characteristics of Clay Minerals ... 271
Origin or Source of Negative Charge ... 274
Isomorphous Substitution ... 274
Exposed Surfaces or Broken Edges ... 275
Adsorption of Ions ... 276
Diffuse Double Layer ... 276
Other Properties of Clays ... 278
Clay Minerals in Different Soil Orders ... 279
Clay Minerals in Indian Soils ... 280
Organic Colloids of Soils ... 281
Soil Fertility and Integrated Plant Nutrition ... 282
Essentiality of Plant Nutrients ... 283
Functions of Plant Nutrients in Plants ... 285
Deficiency Symptoms Of Nutrients On Plants ... 286
Amount of Nutrients in Plants ... 288
Factors Affecting Amount of Nutrients in Plants ... 288
Flow Of Nutrients from Soil to Plant Roots ... 290

Absorption of Nutrients by Plants 291
Entry of Nutrients Ions in Roots 291
Passive Absorption 291
Active Absorption 291
Factors Affecting Nutrients Absorption by Plants from Soil 292
Translocation of Nutrients from Root to Stem and Leaves 294
Soil Fertility Evaluation 294
Deficiency Symptoms of Nutrients on Plants 294
Determination of total Quantity of Nutrients 295
Plant Tissue Analysis 295
Soil Analysis 296
Collection of Soil Samples 296
Analysis of Soil Samples 298
Fertilizer Recommendation 298
Biological Methods of Soil Fertility Evaluation 298
Mineral Nutrition of Crop Plants 300
Nitrogen Nutrition of Plants 300
Phosphorus Nutrition of Plants 301
Potassium Nutrition of Plants 304
Potassium Fixation in Soil 305
Calcium Nutrition of Plants 306
Magnesium Nutrition of Plants 307
Sulphur Nutrition of Plants 308
Zinc Nutrition of Plants 306
Copper Nutrition of Plants 309
Iron Nutrition of Plants 309
Manganese Nutrition of Plants 309
Molybdenum Nutrition of Plants 310
Boron Nutrition of Plants 310
Chlorine Nutrition of Plants 311
Time of Fertilizers and Manure Application 311
Right Time of Fertilizer Application 312
Factors Influencing Fertilizer Placement 318
Methods Of Fertilizer Application in Fruit Trees 320
Preparation of Multinutrient Fertilizer Mixture 324
Calculation of Quantities of Fertilizers for Fertilizer Mixture 325
Integrated Plant Nutrient Management System (IPNMS) 327

Components of the IPNMS 328
Procedures of IPNMS 329
Processes and Cycles of Molecules and Elements in Soil 330
Carbon Cycle 331
Nitrogen Cycle 332
Ammonification 334
Factors Affecting Nitrification 336
Biochemical Process Of Nitrification 337
Nitrification Inhibition 338
Denitrification 339
Factors Affecting Denitrification 339
Chemical Denitrification 341
Volatilization of Ammonia 342
Factors Affecting Ammonia Volatilization 342
Phosphorus Cycle in Soil 344
Phosphorus Immobilization 344
Phosphorus Mineralization 345
Factors Affecting Phosphorus Mineralization 346
Biochemistry of Phosphorus Mineralization 347
Potassium Cycle in Soil 348
Sulphur Cycle in Soil 349
Biochemistry of Sulphur Mineralizatio 349
Factors Affecting Sulphur Mineralization 350
Sulphur Immobilization 351
Oxidation of Inorganic Sulphur 351
Biochemistry of Oxidation of Inorganic Sulphur 352
Reduction of Inorganic Sulphur 354
Biochemistry of Inorganic Sulphur Reduction 354
Cycling of Micronutrients 356
Soil Organic Matter 357
Factors Affecting Organic Matter Decomposition 365
Maintenance of Organic Matter in Soil 368

4. Vulnerability of Soils 371

Vulnerability of Soils 371
Soil Vulnerability, Soil Degradation, Soil Vulnerability Index 371
Soil Erosion By Water And Wind 374
Effect of Soil Erosion 375

Types of Soil Erosion 379
Classification of Soil Erosion 379
Water Erosion 379
Areas Affected by Gully Erosion 388
Components of Hydrological Cycle Related to Soil Erosion 393
Measuring of Water Erosion 396
Wind Erosion 403
Mechanism of Wind Erosion 404
Measurement of Wind Erosion 407
Estimation of Soil Loss By Wind 408
Revised Wind Erosion Equation (RWEQ) 410
Factors Influencing Wind Erosion 411
Depletion of Organic Matter In Soil 416
Loss Of Plants Nutrients 417
Soil Salinity, Sodicity, Acidity 418
Soil Pollution and Soil Contamination 425
Acid Rain 429
Heavy Metals 431
Behaviour of Heavy Metals in Soils and Plants 431
Agricultural Chemicals 433
Pollution of Soil by Fungicides 434
Pollution of Soil by Weedicides 435
Pollution of Soil by Rodenticides 436
Behaviour of Plant Protection Chemicals in Soil 436
Pollution of Soil By Radioactivity 440
Soil Compaction 441
Factors Affecting Soil Compaction 442
Soil Water Logging 443
Characteristics of Water Logging 444
Soil Aridity 446
Landslides 449
Factors Affecting Landslides 450
Effects of Landslides 451
Abandoned Mines 452
Effects Of Abandoned Mines 452

5. Restoration and Management of Soil Vulnerabilities........................453

Importance of Remote Sensing (Rs) and Geographical System (*Gis*) in Soils 453
Importance of Remote Sensing in Soils........ 453
Remote Sensing System 454
Principle of Remote Sensing 454
Stages of Action in Remote Sensing........ 454
Basic Concepts of Remote Sensing 456
Characteristics of Solar Radiant Energy 456
Electromagnetic Radiation........ 456
Thermal Emission Of Radiation 456
Interaction of Emr With The Earth's Surface 457
Types of Remote Sensing 457
Sensors used in Remote Sensing 458
Types of Remote Sensors........ 458
Spectral Signature 458
Platforms for Remote Sensing........ 459
Atmospheric Effects........ 459
Atmospheric Windows........ 459
Orbits 459
Importance of Geographical Information System (GIS) in Soils......... 460
Applications of GIS in Soil Science 461
Soil Erosion Control 463
Water Erosion Control 466
Drainage Channels........ 466
Bunding........ 467
Ditches 468
Terraces........ 468
Types of Terraces 468
Puertorican Terrace 470
Contour Dam 472
Besin Listing........ 473
Subsoiling 474
Diversion Channel 474
Grassed Waterways........ 474
Plantation 476
Wind Barrier 476

Plants for Shelterbelts and Wind Breaks 477
Pasture Development 478
Stabilization of Sand Dunes 479
Strip Cropping 486
Cultural Practices 481
Other Measures of Wind Erosion Control 482
Erosion Control in Gully Or Ravine 482
Ravine Reclamation Programme 483
Restoration and Management of Soil From Depletion of Organic Matter 485
Composting 485
Compost Tea 487
Conservation Agriculture 487
Reduced or Zero Tillage 488
Cover Cropping 489
Mixed Farming With Trees and Livestock 489
Restoration And Management of Soil From Salinity, Sodicity and Acidity 491
Soil Salinity 492
Suitable Time For Leaching 492
Preparation of Field For Leaching 492
New Life for Saline Soil 495
Soil Sodicity 497
Gypsum 497
Pyrite 500
Sulphuric Acid 502
Programme of Restoration of Soil from Sodicity 502
Management of Sodic Soils 504
Soil Acidity 504
Agricultural Lime 504
Lime Requirement to Reclaim Soil Acidity 505
Restoration and Management of Polluted and Contaminated Soils 507
Agricultural Wastes 507
Manure of Cattle Dung 508
Gobar Gas or Biogas 509
Method of Construction of Gobar Gas Plant 509
Compost Making on Land Surface 514
Prevention of Acidity of Rain Water 515

Heavy Metals 517
Agricultural Chemicals 518
Restoration and Management of Soil from Compaction, Sealing, Crusting 519
Reclamation Technology for Abandoned Mines 532
Watershed Management 537
Types of Watersheds 538
Watershed Management Programme 540
Integrated Watershed Management Programme (IWMP) 540
Pradhan Mantri Krishi Sinchayee Yojana 540
Steps in Watershed Management 540

References 543

Index 557

1

Introduction

This chapter presents a birds' eye view of the soil to develop the interest among the readers to know more about the fundamental concepts of soil.

Soil is a major determinant of terrestrial ecosystems and helps in the distribution of land flora and fauna. Soils are sources, transformers and stores of plant nutrients. Soils are the recycling places of plant and animal residues and wastes. Soils are also the screen and filter of pollutants in many ways, forms and degrees. Soils are the main link and recharger of hydrological cycle. Soils influence radiant heat exchange, temperature change and land surface reflection characteristics. Soils are important sources and sinks of biogeochemical cycles of carbon dioxide, methane, nitrogen oxides and sulphur dioxide. Soil is used in buildings, roads, brick kiln, naturopathy etc. Soils are good record of past climatic conditions and human activities, as they have some stable features originated from the interactions of climate, organisms, relief, parent material and time. Soil provides nutrients, water, air, heat and mechanical support to arable plants to produce the food, fodder, fibre, fuel and other plant products for human and animals. Mainly the soil is studied as a natural body and as a medium for plant growth.

Soil as a Natural Body

Soil is a biogeochemically synthesized product of nature. Our solar system has eight planets around the sun. These are Mercury, Venus, Earth, Mars, Jupiter, Saturn, Uranus and Neptune. Earlier Pluto was also in this list but later it was classified as miner or dwarf planet. Mercury, Venus, Earth and Mars are terrestrial (Latin, Terra means Earth) planets as they have hard surface made up of rocks, minerals and metals.

Rocks and minerals present on earth's surface are disintegrated and decomposed by physical, chemical and biological weathering processes to produce asymmetric and heterogeneous parent material. This parent material is converted into the soil-by-soil forming processes in the presence of soil forming factors. Normally the soil formation takes place after weathering of rocks and minerals. But in the dynamic system of nature, it is very difficult to separate these two. Thus, weathering and soil forming processes act concurrently to develop soil profile. All these processes and activities are performed in nature and end-product is soil, therefore soil is a natural body or entity.

Earth

Earth is third planet from Sun and larger than Mercury, Venus and Mars. It is an oceanic planet as oceans cover 70 percent of the Earth's surface. Age of Earth is supposed to be 4.55 billion years. The earliest life on Earth arose at least 3.5 billion years ago. Earth is the perfect place for life. The central portion of the earth is called as core, middle as mantle and outermost as crust. Outer side of the earth on which we are residing is quite cool, but inside there is much hot. The earth started to revolve around the sun after its birth. At this stage it also started to cool down by radiating the heat. Gases started to cool first and converted into the liquid. This liquid went deep into the earth, where it again converted into gas due to heat and come again on the surface. Thus, inner heat of earth comes repeatedly on the surface and ultimately the surface became the solid mass. The repeated conversion of gas into liquid and liquid into the gas and entry into the earth and exit from there caused deposition of heavier elements in the centre and lighter on the surface of the earth. Atmosphere, hydrosphere, and lithosphere are three spheres of earth. These have 0.03, 6.91 and 93.06% of the total weight of the earth, respectively.

Atmosphere has few layers namely Troposphere (0-16 km), Stratosphere (16-50 km), Mesosphere (50-85 km), Thermosphere (85-500 km), Exosphere (500-1000 km) and outer space (above 1000 km). In the troposphere there is gradual decrease in the temperature as we go up. Air gets less and less dense with the increase in height. At the height of about 30 km, there is barely any air. However, gases are present up to the height of 322 km. Nitrogen (78.084%) and oxygen (20.946%) are main gases of air. The other gases are argon (0.934%), carbon dioxide (0.0314%), neon (0.0018%), helium (0.0005%), methane (0.0002%) and krypton (0.0001%). Smaller amounts of hydrogen, xenon, ozone, ammonia, nitrogen oxide, iodine and carbon monoxide are also found in the atmospheric air. Water vapour also found in variable quantities. The Earth's atmospheric nitrogen, oxygen and other gases is the perfect balance for us to breathe and live. Earth's atmosphere protects us from incoming meteoroids.

About 75% area of the earth surface is occupied by the water in the forms of oceans, seas, rivers, ice caps, glaciers, lakes, ponds, ground water etc. This is hydrosphere. The depth of hydrosphere may be up to 10 km. We are dependent on water for irrigation, industry, domestic needs, transport, and recreation. Water as a universal solvent contains various amounts of soluble salts and impurities. The electrical conductivity of oceans water may go beyond 60000 dSm / meter, showing about 1.14% sodium, 0.14% magnesium, 0.05% calcium, 0.04% potassium, 2.07% chloride, 0.27% sulphate etc. Both north and south poles of the Earth are covered by ice in form of ice caps, sheet, and packs.

Lithosphere is Solid part of the earth. This is divided into several rigid tectonic plates. These plates migrate across the surface over periods of Lakhs and Crores of years. This is about one fourth of the total surface area of the earth. Mountains, hills, rocks, plateau, valleys, dunes and plains are come under lithosphere. The outer most stratum of the earth is called as crust. Thickness of crust is 5-56 km. The composition of crust can be indicated by rocks, minerals or elements. The upper 5 km portion of the crust contains 74% sedimentary rocks, 18% igneous rocks and 8% other rocks. The amount of sedimentary rocks decrease and amounts of igneous rocks increase with the increase in the depth of crust. The earth's crust contains all the natural elements and their combinations. However, upper 16 km depth contains mainly eight elements which are more than 98% of the total weight of elements. Gravimetric elemental composition is given in Table 1.1.

Table 1.1: Elemental composition of earth's crust

Element	**Per cent weight**	**Element**	**Per cent weight**
Oxygen	46.60	Hydrogen	0.14
Silicon	27.72	Phosphorus	0.12
Aluminium	8.13	Manganese	0.10
Iron	5.00	Carbon	0.09
Calcium	3.63	Sulphur	0.05
Sodium	2.83	Chlorine	0.05
Potassium	2.59	Barium	0.03
Magnesium	2.09	Strontium	0.02
Titanium	0.44	Others	0.50

Courtesy: www.thougtco.com>chemical composition of earth's crust with some modifications.

In the Vedic context (about 5000 years BC) the earth is supposed to be 'mother' in two senses. When there is mutual relationship between sun and earth, the sun is called as father and earth as mother. When its mutual relationship is described with its animate objects like human, it has been called as their mother and inmate beings have been called as its 'sons. Mother is the earth and I am its son (Atharv Veda. 12.1.12).

Land

Land is the solid part of the earth's crust or surface including its all-natural resources. Land has a geographical area. Length and width are the two dimensions or boundaries of land. Land includes relief, soils, plants, animals, underlying geology, surface and subsurface hydrology, results of past and present human and climatic activities etc.

Landscape

The portion of the land along with its physical and natural features that can be seen by eye in a single view is called as landscape. The soils come in the landscape. The landscapes are subjected to change day by day due to climate, organisms, human activities etc. Figure 1.1 shows a hypothetical landscape.

Fig. 1.1: Landscape

The trees and shrubs disappear and pasture fade away in the industrial environment as more and more land is required for construction of utilities and infrastructures like houses, roads, canals, dams, power plants, various buildings etc. All these encourage the extinction of older natural landscapes and establish new and unnatural or artificial landscapes.

Soil

The soil is a dynamic three-dimensional body of the land or landscape. Length, width and depth are three dimensions or boundaries of any soil. As a body the soil has volume i.e., length, width and depth. Soil is continuously changing and developing due to its dynamic processes. Any land or landscape may have many soil bodies or different types of soils. Our agricultural production mainly depends upon soil and climate to a large extent. The soil is an unconsolidated mineral and organic material on the upper most surface of the earth's crust. It has been formed by climate, organisms, relief or topography all acting over a long period of time on parent material. Soil is a natural medium for plant growth and main component of terrestrial ecosystem and environment.

The geologists first studied the origin and development of soils in field. They indicated that the mantle of loose-weathered material on the earth's surface is called as regolith and upper 100-200 cm part of regolith can be termed as soil. They also classified the soils based on parent rocks and underlying material and given the names like alluvial soil, granitic soil, glacial soil etc.

Soil has various horizontal layers of soil material. Dokuchaev (1883) first observed these layers and indicated their formation due to the joint influence of water, air and various forms of organisms, living or dead. He also considered soil as an independent natural body formed under the integrated activities of climate, vegetation and parent material in which vegetation is most important.

Now these horizontal layers are called horizon. Horizon is a layer of soil or soil material approximately parallel to the land surface. The properties (physical, chemical, biological and mineralogical) of one horizon differ to adjacent or related horizon. These properties are the basis for horizon differentiation or identification. O, A, E, B, C and R are six master or designated horizons. Any soil may have all these or few horizons. The vertical section of the soil from land surface to the C horizon (parent material) is called as soil profile. Solum is the upper part of the soil profile consisting of A, E, and B horizons where eluviation and illuviation are evidenced and recognized. Figure 1.2 A hypothetical soil profile.

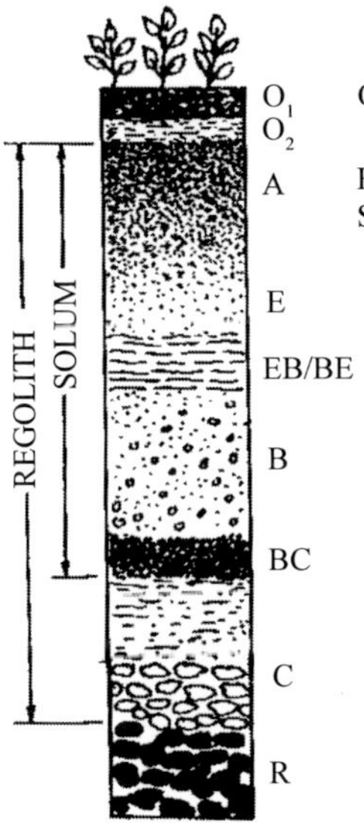

Fig. 1.2: A hypothetical soil profile

Components and Phases of Soil

The soil is a multi-component and multi-phase system. Many soil forming minerals and rocks at different stages of weathering and transformations produce soil particles of different sizes and shapes. These particles are intimately or very closely intermixed with organic matter to create macro and micro pores in soil. The water and air are present in these pores. Thus, four main components of soils are mineral matter, organic matter, water and air. Besides, these living organisms are also present in soils that may be considered as fifth component.

International Society of Soil Science assigned specific sizes for these particles as coarse sand, 0.2 mm to 2 mm; fine sand, 0.02 to 2 mm; silt, 0.002 to 0.02 mm and clay less than 0.002 mm after standard dispersion treatment. The moist mineral soils have about 45% mineral matter, 5% organic matter, 25% water and 25%

air on volume basis. The soil organisms are less than 0.5% which is included in organic matter itself. The relative percentage of water and air depends upon weather conditions. Rainfall and irrigations increase the water volume in soil simultaneously decrease air volume in soil. Air dry soils have more air volume (35-40%).

Mineral Matter

Soil mineral matter includes mineral particles up to 2 mm diameter. The main soil particles derived from weathering and weathering products of rocks and minerals are sand, silt and clay. These individual soil particles i.e., sand, silt and clay are also called as soil separates. Numerous soil properties are directly or indirectly influenced by soil separates as given in Table 1.2.

Table 1.2: Influence of soil separate on soil properties

Property	Sand	Silt	Clay
Aeration	Good	Medium	Poor
Buffering Capacity	Low	Medium	High
Compactability	Low	Medium	High
Drainage rate	High	Medium	Very slow
Leaching rate	High	Medium	Low
Suitability for tillage after rain or irrigation	Good	Medium	Poor
Soil organic matter status	Low	Medium	Medium to high
Soil organic matter Decomposition	Rapid	Medium	Slow
Storage of plant Nutrients	Poor	Medium	High
Water holding capacity	Low	Medium	High

The soil particles are comprised of primary, secondary and accessory minerals. Most of them are silicate minerals. Generally, primary minerals are found in coarse fractions (sand and silt) and secondary and accessory minerals are present in fine fraction (clay) of the soil. The primary minerals occur originally in rocks or derived from the weathering of rocks.

The quartz, feldspars, pyroxenes, amphiboles, olivine and micas are abundantly present primary minerals. Secondary minerals are released from the weathering of rocks or formed by the transformations of primary minerals. Various clay minerals (kaolinite, montmorillonite, illite, chlorite etc.) and oxides are the example of secondary minerals. Clay minerals govern numerous physical, chemical and colloidal properties of soils.

Organic Matter

Organic matter is another component of the soils. It is added into the soil by plants and animal residues, wastes, excreta and soil organisms. Plant leaves, leaf litter, stems, twigs, husk, stubbles, roots, dung, organic manures and dead animal tissues, bodies and cells are the common source of organic matter in soil. The main chemical compounds of the organic residues are starch, cellulose, hemicelluloses, sugars, proteins, lignin, fats, oils, waxes, resins etc.

Different macro and micro-organism act upon the organic residues and change their chemical composition. Each organism acts up to the certain stage of decomposition. The end products of earlier decomposer become substrate for another micro-organism. Thus, the chains of decomposition reactions take place in the soil. Therefore, organic matter remains under various stages of decomposition from undecomposed and partly decomposed organic materials to the humus. The well decomposed, transformed and relatively stable part of soil organic matter is called as humus. The humus gets intermixed with soil mineral matter. Humus also regulates various properties and conditions of the soil.

Soil Water

The water is the third component of soils. Water comes into the soil by precipitation and irrigation. This water is retained in the soil pores and adsorbed by the soil particles. Water present in soils contains dissolved salts, nutrient ions and colloidal substances. The water in soils may be under saturated or unsaturated conditions. Saturated condition comes after heavy rainfall or irrigation, when all the pores of soil are fully filled with water. This water moves downward due to gravity. At this stage macro pores becomes empty of water and they are filled with air. This is the unsaturated condition of water in soil where only micro pores are filled with water. This water is utilized by the plants when they create suction or exert a pressure which is sufficient to absorb it. Water is essential for plant growth. Soil is a storehouse of water and main source of water for land plants. Plants meet their water requirement from water stored in soil.

Soil water plays a significant role in several natural processes like evaporation, infiltration and drainage of water, diffusion of gases, conduction of heat and movement of salts and nutrients. All these depend upon the amount of water present in soil. Soil water is a transporter of plant nutrients from soil solution to root hairs and essential for photosynthesis. The movement and retention of water in soils, uptake by plants, translocation within the plants and evapotranspiration from soils and plants, all depend upon energy states of water in soil-plant-atmosphere-continuum (SPAC).

Soil Air

Soil air is also the mixture of gases but vary in composition than atmospheric air. The soils contain more carbon dioxide and less oxygen as compared to atmosphere. This is due to the respiration of plant roots and anaerobic decomposition of organic materials present in soil. Amount of air in soils increases with the decrease in the amount of water and vice-versa. This helps in the exchange of soil air with the atmospheric air. Good aeration with higher amount of oxygen is essential for growth of the plants and aerobic micro-organisms present in soil. More accumulation of CO_2 in soil causes adverse effects on plant roots and atmospheric nitrogen fixation.

Soil Organisms

Soil is a natural habitat of numerous organisms. These may be microscopic or visible with naked eyes. Flora and fauna, macro and micro, all types of organisms are present in soils. Roots of higher plants, algae, fungi, actinomycetes and bacteria constitute the flora of the soil. Snails, annelids, insects, nematodes and protozoan are the part of soil fauna. All these provide organic matter to the soil. The organisms present in soil may be harmful or beneficial to the crops as well as soil properties. Most of the microorganisms remain in the organic matter rich surface layer of the soil or in the rhizosphere. Sometimes, they reside in the nodules formed on the root surface. Decomposition of organic matter, recycling of nutrients elements and fixation of nitrogen are the three basic function of micro-organisms present in soil.

Phases of Soil

All four components of soil are grouped in three phases i.e., solid, liquid and gaseous. Mineral matter and organic matter are the part of solid phase which provide mechanical support to the plants and act as a store house of plant nutrients. Liquid phase in terms of soil solution supply nutrients and water to the plants and dynamically balance the nutrient pools of soils. Soil air is the gaseous phase of the soil. Soil air is utilized by plant roots and aerobic micro-organism. Thus, all the three phases have specific activities to perform in soil for better growth and developments of plants.

Soil as a Natural Medium for Plant Growth

Plants require a suitable medium for their root support and foothold, supply of water, air and mineral nutrients and resistance against adverse changes in pH and temperature. Besides these plants also require light. Main functions of soil in relation to plant growth and development are to provide better and suitably deep seedbed for easy anchoring of roots and erectness of shoots, optimum supply of nutrients, water and air and proper heating of the soil to give warmth which helps in rapid decomposition of plant residues to provide nutrients. Since soil performs all these necessary functions for the plant growth except light. Therefore, soil is considered as natural medium for plant growth since time immemorial.

The use of soil as a medium for plant growth is not new one. Soil is being used since ancient times and according to Rig-Veda, King 'Prathu' started agriculture.

In Sushruta Samhita, soil has been classified in six different types based on colour, texture and growth of plants. The plants were considered as living beings like the animals and human. It is evident by several references available which show that ancients speculated analogy of morphology, anatomy and physiology between plants and human beings. In Shanti Parva of Mahabharata, it is written, 'as the man sucks water upwards with the help of lotus petiole, trees drink water through the roots with the help of air' indicate knowledge of ancients about transpiration process.

The plants can grow without soil in water (hydroponics), aerated nutrient solutions and other media and may be at the space stations (space agriculture). But cost of cultivation of crops at these media will be too much higher as compared to soil where plants are growing naturally from ancient time. Artificial media require sophisticated instruments to create environment analogous to the natural needs of plants which is impossible for hundreds and thousands of hectares of land. Thus, the prices of agricultural products will be so high that most of population of the world cannot purchase them. Therefore, soil is the ultimate natural medium to produce food, feed and fibre economically and easily.

The population of world is increasing exponentially but food production is not increasing to such an extent and to keep the pace with it, the soils are under intensive agriculture to increase food production per unit area per unit time. Plants require light, water, mineral nutrients, air, heat and mechanical support for their growth and development and except light soil provide all these in more or less quantities. Thus, under natural or agrarian system soil is only medium for plant growth.

The improved high yielding varieties, better irrigation facilities, more use of chemical fertilizers and plant protection chemicals are the basic requirements of intensive agriculture. High yielding varieties take more nutrients from the soil. They decrease the nutrient status of soils. Thus, nutrients are supplied through organic manures and chemical fertilizers. The plant protection chemicals protect the crops against insect-pest and diseases. The crops do not utilize whole amounts of these agrochemicals applied to the soil or crop. The unutilized or residual amounts of these chemicals accumulate in the soil. The better irrigation facilities enhance the movement of chemical residues in soil profile. Ultimately these residues come in to the ground water or other water bodies through deep percolation and horizontal movement.

Later, chemical residues come in the bodies of fishes, birds, domestic animals and humans through trophic cycles and cause harmful effects to them, as numerous food chains operate in the environment. Thus, conservation of whole environment in general and soil in particular is of utmost need. Therefore, judicious use of fertilizers, agrochemicals and irrigation and use of bio fertilizers, organic manures, crop residue management and waste recycling should be adopted for clean and healthy soil i.e., medium of plant growth. This will provide a pollution free environment in agriculture sector for long sustainable growth and development.

Soil-plant Relationship

Plants grow and soil nurture them is the fundamental of soil-plant relationship. Soil is one of the most important natural resources of any country. The soils grow a variety of food, fibre and fodder crops required for men and animals. Soils also produce raw materials for various agro-industries like, sugar and starch, textile, canning and food processing etc. Soil and plant both have direct relation with each other because soil is a natural medium for plant growth and gives mechanical support to them. Simultaneously plants provide their biomass and residues to soil which help in the formation, development, and conservation of soil.

Soil bears certain physical, chemical and biological properties. These properties determine degree of workability, suitability to the specific crops and varieties, fertility as well as productivity of soil. These properties of soil need careful attention in context of their relationship with plants.

Some of the important physical properties of soil showing relationship with plants are texture, structure, density, porosity, consistence, temperature, moisture and colour. Similarly chemical properties of soils like chemical composition, potentials, adsorption, reaction and buffering, ion exchange, base saturation etc. and biological properties as soil flora, soil fauna, and activities of micro-organisms in soil, biochemical cycles of elements in soil and biofertilizers have direct or indirect relationships with plant growth and development.

Soil texture refers to the relative amounts of inorganic soil particles viz. Sand, Silt and Clay. Sand particles are large and coarse, clay particles are very fine and smooth and silt particle are intermediate. Many of the important soil properties are related to texture. Clayey soils show high water holding capacity, poor aeration, high plasticity, and stickiness and swelling whereas sandy soils are conspicuous by the absence of these properties. The soil texture affects plant growth by water and nutrient supply. The available water holding capacity of soil is related to soil texture.

The way in which soil particles are grouped or arranged together to form certain structures or aggregates is known as soil structure. Soil structure can be changed or modified and improved by adopting various soil management practices like tillage, manuring, liming, rotation of crops, irrigation, drainage etc. It may also be damaged by higher sodium etc. The pores control water, air and temperature in soil, which in turn, govern plant growth.

Soil structure indirectly influences plant growth. One of the best example of the effect of soil structure on plant growth is the emergence of seedlings in the seedbed. The seedlings are very sensitive to soil physical condition so that there should not be any hindrance to the emergence of tender seedlings and there should be optimum soil water and soil aeration. The soil in the seedbed should have a crumb structure so that the peds are soft and porous and roots of the seedling can penetrate it easily. The hard compact layer impedes root growth.

The density of soil that is weight per unit volume can be expressed in two ways viz; the density of solid particles of the soil (particle density), and the density of the whole soil (bulk density) that is inclusive of pore space. Generally, soils with low bulk density have better physical condition than those with higher bulk densities. Texture and structure of a soil, its total pore space and organic matter content are all related to bulk densities. In between the soil particles, there are empty spaces, termed as pore spaces, which are occupied, by air and water. Pore spaces between the aggregates of soil particles are macro-pores and those between the individual particles of the aggregates are micro-pores. The percentage of macro pores is more in sandy soils. These soils never remain water logged and allow water to percolate down so rapidly that their moisture content is very low. Thus, crops suffer due to shortage of soil moisture. The percentage of micro pore is more in clayey or fine texture soils. In these soils water logging condition causes adverse effect on respiration of roots and bacterial activities. A proper balance between the macro and micro pores is essential for better soil plant relationship. This can be maintained by timely cultivation and addition of organic matter so that crop suffers from neither shortage nor excess of water.

Soil aeration also has relationship with plant growth. Microbes and plants require oxygen for respiration. Oxygen taken up and carbon dioxide evolved are stoichiometric. Under anaerobic conditions, gaseous carbon compounds other than carbon dioxide are evolved. Root elongation is particularly sensitive to aeration. Oxygen deficiency disturbs metabolic processes in plants, resulting in the accumulation of toxic substances in plants and low uptake of nutrients. Certain plants such as rice are adapted to grow under submerged condition. These have large internal air spaces, which facilitate oxygen transport to the roots.

Soil consistence is a property that determines the resistance of the dry soil for crushing or pulverizing action by implements and when wet its ability to be moulded or changed in shape. All soils have cohesive and adhesive properties. Sandy soils have these properties to a much lesser extent than fine textured clay soils. A good tilth soil has both the macro and micro pores in approximately equal proportion. This ensures adequate retention of water, drainage of excess water and good aeration. The effect of such physical condition of soil on germination of seed and growth of plant is very beneficial.

Soil compaction is the process of increasing dry bulk density of soil, reducing the pore space by expulsion of air through applied pressure on a soil. Soil compaction creates problems for seed germination, water transmission and aeration. Crusting of soil is a form of soil compaction. The crusts present a serious barrier for seedling emergence. Lowering the exchangeable sodium percentage and incorporation of organic matter prevent crust formation.

Soil temperature has relationship with plants. All crops practically slow down their growth below the temperature of about 9^0 C and above the temperature of about 50^0 C. Soil microorganisms show maximum growth and activity at optimum soil temperature range. The biological processes for nutrient transformations and nutrient availability are controlled by soil temperature and soil moisture. Soil temperature has a profound influence on seed germination, root and shoot growth, and nutrient uptake and crop growth. Seeds do not germinate below or above a certain range of temperature but micro- organisms functioning in the soil are very active in a temperature range of about 27^0 to 32^0C.

The temperature should suitably control and modified for better soil plant and microorganism relationships. The various factors that control the soil temperature are soil moisture, soil colour, slope of the land, vegetative cover and general tilth of the soil. Soil temperature can be controlled by regulating soil moisture, proper soil management practices viz. good drainage, proper mulch, good crumby structure, addition of sufficient organic matter etc. These help in keeping the soil sufficiently warm for better chemical and biological activities in the soil.

Soils have various shades of colour. Hematite gives red colour, limonite-yellow colour, and predominance of silica or lime provide light colour to the soils. Soil colour indirectly indicates many other properties of soils e.g., a dark brown or black coloured soil indicates its high organic matter content and fertility. A red or yellowish soil shows good aeration and proper drainage. A white colour due to accumulation of certain salts or alkali indicates deterioration of soil fertility and its unsuitability for normal growth of many crops.

Soil organisms are also having relationship with plants. Different groups of organisms like bacteria, actinomycetes, fungi, algae, protozoa, nematodes, worms, insects and rodents are found in the soil. These groups of organisms are responsible for specific activities in the soil. Such activities may be beneficial or harmful to the crop and yield. These organisms decompose organic matter, take part in atmospheric nitrogen and carbon cycles, and cause plant diseases.

Bacteria decompose the organic matter and release plant nutrients. These are Generally, confined to the surface 20 to 30 cm layer and work best when there is good aeration, a neutral reaction, soil moisture content at about fifty percent of the water holding capacity and temperature between 25^0 C and 38^0 C. Fungi produce microscopic threads called mycelia. Fungi help in breaking down the somewhat resistant parts of the organic matter like cellulose, lignin, gums etc. A large part of slowly decomposing soil humus is made up of the dead remains of fungi. Actinomycetes can grow in deeper layers even under dry conditions of soil, and require less nitrogen. Their main function lies in decomposing the resistant parts of organic matter like cellulose. Algae are microscopic or very minute sized plants having chlorophyll and usually found on the surface of wet soils of paddy fields. They help in adding organic matter to soil, improving the soil aeration and fixing atmospheric nitrogen e.g., blue-green algae.

Definitions of Soil and Branches of Soil Science

Definitions of Soils

The above description will help to derive a particular definition of soil. However, many scientists have defined the soil as per their concepts. Some important definitions of soil are given below.

Whitney (1862): Soil is bin of plant nutrients.

Hilgard (1892): Soil is a more or less loose and friable material from which plants get support, nutrition and other conditions of growth through their roots.

Dokuchaev (1900): Soil is natural body made up from minerals and organic matter which has distinct origin and nature.

Jenny (1941): Soil is a naturally occurring body which originated due to long term integrated effects of climate and organisms on parent material situated at different relief.

Simonson (1957): Soil is a three-dimensional natural body which makes a continuum on land and gradually changes with place to place.

Kanwar (1982): Soil is the soul of infinite life.

Soil Taxonomy (1999): Soil is a natural body comprised of solids (minerals and organic matter), liquid and gases that occurs on land surface, occupies space, and is characterized by one or both of the following; horizons or layers that are distinguishable from the initial material as a result of additions, losses, transfers and transformations of energy or matter, or ability to support rooted plants in a natural environment. (Ref. Soil Survey Staff Soil,1999, Taxonomy: A Basic System of Soil Classification for Making and Interpreting Soil Surveys, Second Edition, Natural Resource Conservation Service, US Department of Agriculture, Hand Book,436).

The WRB (2014 Update 2015) defined soil as a part of the epidermis of the earth i.e., any material within 2 m of the Earth's surface that is in contact with the atmosphere, excluding living organisms, areas with continuous ice not covered by other material and water bodies deeper than 2 m. If explicitly stated, the object classified in the WRB includes layers deeper than 2 m. (Ref. IUSS Working Group WRB. 2015. World Reference Base for Soil Resources 2014, update 2015 International soil classification system for naming soils and creating legends for soil maps. World Soil Resources Reports No. 106. FAO, Rome.)

However, based on all these, (Venkatesh Bharadwaj) defined soil as 'Soil is a three-dimensional natural body present on the upper surface of earth whose all specialties are the result of long-time activities of climate, organisms and relief on its parent material. It is the natural medium of plants to grow them by providing essential nutrients, water and support for their growth and development.'

Branches of Soil Science

The soils are the natural resource on the uppermost surface of the earth. Their origin, nature, properties, uses and management are studied under Soil Science. Soil Science has two main branches Pedology and Edaphology. Pedology studies the origin, formation and geographic distribution of soils as a natural body. Pedology considers the soil as a natural body. Edaphology inter relates the soils with plants and their uses by human beings. Edaphology studies the soils as a natural medium for plant growth.

Later, according to soil properties and specializations many branches of soil science were developed. Soil Physics deals with the physical properties, processes and mechanical behaviour of soil. Soil Chemistry is related with the chemical composition, properties and processes of the soil. Soil organisms, their behaviour and transformations, biological reactions, processes and products are studied under Soil Biology and Biochemistry. The Soil Ecology is related with distribution of soil flora and fauna and their relationship with their environment. Soil Environment studies the relationship between soil with other natural resources and soil pollution management. Soil Fertility refers to inherent capacity of soil to supply nutrients to plants in adequate amount and suitable proportion. Soil Genesis establishes influence of soil forming factors and processes on the development of soil from parent material. Soil Geology refers to the rocks which develop the soil and processes changing rocks including dynamic and physical history of earth. Soil Mineralogy studies about the minerals present in soil and their influence on the physical, chemical, biological properties along with the suitability and management for crop. Soil Geography indicates about the pattern of distribution of soils on the earth surface. Soil Technology deals with the conservation and management of eroded, degraded and problematic soils.

Soil Vulnerability

Soil is very much complex natural resource as compared to water and air. Hence it is more susceptible to vulnerability. Ecological functions of soils are damaged or harmed by many environmental factors. Therefore, soil vulnerability is the susceptibility of soils to be harmed by single or multiple environmental factors. If any soil vulnerability remains unattended it causes soil degradation. The very basic example of soil vulnerability causing soil degradation is soil erosion. The unattended sheet erosion at fertile and productive soils changes them in to wastelands and unproductive ravines and gullies. Mainly soils are vulnerable to Soil erosion, Contaminants and pollutants of soils, Loss of organic matter and plants nutrients from soils, Salinity, alkalinity, acidity of soils, Compaction, sealing, crusting, hard layers of soils, water-logging, aridity, abandoned mines, land slide etc.

2

Soil in Nature and Environment

Genesis of Soil

World's geography is very widely diversified in relief showing various features like mountains and hills covered with snow and trees, glaciers, higher altitudes to plain to lowest valleys, forests to barren lands, fertile fields to waste lands, plateaus, deltas, swamps, lowlands, deserts etc. Temperature varies from arctic cold to scorching hot of equator, Rainfall from aridity to per humidity. All these situations and conditions under different sets of combinations generate and develop numerous types of soil from a single type of parent material. Therefore, many types of soils may develop at a smaller area.

Physical, chemical and biological weathering processes disintegrate and decompose various rocks and minerals to make asymmetrical soil parent material. This parent material is converted in to the soil profile by soil forming processes in the presence of soil forming factors. Soil genesis is defined as mode of formation of the soil from the parent material due to the action of soil forming factors and processes. Parent material, Factors of soil formation and Processes of soil formation, are three main part of soil genesis. These are described here:

1. Parent Material,
2. Classification of parent material,
3. Factors of soil formation,
4. Processes of soil formation,
5. Types of soil forming processes.

1. Parent Material

Soil parent material is the unconsolidated and more or less weathered mineral matter or decomposed organic material from which Solum has developed by soil forming factors and processes. Generally, 'C' horizon is considered as parent material. Since longer periods various activities of earthquakes, volcanoes, water logging, denudation etc. moved, mixed and deposited the weathered materials of rocks and minerals and due to this reason, most of the soils are formed from transported parent material. However, some soils are formed from the weathered residual parent material lying over the bedrock (R). This residual parent material is also termed as sedentary parent material.

2. Classification of Parent Material

The parent material is classified according to its origin and position. The parent material may be originated from rocks, minerals (inorganic parent material), plants, and animal residues (organic parent material). The position of parent material may be sedentary/residual and transported. Types of parent material are described here.

Organic Parent Material

Generally, organic parent material is found in the areas of heavy vegetation growth with ample supply of water and low temperature. These conditions enhance accumulation of a huge amount of partially decayed vegetation material. This organic parent material is called as cumulose. Cumulose can be found at any place but their accumulation is more prominent at north latitude. Cumulose is divided in to peat and muck. Peat is unconsolidated organic parent material having undecomposed or slightly decomposed organic matter accumulated under waterlogged conditions whose plant source can be identified. Muck is highly decomposed organic parent material with more mineral matter and dark colour than peat whose plant source cannot be identified. Sometimes peaty muck and mucky peat terms are also used to indicate the stage of decomposition of peat and muck.

Sedentary Parent Material

Sedentary parent material is originated mainly from igneous, metamorphic or sedimentary rocks at their places. Flat topography, low wind velocity and slow-moving glaciers enhance the origin of sedentary parent material. Residual term is also used for sedentary. Residual parent material is developed from unconsolidated and partly weathered rock or mineral material still present at its original place. Red soils of India are developed from residual parent material.

Transported parent material

Transported parent material is developed from the weathered materials of rocks and minerals, which are transported by water, wind, snow or gravity from their original places and deposited at newer or other places. Water, wind, ice and gravity are the agents for transportation. Different terms are given for the parent material transported by each agent. The brief description of transported parent material is given here:

Parent Material Transported by Water

Alluvial

Materials carried and deposited by flowing water i.e., rivers and streams are alluvial parent material. The particles deposited may vary in size from clay, silt, sand to gravel. Most of the fertile soils of world are formed from the alluvial parent material. In India Indo-Gangetic plain is a good example of alluvium. This occupies more than five lakhs square kilometres (1600 km x 320 km) of area from arid to semi-arid regions of Rajasthan and Punjab to humid and per humid deltaic regions of Bengal. The alluvial parent material is sub divided in to Alluvial fan, Flood Plain and Delta.

Alluvial fan: Fan shaped deposition of alluvial parent material at less steeply sloping place of hilly river is called as alluvial fan.

Flood Plain: Flood plain is the land of river banks frequently flooded by flood water during the flood in the river.

Delta: It is the somewhat flat plain of alluvial deposit at the mouth of river. The fine sediments flow along with the water of river and deposit at the place where river enters to quiet water.

Lacustrine

Fine sediments deposited in the bottom or along the shore of a lake and exposed due to the elevation of land or change in the drainage conditions.

Marine

Sediments deposited in the bottom or along the shore of seas, oceans and beaches and exposed due to the change in sea level are called as marine parent material. The coarser fragments of the sediments deposit at the shore and finer fragments deposit away from the shore. This deposition is changed in to soil parent material by weathering. Most of the soils of Andaman and Nicobar of India are formed from the marine parent material.

Parent Material Transported by Wind

Parent materials can also be transported by wind which include aeolian and loess.

Aeolian: Blown by wind and deposition of coarser material mainly sand in the form of sand dunes, sand ridges and sand sheets are called as Aeolian.

Loess: Silty material transported and deposited by winds of moderate velocity is called as loess. The deposition of wind-transported material is mainly occurring at arid regions such as Thar and Rajasthan deserts.

Parent Material Transported by Snow

Moving snow transports enormous amount of rocks materials since modern periods to the Pleistocene (about 1.8 million years) and Holocene (11700 years) ages. The action of moving snow or melting glaciers can be seen even now a day at Himalayan region of India.

Glacial Till: Generally, the materials directly deposited by ice consisting clay, silt, sand, gravel and boulders in any proportion are called as glacial till or till. Till is a poorly sorted parent material. Till has many subdivisions but moraine, till plain and out wash plain are most important.

Moraine: Moraine is the irregular accumulation of stony material transported and deposited by a glacier. Ground moraine and end moraine are kinds of moraine.

Till Plain: Till plain is the level or undulated land surface covered by till. It generally, contains finer particles, which are deposited when melting rate of ice front is more than its velocity.

Outwash Plain: Out wash plain is the deposition of coarse fragments like sand and gravel at the banks of glaciers at gently sloping places.

Parent Material Transported by Gravity

The material transported by the action of gravitational forces and deposited at the foot of the slope is called as colluvium. This contains heterogeneous parent material. Talus, scree and creep are types of colluviums.

Talus: Talus is the loose and fragmented rocky or soil material deposited at the foot of a cliff.

Scree: The loose and fragmented rocky or soil material deposited at the foot of a steep slope is called as scree.

Creep: It is the slow movement of soil or rocky material down the steep slope due to the gravity and enhanced by water saturation and alternate freezing and thawing. Solifluction and slump terms are also used in this context. Solifluction is very slow movement as a viscous liquid down slope at very low slope angles (2-3 degrees). Slump has relatively fast movement and landslide, mudslide and avalanche are examples of slump.

3. Factors of Soil Formation

The parent material may be present in situ over the bedrock or deposited by streams, glaciers, wind and gravity at another place. Generally, when there is a slope there is movement of soil material. However, the rate of movement may vary from very slow to very rapid. Thus, parent material and slope can be considered as two factors of soil formation. But there are differences in temperature and rainfall, vegetation and organisms and time available for the soil formation, which directly or indirectly influence the soil formation. Therefore, factors of soil formation may be more than two.

These "factors of soil formation" was first identified and discussed by the Russian geologist Vasily Dokuchaev (1846–1903, ref. Pedogenesis wikipedia.org) at the end of 19th century. He is considered as the father of soil science. He indicated that soil is not inert and stable. It develops and evolves under the influence of climatic agents and vegetation that operate over time on a given geological substrate. Dokuchaev wrote the first equation of soil formation:

S = f (p, cl, o) tr where

S = any soil property

f = is the cumulative function of

p = parent material

cl= climate and

tr = relative time available for soil formation

Soil forming factors are changeable and interrelated natural agents, which are responsible for soil formation. These factors are parent material, climate, organisms, relief or topography and time. Jenny (1941) proposed an equation to establish the relationship between any soil property (S) and factors of soil formation as:

S = f (cl, o, r, p, t…), where

S = any soil property

f = is the cumulative function of

cl = climate

o = Organisms

r = relief

p = parent material and

t = time available for soil formation.

The dots indicate some unimportant soil forming factors. Later on, another equation indicating the importance of a particular soil forming factor was proposed as:

S = f (Cl), o, r, p, t.

This, f (Cl), indicates that only climate is responsible for the formation of soil i.e., climofunction. A sequence of related soils which are differing to each other mainly because of difference in climate as a main factor of soil formation is called as climosequence. Similarly, the terminology for other sequences such as clinosequence is the sequence of related soil differing each other due to the variation in slope on which they are formed. The topsequence, biosequence and chronosequence indicated the influence of topography, organisms and time, respectively as the main soil forming factors when other soil forming factors remain same.

However, Joffe (1949) grouped all these five factors in to two; Active factors (climate and organisms) which provide energy for the soil formation and Passive factors (parent material, relief and time) which are the sources of soil forming material and conditions affecting soil formation. These factors, in short, are described here.

Climate

There may be difference in atmospheric and soil climate, but it directly and indirectly influences the soil formation. Directly climate provides water and heat for the reactions within the parent material. Indirectly climate influences the biomass particularly vegetation production. The organic matter provided by the biomass, produces many organic acids during its decomposition. These organic acids increase the solubility of parent material.

The climate is the normally prevailing weather conditions of an area. Climate may be arid (extremely dry without moisture, precipitation is too less than the evapotranspiration), humid (wet or moist, precipitation is too more than the evapotranspiration), tropical (warm temperature throughout the year with a significant amount of precipitation), temperate (warm during summer with cold winters) etc. Rainfall and temperature are two main components of climate, which actively take part in soil formation.

Rainfall: Intensity of rainfall, evapotranspiration, slope, permeability, daily weather and microclimate of the area decide the effectiveness of rainfall in the soil formation.

In humid areas, the precipitation is greater than evapotranspiration. Thus, a significant amount of rain water is available to pass through soil. More percolation and leaching of water through soil continuously alter the properties of parent material, rock and mineral weathering, genesis of clay and movement of soil colloids and salts. Also, production of organic matter, accumulation of nitrogen and exchangeable hydrogen, aggregation and saturation capacity increase remarkably. All these conditions help in the soil formation. More rainfall increases depth of $CaCO_3$ layer in the calcareous soils.

In contrast of this, in arid regions due to low rainfall and high evapotranspiration, very little amount of water for a shorter period is available to pass through the soil. This reduces the weathering of rocks and minerals, movements of salts from upper layer to lower layer, production of biomass and deposition of organic matter and nitrogen. All these conditions hinder the horizon differentiation and profile development. The low rainfall favours the formation of gypsic or calcic horizon in the soil.

Temperature: Temperature affects the rocks and minerals weathering, biomass production and decomposition of organic matter. Each 10^{0}C increase in the temperature doubles the rate of chemical reactions. Thus, at equal moisture levels, the depth and intensity of weathering and amount of clay will be more whereas amount of nitrogen, organic matter, silica/alumina and base/alumina ratio will be low at warmer places as compared to cooler places. The microorganisms require an optimum temperature for their activities. At cold regions due to the low activities of microorganisms, decomposition of organic matter is low. Thus, accumulation is more.

Organisms

Soil is the habitat for microorganisms, plants, animals and human being is also dependent on it. All these help in the soil formation by mineral dissolution, organic matter production, accumulation, decomposition and mixing of soil material, aggregation and aggregate stability, nutrient cycling and agricultural activities.

Bacteria, actinomycetes, fungi, algae and protozoa come under microorganisms. All these decompose organic matter, solubilize the mineral matter, for example oxidation of Sulphur and dissolution of apatite, fix atmospheric nitrogen and ultimately increase soil fertility.

Role of vegetation in soil formation is multifaceted. The roots of plants enter in the cracks and fissures of rocks and make them wider and accessible for air and water. Various root exudates and organic acids liberated by these roots help in the solubilization of rocks and minerals. Roots, leaf litter, small branches and twigs also produce organic matter. The organic matter as a component of soil is essential for soil formation.

The effect of vegetation and soil formation can be very well seen at the proximity of forest and grass lands when other soil forming factors are same. The forest vegetation and grasses have different rooting pattern and foliage. Thus, different types of soil are formed under these two ecosystems.

Forest produces more organic matter than grasses. But fibrous roots of grasses do not go much deeper in soil material. However, they decomposed and humified easily. Thus, more organic matter accumulates on soil surface which gradually decreases with the depth. Hence grass land soils have more aggregation, aggregate stability, water holding capacity and cation exchange capacity and less distinct horizonization. The decomposed and humified grass roots provide more bases to the soil, which develop neutral pH. Generally, Mollisol (dark coloured, organic matter rich and highly base saturated soil) are developed under grasslands.

Woody tap roots of forest vegetation are hard to decompose. But after decomposition these create channels for water and air movement and supply some organic matter to lower depths of soil material. Leaf litter of forest accumulates much more organic matter on soil surface. However, this organic matter is less humified. There is more development of horizons. Leaching of Ca, Mg and K causes acidic pH and leached A horizon. Inceptisols (beginning of soil formation with few diagnostic features) develop under mixed or oak forest and Spodosol (leached A horizon with sub surface accumulation of sesquioxide and humus) are formed under pine forest.

Appreciable population of rodents, burrowing animals, earthworms, ants, termites etc. actively take part in soil formation. Rodents and borrowing animals continuously mix the soil material within the profile. Crotovinas are formed due to the activities of these animals. Crotovina is a burrow of an animal in soil horizon filled with excreta of that animal, organic matter or material from another horizon.

Earthworms thrive well in neutral pH. The acidic pH developed by pine forest in is not good for them. Earthworms do not ingest the acidic litter of the pine and that remains on the soil surface for longer periods in unhumified form. The litter of alder (genus Alnus) is good for earthworms. They ingest that and mix with soil material. Casts of earthworms develop fine blocky to granular structure at A horizon.

Termites bring fine soil material, water and salts from the depths of lands and deposit them on soil surface in the form of mounds called as termitaria. The size of termitarium may be as big as 2.5 meters wide, 8 meters deep and 2-4 meters above the surface. Such a huge amount of soil material can change the properties of soil surface. Figure of a mound or termitarium is given here. Likewise, ants also bring soil materials from lower depths and deposit it on soil surface (Figure 2.1).

Fig. 2.1: Soil material deposited by termites as Termitarium and ants
(Photo by Dr. Venkatesh Bharadwaj)

Human activities also help in the formation of soil. Removal of vegetation as crops and other agricultural use of soil alter the profile development. Terracing, bunding, channelling, drainage, use of fertilizers, manures and amendments, crops and cropping pattern directly influence the soil formation. Provision of irrigation facilities in arid region enhances the soil formation.

In general, low rainfall, high water table, very steep slope, continuous denudation and deposition of soil material, more activities of burrowing animals and lime and silica rich resistant parent material reduces the soil formation and ultimately profile development.

Relief and Topography

Relief is the elevations, height or inequalities of a land surface. Land with even surface or equal elevations is called level, undulating with gentle unevenness, rolling with strong unevenness and hilly with very strong unevenness or elevations. Relief includes hills, mountains, terraces, valleys, cliff, delta, flood plains etc. Micro relief is localized small-scale (less than one meter) differences in the elevation. Micro relief includes mounds, swales, hollows, gilgai etc. Topography is the features of any area that can collectively be shown on a map. Toposequence is the sequence of related soils that differs only due to the variations in topography.

Relief affects the soil formation by influencing water infiltration, retention, movement and drainage, soil erosion and solute movement at any area. Soils formed on flat or level lands are deep with distinct horizons as appreciable amount of rainwater enters in to the soil. But on sloppy lands different type of soil may develop according to the amount of surface runoff, soil erosion and entry of water in to the soil. The term clinosequence denote the group of related soils that differ to each other mainly because of variation in degree of slope in which they are formed. Slope classes as described by Sehgal et al (1987) are given in Table 2.1. Two slope classes Level to nearly level and steeply sloping, respectively are given Figures 2.2.A and 2.2.B.

Table 2.1: Slope classes

Slope class	**Symbol**	**Percent slope**
Level to nearly level	A	0-1
Very gently sloping	B	1-3
Gently sloping	C	3-8
Moderately sloping	D	8-15
Moderately steep sloping	E	15-30
Steeply sloping	F	30-50
Very steeply sloping	G	More than 50

Fig. 2.2: A. Level to nearly level slope (*Photos* Dr. Venkatesh Bharadwaj)

Fig. 2.2.B. Steeply to very steeply sloping land

As soon as the steepness of slope increases the surface run off, soil erosion and soil creep also increase. This reduces availability of water for chemical reactions and plant growth. Therefore, the organic matter status of soil, weathering of minerals, movement of salts and clays, horizon differentiation and thickness of soil profile are also reduced. Thus, soils formed on steep slopes are generally, shallow, coarse textured with stones, without clear-cut horizon differentiation and poorly developed soil profile. When length of the slope increases more runoff water infiltrates in to the soil and there is more accumulation of eroded soil material at foot of the slope. This forms shallow and immature soil with weaker horizon differentiation.

Lowland and valleys are more wet than flat land and at equal rainfall due to more accumulation and storage of runoff water there is more plant growth. Ultimately, there is more accumulation of organic matter. Thus, dark coloured organic matter rich soils are formed. The example is soils of Tarai region of Uttar Pradesh and Uttarakhand.

Too much low-lying areas of humid region with high rainfall, low runoff, more stagnation of water and poor drainage conditions develop hydromorphic soils. These soils show fine texture, impeded drainage and gley horizons. However, the concave areas of arid and semiarid zone may develop saline sodic soils due to the capillary rise of brackish water and deposition of salts at or near the surface due to higher evaporation.

Slope aspects (exposures) also play an important role in the formation of different types of soil. The soils formed at south exposure (facing to sun) are less developed as compared to soil formed at north exposure. This is mainly because of more evaporation of water from the sun facing areas leaving lesser water for growth of vegetation. This reduces the organic matter accumulation and increase the erosion of soil. However, in north exposure due to cooler conditions water remains in soil for longer period. This increases the vegetative growth; ultimately organic matter status of soil is increased and erosion is reduced which form well developed soils.

Parent Material

The parent material is the source of mass required for the soil formation. Its role in the soil formation is passive. Thus, dissimilar parent material may generate somewhat similar soils if other soil forming factors are same.

Soil formation may take place in situ from the weathering material of rocks and minerals or it may develop from the deposition of transported parent materials. In these depositions, there are chances of mixing of different rocks and mineral material. Sometimes materials of different rocks are deposited one by one. Therefore, different layers of parent material show heterogeneity in nature and properties. This is called as lithologic discontinuity of parent material.

Similarly, the lithosequence is a group of similar soils that differ with one another primarily because of differences in parent rocks. Sometimes transported materials deposit upon the soil and hide the actual soil. This hidden soil is known as buried soil. The buried soil is defined as the soil covered by other transported materials up to the depth more than the thickness of the solum.

Effect of parent material on soil properties is quite recognizable at initial stages of soil formation. But the effect of parent material is remarkably reduced at the later stages of soil formation because of more influential soil forming factors like climate and organisms. Soil texture, structure, water holding capacity, colour etc. are greatly influenced by soil parent material.

High silica content rocks like granite, quartz and sand stone form coarse textured sandy soils. Basic rocks (basalt) and impure limestone generate fine textured soils. Amount of clay in soil is directly proportional to the feldspars and hornblende content of parent material. If feldspars in parent material are greater (50-70%) the clay content in soil will be about 50-65%. Clay content in soil will be about 15-20% if feldspars in parent material are about 30-40%. Type of clay formed is also governed by the permeability of the parent material. Sandy parent materials have high permeability which enhances removal of basic ions (Ca and Mg) from upper layer and deposit them at lower layers. This creates acidic state at upper layer which tends to generate 1:1 type (Kaolinite) clay. At the same time in the Ca and Mg rich lower layers there are chances of formation of 2:1 type (montmorillonite and illite) clays. Also, permeable parent material and deep-water table favour the formation of well drained soils. Impermeable or semi permeable parent material along with shallow water table form the poorly drained soils with higher clays, salts, organic matter with reduced (anaerobic) conditions.

The parent material also governs fertility status of the soils. Basaltic soils (Vertisols) are more fertile as compared granitic soils (Alfisols). There is more growth and development of vegetation in the fertile soils. Vegetation provides organic matter to the soil in due course of time that helps in the formation and development of soil profile.

Time

Time refers to the period required for the conversion of weathered rock and mineral materials in to the matured soil. Mature soil is a well-developed soil having all (A, B, and C horizons and distinct horizonation.

Soil formation is a very slow process. Nature requires hundreds and thousands of years to form a particular soil. The time required by nature to form a soil is called as pedologic time. Buol, et al. (1975) indicated rate of soil formation as 1.3 year/cm for Entisol (recent soil with no horizon differentiation in soil) to 750 years/cm for Oxisol (highly weathered soil). Thus, maturity periods for different soils are quite variable.

Soils under warm and moist climate, forest vegetation, permeable, incoherent and sandy parent material, good drainage and flat topography attain maturity in lesser time. The soils under very hot or cold and arid climate, grassy vegetation, impermeable, coherent and clayey and lime rich parent materials, poorly drained, eroded and situated at steep slopes require much more time for maturity.

Soils are classified into zonal, intraoral and azonal according to time and degree of maturity. In India the soils are classified in to Khadar (new alluvium) and Bangar (Old alluvium) according to age of alluvium. Radio carbon dating technique is mainly used to assess the age of soil. Mor and van Warren (1954) indicated five stages from rock and mineral weathering to soil formation. These are given in Table 2.2.

Table 2.2: Different stages of soil formation

Stage	Description
1. Initial	Unweathered patent material.
2. Juvenile	Start of weathering with much of the Unweathered parent material.
3. Virile	Decomposition of easily weatherable minerals has been completed; clay content increased and slowly weatherable minerals are appreciably present.
4. Senile	Decomposition has been completed only highly resistant minerals are left.
5. Final	Under existing set of atmospheric conditions, soil formation completed.

4. Processes of Soil Formation

The parent materials produced due to the weathering of rocks and minerals or by other means start to generate the soil just after growth of plants. Roots absorb nutrients from the parent material and translocate them into shoots and leaves. Leaves provide organic matter and nutrients to the surface of parent material after falling there. Roots present in parent material also provide nutrients and organic matter. This organic matter is the substrate for numerous microorganisms. Microorganisms and other soil organisms also contribute organic matter.

Plants grow again and again. They repeatedly left remarkable amount of organic matter in to the parent material and microorganisms also decompose it simultaneously. Thus, in a cyclic manner repeated inclusion of organic matter changes the characteristics of parent material. Addition of rainwater and ions and atmospheric gases mostly CO_2 to the parent material boost up the changes. All these activities start horizonation in the parent material. Horizonation is the indication of soil formation.

The soil forming processes are also called as pedogenic processes. These are the processes responsible for the development of solum or true soil from unconsolidated parent material. Soil forming processes are very complex biogeochemical activities or reactions which run simultaneously or one by one to form horizons in the parent material. Simenson (1978) summarized all the reactions or activities of soil formation in to four categories namely addition, removal, transfer and transformation. It means that water, mineral matter, organic matter and solar heat may come in to the soil. These may go out from the soil. Mineral matter and organic matter may be transformed in to the soil along with the change in soil structure and the soil material may be translocated or moved in the form of solution and or suspension. All these reactions enhance the horizon development in the soil material. However, burrowing by animals, erosion, silting, sedimentation etc. reduce the horizon formation.

5. Types of Soil Forming Processes

Fundamental and specific soil forming processes are two types of soil forming processes. Transformation and transfer of parent material come under fundamental soil forming processes that generate conditions to work by specific soil forming processes. Humification, eluviation and illuviation are fundamental soil forming processes. The long continued and combined effects of these processes develop wide variety of soil on earth surface.

Specific soil forming processes develop some definite, distinct or categorical characteristics in soils by which one soil can be recognized than the other. Podzolization, laterization, calcification, decalcification, Salinization, desalinization, solonization, desolonization, gleization, pervection and pedoturbation are some important specific soil forming processes. Some fundamental and specific soil forming processes are described here:

Humification: This process is the transformation of organic residues or raw organic matter in to the humus by biological activities, microbial synthesis and chemical reactions. Various soil macro and microorganisms take part in this activity and decompose or transform the raw organic matter or organic residues in to humus. This transformation i.e., Humification generates organic horizon (O). The characteristics of this horizon depend upon nature of organic residues, their type of decomposition and nature of synthesized new organic compounds.

Eluviation: Eluviation is related with the loss or removal. There is the movement of soil material in the form of solution or suspension with in soil mass or from upper layer to lower layer by the downward or lateral movement of water. Movement of soil material in the form of solution is called as leaching.

Eluviation may be mechanical or chemical type. Mechanical eluviation is also called as lessivage in which clay and other fine particles move without any chemical change. In chemical eluviation organic matter, silicic acid, exchangeable bases, salts, hydrated aluminium and ferric oxide and other ions move along with percolating water. The soil layer or horizon from which the material has been transported or moved is called as eluvial layer or horizon and denoted as 'E'. However, A2 and EB are also eluvial horizons.

Illuviation: Illuviation is related with the gain or addition. It is the deposition of soil material transported or moved from upper horizon in to the lower horizon of the soil body. The horizon in which deposition of transported material takes place is called as illuvial horizon and denoted as 'B'. Bt is also an illuvial horizon.

Podzolization: Podzolization is a specific process of soil formation, which forms podzol, podzolic soil or Spodosol. Podzol is Russian term meaning ash like or ash grey. Podzolization is related with loss and translocation of soil material.

The favourable conditions for the Podzolization are temperate, cool and humid climate; sandy, coarse, low bases and silica rich parent material and coniferous and acid producing vegetation.

High rainfall and acidic conditions leach down the cations from the soil. Coarse sandy parent material enhances the percolation of water. Coniferous vegetation after decomposition returns less bases to the soil and reduce the pH less than 5.0 due to the production of organic acids. All these organic acids react with sesquioxide and clay to make them soluble and movable to form organo -sesqui oxide and organo-clay complexes.

These two complexes move downward (eluviate) from A horizon along with acidic percolating water and accumulate (illuviate) in B horizon. Thus, silica rich A horizon shows bleached grey or ash grey colour whereas fine textured B horizon attains rust colour due to precipitation of insoluble ferric ion.

Laterization: The specific process of soil formation generating laterites, Latosol or red and red yellow soil is called as laterization. 'Laterite term is derived from word 'Later' meaning brick. It is related with loss and translocation of soil material.

Favourable conditions for laterization are warm and humid tropical climate; basic parent material comprising iron rich minerals like pyroxenes, amphiboles, biotite and chlorite and deciduous rain forest vegetation.

High temperature and rainfall enhance the weathering of parent rock and leaching of weathering products. The silica reacts with bases of the weathering products which get solubilize and eluviated with percolating water. The prevailing high temperature oxidizes the organic matter at higher rates. Thus, there is very low accumulation of organic matter in the soil. In these conditions the sesquioxide is more stable and remain in the profile.

The Fe_2O_3 and Al_2O_3 of the sesquioxide provide red and grey colour to the soil profile, respectively. B horizon form due to the laterization may be several meters thick below from the surface. It is a cemented horizon and very hard upon drying. Thus, it can be cut into brick shaped pieces and used in the construction of buildings etc.

Calcification and decalcification: Calcification is the process of accumulation of $CaCO_3$ in any horizon or part of the soil profile. The accumulation of $CaCO_3$ may develop calcic horizon.

Arid to semiarid climate, calcium rich parent material and xerophytic vegetation varying from grasses to bushes are common condition favouring calcification.

Under calcification process there is dissolution, movement and reprecipitation of $CaCO_3$ in the soil. The CO_2 gas released due to root respiration and decomposition is dissolved in water to form H_2CO_3. This H_2CO_3 reacts with native calcium to form soluble calcium bicarbonate that move down ward along with water. The calcium bicarbonate reprecipitates as calcium carbonate in the lower layers of soil due to high temperature and lack of water and release of CO_2 as:

$$CO_2 + H_2O \rightarrow H_2CO_3$$

$$CaCO_3 + H_2CO_3 \rightarrow Ca(HCO_3)_2$$

(Insoluble) (Soluble)

High Temperature

$$Ca(HCO_3)_2 \rightarrow CaCO_3 + H_2O + CO_2$$

Formation of calcic horizon in soil profile may take at 50 to 150 cm depth of soil at arid to sub humid regions, respectively. Thus, increase in amount of percolating water increase the depth of calcic horizon. Depth of calcic horizon is also governed by amount of rainfall, depth of ground water table and soil texture.

Decalcification is the reverse of calcification in which $CaCO_3$ or Ca^{2+} ions are removed from the soil profile by eluviation or leaching.

Salinization and Desalinization: Salinization is the accumulation of salts in soil. The soluble salts of sulphates and chlorides of calcium, magnesium and potassium are accumulated in soil profile by translocation. The horizon in which salts are accumulated is called as salic horizon.

Arid and semi-arid climate with high temperature and low rainfall causing more evapotranspiration of water, salt rich parent material, shallow water table, salt rich (saline) irrigation water, poor to imperfect drainage, lower topographic position and coastal areas enhance the process of Salinization. Higher evapotranspiration causes upward movement of saline ground water through capillaries. This causes concentrated accumulation of salts on soil surface or below it. The salic horizons formed due to salts accumulation are designated as Asa or A3 and Bsa or Bz.

The removal of excess soluble salts from the profile or from salic horizon by leaching is called as desalinization. The removal of excess soluble salts from the soil may change the soil order also.

Solonization and desolonization: Solonization is also called as alkalization. Accumulation of the sodium ions on the exchange complex of the clay fraction of soil is called as solonization. Solonization form natric horizon in soil. Bn represents Natric horizon. If the accumulation of exchangeable sodium is 15% or more then sodic soils are formed.

Desolonization is also called as dealkalization or solodization. In this process the exchangeable sodium is removed from the soil profile. The removal of sodium cannot be done by leaching alone. The amendments are also needed. Generally, calculated quantity of gypsum ($CaSO_4 . 2H_2O$) is added and mixed in to the soil before leaching.

Pervection: Pervection is the mechanical eluviation of clay from A horizon to B horizon. This process increases the clay content in B horizon. Therefore, Argillic horizon is formed. Argillic horizon is denoted as Bt. Pervection is also called as lessivage.

Pedoturbation: Pedoturbation is the process of mixing of the soil, which develops homogeneity in the soil. Many agents cause Pedoturbation. When ants, earthworms, moles, rodents and human beings cause mixing of soil, the type of pedoturbation is called faunal pedoturbation. Floral pedoturbation is caused by plants and their roots growing in the soil. Upon decomposition of roots narrow, deep and vertical channels are formed in soil profile which is filled up by soil material of other horizons. Argillipedoturbation is due to presence of more amounts of swell-shrink clays in soil. In this process, soil is churned as in case of Vertisols (Black cotton soil of India). Besides these pedoturbation is also carried out by freezing and thawing, gases, crystallization and earthquakes.

Gleization: Gleization or gleysation is a specific soil forming process mainly related with transformation. This process forms gleyey soil or Gleysol. Humid climate with poor drainage, anaerobic, waterlogged and lower topographic areas and impervious parent material are some conditions that favour gleization.

In these conditions, reduction of iron and other elements takes place resulting greyish, blue-greyish or green-greyish colours in to the soil. This process may take place in whole profile or part of that. Some well aerated place like channels of plant roots and macro pores show orange-red mottles. These mottles contain more amount of iron than blue or grey mottles. Ferrous Sulphide may also form during gleization, which turns in to H_2SO_4 due to oxidation. Gleying word is also used in context of the reduction of iron and manganese compounds by microorganisms under anaerobic and submerged conditions.

Survey of Soil

Soil survey, Utility of soil survey, Objective of soil survey, Types of soil survey, Soil survey methodology, Soil maps, Soil survey interpretation, soil survey report.

Soil is an essential and important natural resource for agriculture. Soils provide nutrients and other prerequisites for the growth of crop plants and trees. Our numerous needs are fulfilled directly or indirectly by soils as they provide food, fibre, feed and fuel to us and to our animals. However, due to differences in soil forming factors and processes various types of soil are formed which are unique in characteristics, divergent in productivity and uses. Therefore, the understanding or knowledge about the type, nature, properties and geographical distribution of soils of any area is essential for correctly utilizing them in agriculture, forestry, irrigation, drainage, housing, industries, other works etc. The suitability of soils for their appropriate and befitting uses can be ascertained by their survey from soil material to farm levels.

Soil Survey

Soil survey can provide useful information for allocating the areas at various levels, as at national level for agricultural, rural, urban, industrial, recreational, wild life sanctuaries etc.; at city level for housing, parks, green belts, roads, railways, aerodrome, industries and waste management sites; at village level for housing, agriculture, pasture, grazing sites and common utility centres and at farm level for the cultivation of different crops, equipment shed, store etc. Besides this soil survey is essential for laying down different field experiment at a research centre and even to collect a representative soil sample of area for the laboratory analysis. In general soil survey provides all the information about the soil as what is the soil? How are they? Where they are located? What are their characteristics etc? Soil Survey is defined as the systematic examination, description, classification and mapping of the soils of any area are called as soil survey.

Utility of Soil Survey

Soil surveys are very much useful in the avoidance of various difficulties related with soil. Now a days requirement of soil survey is increasing continuously to avoid collapse of building due to poor foundation soil, breaking of water supply pipelines due to presence of swell-shrink clays and poor growth of orchards due to the presence of hard pan etc. Thus, use of soil survey is remarkably increasing in the development of command areas, soil conservation, rainfed irrigation, watershed management and reclamation of problematic soils.

Very precisely soil surveys are useful in the Land capability classification and suggesting the best suitability of soil or land for its appropriate and best utilization, Suggesting suitable crops and cropping systems for particular area, Development of pasture, range lands, parks, sanctuaries etc, Identification of the soil problems and suggestions for their reclamation and management, Planning and execution of different soil development strategies for various agroclimatic conditions, Planning and laying of field experiments for irrigation, nutrients, crops etc, Valuation of soil (land) and its usefulness for loaning and taxation and Selection of sites for roads, housing, irrigation projects, public works etc.

More than seventy seven percent area of our country (India) has been surveyed because of its usefulness and numerous field parties of central and state governments, universities and other organizations are continuously working hard to achieve the objectives of soil survey. An All-India Soil Survey Scheme was launched in the year 1956 which converted in to All India Soil and Land Use Survey Organisation (AIS&LUS) in the year 1959. In the Year 1969 bifurcation of AIS&LUS took place and reconstituted as a Directorate in the year 1973. A head quarter of National Bureau of Soil Survey and Land Use Planning (NBS & LUP) was established at Nagpur (India) in the year 1976 as one of the fourteen NRM Institutes of ICAR, New Delhi. At present there are five regional centres of NBSS&LUP at Bengaluru, Jorhat, Kolkata, New Delhi and Udaipur.

Objectives of Soil Survey

Soil survey provides information about soil of any nation for more skilled planning and execution of public policies related with soil and land. The objectives of soil survey may be both fundamental and applied.

Fundamental objectives: Fundamental objectives of soil survey are it increases our knowledge about the main component of the environment i.e., soil, it informs about the nature, properties and classification of soils and it proposes modals and principles regarding the genesis and development of soils, soilscapes and landscapes.

Applied objectives: Applied objectives of soil surveys are mainly related with agriculture, forestry and engineering.

In Agriculture, Selection of suitable soils for sustainable and eco-friendly agriculture, designing of suitable land use-cropping system patterns, assessment of yield and quality of crop produce at different management practices, assessment of soil behaviours under irrigated and rainfed conditions, identification of areas requiring soil conservation and identification of soil problems and suggesting their reclamation measures.

In case of forestry, Selection of areas for forests, suggestion of effective nutrient application and soil conservation programmes for efficient forest management and identification of soil-borne causes of forest degradation.

Similarly in engineering, Selection of site and assessment or quality of soil for housing, high ways, railways, aerodromes etc., Designing of irrigation canals, Selection of sites for septic tanks, river beds and central sewage system, Selection of sites for potable water, water harvesting and storage tanks and Tour planning are few applied objectives.

Types of Soil Survey

Soil survey is done according to agriculture, climate and topography of the area and objectives of soil survey. The soil surveys are of many types according to their objectives. Here five most important soil surveys i.e., Exploratory soil survey, Reconnaissance soil survey, Semi-detailed soil survey, Detailed soil survey and Intensive soil survey are described.

Exploratory soil survey: Exploratory soil survey is mainly done to depict dominant soil groups of large areas. This survey provides a basic inventory of soil resources at regional and national levels and correlate soil data of international territories. It is generally, used for education and land use planning. The scale of this survey is 1:50, 00,000 to 1:5, 00,000 and area covered by 1 cm^2 of map is 625 to 25 km^2. Approximately number of observations per 0.5 cm^2 of map is one per 250 to 50 km^2. Soil order is mapping unit.

Reconnaissance soil survey: Reconnaissance soil survey is mainly done to prepare the resource inventory of soil of a large area. It informs about the types and distribution of soils of any area. It indicates about the less fertile, eroded and salt affected soil along with the soil productivity and identification of the areas for intensive agriculture. Reconnaissance soil survey generates primary information for detailed soil survey, agricultural development and land use planning. This survey is mainly for state or national planning.

Base maps for this survey are topo sheets of survey of India or aerial photographs in the scale of 1:5, 00,000 to 1:1, 00,000 and area covered by 1 cm^2 of map is 25 to 1 km^2. Approximately number of observations per 0.5 cm^2 of map is one per 1/50 to ½ km^2. Soil morphological studies like texture, presence of carbonates, mottles etc. are studied with auger up to 1 meter depth at an interval of 1-2 kilometres (1 km for agricultural land). The drainage conditions and topography of the area are also recorded. Profiles are opened at some specific locations for detailed studies. The soil boundaries are delineated by extrapolation instead of complete traversing. Soil great group is mapping unit.

Semi detailed soil survey: This survey provides detail information on various types of soil for different type of objectives. This survey is done to obtain correlation between physiographic units and soils or soil properties of specific interest or general-purpose surveys for assessing the potential for agricultural development, location for new infrastructure, settlement etc. The validity of such correlation is assessed by random checking of the area. If correlation is found valid, the soil boundaries, based on random checks of physiography, are delineated. This survey is mainly done for regional planning. The scale of this survey is 1:1,00,000 to 1:25,000 and area covered by 1 cm^2 of map is 100 to 6.25 ha. Approximately number of observations per 0.5 cm^2 of map is one per 200 to 12.5 ha. Soil series and association are mapping units.

Detailed soil survey: Detailed soil surveys are done on priority basis like intensive agriculture, experiment station, intensive nutrient application, irrigation projects, tour planning etc. to obtain detailed characteristics of a particular area. The details about texture, depth, salinity, sodicity, erosion and soil formation are also obtained by this survey. This survey is mainly done for watershed planning.

The scale of this survey is 1:25,000 to 1:10,000 and area covered by 1 cm^2 of map is 6.25 to 1.0 ha. Approximately number of observations per 0.5 cm^2 of map is one per 12.5 to 0.5 ha. Sometimes the detailed base maps for detailed survey are cadastral maps in the scale of 1:3000 to 1:8000 and aerial photographs in the scale of 1:10000 to 1:15000. In this survey the main emphasis is given to soil profiles and surface features to differentiate soil types. Soil profiles are selected because of the heterogeneity of the soil and size and shape of the landscape. Morphological study of soil is done at each 100-meter interval with the help of auger and soil boundaries are delineated with the help of actual traverses. Soil mapping unit of detailed soil survey is soil series.

Intensive soil survey: Intensive soil survey is very much time, labour and money consuming. Therefore, it is done under specific conditions requiring very detailed mapping of smaller areas. Intensive soil survey is done for experimental fields, irrigation schemes, civil engineering works, specific purposes etc. to have the direct bearing on the objective. The scale of this survey is 1:10,000 and larger. Area covered by 1 cm^2 of map is 1.0 ha and less. Approximately number of observations per 0.5 cm^2 of map is one per 0.5 ha and less. Soil Type or Soil Phase is mapping unit

Equipments and Instruments Used in Soil Survey

A soil surveyor requires numerous equipments and instruments for location, soil sampling, basic physical and chemical properties of soil, preliminary soil classification etc., during soil survey. These are hand held GPS and compass to know the longitude, latitude and direction etc., spade, khurpi and kudal for digging the profile, different types of augers with extension rods to collect soil samples, polythene and cloth bags and tags for storing soil samples, Abney level meter to know the slope, Munsell's colour book to know the colour of soil, Measuring tape to measure thickness of horizons and depth of soil profile, Soil penetrometer to know hardness and compaction of soil layer, Magnifying lens to identify soil structure and to study clay layer etc. Dilute hydrochloric acid bottle to test the carbonates, distilled water bottle, various maps, stationery and field note book, First Aid Kit etc.

Soil Survey Techniques

Generally, three techniques, Free survey, Grid survey and General-purpose Aerial Photo Interpretation (API) Survey, are used for soil survey.

Free survey: In free surveys, the surveyor chooses observation points assumpting the changes in physiographic features observed in aerial photo or satellite imagery interpretation. Other soil surface features like colour, vegetation, land use etc., are considered to differentiate among soil characteristics. The number of observations may vary as the surveyor confirms the inferred boundaries and checks the uniformity of soil within each boundary. On small scales, the inferred boundaries are often accepted as soil boundaries. However, on a large scale, several new boundaries within the physiographic boundaries are recognized depending upon the scale mapping.

Grid survey: Grid survey is adopted for making small areas such as a micro watershed or area of agriculture research station. The traverse lines are located on a grid pattern and the density of mapping is adjusted according to the area surveyed so that the number of observations per cm^2 of the final map is independent of the scale. Generally, four or five observations per hectare are recommended. Points of comparable observations are drawn. Now a days with the help of computer, grid surveys are done at geo reference points (with latitude and longitude) to digitize the database and generating several thematic maps of practical value. Grid surveys are expensive.

General purpose API survey: General purpose API survey is used for low intensity surveys in which mapped boundaries (soils or land resources) are largely or entirely referred from Air Photo Interpretation (API) with a free survey at low intensities of field observations to characteristics physiographic units. Sehgal (1996) developed the three-tier technique of surveying and mapping of soils and physiography by using Landsat satellite images in the scale of 1:250,000. The three tiers of this technique are image interpretation and land form analysis, field surveys and laboratory analysis and cartography and printing. This whole system is digitized to form a database.

Soil Survey Methodology

The methodology of the soil survey is basically similar as earlier days but the trends have been changed due to the advancement in the technology. In the earlier days soil scientist used blank sheet or cadastral map as base map and his foot step to measure distance. Now a day aerial photographs and aerial photo interpretation (API) technique are widely used. However, soil survey methodology has six main steps as

1. Collection of necessary preliminary information of area to be surveyed,
2. Verification of collected information by ground truth,
3. Actual soil survey, soil sampling and detailed soil profile observation,
4. Analyses of collected soil samples for their properties in laboratory,
5. Soil boundary verification, extrapolation and delineation,
6. Data processing, interpretation, soil map preparation and report presentation. These are briefly described here:

1. Collection of necessary preliminary information: All the necessary preliminary information should essentially be collected before starting actual soil survey. This information may be related with the landform, landform vegetation relationships, land use, geology, climate, topography, erosion, hydrology and irrigation facilities, agriculture, forestry and horticulture status, their past and present productivity etc. and all relevant existing data.

2. Verification of collected information: After collecting all the necessary preliminary information one multidisciplinary pilot team including local residents should go to the particular area for ground truth and to verify and correct the information.

3. Actual soil survey, soil sampling and detailed soil profile observation: On the basis of corrected information actual soil survey work is started to establish distribution of soils and their units. Actual soil survey must be flexible to accommodate changes in definitions based on additional information gathered during survey. Preliminary soil classification and mapping is done simultaneously.

In actual soil survey first of all a base line is selected on the base map. The base line may be the main road or main path of that area. Now the inspection points are selected at both sides of base line and marked by the pencil and scale. Inspection points will be on the roads or paths where one can go by jeep for reconnaissance survey and at every 100-meter interval for detailed survey.

Soil sampling and detailed soil profile observation are done by starting survey from point number one. Bore a hole in the soil with the auger and note texture, structure, moisture, colour, carbonates etc. in the surface and sub-surface soil in the field and collect the soil samples for laboratory analysis. Other features of soils like thickness of horizon, eluviation, and illuviation, presence of mottles, gravel, stones or nodes and depth of bed rock are also recorded from the points inspected. By this way all the points are inspected, their character-features are note down and soil samples are collected for the analysis in the laboratory.

Simultaneously representative soil site is identified and selected for detailed profile studies and soil profile pit is opened with the dimensions approximately of 2.5x2x1m in length, depth and width, respectively. Depth may be more up to bed rock, hard pan, salts, water table etc. In this pit stairs are made opposite of observation site for easy movement. In this pit vertical variations like colour, texture, structure, resistance, consistence, organic matter, clay coatings salts are recorded. Concretions, mottles, rooting pattern, burrows etc. among different horizons are also carefully observed and recorded.

Soil sampling requires highest accuracy and care for the reproducibility of soil properties studied so far. Generally, more than 10 percent variation in the duplicate analysis of soil samples is unacceptable. Therefore, the surveyor must take great care that the samples collected are true representative of the defined soil types and interpretation of the results should not differ due to the errors in sampling procedure.

Soil samples may be disturbed and undisturbed type. Disturbed soil samples are usually taken to assess various chemical, mineralogical, microbiological properties etc. and undisturbed samples are collected for soil physical properties like pF, bulk density etc. The depth of soil sampling in soil surveys may be 0-15, 15-30, 30-45, 45-60, ……,60-90, 90-120 or 0-10, 10-20, 20-30, ……., 90-100…., 140-150 cm or according to the actual depth of soil horizons in the soil profile. After finalizing above profiles of identified soil are opened, detailed studies of each horizon are done and soil samples from each horizon are collected for laboratory analysis.

4. Analyses of collected soil samples for their properties in laboratory: Analyses of soil samples provide data for soil classification and purpose of soil survey decides method of analysis. Analyses of collected soil samples are done after their proper processing. Suitable analytical methods as outlined in soil survey manuals are adopted for particle size distribution, EC, pH, organic carbon, nitrogen, phosphorus and potassium contents, cation exchange capacity, exchangeable bases, moisture holding capacity, saturated hydraulic conductivity etc., of each horizon.

5. Soil boundary verification, extrapolation and delineation: Based on data collected soils are classified in to soil series, type and phases. Actual boundaries of soils are delineated by recording the homogeneity and heterogeneity in the soil characters with in the distance less than 100 meters. Any soil having area not more than the 15 percent of the area of main soil is not shown separately but shown with in the area of main soil.

6. Data processing, interpretation, soil map preparation and report presentation: Data processing is very much essential as during soil survey huge data of various types and formats are collected. These data may be managed, maintained and distributed as hard copy, electronically digitized form or both. Various methods are used to process the collected data but now a day Automated Data Processing System (ADPS) is utilized everywhere including soil survey. ADPS can perform time consuming tasks and retrieve data in no time. However, it requires computers with word processors, initial database, spreadsheets, Geographical Information and Global Positioning Systems, specific softwares, highly skilled personnel etc.

Soil Survey Interpretation

One important aspect of soil survey is to interpret the collected soil data correctly for various agricultural and non-agricultural uses or end users. Many properties of soils like texture, structure, swelling and shrinkage, presence of $CaCO_3$, salts, organic matter, nutrients status, concretions, hardpans, water table, water movement, depth of soil etc. directly and indirectly influence the uses of soil for crop production, selection of crops, water supply, construction of rural, urban and industrial structures, waste management etc.

A correct interpretation of soil data or properties will suggest the befitting use of soils. Some examples are soils having hard pan or bed rock above one meter depth are not suitable for laying orchard and septic tanks. A soil with deep water table, high infiltration rate and rapid permeability will be suitable for constructing septic tanks. Soils having high amount of swell- shrink clays are not good for construction and laying pipe lines. Soils with high salt contents are not suitable for plant growth as well as laying the pipe lines and iron structures.

Soil Map Preparation

Various countries have map of their soil resource as they provide useful information about soil according to objectives and types of soil survey. Soil map depicts the soil series or other soil mapping units in relation to the prominent physical features and significant agricultural systems of land surface. The soil mapping units are soil series, soil type, soil phase, soil association and soil complex. Soil maps are prepared on the basis of type of soil survey or base map and actual survey done.

Generally, the base soil maps are of three types i.e., Reconnaissance soil map, detailed soil map and reconnaissance-detailed soil map. Other soil maps, like generalized, climatic and inventional soil maps can be prepared with the help of these base soil maps. Soil maps are most used for land evaluation and planning, agricultural extension, environmental management etc. Traditional soil maps typically show only general distribution of soils, accompanied by the soil survey report. Many new soil maps are derived using digital soil mapping techniques. Such maps are more informative in context and show higher spatial detail than traditional soil maps. The legends of maps should be self-explanatory and serve the purpose up to maximum extent.

Like other cartographical maps soil maps also have scales. The scale determines how much details can be shown on the map. Small-scale soil maps (e.g., 1:100,000 to 1:250,000) cannot show as much detail on the ground as large-scale maps (e.g., 1:12,000 to 1:50,000). Because of this, in small-scale maps, individual soil types may often be grouped together to make soil association. A large-scale map could show boundaries for actual soil types as they can fit into the map more easily. However, map scales vary country to country. The use of computer mapping tools such as a Geographical Information and Global Positioning Systems allow maps of different scales to be displayed on the computer screen.

Soil Survey Report Presentation

Lastly soil survey report is prepared and published along with maps and diagrammes. A good soil survey report provides information on soil, land and physiography of the area surveyed under various sections. First introductory section deals with the importance and objective of the survey. Second section informs about general geography and land use pattern of the area. Next section is related with soil survey techniques.

Fourth section is the main section of the report. It provides all the information upon soil like texture, structure, characters of horizon, thickness of profile, slope, erosion, salinity and sodicity, present land use and condition, stone, rocks, percent sand, silt and clay, water holding capacity, pH, organic matter status, carbonates, exchange capacity, nutrient status, salt etc. in horizons, soil forming process, nature and type of bed rock, ground water table, soil fertility and productivity etc. The fourth section also contains information on soil physiographic relation, geographic position of soils, morphological, physical and chemical properties of soils and classification of soils.

The information on degraded and problematic soils is furnished in chapter five. Sixth chapter indicates about present land use, suitability of soil for different crops and uses. Mainly this chapter indicates soil survey interpretations. The seventh and last chapter relates with the summary and conclusion.

Morphology of Soil

Soil profile, Soil horizons and layers, Main characteristics of master horizons, Sub ordinate distinctions of Master Horizons, Vertical sub divisions and discontinuities, Soil micromorphology, Soil monolith.

The long-continued action of soil forming process on parent material for hundreds and thousands of years in the presence of other soil forming factors create horizontal layers in the parent material and a new natural body called as soil is formed. This newly formed soil body has quite different characteristics and morphology than its parent material. Study of soil under field conditions comes under soil morphology that indicates physical characters of a soil profile in terms of kinds, thickness and arrangements of horizons in it and the colour, texture, structure, consistency and porosity of each horizon. Generally, soil colour and mottling, texture, consistency at different moisture conditions, structure, cutans, nodules and concretions, voids, horizon boundaries, master horizons and diagnostic surface and subsurface horizons are studied in the field under soil morphology. This chapter is mainly based on Soil Taxonomy.

Soil Profile, Soil Horizons and Layers

Soil profile is the vertical section of the soil through all its horizons up to the parent material. In other words, soil profile is the vertical section of soil from its surface to bedrock. A soil profile has different layers of soil or soil material, which are called as soil horizon. The soil horizon may be defined as relatively uniform layers of soil or soil material which is approximately parallel to land surface and differ from adjacent related layer in physical, chemical, biological and mineralogical properties. The horizons are identified due to the differences in their properties. Horizons are designated by capital letters as O, A, E, B, C, and R. These are called as master horizons. Sometimes few master layers are also present in soils. These are designated as L, M and W. L indicates coprogenous earth i.e., sedimentary peat, diatomaceous earth and marl. M layer is root limiting, horizontal continuous layer below the soil surface which is made up of geo textile liner, asphalt, concrete, rubber, plastic etc. W is water layer within or below the soil.

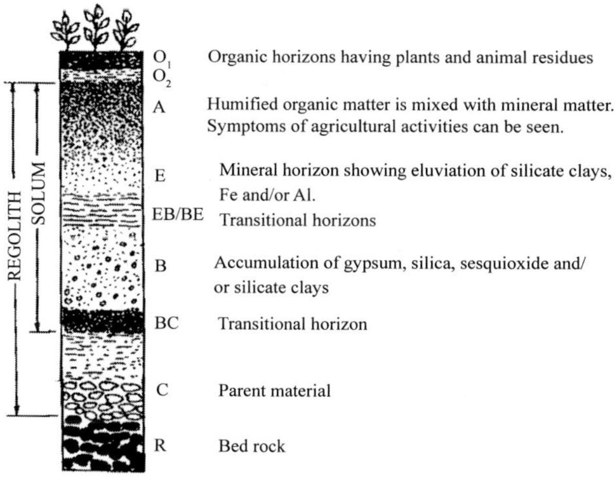

Fig. 2.3: A hypothetical soil profile showing sequence of its all horizons

Soils can have all these horizons or less. Any soil may lack in A, E, B, C or R. Sometimes transitional horizons are also formed which show main characteristics of one master horizon and secondary characteristics of another horizon. Transitional horizon is designated by two capital letters, such as AB, EB, BA, BC or CE. AB indicates that the characteristics of horizon are more like A than B.

Sometime distinct portions of two different horizons are also founds, in which one portion is covered or surrounded by other. These are also called as transitional horizons and designated as E/B, B/E or B/C. E/B indicates that horizons have different part of E and B horizon but E is more effective and covering B. The reverse is true with B/E.

Main Characteristics of Master Horizons

O is organic horizon and A, E, B, C are mineral horizons. Capital letters are base to which other symbols are attached. These horizons are formed due to long and continued activities of soil forming process. Master Horizons are described here in short.

O Horizon: This is an organic horizon. This horizon is dominated by the deposition of undecomposed or partially decomposed organic materials such as leaves, needles, twigs, moss and lichen. The deposition of organic material may be on the top of mineral or organic soil. O horizon is found in forest lands and not in pastures, range and agricultural lands.

A Horizon: A horizon is mineral horizon which is formed at soil surface or below the O horizon. Main characteristics of A horizon are accumulation of humified organic matter intimately mixed with mineral fraction, dark colour than just lower horizon, mostly absence of properties of E and B horizons and presence of properties originated due to ploughing, pasturing, grazing or similar agricultural activities.

E Horizon: This is a mineral horizon showing eluviations. Main characteristics of E horizon are the loss (eluviation) of silicate clays, iron, aluminium or combination of these. This causes dominance of sand and silt fractions of quartz and other dominant minerals. The E horizon is lighter in colour and has quantitatively less organic matter than A horizon.

B Horizon: B horizon is formed below the O, A or E horizon. Generally, the signs of original rock structure are not found in B horizon. It has some specific characteristics like illuvial accumulation of silicate clays, iron, aluminium, humus, gypsum or silica alone or in combinations; removal of carbonates; residual accumulation of sesquioxides or coatings of sesquioxide that change the colour of horizon and presence of alternations that form silicate clay or release oxides. Generally, structure of B horizon is blocky or prismatic. Sometimes granular structure may also find. Organic matter status is more than E horizon but clearly less than A horizon.

C Horizon: C horizon is the unconsolidated layer below the solum and does not have the characteristics of O, A, E, or B horizons. It excludes hard bed rock. The effect of soil forming process in this horizon is very less. The material of C horizon may be similar or dissimilar to the parent material from which solum is developed. The upper portion of C horizon may be the part of solum due to the continuous action of weathering, erosion and soil forming processes.

R Horizon: This is the horizon of hard bedrock consisting granite, basalt, quartzite and indurated sandstone or limestone. The structure of this horizon is so coherent that can only be chipped or scraped by spade but not dug. R horizon may have some small cracks which can be penetrated by few finer roots.

Subordinate Distinction of Master Horizons

Master horizons and layers of profile also have some subordinate distinctions. The subordinate distinctions inform about the operation of specific soil forming process, presence of some specific material in the horizon or any specific character of the horizon. These are indicated or designated by suffixing small letters with the master horizon. The examples are Oa, Ap, Bh, Cr etc. indicating highly decomposed organic material in O horizon, ploughing in A horizon, illuvial accumulation of organic matter in B horizon and layer of soft bed rock in C horizon, respectively. All the sub ordinate distinctions are described here:

a	=	Highly decomposed organic material used with 'O'
b	=	Identifiable buried genetic horizon
c	=	Concretions of Fe, Al, Mn or Ti or hard non-concretionary nodules cemented by material other than silica. co = Coprogenous earth, related with sedimentary peat
d	=	Physical root restriction, related with non-cementing natural root restricting layer di = Diatomaceous earth, related to limnic layer and used with L layer
e	=	Intermediately decomposed organic materials. Used with 'O'
f	=	Frozen soil or water having permanent ice in horizon. ff = Dry permafrost, related with horizon or layer continually colder than 0°C and not cemented by ice
g	=	Strong gleying showing reduction and removal of iron during soil formation or preserved a reduced state due to saturation with stagnant water. Generally, used with 'B' and 'C' horizons.
h	=	Illuvial accumulation of organic matter. Used with 'B' horizon.
i j	= =	Slightly decomposed organic material. Used with 'O' horizon Accumulation of Jarosite, $KFe_3(SO_4)_2(OH)_6$, mineral jj = Cryoturbation
k	=	Accumulation of calcium carbonate. kk = Accumulation of pedogenic calcium carbonate.
m	=	Continuous or nearly continuous pedogenic cementation or induration of soil. ma = Marl, used with L layer
n	=	Accumulation of exchangeable sodium.
o	=	Residual accumulation of sesquioxides.
p	=	Ploughing or other agricultural activities. Used with 'O' or 'A' horizons.
q	=	Accumulation of secondary silica.
r	=	Weathered or soft bed rock. Used with 'C' horizon
s	=	Illuvial accumulation of sesquioxides and organic matter, used with 'B' horizon se = Presence of sulphides in mineral or organic horizon ss = Presence of slickensides

t	=	Accumulation of silicate clays that has formed in the horizon or moved in to the horizon by illuviation.
u	=	Presence of artifacts
v	=	Plinthite formed with high iron, low humus and reddish material.
w	=	Development of colour and or structure. Used with 'B' horizon.
x	=	Showing the presence of fragipan with genetically developed firmness, brittleness or high bulk density.
y	=	Accumulation of gypsum. yy = dominance of gypsum in horizon
z	=	Accumulation of salts having more solubility than gypsum.

Vertical Sub Divisions and Disconutinuties

Besides the subordinate distinctions of master horizons, the international numbers are also suffixed to indicate vertical subdivisions and prefixed to indicate discontinuities. Vertical subdivisions indicate changes in the morphological characters, texture, structure or colour with in a master horizon i.e., with in B horizon the successive layers may be B1, B2, B3, B4 etc. A discontinuity indicates the significant change in particle size or mineralogy. The numbering starts with the second layer of different material and '2' number is given. Below this the number '3' is given. These prefixes or international numbers indicate only change in material with horizons or layer and can be shown as A, 2B1, 2B2, 2B3. Here prefix 2 with 'B' show the discontinuity and suffix 1, 2 and 3 with 'B' are the sub divisions of 'B'.

Solum: The upper most part of the soil profile consisting upper horizons (A, E and B) in which the parent material has been modified or altered and which is under the influence of plant roots.

Regolith: Generally, engineers use this term, as the synonym of soil. Regolith is the unconsolidated layer of weathered rocks, minerals and soil material overlying solid rock. By this way, all the horizons above 'R' come under regolith.

Pedon: Pedon is the smallest, three dimensional, and somewhat hexagonally shaped soil body whose area is 1 to 10 square meters. The lateral dimensions of pedon should be so wide that allow the study of properties of horizons present in it. A pedon is considered as a soil.

Polypedon: Polypedon is the group of many similar and adjoining (contiguous) pedons bound together by the pedons of dissimilar characters. The area of polypedon may be quite variable from few square meters to many square kilometres. Polypedon is the real soil body that is classified in to soil series and higher categories of soil classification.

Soil individual: Unlike to human, animals or plants, soil cannot be seen as an individual unit. The presence of soil under a continuum creates the problem to define the soil individual as a basic entity. Cline (1949) defined soil individual as the smallest natural body that can be defined as a thing complete in itself. By this way polypedon is considered to be a soil individual which is the smallest natural unit and physical body of the soil that differs from its adjoining units on the landscape. The boundaries of soil individual are up to the place where significant differences in one or more soil forming factors are observed.

Soil control section: The soil control section is the portion of soil profile confined to arbitrary depth or depth range. Generally, the layer of soil and parent material in the depth range of 25 to 100 cm is considered as soil control section. Soil control section is used in soil survey.

Soil Micromorphology

Soil micromorphology is study of soil morphology with the help of some optical instruments. The instruments may vary from simple hand lens to the electron microscope. Micro morphology of soil helps to know the soil genesis, soil plant relationship, nutrient uptake etc.

Soil plasma, soil skeleton, soil voids and micro pedological features are few general terms of soil micromorphology. Soil plasma is the soil material that can move or have moved, rearranged and assembled by the soil forming processes. Soil skeleton is the soil material that cannot easily move, rearranged and assembled by the soil forming processes. Void is the spaces between solid soil materials.

Micro pedological features are some definite microstructures like cutans, pedotubules and slicken sides. Cutans are the coating or deposition of soil material on the surfaces of peds or natural surfaces of other soil materials. Most common example of cutans is clay cutan in which translocated clay particles deposit on the ped surface. Cutans are classified according to their position, composition and boundary conditions. Pedotubules are the external single or branched tubular system of soil material. Pedotubules are somewhat like to channels present in soil. The large size of pedotubule that can be seen with naked eye is called as Crotovina (Krotovina). Coatings on soil materials are also considered a part of micromorphology. Calcium Carbonate coatings in a soil of road cut at Chakrata, Dehradun Road can be seen in following figures.

Fig. 2.4: Calcium Carbonate coating on soil material
(Photo Dr. Venkatesh Bharadwaj)

Fig. 2.5: Downward movement of Calcium Carbonate coating
(Photo Dr. Venkatesh Bharadwaj)

Soil Monolith

Soil monolith is the vertical section of a soil profile removed from the field, mounted in the box and installed in the soil museum or laboratory for display or study. Macro monolith and micro monolith are two types of monoliths. Macro monolith represents the position, thickness and depth of horizons in the same manner as seen in the field during survey. Thus, a wooden or tin box measuring 10 cm in breadth, 5 cm in depth and length equal to the depth of 'C' horizon is used for the removal of profile from the field. Micro monolith is much smaller in size. Generally, an acrylic box of 15 - 30 cm in length, 5 cm in breadth and 3 cm in depth is used for mounting the soil profile. The thickness of each horizon is reduced up to ten times according to the length of the box. Therefore, a15 cm thick horizon of a 150 cm deep profile will be accommodated in 1.5 cm long space in the box. All the horizons of the profile are serially affixed in the box with the help of resin or permanent adhesive.

Classification of Soil

Purpose of soil classification, Systems of soil classification, old soil classification systems, Comprehensive soil classification system, Soil Taxonomy, World Reference Base (WRB) - The Revised Soil Legend of FAO-UNESCO, Numerical soil classification, Universal soil classification and Unified soil classification, Land Capability Classification, Land suitability classification.

Soils can also be classified just like plants and animals in to different categories according to their characteristics and properties. The classification of soils is not a new one. It journeyed from Vedic Era to Numerical classification of soils and very recently up to Universal soil classification system via Dokuchaev etc. In the Vedic period soils were classified on the basis of their colour, place (topography), fertility etc. Soil classification is defined as the grouping or categorization of the soils according to their characteristics.

Purpose of Soil Classification

It is necessary to arrange systemically the information on soil collected after soil survey in to appropriate classes. The main purposes of soil classification are,

1. Organization of collected knowledge about soils,
2. Easy understanding of relationships among different soils,
3. Remembering characteristics of soils,
4. Establishing new relationships and principles among soils being classified and
5. Using soils, their behaviour, productivity, best uses and results of soil-based experiments in to further research and agrotechnology transfer.

Systems of Soil Classification

Many systems of soil classification were developed and tested since long back. Now a day many countries like U.S.A., Africa, Australia, Canada, France, Brazil, Russia, China etc. have their own soil classification system. In India also soils were classified according to their colour, topography, fertility, irrigability and productivity also. FAO and UNESCO also proposed the soil classification systems. The system of soil classification can be grouped in to old soil classification systems and new soil classification systems. In India, soil had been classified from the ancient period itself even though it was not as detail as the new or modern classifications.

Old soil classification systems: These systems were quite simple and generally, based on the use of the soil or any recognizable soil property. These systems help in the taxation and adopted by revenue department. In the ancient India, the classification was based on the soil fertility as Urvara (fertile) and Usara (sterile), colour and position on the land scape. Some old systems are briefly described here,

Crop suitability classification: Soils were classified according to their suitability to grow different crops like wheat soil, groundnut soil, cotton soil, paddy soils, sugarcane soils etc.

Physical classification: This was based on physical properties of soils. Mainly soil texture or soil structure was used and soils were classified as sandy soil, loamy soil, clayey soil, sandy loam soil, aggregated soil etc.

Chemical classification: The chemical properties of soil were also used for their classification and they were classified as mineral soils, organic soils, acidic soil, sodic soil, calcareous soil etc.

Physiographic classification: The soils were also classified as per their topographic location or position on the land scape as hill soils, forest soils, roadside soils, terrace soils, low land soils, upland soils, dry land soil, water logged soil, submerged soils etc.

Geologic classification: Geologists classified soils in to residual or sedentary soils, transported soils. They also gave name like granitic soils, basaltic soils, alluvial soils, aeolian soils, colluvial soils etc. because of rocks and soil parent material.

New soil classification systems: The new soil classification systems were started at Russia where the concept of soil as an independent natural body was established. Some new systems are described here:

Genetic Classification: This system was proposed by Russian scientist Dokuchaev (1900). In this system soils were categorized in to normal, transitional and abnormal soils. Later on, these categories were termed as Zonal, Intrazonal and Azonal soils, respectively. The zonal soils comprise fully developed soil profiles and their main characteristics are depending upon climate and vegetation. Characteristics of intrazonal soil are depending upon topography, parent material, and drainage or soil age. There was no profile development in azonal soils. This system of Dokuchaev and his associates was adopted by most of the countries of the world including India till 1964-65.

Coffey's soil classification system: Coffey (1912) classified soils in to five major classes according to their properties. These classes were arid soils, dark coloured prairie soils, light-coloured timbered soils, black swamp soils and organic soils.

Morphogenetic system: Marbut (1928 and 1935) proposed morphogenetic soil classification system based upon the iron-aluminium and lime content in the soil. He divided soils in two main classes Pedalfers and Pedocals which were supposed to present in areas of high rainfall and high evaporation, respectively.

Comprehensive system: Baldwin et.al. (1938) started comprehensive soil classification system which was revised by Thorp and Smith (1949). In this revised system the soils were grouped in to three orders (Zonal, Intrazonal and Azonal), nine suborders (six for zonal soils, three for intrazonal soils and none for azonal soils) and in all 37 great soil groups.

Since 1951 the comprehensive system was further developed by U.S. soil survey staff and their associating agencies. In each year since 1953 they started publishing and testing 'One or An Approximation' for the classification of soils. After thorough commentary, suggestion and discussion the 'Seventh Approximation' was published in 1960 (Soil survey staff 1960). Later, its two supplements were also published in the year 1964 and 1967. All these were used by various countries including India for classifying the soils.

Soil Taxonomy

Among the comprehensive soil classification systems, the soil taxonomy is most advance and latest one. The seventh approximation and its supplement were intensively examined by the scientist of different countries. Their suggestions were included, comments were rectified and a comprehensive soil classification system was published in the year 1975 entitled 'Soil Taxonomy' (Soil survey staff, 1975). Its second edition named 'Soil Taxonomy: A Basic System of Soil Classification for Making and Interpreting Soil Surveys' was published (Soil survey staff, 1999, United States Department of Agriculture, Agriculture Natural Resources Conservation Service, Handbook Number 436). This national soil classification system of USA (Keys to Soil Taxonomy, Twelfth Edition 2014) has widespread international acceptance, approval and adopted by various countries in Latin America and Asia for their soil survey and classification due to its specialties.

Specialties and Drawback of Soil Taxonomy

Soil taxonomy has many specialties or distinctive features over earlier classification systems. The specialties are,

1. The soil classification is based upon soil properties which exist today and can be measured in field as well as in laboratory.

2. A new nomenclature system is used for soil classification. The names of soil are connotative of soil properties. Thus, one can understand properties of soil by its name. The connotatives are mostly derived from Greek and Latin languages, Thus, they are understandable in other languages also.
3. The soil properties which affect the soil genesis or which are the result of soil genesis are also considered. Therefore, the soils of unknown genesis can also be classified.
4. Soil is at the central place of classification instead of soil forming processes, factors, climate or geology.
5. There is uniformity in soil classification because of large number of scientists of numerous countries has approved and using soil taxonomy.
6. Soil taxonomy is sequential plan of classification for easy remembrance of soil.

The only drawback of the Soil taxonomy is that the terminology below order and sub-order is quite difficult and incomprehensive. Even soil taxonomist cannot understand it without key.

Structure of Soil Taxonomy

Structure of soil taxonomy is like to biological taxonomy. It has six categories from higher to lower level of generalization. The categories are order, sub order, great group, sub group, family and series. Order, suborder and great group are higher categories, whereas sub group, family and series are lower categories. A simplified description of the six categories (order to series) of soil taxonomy and their differentiating characters are given in table 2.3.

Table 2.3: Description of Categories

Category	Number of taxa	Description of differentiating characteristics
Order	12	Orders are based on soil morphology resulted due to soil forming processes and indicated by major diagnostic horizons.
Sub order	64	Suborders are sub divisions of orders. These are related with genetic homogeneity, wetness, climatic change, parent material and vegetational effects on soil properties.
Great group	300	The main emphasis is on presence of diagnostic horizon (except Entisol), and presence or absence of diagnostic layers plinthite, fragipan, base status, soil temperature and moisture regimes.

Category	Number of taxa	Description of differentiating characteristics
Sub group	2000	Central concept or great group and properties indicating, intergradation of other great groups, suborders and orders on the basis for differentiation.
Family	Numerous, more than 4500 in U.S.A. only	Soil properties important for root growth, like texture, mineralogy and mean annual soil temperature at 50 cm depth are considered for differentiating families.
Series	Numerous, more than 1649 in India and more than 19000 in USA	Series is lowest category whose differentiating characteristics are kind and arrangements of horizon, colour, texture, structure, consistence, pH and EC, organic matter, hard layers etc. Series is indicated by the name of geographical place where it was first seen or recognised.

Order

There are twelve soil orders. All the soils of world will come in any order out of twelve orders. Name of these orders can be remembered by an abbreviation (GAUSVIAHOMEA). The names, formative element, derivation of formative element, diagnostic horizon and prominent characteristics of orders are given in Table 2.4

Table 2.4: Brief description of soil orders

Order's name		Formative element	Derivation of formative element	Diagnostic horizon and prominent characteristics
1.	Entisol	ent	Recent	Recent soils without diagnostic horizon, or with ochric or anthropic epipedon, poor or no profile development.
2.	Inceptisol	ept	Inception Latin-Incep-tum-beginning	Embryonic soils of humid region, No spodic, argillic, nitric, oxic, petrocalcic and plinthite. With altered or cambic B horizon or cambric, mollic or plaggen epipedon.
3.	Alfisol	alf	Pedalfer	Soils of humid and sub-humid region with more than 35% base status, argillic, natric or kandic horizon and ochric epipedon.
4.	Mollisol	oll	Latin- Mollus-soft	Dark coloured soils of grassy vegetation region with more than 50% base status and mollic epipedon which is not hard and massive when dry.
5.	Ultisol	ult	Latin-Ultimus-last	Forest soil with less than 25% base status and argillic or kandic horizon.

Order's name		Formative element	Derivation of formative element	Diagnostic horizon and prominent characteristics
6.	Aridisol	id	Latin-Aridus-dry	Soils of arid (dry) region with ochric or anthropic epipedon and either salic, calcic, gypsic, cambic or duripan. Sometimes argillic or natric horizon with aridic soil moisture regime.
7.	Vertisol	ert	Latin-Verta-turn, inversion	Dark coloured soils with swell-shrink clays, deep cracks when dry. 30% or more clay up to 50 cm depth or lithic or paralithic contact, slicken sides.
8.	Oxisol	ox	France-oxide	Highly weathered soils with oxic horizon and without argillic horizon.
9.	Andisol	and	Japan-and-black soil	Andic soil properties in 60% or more thickness of the profile, less than 25% organic carbon.
10.	Spodosol	od	Greek-spodos-wood ash	Spodic horizon within 2 meter, more accumulation Fe, Al, and humus but no plaggen epipedon.
11.	Histosol	ist	Greek-histos-tissue	Peat or bog soil with more than 30% organic matter up to the depth of 40 cm.
12.	Gelisol	el	Latin-gelare-freeze	Soils with Permafrost or Gelic materials within 100 cm of the soil surface and Permafrost within 200 cm of the soil surface.

(*Sources*: Soil Survey Staff, Soil Taxonomy: A Basic System of Soil Classification for Making and Interpreting Soil Surveys, Second Edition, 1999 and Keys to Soil Taxonomy, Twelfth Edition, 2014)

Diagnostic Horizons

Pedogenic processes form the diagnostic horizons which have prominent and measurable soil properties. Diagnostic horizons are used to identify soils and classify them in different categories particularly great group. Soil taxonomy describes numerous diagnostic horizons. These can be grouped in to two, surface diagnostic horizons also called as epipedons and sub-surface diagnostic horizons called as endopedons.

Epipedons (the surface diagnostic horizons): Epipedon is derived from two Greek words 'Epidermis-skin and pedon-soil or Epi-over, upon and pedon-soil'. So, epipedon is skin of soil or over and upon soil i.e., the diagnostic horizon present or formed on the soil surface. This includes the upper part of the soil darkened due to the presence of organic matter or shows evidence of eluviation or both. There are eight epipedons namely mollic, umbric, ochric, histic, plaggen, anthropic, folistic and melanic. The first three epipedons (mollic, umbric and ochric) are commonly observed in Indian soils. The brief description of eight diagnostic horizons is given here:

Mollic: A dark coloured surface horizon, colour darker than 5.5 when dry and 3.5 when moist, 1% or more organic matter in the upper 18 cm depth after mixing, strong structure, thickness 25 cm and base saturation more than 50%.

Umbric: Latin word, umbra i.e., shade or dark. An epipedon just like mollic except base saturation less than 50%.

Ochric: Greek, ochros i.e., pale. It does not meet the requirements any of the other seven epipedons as it is too thin or too dry and hard and massive when dry. Light coloured surface horizon, organic carbon less than 1% or less.

Histic: Greek, Histos i.e., tissue. A surface horizon with more than 20-30% organic matter depending upon clay content, saturated with water for 30 days at some season of the year. Thickness is less than 30 cm if drained or 45 cm thick if not drained.

Plaggen: German, plaggen i.e., sod. A man made more than 50 cm thick surface horizon developed due to long continued application and mixing of manures.

Anthropic: Greek, Anthropos i.e., man. A transformed surface horizon somewhat similar to mollic epipedon containing more than 250 mg kg^{-1} 1% citric acid soluble P_2O_5 or dry more than 10 months if not irrigated. Anthropic is formed under long continuous cultivation and fertilization.

Folistic: A surface layer with two or more horizons, saturated with water for less than 30 days in normal years containing 75% or more Sphagnum fibres or bulk density less than 0.1 g/cm^{-3}

Melanic: Upper boundary at or within 30 cm is mineral soil surface or upper boundary of an organic layer with andic soil properties and a melanic index of 1.70 or less.

Endopedons (sub surface diagnostic horizons): The term endopedon is derived from Greek word Endodermis-inner skin and pedon-soil. Thus, endopedon is inner skin or sub surface or lower part of the soil where soil materials generally, accumulate. These are designated as E horizon or part of A or E horizons.

In all there are twenty-two endopedons (agric, albic, anhydritic, argillic, calcic, cambic, durinodes, duripan, fragipan, glossic, gypsic, kandic, natric, orstein, oxic, petrocalcic, petrogypsic, plasic, salic, sombric, spodic and sulphuric). The short description of these sub surface diagnostic horizons is given here:

1. **Agric:** Latin, Agric i.e., field. Agric is an illuvial horizon formed due to the accumulation of clay, silt and humus below the ploughing depth.
2. **Albic:** Latin, Albus i.e., white. Albic is bleached E horizon with thin or discontinuous coatings on the sand or silt and colour values about 5 when dry and 4 when moist.
3. **Anhydritic:** Anhydrite (Calcium Sulphate) accumulation is 5% or more by weight through neoformation or transformation, thickness is 15 cm or more and hue is 5Y.
4. **Argillic:** Latin Argilla i.e., white clay. This is an illuviated clay horizon in which amount of silicate day is 1.2 times more than tenth of the overlying horizon. The fine clay is moved downward with percolating water and deposited as clay cutans on pad faces and walls of the pores.
5. **Calcic:** Latin calcic i.e., lime. It is a mineral soil horizon of secondary calcium and or magnesium carbonate enrichment. It is more than 15 cm thick $CaCO_3$ equivalent is more than 15 percent and it has at least 5% more $CaCO_3$ equivalent than underlying C or another horizon.
6. **Cambic:** Cambic is a leached horizon by physical movements or chemical reactions. It may develop in the presence or absence of fluctuating ground water and show extremely variable mineralogy.
7. **Durinodes:** Latin Durces i.e., hard and nodes i.e., nodules. Durinodes are weakly cemented to indurated nodules of soil cemented by SiO_2. Durinodes do not break down with concentrated HCl alone.
8. **Duripan:** Latin Durus i.e., hard and pan i.e., layer. Duripan is a sub-surface horizon which is mostly cemented by silica. Its air-dry fragments do not slake in water or HCl but disintegrate in hot concentrated KOH solution after acid washing.
9. **Fragipan:** Latin Fragilis i.e., brittle and pan i.e., layer. Fragipan is a sub-surface horizon of 15 cm or more thickness and high mechanical strength to the horizon above. It is brittle when moist but very hard when dries. The air-dry fragments slake in water. It does not find in calcareous soil material and does not effervescent in dilute HCl.
10. **Glossic:** Greek Glossa i.e., tongues. It is 5 cm or thicker sub surface horizon in characteristics of albic horizons gradually intrudes or encroached into argillic, kandic or natric horizon and from which clay and free iron oxides are removed.
11. **Gypsic:** Gypsic is a calcium sulphate and or magnesium sulphate enriched horizon which is 15 cm or more thick and contains at least 5% more sulphate than underlying horizon.

12. **Kandic:** Kandic is a horizon of low activity clays. It has the exchange activity of clays as ≥ c mol (P+) Kg^{-1} clay or CEC as ≥ 16 C Mol (P+) Kg^{-1} clay at pH 7.0.

13. **Natric:** Natrium i.e., sodium. Natric is an argillic horizon with high sodium clay enrichment. Its 15% or more of the exchange complex should be saturated with sodium or exchangeable magnesium + sodium should be more than calcium + exchange acidity at pH 8.2. Structure should be prismatic or columnar. But blocky structure is also found in the natric horizon of Indo-Gangetic plains.

14. **Orstein:** It consist spodic materials. It is a layer which is 50% or more cemented and 25 mm or more thick.

15. **Oxic:** It is at least 30 cm thick sub surface horizon enriched with Fe and Al oxides and 1:1 type (kaolinitic) clays. Highly weathered and contains less than 10% weatherable minerals in the sand fraction. The texture is sandy loam or finer.

16. **Petrocalcic:** Petrocalcic is an indurated calcic horizon whose hardness is 3 or more (Moh's scale). Its 50% or more dry fragments break down in acid but not in water.

17. **Petrogypsic:** A cemented gypsic horizon. Its clay fragments do not slake in water.

18. **Placic:** It is very thin (2-25 mm), slowly permeable and black coloured iron or manganese pan present within 50 cm of the surface.

19. **Salic:** A horizon with secondary accumulation of 2% or more water soluble salts (NaCl, Na_2SO_4 etc.) other than gypsum, 15 cm or more thickness and product of thickness (cm) and percent salt content is 60 or more.

20. **Sombric:** A darker subsurface horizon common in cool, moist and high-altitude tropical region, formed due to illuviation of humus and lacks in the properties of a spodic horizon.

21. **Spodic:** Greek spodos i.e., wood ash. Spodic horizon has illuvial accumulation of humus and or aluminium oxide with or without iron and generally, formed in cold humid regions.

22. **Sulphuric:** Sulphuric horizon is a mineral or organic soil horizon, where pH is less than 3.5. It shows yellow mottles of jarosite. Sulphuric horizon is highly toxic to plant roots.

Sub Order

Soil Orders are further divided into suborders mainly on the basis of genetic similarities i.e., all the wet soils will be kept in one group and all the dry soil will be kept in another group. Also changes in morphological, physical and chemical properties of soil due to climatic differences and other associated variable like vegetation etc. are also the basis of sub order groupings. The suborders for each order are given in Table 2.5.

Table 2.5: Name of sub orders

Order		Number of sub order	Name of sub orders
1.	Entisol	5	Aquent, Arent, Psamment, Fluvent, Orthent
2.	Inceptisol	6	Aquept, Andept, Ochrept, Plaggept, Tropept, Umbrept
3.	Alfisol	5	Aqualf, Boralf, Ustalf, Udalf, Xeralf
4.	Mollisol	7	Alboll, Aquoll, Boroll, Rendoll, Udoll, Ustoll, Xeroll
5.	Ultisol	5	Aqult, Humult, Udult, Ustult, Xerult
6.	Aridisol	2	Orthid, Argid
7.	Vertisol	4	Torrert, Udert, Ustert, Xerert
8.	Oxisol	5	Aquox, Torrox, Ustox, Udox, Perox
9.	Andisol	7	Aquand, Cryand, Torrand, Xerand, Vitrand, Ustand, Udand
10.	Spodosol	4	Aquod, Ferrod, Humod, Orthod
11.	Histosol	4	Fibrist, Hamist, Folist, Saprist
12	Gelisol	3	Histel, Orthel, Turbel

Name of suborder is written by prefixing formative element of sub order with the formative element of order e.g., Aquent. Here,' Aqu' is the formative element of 'Aqua' meaning wetness indicating property of sub order and 'ent' is the formative element order Entisol. The name of formative elements of sub order are given in Table 2.6. These are short forms of Latin (L), Greek (Gk), German (Gr) etc., words.

Table 2.6: Name of formative elements of sub order

Formative	Derivation	Meaning
Alb	L-albus- white	Presence of albic horizon
And	Ando	Similar to Ando
Aqu	L-Aqua-water	Presence of wetness
Ar	L-Arare-to plough	Mixed horizons
Arg	L-Argilla-White clay	Presence of argillic horizon
Bor	Gr-Boreas-Northern	Cool
Ferr	L- Ferrum-iron	Presence of iron

Formative	Derivation	Meaning
Fibr	L- Fibra-fiber	Fiber, least decomposed stage
Fluv	L- Fluvius-river	Flood plain
Hemi	Gr- Hemi-half	Intermediate stage of decomposition
Hum	L-Humus-earth	Presence of organic matter
Lept	Gr- Leptas-thin	Thin horizon
Ochr	L-Ochros- pale	Presence of ochric epipedon
Orth	Gr- Ochros-true	Common
Plag	Gr- Plagge- sod	Presence of plaggen epipedon
Psamm	Gr- Psammos-sand	Sandy texture
Rend	Rendzina	Rendzina like
Saper	Gr- Sapros- rotten	Most decomposed stage
Torr	L- Torridus-hot, dry	Usually dry
Trop	Gr-Tropikos-solstice	Continuously warm
Ud	L-Udus-humid	Humid climate
Umbr	L-Umbra-shade	Presence of umbric epipedon
Ust	L-Ustus-burnt	Dry climate, Hot in summer
Xer	Gr- Xeros-dry	Annual dry season

Great Group

Soil Great groups are the sub divisions of sub order. These are differentiated mostly on the presence and absence of diagnostic horizons and their sequence in the soil profile, soil temperature and moisture regimes and base status of soil. The great group formative elements indicate the diagnostic properties and prefixed with the sub orders to denote great groups. Thus, names of great groups may have three or more syllables.

Suppose 'Torr' formative element is prefixed with 'psamment' sub order, then great group will be 'Torrpsamment' meaning the Entisol (ent with 'Psammos', Psamm-sand- sandy or coarse texture) and developed at Torridus (Torr-hot and dry) conditions. A list of names formative elements of great groups is given in table 2.7.

Table 2.7: Name of formative elements of great group

Formative	Derivation	Meaning
Acr	Gk-Akros-at the end	Extreme weathering
Agr	L-Ager-field	Agric horizon
Alb	L- albus-white	Albic horizon
And	Ando	Similar to Ando
Anthr	Gk- Anthropos-man	Anthropic epipedon
Aqu	Aqua	Wetness
Arg	Argillus	Argillic horizon
Calc	Calcic	Calcic horizon
Camb	L- Cambiare-exchange	Cambic horizon
Chrom	Gk- Chroma-colour	High chroma
Cry	G k- Kryos- cold	Cold
Dur	L- Dusces-Hard	Duripan
Dystr	G k- Dystrophic infertile	Low base saturation
Eu	G k- eutrophic-good	High base saturation
Ferr	L- Ferrum	Presence of iron
Frag	L- Fragilus	Presence of fragipan
Gibbs	Gibbsite	Presence of gibbsite
Gloss	Gk- Glossa-tongue	Tongued
Hal	Gk- Hals-salt	Salty
Hapl	Gk- Haplous-simple	Minimum horizon
Hum	Humus	Presence of humus
Hydr	Gk- Hydro-water	Presence of water
Luo, Lu	Gk- Louo-wash	Illuviation
Moll	Mollis	Mollic epipedon
Natr	Natrium	Nitric horizon
Ochr	Ochros	Ochric epipedon
Pale	Gk-Palcon-old	Old development
Pell	Gk- Pellos-dusky	Low chroma
Plac	Gk- Plax-flat	Presence of thin pan
Plag	Plaggen	Plaggen horizon
Plinth	Gk- Plinthos-brick	Plinthite
Quartz	Quartz	High quarty content
Rend	Rendzina	Similar to Rendzina
Rhod	Gk-Rodon- rose	Dark red colour
Salt	Salt	Salic horizon
Sider	Gk- Sideros- iron	Free iron oxide
Sombra	Fr- Sombre – dark	Dark horizon
Sphagn	Gk- sphagnous-bog	Sphagnum moss
Torr	L- Torridus – hot and dry	Usually, dry
Trop	Tropikos	Continuously warm
Ud	Udus	Humid climate

Formative	Derivation	Meaning
Umbr	Umbra	Umbric epipedon
Ust	Ustus	Dry climate, hot in summer
Verm	L-Vermes-worm	Wormy or mixed by animals
Vitor	L- Vitrum-Glass	Presence of glass
Xer	Gk-Xeros	Annual dry season

Sub Group

Soil Great groups are divided in to sub groups. Sub groups indicate main concept of great groups as well as gradation of other categories of classification. There are three divisions of sub groups. The sub group completely representing the central concept of great group is called as 'Typic' for example Typic Hapludoll. The sub groups having one or more properties of order, sub order or great group along with the central concept are come under intergrades; example is Ustochrepts i.e., towards Inceptisol and extra grades sub groups have important property not representing great group and do not show transition towards any known soil type; example is Pergillic Cryhorent. The name of sub group is written by prefixing an adjective with the appropriate great group as Aquic Hapludoll i.e., Hapludoll developed under poor drainage conditions or wetness.

Family

Soil family is a category of soil taxonomy between soil series and sub group. It is related with the plant growth and land use. Differentiating criteria for soil family are quite variable and change according to sub groups. However, soil textural class, soil mineralogy and soil temperature regimes are three main criteria for differentiation and on the basis of these, families may be classified into fine, fine loamy, sandy or clayey, kaolinitic, montmrillonitic, gypsic or mixed and frigid, mesic, thermic or hyperthermic etc.

Soil Series

Soil series is the lowest and most specific category of classification. It delineates the properties of a pedon among the polypedon. Series can be identified on the basis of kind and arrangements of horizons and their physical, chemical and mineralogical properties. Soil series are area specific and basic units of soil maps. Series are named after the name of geographical place, town or area where they were first recognized, identified or described. Classification of Valentine loamy fine sand and Beni silty clay loam soils up to soil series according to Soil Taxonomy are as Valentine loamy fine sand, rolling, mixed, mesic, Typic ustissament and Beni silty clay loam fine loamy, mixed, hyperthermic, Aquic hapludoll, respectively. At present more than 1649 soil series have been identified in India.

Soil Phase

Soil phase is based upon some important soil properties like surface texture, slope, erosion, stoniness, soluble salt contents etc.

World Reference Base for Soil Resources (WRB)- The Revised Soil Legend of FAO-UNESCO

After more than hundred years of development of modern soil science a generally, accepted common soil classification system has not yet been universally adopted. Also, Soil Taxonomy of USA was not accepted in whole of the world. This is because of soils constitute a continuum which is not easily separable into classes and identifiable like plants and animals and adherence of many countries to their own soil classification systems. This lack of common soil classification system reduced the creditability of soil science among other basic sciences having their common and universally accepted classification systems.

History of WRB

In the year 1974 FAO and UNESCO jointly published a world soil map of the soil units. In this soil map, there were 28 major soil groupings sub divided into 153 soil units. This legend (map) was very useful for many scientific disciplines of the world besides soil scientists because of its simplicity, universality and serving the base for further developments of soil classification system.

Later in the year 1980 work was started to provide a common and universally accepted soil classification system, named as World Reference Base for soil Resources (WRB). WRB is the successor of IRB (International Reference Base for soil classification) and based on the Legend (FAO- UNESCO, 1974) and the Revised Legend (FAO, 1988) of the Soil Map of the World (FAO-UNESCO, 1971-1981).

In 1980, the International Society of Soil Science (ISSS, renamed as the International Union of Soil Sciences, IUSS in 2002) formed a Working Group named 'International Reference Base for Soil Classification' for further elaboration of a scientific international soil classification system. This Working Group was renamed 'World Reference Base for Soil Resources' in 1992. The Working Group presented the first edition of the WRB in 1998 (FAO, 1998) and the second edition in 2006 (IUSS Working Group, WRB, 2006). In 1998, the IUSS Council endorsed the WRB to officially recommend Terminology to name and classify soils. Lastly World Reference Base for Soil Resources (2014 update 2015) was published by Food and Agriculture Organization of The United Nations as international soil classification system for naming soils and creating legends for soil maps.

In the year 2022, within the framework of the IUSS "International Decade of Soils 2015-2024" and with the firm commitment to offer to the international community a soil classification system to facilitate both the implementation of soil inventories and the interpretation of soil maps as practical tools for decision-making for geologists, agronomists, farmers, engineers, politicians, etc., the International Union of Soil Sciences presents the fourth edition of the World Reference Base (2022).

Importance and Speciality of World Reference Base for Soil Resources: WRB has very wide importance as it is a basic soil classification system as;

1. It facilitates the identification of pedological structures; information of soil inventories and transfer of pedological data.
2. It enhances international use of pedological data by other users of soil and land such as agronomist, architects, botanist, civil engineers, ecologists, farmers, foresters, geologists, hydrologists etc.
3. It helps in the evaluation of soil resources and potential uses of different types of soil.
4. It develops models of the soil development depending on the human use and validates experimental methods of soil use for sustainable development.
5. It suggests improvement of soil potential and transfer of soil related technologies from one region to another region.

The basic specialties of WRB are

1. The classification of soils is based on soil properties measurable and observable in the field which is defined in terms of diagnostic horizons, diagnostic properties and diagnostic materials i.e., diagnostic characteristics.
2. The selection of diagnostic characteristics considers their relationship with soil forming processes. Soil-forming processes contribute to a better characterization of soils. However, soil forming processes are not as such used as differentiating criteria.
3. At a high level of generalization significance diagnostic features are selected up to the possible extent for soil management.
4. Climate parameters are not used in the classification and definitions of soils. Climate parameters in combination with soil properties are used for interpretation only.
5. The name of a particular soil will not become obsolete due to global or local climate change or unavailability of climate data.

Elements of World Reference Base (WRB)

Two levels of the categories and the diagnostic horizons, properties and materials are the Elements of World Reference Base (WRB).

Two levels of categories are First Level Category and Second Level category.

First Level Category has 32 Reference Soil Groups (RSGs). The RSGs classes are differentiated mainly according to characteristic soil features produced by primary pedogenetic process and secondly by any specific soil parent material.

The Second Level Category consist the name of the RSG combined with a set of principal and supplementary qualifiers. Here the soils are differentiated according to soil features resulting from any secondary soil-forming process that has significantly affected the primary characteristics. In many cases, soil characteristics that have a significant effect on land use are also considered.

The diagnostic horizons, properties and materials of the WRB are shortly described here.

The diagnostic horizons are grouped in three, Anthropogenic diagnostic horizons, Organic or mineral diagnostic horizons and Organic diagnostic horizons.

I. Anthropogenic diagnostic horizons are mineral horizons: These are seven in number.

1. **Anthraquic horizon in paddy soils:** This horizon comprises the puddled layer and the plough pan. Both show a reduced matrix and oxidized root channels.
2. **Hortic horizon:** This is dark coloured, rich in organic matter and Phosphorus, high animal activities and high base saturation which result from long-term cultivation, fertilization and application of organic residues.
3. **Hydragric horizon in paddy soils**: This horizon is below the anthraquic horizon showing redoximorphic features and/or an accumulation of Fe and/ or Mn.
4. **Irragric horizon:** This horizon is uniformly structured with at least moderate content of organic matter and high animal activity. This horizon is gradually built up by sediment-rich irrigation water.
5. **Plaggic horizon:** Dark, at least moderate content of organic matter, sandy or loamy; resulting from application of sods and excrements.
6. **Pretic horizon**: Dark, high content of organic matter and Phosphorus, low animal activity, high contents of exchangeable Ca and Mg, with remnants of charcoal and/or artefacts; including Amazonian Dark Earths.
7. **Terric horizon:** It shows a colour related source material, high base saturation; resulting from adding mineral material (with or without organic residues) and deep cultivation.

II. Organic or mineral diagnostic horizons are six

1. **Cryic horizon:** Perennially frozen (visible ice or, if not enough water, ≤ 0°C)
2. **Calcic horizon:** Accumulation of secondary carbonates, non-cemented
3. **Fulvic horizon:** Andic properties, highly humified organic matter, higher ratio of fulvic acids to humic acids
4. **Melanic horizon:** Andic properties, highly humified organic matter, lower ratio of fulvic acids to humic acids, blackish
5. **Salic horizon:** High amounts of readily soluble salts
6. **Thionic horizon:** With sulfuric acid and a very low pH

III. Organic diagnostic horizons are two

1. **Folic horizon:** Organic layer, not water-saturated and not drained.
2. **Histic horizon:** Organic layer, water-saturated or drained

IV. Diagnostic materials related to the concentration of organic carbon: Mineral material < 20%, soil organic carbon organic material ≥ 20% soil organic carbon soil organic carbon organic carbon that does not meet the diagnostic criteria of artefacts.

V. Diagnostic material related to colour: Albic material light-coloured fine earth, expressed by high Munsell value and low chroma.

VI. Technogenic diagnostic materials: It is predominantly understood as parent materials, artefacts created, substantially modified or brought to the surface by humans; no subsequent substantial change of chemical or mineralogical properties, technic hard material consolidated and relatively continuous material resulting from an industrial process.

Classification of Soils According to WRB

There are three steps in WRB Classification of soils.

In the step one diagnostic horizons, diagnostic properties and diagnostic materials are detected by following the Guidelines for Soil Description (FAO, 2006) and the soil and its features are described.

Step two allocates the soil to a Reference Soil Group (RSG). The described combination of diagnostic horizons, properties and materials is compared to the WRB Key in order to allocate the soil to the appropriate RSG.

Step three allocates the qualifiers. Qualifiers are used for the second level of the WRB classification. The qualifiers available to use with a particular RSG are listed in the Key, along with the RSG.

The qualifiers are divided into principal and supplementary qualifiers. The principal qualifiers are ranked and given in an order of importance. The principal qualifiers are added before the name of the RSG without brackets and commas. The sequence is from right to left, i.e., the uppermost qualifier in the list is placed closest to the name of the RSG.

The supplementary qualifiers are not ranked and used in alphabetical order. The supplementary qualifiers are added in brackets after the name of the RSG and are separated from each other by commas. The sequence is from left to right, i.e., the first qualifier according to the alphabet is placed closest to the name of the RSG. There are thirty-two (32) Reference Soil Groups (RSGs) in WRB. (Reference: World reference base for soil resources, 2014, International soil classification system for naming soils and creating legends for soil maps, Update 2015).

Numerical Classification of Soil

The history of numerical classification is quite new. Since late 1950 there is much development in IT and associated numerical classification of many objects including soils. In the year 1963 Sokal and Sneath (Numerical Taxonomy, 1963 and Principles of Numerical Taxonomy, 1973, W.H. Freeman, San Francisco) first developed Numerical Classification of biological systematics. Earlier it was also used in social science studies. In this system taxonomic units because of their characteristics or properties are grouped using numeric algorithms like cluster analysis.

Later on, in the advancement of computer and software numerical classification was rapidly developed and adopted in numerous biogeochemical fields including soil science. Gruijter (1977) described numerical classification of soils and its application in survey in much detail. Presently many software and programmes are available and used in numerical classification and more than 10 lakhs of soil observations are available to increase the feasibility of numerical classification of soils.

In soil science numerical classification is done for various aspects like genesis, properties, specific characteristics like concretions etc. The general classifications indicate about the specific name of the soil or they are qualitative. But numerical classification is a quantitative approach as all the characteristics of soil have digits or numerals. The soil profiles are grouped together according to similarity in them using distance matrix to calculate the distance or height between the pairs of soil profiles. Generally, hierarchical clustering of soil profiles is done where similar soils will be closer than the dissimilar soils and Dendrogram is constructed.

Dendrogram word is derived from Greek Dendro means tree and Gram means drawing, so this is a tree drawing. Dendrogram, often miswritten as Dendogram, is generally, used to depict the relationship among clusters resulted from hierarchical clustering. Height of link at which two soils joined together is most important. If the height of link or distance that joins the soils together is less then soils are similar and height of link or distance is more than soils are dissimilar. This is like family tree in which father is nearer to his son in relation then nephew or grandson.

Universal Soil Classification System and Unified Soil Classification System

Universal Soil Classification System is very necessary as many countries have their own soil classification system and they strictly follow them to classify their soils. This creates very much confusion to understand the findings at international level. Therefore, the 19th World Congress of Soil science (2010) at Brisbane, Australia conducted a conference named Soil Solutions for a Changing World and approved a task group with a time limit of ten years to develop a simple, user friendly, continuous and numeric Universal Soil Classification System considering the human influence also. Human influence on soil is varied and results numerous problems and solutions on agricultural activities, compaction, drainage, erosion, fertility, salt accumulation etc. The work on this system is under progress and will be communicated by IUSS and other related agencies in coming years.

Unified Soil Classification System is used by engineers and geologists to describe the texture and grain size of soil or soil materials. This is adopted by ASTM D-2487-98 and IS: 1498-1970. Permeability, shear strength, consolidation and swelling and shrinkage, grain sizes etc. of soils are few important properties considered for construction etc. This is a two-letter system. First letters are G (gravel), S (sand), M (silt), C (clay) and O (organic). The second letters are P (poorly graded i.e., uniform particle sizes, W (well graded i.e., diversified particle sizes), H (high plasticity) and L (low plasticity). One combination may be GW meaning well graded fine to coarse gravel. Pt term is used for peat i.e., highly organic soils. Numerous useful combinations can be identified with the help of these two letters and utilized by concerned persons.

Land Capability Classification

Soils are also classified for different uses based on information collected through soil surveys. Various limitations present in soil or land are the basis of such classification. The limitations may be permanent like slope, climate and recurring floods or temporary like fertility or nutrient status, drainage etc. The temporary limitations can be reclaimed. The presence of these limitations in soils decides their capability for different uses.

Land capability classification is the grouping of soils in to soil capability classes, soil capability sub classes and soil capability units for their long and sustainable use without adversely affecting the productivity and suggesting the required management practices. Just like soil taxonomy this system was also proposed by U.S. Deptt. of Agriculture in the year 1961 (Klingelbel and Montgomery, 1961).

Limitations Affecting Land Capability

The limitations or factors affecting land capability classification are

1. Depth of soil,
2. Presence of rocks and stone fragments,
3. Movement of air and water through soil,
4. Slope,
5. Erosion,
6. Moisture conditions, wetness and flooding
7. Nature and amount of salts present in soil and
8. Climatic conditions

Soil capability classes: Soil capability classes are eight in number (I to VIII). The intensity of limitations gradually increases from I to VIII. The capability classes are placed in to two groups, land suitable for agriculture and land unsuitable for agriculture.

Land suitable for agriculture: Class I to IV come under this group.

Class I has very good lands with no limitation, then limitations increase gradually and require soil conservation measures. However, class I has no sub class because there is no limitation or problem in this class.

Class II are good cultivable lands on level plains or gentle slope with slight limitations.

Class III are moderately good cultivable lands on moderate slope and moderate limitations.

Class IV are fairly good lands on moderately steep slope and suitable for occasional or limited cultivation and unsuitable for several crops.

Land unsuitable for agriculture: Lands of Class V to VIII come under this group. These lands are suitable for pastures, grazing lands, permanent vegetations, parks, forest, wild life sanctuaries etc.

Class V lands are suitable for grazing and limitations are related with ploughing due to stones and rocks fragments and marshy conditions. Subclass s is not used with class V.

Class VI lands are suitable for grazing and forestry but not for cultivation. The moderate limitations are related with steep slope and erosion etc.

Class VII lands are fairly well suited for grazing, forestry and wild life. Severe limitations related with very steep slope, very severe erosion, rockiness and moisture are common in these lands. These lands require very much careful management practices even for grazing and forestry.

Class VIII soils are non-arable extremely rough, rocky, arid or wet and salt affected. These are well suited for aesthetic, recreation, water supply and wild life. These lands include highly eroded, mountain tops and undulating rocky, plains etc.

The capability classes are shown by specific colour in the soil map e.g., class I (green), II (yellow), III (brown), IV (pink), V (dark grey), VI (orange), VII (red) and VIII (purple).

Soil capability sub class: Soil capability sub class includes the soils having dominant limitations. The dominant limitations or dangers are four;

1. e, limitation of erosion,
2. w, limitations of excess water or wetness,
3. s, limitations of root zone of plants i.e., soil depth, stickiness etc. and
4. c, limitations related with climate.

The symbols of these limitations are suffixed with soil capability class to denote the soil capability sub class e.g., II w. II w land is kept in class II due to excess water or wetness problem.

Soil capability units: Soil capability sub classes are further divided into soil capability units. These are the groups of soil mapping units. A capability unit includes the soils which are sufficiently uniform in their characteristics, potentials, limitations and management. These soils can be used for same crops, produce almost same yield and require similar conservation measures. Soil capability units are denoted by suffixing international numerals with sub class e.g., IIe1 i.e., the soil is kept in class II due to erosion but there is only mild sheet erosion.

Figure of land capability classification

Land Suitability Classification

Land suitability classification indicates the appropriateness or rightness of a land for particular use like agriculture including rainfed cropping, irrigation, livestock production, horticulture, forestry and plantations, resource conservation, tourism etc. All the lands are not right or appropriate for a particular use. Once the suitability criteria are established the land may be assigned for its best use. Therefore, it is necessary to classify lands according to their suitability and suggest the improvements also. The purpose of land suitability classification is the evaluation and grouping of a land to know its suitability for specific uses.

Categories of land suitability classification: There are four major categories of land suitability classification. These are,

1. Land Suitability Orders: Orders show kinds of suitability. Land Suitability Orders indicate the assessment of any land about its suitability or not suitability for a particular use. There are two orders. Order S means Suitable and Order N means Not Suitable which is depicted as S and N symbols respectively. Generally, Land Suitability Orders are based on economics of their use i.e., income over expenditure or investment. If income is economically more than expenditure or investment then land is suitable otherwise not suitable. Any waterlogged or wet land may be Suitable for fisheries, rice or lotus cultivation with sustained production but Not Suitable for maize, wheat or legume cultivation.

2. Land Suitability Classes: Classes indicate degrees of suitability within Orders. There are five classes, three in suitability (S1, S2 and S3) and two in not suitability (N1 and N2). The classes are numbered consecutively in sequence of decreasing degrees of suitability within the Order.

S1 Class: Highly suitable land without significant limitations for sustainable use or with very minor limitations which can not significantly reduce productivity or benefits and will not raise inputs above an acceptable level.

S2 Class: Moderately Suitable Land having limitations which in total are moderately severe for sustained use, appreciably inferior to Class S1 land but still attractive. The limitations will reduce productivity or benefits and increase required inputs to the extent to get overall advantage from the use.

S3 Class: Marginally Suitable Land having limitations which in total are severe for sustained use. The productivity or benefits are so reduced or require increased inputs, that the expenditures are only marginally justified.

N1 Class: Currently Not Suitable Land having limitations which may be overcome in time, cannot be corrected with existing knowledge at currently acceptable cost and the limitations are so severe as to prevent successful sustained use of the land in the particular manner.

N2 Class: Permanently Not Suitable Land having limitations which are so severe as to prevent any possibilities of successful sustained use of the land in the particular manner.

3. Land Suitability Subclasses: Sub classes indicate kinds of limitations e.g., moisture deficiency, erosion hazard etc. and indicate requirement of major improvement measures within Classes. There are no subclasses in Class S1. Subclasses are indicated by lower-case letters with significant patterns like S2e, S2m, S3me etc. where e for erosion and m for moisture.

4. Land Suitability Units: Land suitability units are subdivisions of subclass. Units inform about minor differences in management practices required within Subclasses. All the units within a subclass have the same degree of suitability at the class level and similar kinds of limitations at the subclass level. The units differ from each other in their production characteristics or in minor aspects of their management requirement. There is no limit to the number of units recognized within a subclass. Land Suitability units are indicated by Arabic numerals with a hyphen like S2e-1, S2m-2.

Geography and Cartography of Soil

Soil geography is the branch of geography dealing with the area wise (areal) distribution of soil types. It is the study of soil and its numerous morphogenetic attributes as a function of climate, geology, relief, vegetation, human activities and natural and anthropogenic history. It is a system to transfer knowledge of soils gained through their formation, morphology, classification etc. for agricultural sustainability, food security, lands' carrying capacity, human and animal health and clean environment. Different types of maps in various scales show the areal distribution to soil for above mentioned priorities are also the part of soil geography.

Soil Mapping

The distribution of different soils on the land scape requires their depiction. This may be achieved by preparing soil maps. A soil map is a map which shows the distribution of different soil types or other soil mapping units related to the prominent physical features including agriculture of the land surface. A mapping unit is the smallest area of a map that can be delineated by a single boundary at the scale used.

All maps have scale. The scale indicates how much detail can be depicted on the map. Each mapping unit is depicted by a symbol, colour or name. The depiction of mapping units depends upon the scale of map. Small scale map can not show as much detail on ground as compared to large scale map. A large-scale map has simple mapping units like soil series and depicted up to 80% or more purity and correctness. The small-scale maps (1:250,000 to 1:100,000) use complex mapping units like soil association. In general, the soil mapping units are soil series, soil type, soil phase, soil association and soil complex.

Preparation of Soil Maps

Soil maps are prepared on the basis of soil surveys or based on the abstract of soil surveys and some base maps. The base maps may be cadastral maps, toposheets and aerial photographs. Cadastral maps are also called as village Patwari map. These are prepared for the use of revenue personnels. These show the field of each farmer and other land of village. Scale of cadastral map is very large. Normally 12.5 to 25 cm length shows 1 km distance. In hilly areas 80 or 160 cm shows 1 km. These maps are not absolutely correct. The height and depth of any area can not be known by cadastral maps.

Survey of India prepares toposheets on 1:5,000 and 1:25,000 scales which are traditionally called as 1 inch and 0.25 inch sheets, respectively. In one inch sheet contour interval is 20 meters and covers a geographical area of 10 minutes latitude and 15-minute longitude. In 0.25-inch sheet contour interval is 40 meter and covers 1 degree latitude and 1 degree longitude geographical area.

Aerial photographs are taken by the aerial camera fitted with an aeroplane. In India the scale of these maps may vary 1:15,000 (4" = 1 mile) to 1:60,000 (1" = 1 mile). Aerial photographs record all features of land surface on the basis of resolution of lens and scale of photography. In aerial photography, two photographs of a single point are taken due to two different positions of the camera. The common area of two photographs is called as overlap. End lap and side lap are two types of over lapping which help in the estimation of height and depth of a particular area by using stereoscope. The aerial photo coverage is 5x5 km at a single view. Now a days remotely sensed satellite imageries (earth resource satellite, Landsat imagery) are also used as base maps. Each imagery shows a synoptic view of 185x185 km area.

Types of Soil Maps

There are five types of soil maps. These types are generally, based on types of soil surveys. Their brief description is given here:

Reconnaissance soil map: A map showing general distribution of soils as revealed by traverses at the intervals of 2 to 4 km. The mapping unit is soil association and scale vary from 1:125,000 to 1:250,000.

Detailed soil map: A map slowing the boundaries between all soil types. Traverses are generally, made at ½ km intervals or closer intervals. The scale of map may vary 1:10,000 to 1:25,000 depending upon the purpose of map, the intensity of land use, soil pattern and scale of other cartographic materials.

Detailed- reconnaissance soil map: A map showing distribution of some soils, area or features of land surface in much detail than others. The traversing is done at the intervals of 1 to 2 km and scale may vary from 1:50,000 to 1:1, 25,000 depending upon details required.

Generalized soil map: A small scale soil map showing the general distribution of soils of a large area with less detail. The scale may be 1:15,000 to 1:1, 75,000.

Schematic soil map: A soil map prepared based on available information on soil forming factors due to the inadequate information about the soil of a new and undeveloped area. The scale of these maps may be 1:25, 00,000 or smaller.

Types of Soils of World, Asia and India

Types of Soils of World

Soil forming factors and processes played the key role in soil formation and development. Therefore, soils types or groups very widely differ from one region or zone to another. In the earlier times types The Vedas and Sushruta Samhita, indicated six different types of soils because of their colour, texture and growth of plants. Similar were the information in other ancient civilizations of the world.

Also, types of soils were sandy soil, loamy soil, clayey soil, sandy loam soil, aggregated soil etc., or mineral soils, organic soils, acidic soil, sodic soil, calcareous soil etc., or hill soils, forest soils, roadside soils, terrace soils, lowland soils, upland soils, dry land soil, water logged soil, submerged soils etc. as per their topographic location or position on the landscape. World soils were also grouped in granitic soils, basaltic soils, alluvial soils, aeolian soils, colluvial soils etc. on the basis of rocks and soil parent material.

Russian scientist Dokuchaev (1900) categorized world's soil in to normal, transitional and abnormal soils. Later on the concept of zonality was developed which divided world's soil into zonal. The zonal soils comprise fully developed soil profiles and their main characteristics are depending upon climate and vegetation. There was no profile development in azonal soils. Intrazonal soils are depending upon topography, parent material, drainage and soil age.

Later, Glinka, one of Dokuchaev students, worked on the development of the soil-geographic concepts. Glinka contributed to the development of soil mapping and prepared the first schematic soil map of the world at a scale of 1:80 million in 1908. This map included 18 soil types and a unit for water bodies. These soil types are: 1- podzolic and soddy soils, 2- forest soils and degraded chernozem, 3-chernozem and regoor, 4-chestnut soils, 5- stratified columnar soils of temperate deserts, 6- desert crusts of temperate deserts, 7- terra rossa, 8- laterite, 9- zheltozem (Gehanmgelehm of Rihtgofen), 10- Soils of dry tundra, 11- meadow and meadow-steppe soils, 12 krasnozem, 13- desert crusts, 14- vertical zones of mountain region, 15- large lakes, 16- Bog soils, 17- sands of deserts, 18- solonetz, 19 dark soils of savannas. (Ref. Hartemink, A.E., et.al., 2013)

In another supposition Coffey (1912) placed world's soils in to five major classes as arid soils, dark coloured prairie soils, light coloured timbered soils, black swamp soils and organic soils. But Marbut (1928 and 1935) approximately divided soils only in two main classes Pedalfers and Pedocals which were supposed to present in areas of high rainfall and high evaporation, respectively. Thorp and Smith (1949) revised Baldwin et.al. (1938) and soils of the world were grouped in to three orders (Zonal, Intrazonal and Azonal), nine suborders (six for zonal soils, three for intrazonal soils and none for azonal soils) and in all 37 great soil groups.

Presently all these suppositions or approximations are the basis to describe soils types or groups of the world by World Reference Base for Soil Resources and Soil Taxonomy: A Basic System of Soil Classification for Making and Interpreting Soil Surveys. However, finally described and accepted soils types or groups of the world are yet to come.

Similarly, Soil Taxonomy described 12 soil types in the ice-free land of the world in the form of soil orders as,

Entisols: This is largest soil type, 16% of the land area and found in the Sahara, Arabian Peninsula and Australia.

Inceptisols: This is 10% of the world's land and found in China, Alaska and both sides of the Mediterranean.

Alfisols: This is about 10% of the world's ice-free land found in Western Europe and parts of North and South America, Africa, India and Australia.

Mollisols: This is 7% of the world's land and found in Ukraine, parts of Russia and the USA.

Ultisols: 8.5% of the world's land and found in Southern USA and Asia, Africa and South America.

Aridisols: This is nearly 13% of the world's land and present in the world's deserts.

Vertisols: This is 2.5% of the world's land and found in India, Australia, sub Saharan Africa and South America.

Oxisols: This is 7.5% of the world's land and found in the world's tropical rainforests, including the Amazon and in Africa.

Andisols: This is 1% of the world's and present in pockets in New Zealand, Northwest USA, Chile, Kenya, Indonesia and Japan.

Spodosols: This soil type covers 2.5% area of the world's land. Mainly found in many forests including much of Scandinavia.

Histosols:This soil type occupies about 1% area of the world's land and found in Northern Scandinavia and Canada and give some of Scotland's whisky its unique taste.

Gelisols: This is under permafrost, about 9% of the world's land and present in northern Russia and North America.

Types of Soils of Asia

Asia is the biggest continent in the world occupying an area of Forty Crore Five Hundred Eighty Lakhs square kilometres by its 50 independent countries. Russia the largest Asian country by area occupies about 30% of the total territory of the continent and Maldives is smallest. Asia occupies the eastern part of the single Eurasian landmass and surrounded by the Arctic, the Pacific and the Indian Oceans from the North, East and South, respectively. It is also separated from Africa by Suez Canal and the Mediterranean and the Black Seas from Europe.

Soil is a very important resource for all. Soil forming factors and processes played the key role in the formation and development soils of the world including Asia. However, to ascertain the types of soil, many factors including remoteness of area, climate, topography and relief, temperature, moisture, biosphere, time, political and economic activities etc. play their role. All these factors vary considerably from one part of the Asia to another, in all directions and elevations giving a horizontal zonality and diversity to soils of this vast continent. Even though numerous types of soils based on different soil surveys and classification systems, are identified in world, continents, countries, states, provinces etc. and indicated that not more than 20 percent of the soils are suitable for agricultural production.

Earlier many workers including Dokuchaev etc. provided older classifications of soils including Asia and Tamhane, et.al. (1965) indicated Chemistry and Fertility of Tropical Asia including their classification. Later on, Kovda and Lobova (1968) classified soil of Asia. Now Asian soils are classified on various national classification systems as well as soil taxonomy and WRB of FAO. WRB described Asian soils into twenty-seven soil groups.

Types of Soil of India, their Distribution, Nature, Properties and Uses

The conditions like high mountains, snowy tops and flats, glaciers, high altitude or plain, very dense to scanty forests, fertile fields to barren lands, plateaus, deltas, swamps, lowlands and deserts, temperature variations from arctic cold to scorching hot of equator and rainfall from aridity to per humidness prevail in India under different sets of combinations and forms. All these develop numerous types of soil from a single type of parent material. Therefore, India has many more types of soils than countries of almost equal geographical area.

Thus, the earlier Indian soil scientists, located in different parts of the country, were attracted to a systematic study of soil genesis and classification. Bhattacharya et. al. (2013) indicated that, derived from a wide range of rocks and minerals, a large variety of soils occur in the Indian subcontinent. Soil-forming factors like climate, vegetation and topography acting for varying periods on a range of rock formations and parent materials, have given rise to different kinds of soil.

Earliest investigations on soils of India by Voelcker date back to 1893 and by Leather to 1898. They categorized the soils of the country into four major groups, namely the Indo-Gangetic alluvium, the black cotton soil or regur, red soil and laterite soil. Schokalskaya published a soil map of India in 1932 following the Russian concept and described 16 soil groups based on climate, vegetation, soil-forming materials, salinity, alkalinity, swamps and peats.

In 1935, Wadia and his co-workers compiled a soil map of India with the emphasis on geological formations and classified the soils as red, black (regur), laterite and lateritic soils of Peninsular India, delta, desert, bhabar, terai and alkali soils of the Indo-Gangetic Plains (IGP). Vishwanath and Ukil (Vishwanath, B. and Ukil, A. C., Soil Map of India, Indian (Imperial) Agricultural Research Institute, New Delhi, 1943) published a soil map of India by placing the soils in different climatic zones.

Integrating the effects of climate, vegetation and topography, 16 major and 108 minor soil regions were identified and brought under 27 units by Raychaudhuri et al. (Raychaudhuri, S. P., Aggarwal, R. R., Datta Biswas, N. R., Gupta, S. P. and Thomas, P. R., Soils of India, Indian Council of Agricultural Research, New Delhi, 1963).

Later a revised soil map of India was generated with 23 major soil groups under FAO/UNESCO's scheme on World Soil Map project. This map was refined with 25 broad soil classes represented on a 1:7 million scale map (Govinda Rajan, S.V., Soil Map of India. Ref. Review of Soil Research in India (Eds Kanwar, J. S. and Raychaudhuri, S. P., 1971, 1:7)

Agro-Climatic Zones or Regions of India

Agro-Climatic zone is an extension of the climate classification keeping in view the suitability to agriculture and defined as a land unit in terms of major climates, suitable for a certain range of crops and cultivars to obtain maximum production from the available resources and prevailing climatic conditions. Large numbers of complex agro-climatic conditions prevail in the vast 32, 87, 263 square kilometre geographical area of the India. However, Planning Commission, NARP, ICAR, Niti Aayog etc. because of homogeneity in agriculture, cropping and farming systems, climate, soil, topography, water resources, etc. delineated fifteen main agro climatic zones or regions in our country. These 15 zones were further sub-divided into 73 sub-zones based on more specific soil types, topography, climate and cropping pattern characteristics. These fifteen main agro climatic zones or regions are,

Western Himalayan division: Jammu and Kashmir, Uttar Pradesh and Kumaon-Garhwal areas of Uttaranchal

Eastern Himalayan division: It consists of Sikkim, Darjeeling area (West Bengal), Arunachal Pradesh, Assam hills, Nagaland, Meghalaya, Manipur, Mizoram and Tripura

Lower Gangetic plain region: This region spreads over eastern Bihar, West Bengal and Assam valley

Middle Gangetic plain region: It includes eastern Uttar Pradesh and Bihar (except Chotanagpur plateau)

Upper Gangetic plain region: This region encompasses central and western parts of Uttar Pradesh

Trans-Gangetic plain region: It consists of Punjab, Haryana, Delhi, Chandigarh and Ganganagar district of Rajasthan

Eastern plateau and hill region: It comprises the Chotanagpur plateau, Raajmahal hills, Chhattisgarh plains and Dandakaranya

Central plateau and hill region: This region spreads over Bundelkhand, Baghelkhand, Bhander plateau, Malwa plateau and Vindhyachal hills

Western plateau and hill region: It comprises southern part of Malwa plateau and Deccan plateau (Maharashtra).

Southern plateau and hill region: It includes southern Maharashtra, Karnataka, western Andhra Pradesh and northern Tamil Nadu

East coast plain and hill region: It includes the Coromandel and Northern Circar coasts.

West coast plain and hill region: It extends over the Malabar and Konkan coasts and the Sahyadris and is covered by laterite and coastal alluvial.

Gujarat plain and hill region: It includes Kathiawar and fertile valleys of Mahi and Sabarmati rivers.

Western plain and hill region: It comprises western Rajasthan and west of the Aravallis

Island region: The island region includes Andaman-Nicobar and Lakshadweep which have typically equatorial climate

Agro-ecological Zones or Regions of India

Agro-ecological zone is defined as a land resource mapping unit particular in terms of climate, landform, soils and land cover, with a specific range of potentials and constraints for land use. It is a geographical area having similar climatic condition that determine their ability to support rained agriculture and influenced by latitude, elevation, temperature, seasonality and rainfall amounts and distribution during the growing season.

The National Bureau of Soil Survey & Land Use Planning (NBSS&LUP) indicated twenty agro-ecological zones or regions based on the growing period integrated with effective rainfall, soil groups and district boundaries. These twenty agro-ecological zones or regions were further sub- divided into sixty sub-zones zones or regions.

Western Himalayas

Western Plain, Kachchh, and part of Kathiawar Peninsula

Deccan Plateau

Northern Plain and Central Highlands including Aravalli

Central Malwa Highlands, Gujarat Plains, and Kathiawar Peninsula

Deccan Plateau, hot semi-arid ecoregion

Deccan (Telangana) Plateau and Eastern Ghats

Eastern Ghats, Tamil Nadu Plateau and Deccan (Karnataka)

Northern Plain, hot sub-humid (dry) ecoregion

Central Highlands (Malwas, Bundelkhand, and Eastern Satpura)

Eastern Plateau (Chhattisgarh), hot sub-humid ecoregion

Eastern (Chotanagpur) Plateau and Eastern Ghats

Eastern Plain

Western Himalayas

Bengal and Assam plains

Eastern Himalayas

North Eastern Hills (Purvanchal)

Eastern Coastal Plain

Western Ghats and Coastal Plain

Island of Andaman Nicobar and Lakshadweep.

Physiographical Zones or Regions of India

Physiography is also called as physical geography which is a sub division of geography studying the physical patterns and processes of the earth to understand the forces that produce and change rocks, oceans, weather, and patterns of global flora and fauna.

India is the seventh largest country with 2.4% of total area of the world located in northern hemisphere. The Indian Mainland extends from 8°4′ north and 37°6′ north in length (latitudes) and between 68°7′ East and 97°25′ east in width (longitudes). The North-south extension is 3214 km and East-west extension is 2933 km. At 23°30′ north, the Tropic of Cancer passes through the center of India, dividing the country into two equal Northern and Southern parts. The Standard Meridian is on 82°30′ East longitude in the middle of the country and passes through Mirzapur near Prayag Raj in Uttar Pradesh. The unique physiography of country has developed distinctive features in it including different types of soils besides other factors of soil formation.

Physiographic Regions and Divisions in India

India can be broadly classified into four broad regions namely,

1. Northern mountains
2. Great plains
3. Deccan plateau and
4. Coastal plains and Islands.

In India there are six physical divisions because of physical features namely,

The Northern Mountains

The North Indian Plain

The Peninsular Plateau

Great Indian Desert

The Coastal Regions

Islands

Crustal Segments of India

India has three main crustal segments, namely

The triangular plateau of peninsular India in the Deccan and south of Vindhyan,

The mountain region of the Himalayas bordering the peninsula on the west, north and east which is also called as extra peninsula and

The great plain area between above two i.e., the triangular plateau and mountain, filled with alluvium of Indo-Gangatic River system and extending from the valley of Sind (Pakistan) to the Brahmaputra in Assam (India).

Also based on recent geological, geophysical and borehole petrophysical studies, Om Prakash Pandey (2020) reported that Indian crust can be subdivided into as many as seven distinct layers namely:

1. Sediments,
2. Upper crust,
3. Middle crust,
4. Lower crust,
5. Lowermost high velocity crust,
6. Magmatic lower crust and
7. Differentiated mantle magma layer over the normal ultramafic mantle.

Om Prakash Pandey (29 April 2020) Geodynamic Evolution of the Indian Shield: Geophysical Aspects pp 331-349| Indian Crust 1.CSIR-National Geophysical Research Institute, Hyderabad India. Part of the Society of Earth Scientists Series book series (SESS)

Soils of India

It is reported that structural fabric and composition of the crust have undergone considerable changes since Archean era. Almost 5 to 20 km thick upper portion of the crust is eroded away from several Archean-Proterozoic terrains of India and possibly other global shields and platforms as well. The bulk of the continental crust in stable shields now appears to be made up of mainly metamorphic grade rocks. In these main and subdivided crustal segments two categories of soils are found,

(1) Soils developed over the under-lying bed rock of the area. The nature and properties of these soils show a broad relationship with the genetic processes and characteristics of parent material and climate and

(2) Soil developed from the transported material. The nature and properties of these soils have distinct relationship with mineralogy of the area.

These two categories of soils were divided in to 24 soil groups by Govindarajan (1965). Later, Govindarajan and Gopala Rao (1978) short listed these 24 soil groups in to 7 major soil groups which basically come under genetic classification and considered by Tapas et.al. (2013).

The author now grouped the Indian soils in nine major or broad soil groups with slight modifications in earlier groups. These broad soil groups are,

Alluvial soils

Red soils

Black soils

Desert and arid soils

Laterite and lateritic soils

Tarai and sub- montane soils

Forest and Hill / mountain soils

Saline and alkali/ salt affected soils and

Peat and marshy soils

The brief description of these broad soil groups is given here:

Alluvial Soils

Alluvial soils of India can be classified in to Inceptisols, Entisols, Alfisols and Aridisols under U.S. Soil Taxonomy and Cambisols, Fluvisols, Luvisols, Gleysols, Solonchaks and Solonetz under World Reference Base.

Occurrence: Alluvial soils are mainly found in Indo-Gangetic plain, Brahmaputra valley and coastal region of the peninsula. The states which are having alluvial soils are Jammu & Kashmir, Himachal Pradesh, Punjab, Haryana, Uttar Pradesh, Uttarakhand, Delhi, Gujrat, Goa, Madhya Pradesh, Chhattisgarh, Andhra Pradesh, Telangana, Kerala, Puducherry, Arunachal Pradesh, Nagaland, Manipur, Mizoram, Tripura, Meghalaya, Assam, Bihar, Jharkhand, Odisha, West Bengal, Tamil Nadu, Kerala, Maharashtra and Andaman and Nicobar. Total area is about 10, 00,060 sq. km i.e., 30.4%.

Parent material: Parent material of alluvial soils is alluvium. The alluvia are of different types at different places. Punjab plain which comprises Punjab, Haryana and part of Rajasthan has the alluvium deposited by Ravi, Beas, Sutlej and Yamuna rivers. These rivers brought the sediments from Himalaya and Shivalik. The Aravali, Vindhyan and Gondwana area is also associated with the tributaries of Yamuna.

The Gangatic plain comes under Uttar Pradesh, Uttarakhand, Bihar and West Bengal. This region has alluvium derived from the sediments brought by Ganga, Yamuna, Ghaghara, Gandak and Gomti rivers.

The alluvium of Brahmaputra and Tributaries has other types of deposition. The coastal alluvium is mostly bound along with the coastal belt of peninsular India which lies at Arabian Sea and Bay of Bengal. These are called as Western coastal plain and Eastern coastal plain, respectively.

Another type of parent material is deltaic alluvium. This contains very heterogenous sediments brought by the rivers Ganga at Sundarvan, Hooghly and Howrah in West Bengal, Mahanadi near Cuttack in Odisha, Godavari and Krishna near Eastern coastal plains.

Climate and vegetation: These soils are found under quite different climatic conditions varying from arid and semi-arid to humid and per humid climates. Similarly, the vegetation is also diversified from small grasses to long dense forests.

Nature and properties: The colour of alluvial soils is usually variable due to the nature of parent material, organic matter status and moisture conditions. Texture is from sandy to clayey. These may be calcareous or non-calcareous and pH may vary from acidic to alkaline.

The soils if Indo-Gangatic alluvium is locally termed us Khadar and Bhangar. Khadar soils are developed from recent alluvium. These are low in gravel, sandy in texture, light coloured, lesser lime concretions (Kankar) and show poor profile development. Bhangar soils are developed from old alluvium. These are rich in gravel and clay with a thick Kankar layer, fine texture, dark in colour and show some more profile development. The Kankar layer is Generally, common in the soils of Delhi, Uttar Pradesh, Punjab and Haryana states.

The soils developed from recent and old alluvia are also found in the valley of Brahmaputra and tributaries. Colour of these soils very from light grey to yellow, yellow brown and dark grey. Texture is silty varying from loam to silty clay loam, structure may be angular blocry. There is well developed 'B' horizon. Some time there may be presence of Kankar, pebble beds, calcareous material or hard pan of silica mixed with soils particles. This hard pan hinders easy movement of water and root penetration.

Sub Groups of Alluvial Soils

Calcareous alluvial soil, Coastal alluvial soils, Coastal sand and Deltaic alluvial soils are sub groups of alluvial soils.

Calcareous alluvial soil: These are Generally, developed in north eastern part of the U.P. to north western part of Bihar. These soils contain 10-40% or more amount of $CaCO_3$ which may be distributed in whole of the profile. These soils are faint brown to yellow in colour, sandy loam to loam in texture, alkaline in pH with no distinct horizonation.

Coastal alluvial soils: East coast and West coast are two coastal regions of India. The characteristics of soils of these coastal regions are quite different due to the nature of sediments brought by the rivers flowing there. The colour may be brown, yellow brown, grey, dary grey and red. Soils derived from calcareous parent material are of dark colour, few fine textures and whereas soils derived from granite and granitic gneiss are red in colour and clayey. Although some coastal soils are also salt affected due to the oceanic saline water. Coastal soils are suitable for growing paddy and continuously two crops of paddy are grown in a year.

Coastal sand: In some specific areas of coastal region sandy and deep soils (Regosols) are found with very poor profile development. Poor profile development is mainly due to the coarse parent material. Topography of sandy stretches may be flat to slightly undulating. Low water table and well drained conditions hinders the development of salinity of these soils. However, some low lying and swampy areas may be saline, Kutch and swampy can not be used for crop production. In some areas coconut, plam and fuel wood trees are grown successfully.

Deltaic alluvial soils: Ganga, Brahmaputra, Mahanadi, Godavari, Krishna and Kaveri rivers bring the alluvial material from huge area of Indian sub-continent. This alluvial material is deposited at the mouth of the river (estuary) and delta is formed. The alluvial soils of various delta or deltaic alluvial soils show the characteristics of the sediments and soils of the areas where from the reivers are coming.

Deltaic alluvial soil of Gangatic plains is light in colour and silty to silty clay in texture. Deltaic alluvial soils of Mahanadi are light brown to light in colour and sandy loam, loam and silty loam in texture. This colour and texture are the indication of the colour and texture of eastern central plateau of M.P., Chhattisgarh and Odisha. Likewise deltaic soils of Godavari and Krishna are dark coloured and fine textured of silty clay and clay. This is mainly due to the black soils derived from basalt trap.

Soils of Kaveri delta are dark coloured and silty to silty clay in texture. Except deltaic soils of Ganges other soils have brown and mottled horizons in deeper layers due to imperfect drainage conditions. Organic matter accumulation in A horizon is variable due to prevailing drainage conditions and agricultural operations. Different types of vegetations are naturally grown in the swampy deltaic soils of Ganga. These soils are enriched with organic matter as compared to other deltaic soils which are under crop cultivation.

Uses: Alluvial soils are mainly used for growing different crops like wheat, rice, sugarcane, jute, maize, cotton, sunflower, oilseeds, legumes, fodders etc.

Red soils

Red soils come under Alfisols, Ultisols, Entisols, Inceptisols, Mollisols and Aridisols of U.S. Soil Taxonomy and Luvisols, Cambisols, Leptosols, Lixisols and Nitisols of World Reference Base (WRB).

Occurrence: Red soils are mainly found in Andhra Pradesh, Telangana, Kerala, Puducherry, Uttar Pradesh, Himachal Pradesh, Uttarakhand, Andaman and Nicobar, Gujrat, Goa, Madhya Pradesh, Chhattisgarh, Arunachal Pradesh, Nagaland, Manipur, Mizoram, Tripura, Meghalaya, Assam, Bihar, Jharkhand, Odisha, West Bengal, Tamil Nadu, Kerala and Maharashtra. Total area is about 8798 sq. km i.e., 26.8%.

Parent Material: Parent material of red soils is mainly ancient granite, gneiss and schists. Red soils are also developed from sand stone, mica, acid trap, quartzite, lime stone and shale.

Climate and vegetation: Warm semi-arid to humid climate and broad-leaved forest vegetation.

Nature and Properties: Colour of red soils may vary from red to yellow due to the presence of hematite and limonite, respectively. The dehydration converts limonite to hematite by which yellow colour of soil changes in to red.

Texture is Generally, loam which may vary from sandy loam to clay loam. The soil formed from limed feldspars may have nodules or thick veins of lime. pH is Generally, neutral in the range of 6.0 – 7.5 but calcareous red loamy soils may show pH up to 8.5.

There is high amount of Fe and A1 and low CEC and percent base saturation. Kaolinite and hydrous oxides of Fe and Al are main clay minerals along with illite and other 2:1 type clay mineral. These soils have phosphorus fixing capacity. A horizon is 25-40 cm deep, crumb structure, dark reddish grey in colour and loamy in texture. B horizon is 70-80 cm deep from the surface, angular blocky structure, colour dark brown and loam to clay loam in texture. Below B horizon there may be presence of weathered gneiss or thick veins or layers of hard lime or hard lime stone rocks.

Sub Groups of Red Soils

Red sandy, red loams and red and yellow soils are three sub groups of red soils.

Red sandy soil: Red sandy soils are derived from granite, gneiss, quartzite and sand stone. These soils have fine and coarse fractions of sand. Clay minerals, mostly kaolinite are covered by the layer of hematite (red), limonite (yellow) or the mixture of both. Thus, show red yellow or reddish yellow colour. pH of these soils is Generally, acidic (4.5-6.5) but some soils show alkaline pH. A horizon of red sandy soils has fine sandy, coarse sandy or loamy sand texture with light reddish brown or red colour. B horizon is loam or sandy clay loam in texture and dark reddish brown in colour. C horizon has weathered parent material of coarse grained granite or granitic gneiss.

Red loamy soils: Red loamy soils are derived from granite, gneiss, quarts, feldspars and mica. The pH is almost neutral (6.7-7.5). Base saturation is 88-93%, organic matter status is low and low to medium in nitrogen. Some red loamy soils of Meghalaya are high in available nitrogen status.

Red and yellow soils: These soils are also called as yellow soils because the main colour is yellow which may vary from reddish yellow to yellowish brown. Yellow soils are developed from micaceous quartzite, schist, gneiss and phyllite. These rocks found 100-200 cm below soils profile. The soils are fine textured and show some relationship with laterites. A horizon is slightly rich in humus and show loam to silty loam texture. B horizon is single textured, blocky in structure and yellow, sometime red, in colour. B horizon also shows mottles and pigmentation of iron and upon drying cracks may develop.

Uses: Different types of agro-horticultural crops can be grown in these soils due to their presence in different agro climatic zones. Thus, smaller millet, paddy, maize, jute, soybean, gram, moong, groundnut, potato, tea, cocoa, grapes, banana, papaya, mango, cashew etc. can be grown satisfactorily after managing soil, water and nutritional conditions properly.

Black Soils

Black soils are also called as black cotton soils of India and come under Vertisols, Inceptisols and Entisols of U.S. Soil Taxonomy and Vertisols, Cambisols, Leptosols and Regosols of World Reference Base of Soil Classification.

Occurrence: Black soils are occurring mostly in Maharashtra and Malwa plateau. Narmada valley and parts of Satpura range of Madhya Pradesh and some parts of Karnataka, Andhra Pradesh, Tamil Nadu and Gujrat.

Parent material: Black soils are developed mainly from basalt but at some places these are also derived from granite, schist and gneiss with soda lime feldspars.

Climate and Vegetation: Black soils are formed under warm semi arid to sub humid tropical climate and tropical deciduous and thorny forest and dense grasses.

Nature and Properties: Black soils are swell-shrink soils. The texture of black soils is fine varying from clay loam to silty clay loam and clay. The clay content ranges from 35-60% which may go up to 80%. Silt may also high but sand is very less. Sometimes gravel may also present in the form of weathered fragments of parent material and concretions of lime and ferro- manganese minerals.

Clays in the black soils is mainly of 2:1 smectite type. This clay shows higher expansion and contraction properties due to which a continuous churning takes place in the soil. The continuous churning develops some specific features in black soils like wide (up to71 cm) and deep (up to 50 cm) cracks and fissures, Gilgai micro relief and coarse intersecting slicken sides.

Cracks and fissures in black soils develop due to higher swelling at high moisture content and higher shrinkage at low moisture content. These cracks become wider and deep upon drying of soils. There is a physical movement of surface soil towards the subsoil through these cracks and fissures. Hence these soils are also called as self-mulched or self-ploughed soils.

Dark grey to black colour of black soils is due to the presence of clay-humus complex and /or titaniferous magnetitic minerals. Black soils are low in organic matter (0.4-0.9%) and calcareous in nature with high $CaCO_3$ content. pH is neutral (7.2) to alkaline (8.5) and bulk density is high (1.5 to 1.8 mega gram per cubic meter). Cation exchange capacity of black soils is high (34-54 cmol (p+) per kg soils) which is due to fine 2:1 type clays.

Sub Groups of Black Soils

Shallow, medium and deep black soils are three sub groups of black soils according to their depth and one more sub group is mixed red and black soils.

Shallow black soils: Shallow black soils are 30-35 cm deep and derived from basalts of Deccan trap. Colour of surface may vary from dark grey to dark yellowish grey. Structure is granular or weak blocky. Lime is present in the form of fine grains or nodules. Drainage condition is good. The solum joins with the decomposing basaltic bed rock.

Medium black soils: Medium black soils are 50 to 120 cm deep and derived from basaltic traps, schists, alkaline granite, gneiss and hornblende. Texture may be up to silty. Clay with equal amount of clay is in upper and lower horizons. Organic matter status is somewhat higher and drainage condition is good. Lime is present in variable quantities and calcic horizon has 20-30 cm thick layer of calcareous nodules.

Deep black soils: Deep black soils are more than 120 cm deep. Sometimes the depth may be more than 240 cm. These soils are also called as Regur in central India. These are mostly derived from basaltic traps. The cotton is main crop in these soils to denote them as "Black cotton soils" of India. Black cotton soils show almost all characteristics described for Black soils.

Mixed red and black soils: In some of the regions mixed red and black soils are also found. Generally, in the region of red soils, black soils are found at lower topographic levels. Similarly red soils are found in the region of black soils. In these regions, black soils due to fine in texture, higher in permeability and more susceptible for erosion cover the red soils. Therefore, mixed black and red soils are formed.

Uses: Most of the black soils are inherently well fertile and mainly used for growing cotton. However, millet, sorghum soybean, pigeon pea etc. are also grown. Sugarcane is also cultivated under irrigated conditions.

Desert and Arid Soils

Desert soils come under Aridisols, Inceptisols and Entisols of U.S. Taxonomic System and Cambisols, Leptosols, Calcisols, Arenosols and Fluvisols of World Reference Base system.

Occurrence: There is a desertic stretch in between Indus River to Aravalli ranges of the north western part of the country having area about 30 million hectares. The states coming in this part are, Rajasthan, Southern part of Haryana, south west Punjab and some part of north Gujrat. This area comes under hot arid region. Besides this cold desert of Jammu and Kashmir and tropical desert between Karnataka and Andhra Pradesh also have desert soils.

Parent material: The main parent material of desert soils is alluvium, sand and weathered residues of different rocks and minerals. The quartz, feldspars, hornblende and calcite grains are main component of desertic sand. This desertic sand forming sand dunes may be derived from Run of Kutch, lower Indus basin, ocean, weathering of localized rocks and/or disintegration of light textured soils.

Climate and vegetation: The climate is mostly hot and arid, rainfall is less than 100 mm to 500 mm per annum whereas potential evapotranspiration (PET) is 1620 mm to 2060 mm. In case of cold arid rainfall is less than 150 mm and PET is around 800 mm. The tropical (Deccan) arid region also has more PET (around1850 mm) than rainfall (less than 500 mm). Thus, there is general scarcity of water which hinders the growth of vegetation. Mostly xerophytes and some thorny bushes and grasses are naturally grown in this area.

Nature and properties: There is very poor profile development in the desert soils. 'A' horizon is absent or very poorly developed. Similar is the case with 'B' horizon' and sand is the main soil separate up to the depth of 50 cm. Texture is fine sandy to loamy fine sand with less than 10% clay. Soil colour is pale brown to dark yellowish brown. Structure is Generally, weak sub angular blocky. pH of desert soil is high (7.2-9.0). Some times between 8.0-8.6 due to calcareousness. Concretions of $CaCO_3$ are present in variable quantities in the profile. Occasionally gypsum also found in appreciable quantities to form gypsic horizon. Soils are very low in organic matter content.

Uses: Soils are productive if irrigation facilities are provided and two crops in a year can be grown. Arid fruits like Ber, bel and pomegranate, pearl millet, mustard and legumes can be grown successfully with short supply of water. In rainy season natural vegetation grow remarkably which provide fodder and fuel wood.

Laterite and Lateritic Soils

Laterite and lateritic soils come under Alfisols, Ultisols, Inceptisols, and Oxisols of U.S. Soil Taxonomy and Cambisols, Alisols, Leptosols and Plinthosols of W.R.B. (World Reference Base). Laterite word is derived from Latin word, 'Later' which means brick. Thus, soil can be dig in to brick like pieces and used for construction purposes.

Occurrence: Well developed laterite and lateritic soils are found in Karnataka, Maharashtra, Tamil Nādu, Goa, Andhra Pradesh, Eastern Ghats of Orissa, Malabar Coast of Kerala, Parts of West Bengal and Assam and some parts of Madhya Pradesh.

Parent material: Main parent material of these soils is basic igneous rocks like basalt, norite and diabase and minerals like pyroxene, amphiboles and biotite. Sometimes these soils are found overlying shale, sandstone and other sedimentary rocks. These overlaying soils are carried down to lower places by flowing water and again cemented in to compact mass in due course of time.

Climate and Vegetation: These soils are formed in the areas of tropical and subtropical region where there is immediate dry spell after the rainfall. Rain forests of tropical areas are favourable for the formation of laterites.

Nature and Properties: Soils are highly weathered and weathering depth may be numerous meters. The intermittent wetting and drying enhances the eluviation of silica to the lower horizon. This increases the amount of sesquioxide in the upper horizon of the soil.

The main characteristic of lateritic soil is the compact and vesicular structure in the subsoil horizons. This structure is essentially made up of hydrated oxides of iron and aluminium. Below this the soil material may be brick red or brown to yellowish in colour. This material is soft when moist but becomes hard like granite upon drying. The quarrying of hard and massive laterite is done in the form of bricks or tiles for construction etc.

Kaolinite is the main clay mineral which is present in abundance. Thus, plasticity, cohesion, swelling and shrinkage and cation exchange capacity are very low. The silica: Sesquioxide ratio is low, combined and constitutional water is more and Ca, Mg, Na and K are also low due to pronounced leaching. Thus, causing deficiency of Ca, Mg, and K. Plant nutrient status is low, high phosphate fixation as iron and aluminium phosphates also causes phosphorus deficiency. Lower pH in the range of 6.0-6.8 causes toxicity of aluminium and manganese.

Uses: Although lateritic soils are deficient in plant nutrients even through successfully used for growing different types of crops. The low-lying lateritic soils are used for the cultivation of paddy, banana, areca nut and coconut. The laterite at higher topographical places is under the cultivation of cocoa, cashew, tea, rubber, coffee, cinchona and mangoes.

Tarai and Sub- Montane Soils

Tarai soils come under Mollisols order of U.S. Soil Taxonomy and Chernozems and Kastanogems of World Reference Base.

Occurrence: Tarai soils are found in the northern part of Uttar Pradesh, Uttarakhand, Bihar and West Bengal under foot hills of Himalayan ranges. These occupy area about 29000 sq. kilometres. These soils remain moist during most of the part of the year due high-water table and wet moisture regime. Therefore, these soils are called as Tarai soils.

Parent material: Tarai soils are derived from alluvial parent material. Alluvial sediments of Tarai belt have interstratified coarse sand and gravel with hard clay. The new deposits as well as older sediments of Pleistocene age are derived from mica, chloritic schists, gabbro, gneiss, sand stone, shale and slate.

Climate and vegetation: Tarai soils are developed under hot and humid climate and tall grasses and dense bushy vegetation.

Nature and Properties: Humid climate of Tarai region enhance the fast and more growth of vegetation and facilitates greater organic matter accumulation in the soils. Thus, Tarai soils are rich in organic matter. Dark coloured 'A' horizon contains 1-4% organic matter which gradually decreases with depth. Horizons can be differentiated on the basis of their organic matter content.

Texture of Tarai soils vary from sandy loam to silty clay loam. In some places lower horizon (100-300 cm from the surface) may have fine nodules of lime or a layer of rounded pebbles. This pebble layer makes the soils more permeable. The texture of parent material directly influences the thickness of profile and leaching conditions of soils. Soils developed from coarser parent material are thicker with greater leaching. Tarai soils are imperfectly drained to well drain.

Uses: Tarai soils are highly fertile. But drainage conditions determine the productivity of these soils. Well drained Tarai soils are most productive. Rice, sugarcane, wheat and summer sunflower are commonly grown in Tarai soils. There is loss of organic matter in the newly cultivated Tarai soils due to intensive agriculture. This is adversely changing the fertility and productivity of these soils.

Sub-montane Soil

Such soils are distributed in the dry subhumid environment of Hoshiarpur in Punjab and Rakh Dhiansar in Jammu and Kashmir and in the humid tract of Dehradun. In J&K area of Alfisols is104000ha. The lands are sloping, the soils range from loamy sands to sandy loams, silty loams and clay loams with soil moisture storage capacity improving in that order. Soil crusting occurs in soils of dry and sub-humid regions. These soils are formed in high rainfall regions of sub-Himalayan under coniferous forests. The soils are acidic in reaction. The organic matter accumulation is high and there is absence of free lime.

These are formed by the deposition of scoured material from Shivalik and also the lesser Himalayas. These are found within the Tarai region of the submontane stretching from Jammu and geographic region to Assam. The soil supports a luxuriant growth of forest and a lot of susceptible to soil erosion.

Forest and Hill / Mountain Soils

Forest Soils

The recorded forest area of India is about 708273 sq. km which is 21.54 % of country's total geographical area (India State Forest Report, 2017, Forest Survey of India, MOEFCC New Delhi). Total forest and tree area is eight crore nine lakh hectares which is 24.62 percent of total geographical area of India.

Forest soils are commonly found in Andhra Pradesh, Telangana, Assam, Tripura, Mizoram, West Bengal, Odisha, Madhya Pradesh, Maharashtra, Chhattisgarh, Uttarakhand, Gujrat, Himachal Pradesh, Karnataka, Kerala, some part of Uttar Pradesh, Jharkhand, Tamil Nadu, Jammu and Kashmir etc.

Fig. 2.7: A view of mixed coniferous forest at way to Srinagar, Jammu & Kashmir (by Dr. Venkatesh Bharadwaj)

Parent material: Forest soils of Himalayan areas are developed on sedimentary, igneous and metamorphic rocks, whereas forest soils of southern area and Deccan plateau are developed on igneous or metamorphic rocks.

Climate and vegetation: Forest soils of India are developed under varied climate and forest vegetation. The climate ranges from hot and dry subtropical, subhumid to humid and cold temperate. Species of forest vegetation directly influence the physical and chemical properties of the soils under their development. The tropical deciduous, tropical evergreen and conifers are major forest types under which forest soils are developed.

Nature and properties: Generally, forest soils are developed under acidic conditions in the presence of acidic humus and low base status are podzolic type and these developed under slightly acidic to neutral condition with higher base status are brown forest soils and non calcic brown soils. The properties of forest soils are quite variable. These soils are calcareous to non calcareous type. The texture varies from sandy in littoral forest to silty clay loam in coniferous forests. pH is slightly acidic to slightly alkaline. Accumulation of organic matter is moderate to higher (1-3.5%). CEC of forest soil also vary from 10-20 cmol (p+1)/ kg soil. Most of the forest soils are quite fertile and rich in plant nutrients, but some podzolic forest soils are deficient in available phosphorus.

Uses: Forest soils are used to produce timber, fuel wood, fodder, other minor forest products and support shifting cultivation by tribal community of the country

Hill / Mountain Soils

Hill soils are Generally, classified into Entisols, Inceptisols, Mollisols, Alfisols and Ultisols of U.S. Soil Taxonomy and Luvisols, Leptosols, Cambisols and Phaeozems of World Reference Base.

Occurrence: Hill soils are found in Himalayan ranges of Jammu and Kashmir, Himachal Pradesh, Uttarakhand, Nagaland, Assam, Meghalaya, Arunachal Pradesh, Mizoram, Manipur, hilly region of West Bengal, Satpura and Nilgiris ranges of southern states namely Tamil Nadu and Kerala, Total area under hilly soils is about 13,300 square kilometre

Parent material: Parent material of hilly soil is quite diversified mostly these soils are developed from sand stone, shale, charnokite, norite, conglomerate, schist, quartzite, mica etc.

Nature and properties: Most of the hill soils show the nature and properties of brown forest soils and podzols. Soils are sandy loam to loam in texture. At some places clayey texture of B horizon may vary form silty clay to clay and below this the soils of horizons at the depth of 100-200 cm brown to dark brown coloured compact clay. Hill soils are mostly acidic in nature. pH may vary from 4.8-6.1. Upper horizons contain more amounts of organic matter and phosphorus and rich in available potassium. CEC varies from 9.4-44.0 cmol (p+1)/kg soil. There may be more amount of exchangeable calcium in lower horizon due to the presence of lime stone in bed rock.

In the Himalayan region, under spruce, pines and deodar forest brown podzolic soils are found. Soils are non-calcareous and acidic in nature. Surface horizon of these soils is rich in silica and there is illuviation of sesquioxide at the depth of about 60 cm.

Very shallow soils with very poorly developed profile are found at higher ranges of Himalayas. These soils are formed on sand stone and shale parent material under lesser rainfall, scanty vegetation and lower temperature. These soils are also called as skeletal soils due to their shallowness and poor profile development.

Soils of Nilgiris hills are derived from granite, gneiss, basalt, and charnokite and norite parent material. Soils are mostly red and black in colour. Texture is sandy loam, clay loam to clay. pH is acidic, drainage condition is good. Mostly soils are poor in organic matter, nitrogen and phosphorus and rich in Ca, Mg, K, Fe and Al.

Uses: Uses of hill soils depend on their low, medium and high height. Production of timber, temperate and other fruits, sheep and goat herding, pastures and grazing, ornamental, medicinal plants and herbs, fodder, agricultural crops etc. are common uses of hill soils

Saline and Alkali/Salt Affected Soils

Saline and alkali/Salt affected soils come under Aridisols, Inceptisols, Alfisols, Vertisols, Entisols and Mollisols orders of U.S. Soil Taxonomy and Leptosols, Vertisols, Solonchaks, Solonetz, Gypsisol and Calcisols of World Reference Base.

Occurrence: Salt affected soils occupy about 12-million-hectare area of India in which major portion falls under arid and semi-arid region of Indo-Gangatic plain followed by Deccan Plateau and Coastal and Deltaic regions. Salt affected soils are Generally, found along with normal soils. Most commonly salt affected soils are found in Uttar Pradesh, Bihar, Delhi, Punjab, Madhya Pradesh, Rajasthan, Gujarat, Haryana, Maharashtra, Andhra Pradesh, Orissa, Tamil Nadu, Kerala and West Bengal.

Parent Material: Parent material of salt affected soils is mostly alluvium and also quartz, feldspars, calcite and basalt.

Climate and vegetation: Salt affected soils are mostly formed under hot arid and semi arid areas where annual rainfall is below 500 mm or between 500-800 mm except some coastal areas. Vegetation in the salt affected areas is very rare. Only some halophytes and sodium tolerant grasses and bushes are seen as natural vegetation.

Nature and properties: Saline and alkali or Salt affected soils differ from normal agricultural soils in their soluble salt content and pH. Increase in soluble salt content of soils changes the balance of cations, soil reaction, physical properties, and osmotic pressure and enhances the specific ion toxicity. All these adversely influence the crop production.

Texture of salt affected soils of northern India is loamy with more amounts of illite and mixed minerals. These have 30-40% exchangeable sodium precent. pH between 9.0-12.0 and salt content between 10-100 dSm^{-1} as EC. Soils also have platy at surface and blocky structure at sub surface. Semi-arid tract of Gujarat, Maharashtra, Karnataka, Andhra Pradesh and Tamil Nadu show 10-15% exchangeable sodium percentage due to more montmrillonitic minerals. These soils also have lesser permeability and situated at lower topographical position. Therefore, productivity is also too much less.

Salt affected soils of arid regions of Gujarat and Rajasthan have very low permeability and poses very poor drainage conditions. Coastal salt affected soils are always rich in salts due to tidal inlet and lower topographical situation. They have high soluble salts of chlorides and sulphates of Na, Ca and Mg as EC ($4dsm^{-1}$), less than 15% exchangeable sodium and pH more than 8.2. Some soils of Kerala and West Bengal have more amounts of acidic salts. Therefore their pH goes down to below 4.0.

Uses: Reclamation of salt affected soils is essential to use then for agricultural production. After reclamation these soils can be used for growing rice, sugar beet, wheat, barley, bajra, sugar cane, cotton, various grasses etc.

Peaty, Marshy and Skeletal Soils

Peaty Soils

Occurrence: Peaty soils are found in southern Uttarakhand, northern part of Bihar, coastal areas of West Bengal, Odisha, Kerala and Tamil Nadu. These soils mainly occur in the western parts of Kottayam district and some parts of Alappuzha district (Kerala) where it is called Kari.

Parent material: Parent material of peaty soils is organic matter enriched with soluble salts.

Climate and vegetation: Peaty soils are developed under heavy rainfall and humidity climate and plenty of mangrove vegetation.

Nature and properties: Peat soils are almost black or dark in colour. There is more accumulation of partly decomposed organic matter, more than half kg per kg soils, under wet or water-logged condition. The source of organic matter can be recognized due its partly decomposition. Texture of peat soils is fine with more amount of smectite under water logged (anaerobic) condition reduces the pH up to 3.5-4.0. Base saturation percentage is also low.

Saline peat soils also called as Kari soil are found in Kerala. These are developed from sediments of brackish water in the tidal swamp areas. Kari soils contain pyrite (FeS_2) and other sulphides, which upon oxidation and decomposition produce sulphuric acid (H_2SO_4). This acid reduces the pH of the soils from 4 to 3 and up to 2.0. Reactions with acid decompose the clay minerals which increase the iron and aluminium concentration in the soil.

Uses: Peat soils are used for cultivation of rice and pineapple after proper care and management. Peat soils of Sunder ban (W.B.) are also under mangrove forests and help in the maintenance of ecological balance there.

Marshy Soils

Occurrence: Marshy soils also occur in the coastal areas of Orissa and Tamil Nadu, Sundarbans of West Bengal, Central part of Bihar and Almora district of Uttarakhand

Parent material: Marshy soils are mostly derived from alluvial parent material such as mica kaolinite, dried loess or coastal plains.

Climate and vegetation: Marshy soils are developed under humid climate and vegetation.

Nature and properties: Marshy soils show bluish and greyish colour as these are developed under anaerobic water-logged conditions. Marshy soils are Generally, found at depressed lands of alluvial and coastal areas. These soils are dominated by mica and kaolinite.

Uses: Generally, in enclosed area, marshy soils are used to grow grasses, reeds, sedges, wild rice, aquatic medicinal plans, shrimp, fish etc.

Skeletal Soils

Occurrence: Soils are, generally, found at Ladakh, Himachal Pradesh and Vindhyachal ranges.

Parent material: Skeletal soils are developed from sand stone and granite types igneous rocks.

Climate and vegetation: Hot or cold semi-arid to arid climate helps in the development of skeletal soils. Skeletal soils are developed under scanty vegetation.

Nature and properties: Skeletal soils have some thickness. But due to lack of profile development these soils are called as skeletal soils. Thickness of skeletal soils may very 7 to 15 cm. Soils developed over sand stone are thicker than those developed on granite and other igneous rocks. Granite is more resistant for weathering. Low rainfall, lesser vegetation, adverse condition for weathering and more erosion help in the formation of skeletal soils. Skeletal soils are horizon less soils. Soils show light grey to dark grey colour. Texture is sandy loam to loam with more amount of silica.

Uses: Skeletal soils in lowlands are used to cultivate rice and well-drained skeletal soils in high rainfall areas are mainly used to grow para rubber, tropical fruits and oil palm. Skeletal soils of drier areas are suitable for cassava, sugarcane, beans, maize and sorghum cultivation.

Soils of Indian States

Soil survey done by central and state agencies and agricultural universities indicate that Indian soils mainly come under Inceptisols (40.0%), Entisols (28.0%), Alfisols (13.5%), Vertisols (8.5%), Aridisols (4.0%), Ultisols (2.5%) and Mollisols (< less than 1%). State wise distribution (Yadav and Singh 2000, Bhattacharyya and Pal 2015) of various soil orders is given in table 2.8:

Table 2.8: Soil orders of soils of Indian states

State	Soil orders (Approximately area in square km)
Andaman & Nikobar	Inceptisols (93), Entisols (8), Alfisols (118), Mollisols (96)
Andhra Pradesh	Inceptisols (103080, Alfisols (83550), Entisols (290), Vertisols (22387), Aridisols (8170), Mollisols (5103)
Arunachal Pradesh	Inceptisols (31264), Entisols (29855), Ultisols (11876), Alfisols (228)
Assam	Inceptisols (32490), Entisols (26401) Alfisols (8869), Ultisols (4365)
Bihar	Inceptisols (60600), Alfisols (59920), Entisols (47413), Vertisols (1068)
Delhi	Inceptisols, (1295), Ultisols (583),
Goa	Inceptisols (116674), Entisols (23288), Aridisols (24399), Vertisols (20475), Alfisols (1188)
Haryana	Inceptisols (25591), Entisols (12753), Aridisols (4158), Alfisols (298)
Himachal Pradesh	Entisols (286541), Inceptisols (10999), Mollisols (400), Alfisols (298)
Jammu & Kashmir	Entisols (206470), Inceptisols (14272), Alfisols (1005), Mollisols (417),
Karnataka	Alfisols (51100), Inceptisols (45431), Entisols (31335), Vertisols (28970), Aridisols (18894),
Kerala	Ultisols (24033), Inceptisols (11563), Entisols (2254), Alfisols (565), Mollisols (448)
Lakshadweep	Entisols (320)
Madhya Pradesh	Inceptisol (179153), Vertisols (91896), Alfisols (8500), Entisols (84671), Mollisols (2602)
Maharashtra	Entisols (44480), Inceptisols (94636) Vertisols (81062), Alfisols (1731), Mollisols (194)
Manipur	Inceptisols (8599), Ultisols (8110), Entisols (5106) Alfisols (38)
Meghalaya	Inceptisols (10238), Ultisols (8983) Entisols (2401), Alfisols (817)
Mizoram	Ultisols (8147), Inceptisols (7857), Entisols (4522), Alfisols (555)
Nagaland	Inceptisols (14888), Entisols (1213), Alfisols (478),
Odisha	Inceptisols (79099), Alfisols (52190), Entisols (15825), Vertisols (8593),
Pondicherry	Inceptisols (215), Vertisols (96), Entisols (95), Alfisols (24)
Punjab	Inceptisols (29752), Entisols (1042), Andisols (808), Alfisol (211)
Rajasthan	Entisols (144740), Inceptisols (86112), Aridisols (76987), Vertisols (11156), Alfisols (2883)
Sikkim	Inceptisols (3908), Entisols (2371), Mollisols (817)
Tamil Nadu	Inceptisols (65983), Entisols (2371), Mollisols (817)
Tripura	Inceptisols (8460), Ultisols (691), Entisols (840), Alfisols (417), Histosols (24)

State	Soil orders (Approximately area in square km)
Uttar Pradesh	Inceptisols (209083), Entisols (57788), Vertisols (4791), Mollisols (1033)
West Bengal	Inceptisols (48996), Entisols (19723)
Total 3280565 sq. km.	Inceptisols (1303730), Entisols (921320), Vertisols (279600), Andisols (140690), Ultisols (82500), Mollisols (13200), Histosols (24).

3

Properties and Processes of Soils

Physical Properties of Soil

Physical properties of soil have many important roles in the crop production. The basic requirements for germination, seedling emergence and root development are adequate moisture and aeration, optimum temperature and absence of mechanical hindrance. After germination the further development of plant is also governed by physical conditions of soil. Soil texture, soil structure, soil density, soil consistence, soil water, soil air, soil temperature and soil colour are main physical properties of soils also affecting mechanical behaviour of soils. These are described here.

1. Soil Texture

The weathering divides larger particles of rocks and minerals in to the smaller ones. This process goes continuously. Thus, particles of different sizes ranging from large fragments of rocks to the very fine clay minerals are found in soil. However, for texture particles less than 2mm diameter is considered. The relative amount of these particles indicates the soil type also. Such as sandy soils contain more amounts of coarse particles i.e., sand and clayey soils are rich in fine particles. The rock and mineral particles having diameter less than 2.0 mm and come between specifies size are called as soil separates. International society of soil science (ISSS) and United States Department of Agriculture (USDA) systems are Generally, followed to classify the soil separates. This classification is given in table 3.1.

Table 3.1: Classification of Soil Separates

Soil separates	Diameter(mm)	
	ISSS System	USDA system
Very coarse sand	--	2.0-1.0
Coarse sand	2.0-0.2	1.0-0.5
Medium sand	--	0.5-0.25
Fine sand	0.2-0.02	0.25-0.1
Very fine sand	---	0.1-0.05
Silt	0.02-0.002	0.05-0.002
Clay	Less than 0.002	0.002-0.0005
Colloid	--	Less than 0.0005

The relative amount, proportion or percentage of the soil separates i.e., sand, silt and clay present in a soil is called as soil texture. This indicates mechanical or physics composition of soil. Numerous textures of soils can be identified according to relative percentage of sand, silt and clay present in them. Therefore, to overcome this problem few soils textural classes were designated. The soil textural classes are the groupings of soil textural units based on relative percentage of soil separates i.e., sand, slit and clay. Soil textural classes based on USDA are given in table 3.2 and shown in figure 3.1

Table 3.2: USDA based Soil textural classes

Soil textural classes	Percentage range (weight basis) of		
	Sand	**Silt**	**Clay**
Sand	85-100	0-15	0-10
Loamy sand	70-90	0-30	0-15
Sandy loam	43-80	0-50	0-20
Loam	23-52	28-50	7-27
Silty loam	0-50	50-88	0-27
Silt	0-20	80-100	0-12
Sandy clay loam	45-80	0-28	20-35
Clay loam	20-45	15-35	27-40
Silty clay loam	0-20	40-73	27-40
Sandy clay	45-65	0-20	35-45
Silty clay	0-20	40-60	40-60
Clay	0-45	0-40	40-100

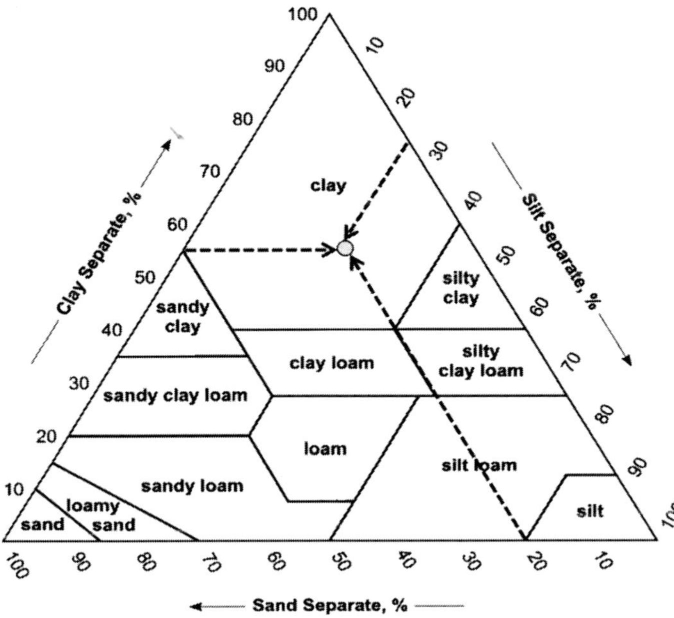

Fig. 3.1: Soil Textural Classes

Soil texture is stable and permanent property of soil. It cannot be changed even after longer periods of soil use. However, in smaller areas like nurseries, glass houses, poly houses and green houses desired soil texture can be obtained by adding required quantities of sand or clay to the soil.

Importance of Soil Texture

Soil texture can help in prediction of many characteristics of soil. The several textural classes can be grouped in three major groups in sandy, loamy or clayey soils. The sand and loamy sand soils come in sandy group which has coarse texture. The group of medium textured soil is too large. This group has sandy loam, loam, silty loam, silt, sandy clay loam, clay loam and silty clay loam soil. The fine textured soils are silty clay and clay.

The sandy or coarse textured soils have rapid water movement, lesser water holding capacity, good aeration, lesser nutrients, and more leaching losses of nutrients, well drained, easily tilled and require more irrigation to grow crops. The medium textured soils have optimum physical and chemical characteristics for plant growth. Most of the crops can be grown in these soils. These soils are good in moisture holding, nutrient retention and supply, moderately drained and aerated, responsive to fertilizer applications and amiable for tillage practices.

The fine textured clayey soils are less permeable, slow movement of water, imperfectly drained, water can stagnate on soil surface, poor aeration, more water holding capacity, provide water to the crops for longer periods, lesser irrigation is required, more adsorption/absorption and less leaching losses of nutrients and difficult for tillage practices as wet clay soils stick with implements and dry clayey soils produce large and hard clods upon ploughing. Therefore, these should be ploughed at an intermediate soil moisture condition. Clayey soils are inherently productive, if properly managed for tillage and drainage conditions.

Determination of Soil Textures

Numerous physical and chemical properties of soils are associated with the size and surface area of the soil particles. Determination of size of soil particles is necessary to know the texture of soil. Estimation of amounts of different soil particles i.e., soil separates by sieving, sedimentation, microscopic or combination of these techniques is called as particle size analysis or mechanical analysis of soil.

Soil particles, ranging from fine gravel to colloidal clays, Generally, cluster together as aggregates. They can be separated from one another by mechanical and chemical means to know the amount of specific sized particles. In the laboratory separation of soil particles is done mainly by two methods,

1. The hydrometer method also called as Bouycous method and
2. The pipette method.

1. The hydrometer method measures the density of soil suspension at various times. In this method particles suspended in water, settle differentially according to their surface area per unit weight. The clay particles having high surface are settle slowly as compared to silt and sand which settle rapidly because of their low surface areas. The rate of which particles settle can be calculated by using Stoke's law. This method is rapid and less accurate than pipette method but can be used for routine analysis.

2. The pipette method depends upon the concentration of suspended soil particles at particular depth and time. The suspended soil particles take time (t) to settle at the depth (d). At this stage the concentration of such particles will be more which have settling velocity less than d/t. This procedure also depends upon Stocks law. This law was proposed by G.G Stokes' in 1851. The law indicates the relationship between the radius (size) of particles and their rate of fall in liquid medium. The formula of Stoke's law is given below,

$$\vartheta = \frac{2r^2 g(DP - DM)}{9\eta}$$

ϑ = velocity of fall of particles i.e settling velocity (cm sec^{-1}), g = Acceleration due to the gravity (980 cm sec^{-1} sec^{-1}),

r = Radius of spherical particles (cm),

DP = Density of particles (g cm^{-3}),

DM = Density of medium (g cm^{-3}) and

η = Viscosity of medium (dyne sec cm^{-2})

The freely falling particles acquire accelerated motion due to gravity which will be equal to 4/3 π r³ g (DP-DM). But in soil suspension falling particles face resistance offered by viscosity of the medium (water). This resistance will be equal to 2 πr η 3 φ. These two forces provide a relationship as 2 πr η 3 φ = 4/3 π r³ g (DP-DM).

$$\vartheta = \frac{\frac{4}{3}\pi r^3 g(DP - DM)}{2\pi rn3}$$

$$\vartheta = \frac{2r^2 g(DP - DM)}{9\eta}$$

If any particle falls from X to Y i.e., from height 'h' in time't', the velocity will be equal to h/t.

Thus, $h/t = \frac{2r^2 g(DP - DM)}{9\eta}$ and

$$t = \frac{9\eta h}{2r^2 g(DP - DM)}.$$

The mechanical analysis or particle size analysis is done after removing the organic matter, $CaCO_3$, soluble salts from the soil and dispersing it with various chemicals.

The amount of sand, silt and clay obtained either by hydrometer or pipette method can be fitted in textural triangle table or in figure, indicated earlier, to know the texture of soil.

Fingers Feel method to Determine Soil Texture

The texture of soil can be known by fingers feel method also. This method is used to good advantage by soil surveys, soil conservation and farm planning personnel. It requires considerable practice to be highly proficient. The balls and ribbons of soil are made in this method.

One tea spoonful of soil is taken in the palm; the water is added drop by drop. Kneading of soil is done simultaneously and soil is brought to a workable consistency. During this process feeling of soil between thumb and fingers can indicate its texture. More gritty feeling will be due to more sand in soil and smoothness will be the indication of more amount of clay. The feeling of loam soil will be slightly gritty and fairly smooth.

Very weak balls and ribbons are formed by sandy soils. Fairly firm balls and ribbons are formed with loam soil. Very firm balls which are hard upon drying form very hard balls and long flexible ribbons are formed with clayey soils.

Soil texture and Plant Growth

Soil texture has more agronomic importance. It decides the selection of crop to grow in particular soil. Crops like sugarcane, paddy etc. are grown on fine textured soil whereas groundnut, potato and root crops grow very well on coarse textured soils. Most of the crops can be grown successfully on medium textured i.e., loamy soils.

Sandy soils have less available water and organic matter as compared to clayey soil. Organic matter contains essential plants nutrients. Therefore, sandy soils are inherently weak in the plant nutrient status. Low nutrient and available water status of sandy soil directly reduce the plant growth. In contrast of this clayey soils contain more amounts of organic matter and available water. But due to the water stagnation few crops can be grown in clayey soils. Loamy soils have proper amounts of organic matter, plant nutrients and water. Therefore, most of the crop plants grow better in loamy soils.

2. Soil Structure

Soil structure influences the aeration, thermal, water holding, water movement, root penetration and microbial properties of soil. The sand, silt and clay (primary particles) of soil by nature do not remain separate with each other. These primary particles combine together and arrange themselves in different shapes and pattern to form secondary particles or groups of aggregates.

The combination or arrangement of primary particles to the secondary particles is called as soil structure. Thus, soil structure is the natural arrangement of soil particles. Secondary particles are also called as ped. Ped is a naturally formed unit of soil structure. Natural processes, forces and substances of plant and microbial origin bind the primary particles to form a structural unit called aggregate which has diameter less than 10 mm or 1 cm. Process of aggregate formation is known as aggregation. Clod is a compact and coherent unit of soil particles which is obtained due to the ploughing or digging of highly moist or dry soil. The term clod is reverse of aggregate and ped.

Categories of soil structure: The soil structure is categorized into three,

1. Types of soil structure i.e., Shapes of peds,
2. Class of soil structure i.e., Size of peds and

3. Grade of soil structure i.e., Stability of peds.

Types of Soil Structure

Platy, Prism like, Blocky and Spheroid are four main types of soil structure. These types are based upon the shape of the peds. The types of soil structure are identified or described under field conditions.

1. Platy: Soil particles are arranged in thin plate like structure. Their horizontal axes are too much long than the vertical axis (Fig. 3.2). Plate like soil structure is found near mines of mica, above the impermeable B horizon and in the upper layer of heavily compacted soil. This structure is not good for plant growth as there is no enough movement of air and water. Lenticular is another type of platy structure where edges of peds are thin than the middle portion.

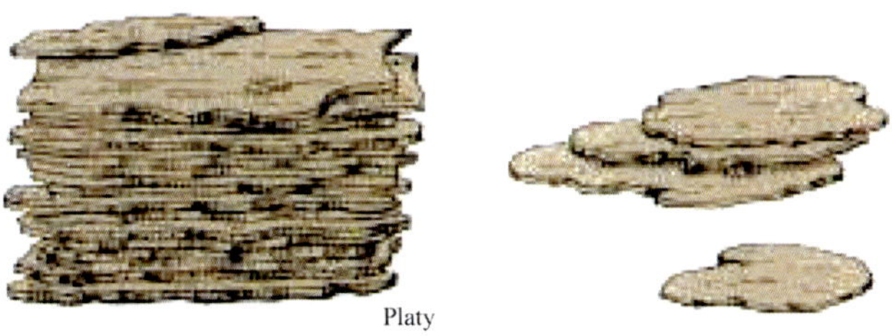

Fig. 3.2: Please redraw in b/w with changes in drawing

2. Prism like: The vertical axis is larger than the other axes. The height of peds may be up to 15 cm. The peds have the appearance of a prism. This structure is divided into two.

A. Prismatic: In the prismatic structure the top of ped did not have caps i.e., they are not rounded in shape. The level of tops of ped remain flat or plane (Fig. 3.3). Prismatic structure found in Bt horizon or sub surface layer of clayey soils.

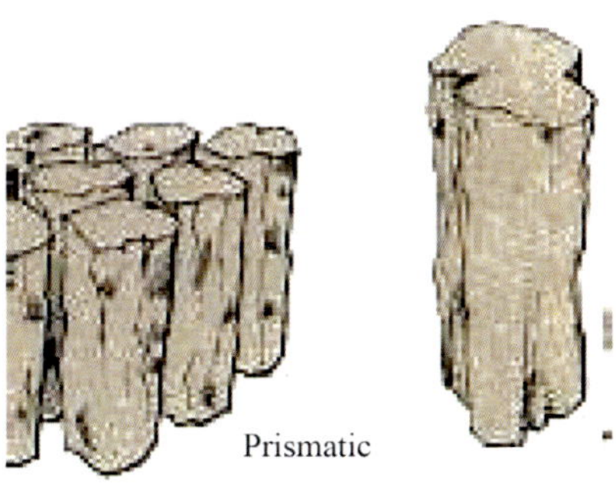

Fig. 3.3: Prismatic

B. Columnar: The peds have rounded caps on their tops i.e., the tops are not flat or in plane level (Fig. 3.4). Columnar structure is found in sub soils of arid and semi-arid regions and B horizon of sodic soils. These two structures are not good for crops due to the imperfect in air and water movement.

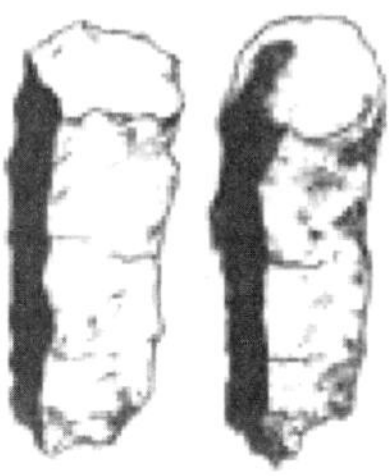

Fig. 3.4: Columnar

3. Blocky: Blocky structure shows almost equal axes i.e., length, width and thickness or height. Thus, there is formation of somewhat cubical structures. Blocky peds are generally, found in B horizon and common in sub surface horizon of black cotton soils of India. Based on stability and acuteness of angles or corners, blocky structure is sub divided into angular blocky and sub angular blocky.

In the angular blocky the surfaces of peds remain plane or flat and vertices show acute angles (Fig. 3.5). In case of sub angular blocky, the peds are roundish in shape and vertices have subdued angles. The appreciable number of voids among the peds provides free movement of air and water and ease to root penetration. Thus, blocky structure is good for plant growth.

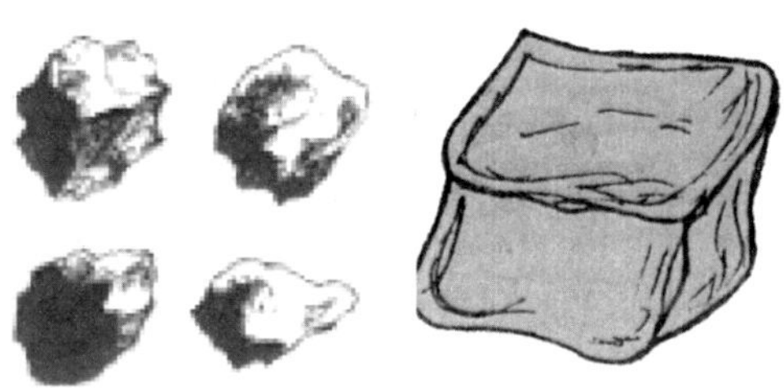

Angular Subangular

Fig. 3.5: Subangular

4. Spheroid: All the axes of peds are almost equal with roundish surface. The peds are not in close contact to each other. Thus, have plenty of pore spaces. The peds are formed due to the interaction of soil mineral matter with plant roots and soil organisms. The spheroid soil structure is found in organic matter rich A horizon of soils. The diameters of the peds remain less than 10 mm.

The spheroid soil structure is divided into granular and crumb on the basis of porosity of peds. Granular is relatively less porous in which numerous small soil particles join together to make a large granule (Fig. 3.6). Granular structure is good for plant growth. Crumb soil structure, just like bread crumb, is too much porous in which finest particles make a larger granule (Fig. 3.7). The peds are small less than 5 mm in diameter, not fitting with other, and rich in plant nutrients. The higher porosity of peds helps in free movement of air, more retention of water and easy penetration of plant roots. Thus, crumb structure is appreciably good for plant growth and development.

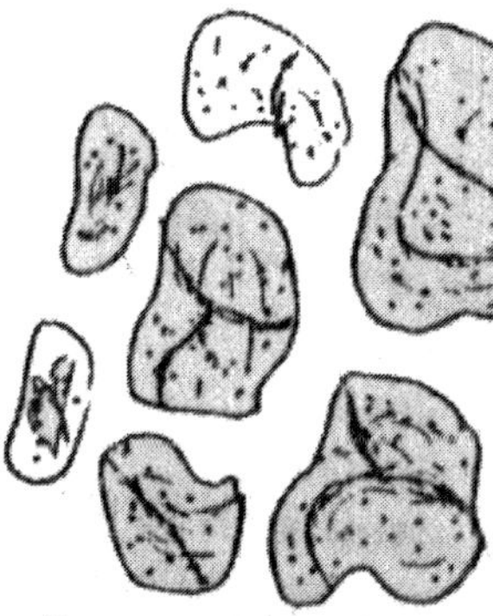

Fig. 3.6: Granular

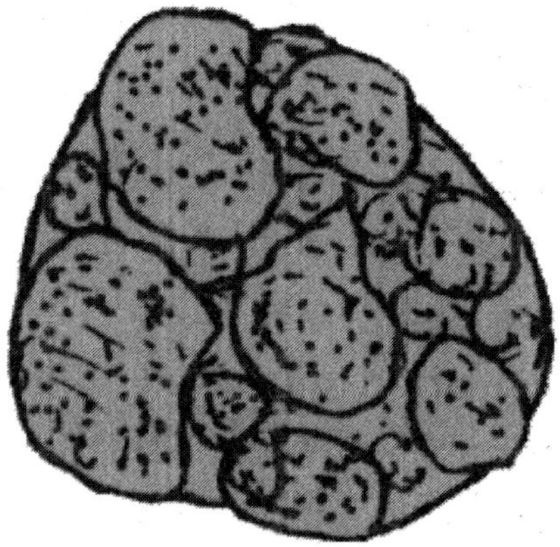

Fig. 3.7: Crumb

Class of Soil Structure

Class of soil structure is the grouping of peds on the basis of their average size in mm. This is expressed in thickness of platy, height, width and thickness of prismatic and diameter of spheroid structures. The numeral values for these are given in table 3.3 (Soil Science Society of America, 1984.

Table 3.3: Class of soil structure

Class of soil texture	Type of soil structure				
	Plate	Prism	Block	Spheroid	
	-	-	-	Granular	Crumb
Very fine or very thin	<1	<10	<5	<1	<1
Fine or Thin	1-2	10-20	5-10	1-2	1-2
Medium	2-5	20-50	10-20	2-5	2-5
Coarse or thick	5- 10	50-100	20-50	5-10	-
Very coarse or thick	>10	>100	>50	>10	-

Grade of Soil Structure

Grade of soil structure indicates the distinctness and stability of peds and ratio between aggregated and unaggregated soil material. Grade of soil structure depends upon the moisture content of soil and strongness of ped increases with decrease in moisture content of soil.

Grade of soil structure expressed under moist condition of soil. There are four grades of soil structure,

1. Structure less: No peds are observed in undisturbed soil. Puddled soil or compact subsoil is structure less. The coherent structure is called as massive and non-coherent structure like sand is known as single grain.
2. Weak: The undisturbed soil contains poorly formed indistinct peds which are not much stable. When the soil is disturbed by hands, it contains few complete peds, many broken peds and a large amount of unaggregated soil material.
3. Moderate: The well-formed peds which are fairly stable. The disturbed soil contains many distinct and complete peds, few broken peds and little unaggregated soil material.
4. Strong: The undisturbed soil contains highly distinct and stable peds which are very weakly attached together. The peds can tolerate displacement and can be separated easily. The disturbed soil shows many complete peds, very few broken peds and almost no unaggregated soil material.

The grade, class and type of the structure is the sequence to denote the structure of any soil, example is moderate, fine, columnar prism like structure.

Ref. Soil Structure - Food and Agriculture Organization
https://www.fao.org/fishery/docs/CDrom/FAO_Training/FAO_Training/.
Ref. Soil structure Classification - Indian Agricultural Statistics ...
ecoursesonline.iasri.res.in/Courses/Introduction to Soil Science...·

Formation and Development of Soil Structure

The main processes in the formation of soil structure or aggregates are cohesion and adhesion by which soil colloids join or stick together and attach themselves with other soil components. This is mainly due to the electrical imbalance or presence of electrical double layer in the system.

The lowering of electro kinetic potential of the system develops the attraction among particles due to their collision. This forms floccules. These floccules get cemented to form aggregates or to develop soil structure. The main cementing agents are organic matter, silicate clays and hydrous oxide clays, calcium carbonate and cations. Under natural conditions development of soil structure is a complex process in which physical, chemical and biological activities, processes and forces take part along with the climate

Physical activities: Evapotranspiration removes water from the soil. The removal of water from the soil cannot be uniform if the dehydration process is rapid. Therefore, there will be uneven shrinkage in soil which results into the development of cracks of various size and shapes. Columnar structure is formed due to drying of soil.

When there is rainfall or irrigation is done, there is rapid entry of water in to the soil which causes uneven swelling of soil material. This produces fragments and fractures in soil. Rapid intake of water causes entrapment of considerable amount of air in capillary pores which creates an explosive condition with in the ped due to the higher pressure. Thus, there is disintegration and disruption of larger aggregates and peds in to the smaller ones.

There is about 9% increase in the volume of water when it changes into the ice. Thus, freezing of water present in the larger pores of soil creates pressure on the soil particles due to this water come out from the smaller pores and particles come closer and closer. Ultimately they attach each other to form large aggregates.

Higher temperature and low rainfall at arid regions enhance the localized accumulation of salts in profile. This is called as concretion which is a form of aggregate. Movement of water vapours, decomposition of organic matter and weathering of minerals are activated by temperature. These are helpful in soil structure formation.

Chemical activities: Dipolar nature of water, presence of organic and inorganic colloids, salts and cations promote the formation of soil structure. Water has positive charge at one side and negative charge at another side. The negative side of the water molecule is attached with the positive side of other side molecule to form a long chain of water molecules.

But in the soil system clays, salts etc. are also present. The positively charged cations may attach with negative side of water. This forms a bridge in which cation is in the middle, many water molecules are attached at both the sides of cation and lastly negatively charged clays are attached with the positive side of water. This forms a clay- water- cation- water- clay complex. Many such complexes are present in soil, when there is removal of water from the soil, the chain of water molecules become shorter and shorter and ultimately clay particles unite to form the aggregate.

The aggregation is much pronounced in calcium rich soils as compared to sodium rich soils. Therefore, sodium is replaced by the addition of calcium for the reclamation of sodic soils. The ferric hydroxide produces stable aggregates in lateritic soils. The Fe^{3+} iron is a flocculating agent whereas gelatinous iron acts as a cementing agent which binds up the particles. Iron oxide precipitates as hydrated gel which upon dehydration firmly binds the soil particles. Similarly, coatings of hydrous oxides of iron and aluminium on quartz increase the strength of quartz-clay bonds in the soils. This increases the aggregation.

There is more stable aggregate formation in acidic soils which are rich in 2:1 type clay due to aluminium bonding. Clay - aluminium bonding is facilitated by polyvalent phosphate ions. The precipitated Fe and Al phosphates cause aggregation due their sorption on soil particles.

Biological activities: Activities of soil flora and fauna have a significant role in the development of soil structure. Earthworms develop good soil structure. They grind and inter mix the ingested silt and clay particles and partially decomposed organic matter into their gut and add extra calcium. They excrete this enriched material and deposit as granular surface or sub surface casts. The movement of earthworms also makes burrows in soil which increases the permeability of soil. More population of earthworm in soil greater will be water stable aggregates.

Plants also help in the genesis of soil structure. Roots hairs bind together smaller particles. Roots exert the pressure on soil particles to group them in larger aggregates. The decomposition of plant residues produces energy for microorganisms. The microbial cells and fungal hyphae bind up smaller soil particles in the large ones.

Microbes also produce polysaccharides, slimes, mucilages and gums during organic matter decomposition. These products act as cementing agent for soil particles. Thus, the use of organic residues and green manure enhance aggregation and aggregate stability. A good vegetative cover on the soil surface protects the aggregates.

Soil Structure and Plant Growth

The soil structure plays indirect role in the plant growth. The germination of seed, entry of radical into the soil and emergence of plumule from the soil are governed by the soil structure. A good germination requires optimum water, air and temperature conditions along with the rightly prepared seed bed. These requirements are provided by a soil having good structure. Pores of different sizes and shapes are formed during the formation of soil structure. These pores regulate the entry, movement and retention of water and exchange of air and temperature in soil.

The soil structure which permits rapid infiltration and sufficient retention of water as well as adequate aeration is good for seed germination and plant growth. The physical conditions of soil at granular structure provide all these. Granular structure is also suitable for root penetration and seedling emergence as compared to fine and compact soil.

The hard and compact soil impedes root elongation and development. Such soil also decreases the infiltration and increase the runoff water. Like wise to granular structure crumb structure is most suitable for plant growth as it shows soft and porous peds which permit easy penetration of roots and emergence of seedlings.

Management of Soil Structure

Improvement and maintenance are two main criteria of soil structure management. Application of organic manures and plant residues, application of fertilizers, application of soil amendments, and soil conditioners and cultivation of crops are main practices adopted to manage the soil structure.

Aggregation may be improved by the application of FYM, compost, green manure and even by organic residues like cow dung, sugarcane trash, rice husk, wheat straw, saw dust etc. All these materials provide organic matter to the soil. The relation between organic matter application and structure improvements depends upon quality and period of application, climate, and properties of soil and production of compounds during the decomposition. In sandy soils application of organic matter increases the number of micro pores and decreases the number of macropores. This increases the aggregation and water holding capacity of soil.

There is about 1.3 to 1.7 times increase in the structural index of Indian soils due to the addition of variable quantities of organic matter. But aggregation persists only for 3-4 months. Thus, judicious and continuous use of organic residues is necessary. The organic matter either containing or producing soluble gums, polysaccharides, fulvic acid, lignin and uronic acid increases the aggregation and aggregate stability.

Application of single super phosphate (SSP) increases the water stable aggregates in soil. Application of SSP with or without FYM has somewhat similar effect on the aggregation. The phosphate ions have bonds to attach clay particles. The effect of Al-P and Fe-P is more pronounced as compared to Ca-P in soil structure formation and maintenance. Nitrogen alone is not good for aggregates but nitrogen along with phosphate is significantly better for aggregation and aggregate management.

Applications of lime to the acidic soil and gypsum to the alkali soils improve the soil structure. This is because of the flocculation created by the Ca ions present in lime and gypsum. Application of polyvinyl alcohol, polyvinyl acetate and poly acrylamide gel increase the mean weight diameter (MWD) of the soil aggregates.

Long continuous cultivation of crops without the application of organic manures or residues destroys the soil structure. The cultivation of crops or even growths of the grasses on the soil surface improve soil structure. The role of vegetation in the improvement of soil structure is due to their canopy and root system. The dense vegetative canopy keeps the soil surface moist and cool where microorganisms flourish well. The canopy also protects the soil surface from the beating action of rain drops which destroys the soil structure.

The organic matter produced by the canopy, microbial activities during the decomposition of organic matter and pores formed by the decomposed roots improves the soil structure. Legumes in crop rotation create more aggregation and prevent the structure from destruction as compared to the cereals. This is not only due to the provision of organic matter and nitrogen by the legumes for the microbial activities but also due to the cementing agents produced during the crop growth (Ghildiyal, 1969).

3. Soil Aeration

Respiration starts with the germination of seeds. During respiration plants roots as well as organisms present in soil take oxygen from soil and release carbon dioxide in to it. Therefore, carbon dioxide content in soil increases many times as compared to atmospheric air. Soil aeration is the process to replace the soil air to the atmospheric air. The composition of air of well aerated soil is somewhat like to the composition of the atmospheric air at soil surface. But poorly aerated soils have more amount of CO_2 gas. Soil aeration mainly depends upon the volume and continuity of air-filled pores. The general composition of major gases in soil air and atmospheric air are given in table 3.4 which clearly indicates this fact.

Table 3.4: General Composition (percent volume) of soil air and atmospheric air

Air	Gases		
	Nitrogen	Oxygen	Carbon dioxide
Soil	79.2	20.6	0.25
Atmosphere	79.0	20.97	0.03

If enough oxygen is not present in soil, then normal activities of plants and aerobic organisms are adversely affected. Therefore, continuous contact and exchange of soil air to that of atmospheric air is essential. This will reduce the CO_2 and increase O_2 in the soil. This is possible when movement of air in soil or aeration is good enough.

Composition of Soil Air

Composition of soil air is quite variable due to various factors affecting it. These are described here,

Activities of roots and organism- Activities of roots and organisms are not similar for all the times. They utilize different amounts of oxygen at different times and release carbon dioxide accordingly. This changes the composition of soil air. When aerobic microorganisms are more active, they decrease oxygen status of soil air.

Soil water content- More than 40% of water content or less than 10% air content in soil decrease the air exchange and increase CO_2 status of soil air. Also, anaerobic conditions increase the activities anaerobic microbes, which increase the concentration of methane, hydrogen sulphide, ammonia etc. gases in the soil.

Addition of fertilizers, manures and soil amendments- Application of fertilizers, manures, lime, gypsum etc. reduces the oxygen status of the soil air. Because these enhances the activities of roots and microorganisms.

Diffusion of air: The restricted diffusion of air increases the CO_2 status of soil air. Fine texture, poor aggregation, wet, muddy, water logged, compact and subsurface soil has more CO_2 concentration as compared to coarse textured, better aggregation, dry and crumb structure, loose and surface soil.

Organic matter: Presence of easily decomposable organic matter enhances the microbial activity to utilize more O_2 from soil air. This increases the CO_2 status of soil air.

Soil Aeration and Plant Growth

The concentration of oxygen and carbon dioxide gases in the soil air are most important for growth and development of crop plants. The effect of soil aeration on plant growth is summarized here,

Water and nutrient absorption from soil- Under more moisture or aquic conditions, oxygen supply is reduced. This adversely affects the plant growth

Poor growth: Poor acration reduces the developments of roots as adventitious roots grow on stem. There is no adjustment between uptake of water and nutrients. Leaves curled down and become yellow. Even under water logged conditions plants suffer from shortage of water and develop wet wilting. However, this also depends upon plant species, duration of water logging and prevailing temperature.

Production of toxic substance: Glycolysis process increases in the sensitive plants for poor aeration. There is increase in production of toxic substances like ethanol and acetaldehyde in the plants. Ethanol kills the roots cells.

Plant growth: Aeration requirements for crops are different. Tomato, potato, sugar beet, pea, wheat, oats and barley require good aeration. Maize, soybean and millets grow well under moderate aeration. Jowar, mustard can tolerate poor aeration up to some extent. However, there is no adverse effect of poor aeration on paddy. Paddy takes downward diffused oxygen through the air spaces present in roots and stem.

Concentration of oxygen in soil air: There is normal growth of plants if soil air contains more than 10% oxygen. But less than 2% oxygen in soil air completely stops the plant growth.

Oxygen Diffusion Rate (ODR): Oxygen diffusion rate is the rate at which oxygen in the soil air is exchanged with the oxygen in the atmospheric air. Aeration requirement of the crops as ODR is about $20\text{-}39 \times 10^{-8}$g/ cm^2/ minute. The rate of growth of roots of different plants stops when ODR is less than 20×10^{-8}g/ cm^2 / minute. Oxygen diffusion requirement of peas is more, wheat and soybean medium and maize and Toria is less.

Renewal of Air in Soil

Entry of air in soil is through the air-filled pores from surface to lower layers. However, many factors govern the entry and movement of air in soil. These factors are,

Water Logging: Water logging and water filled in pores hinder the movement of air in soil and check the renewal of soil air.

Soil Temperature: Soil temperature helps in the renewal of soil air. Temperature difference between soil and air helps in the movement of air. Soil air expands due to increase in soil temperature and become light in weight. This air moves upward. This process helps in the exchange of air present in different horizon with the atmospheric air to renew the soil air.

Atmospheric Pressure: Decrease in atmospheric pressure reduces the volume of air in soil. Reduction in air volume is balanced by the entry of air in the soil.

Wind Velocity- Wind velocity also helps in the renewal of soil air. Air blowing on the soil surface renews soil air up to 1-2 cm of soil depth.

Irrigation and rainfall: Rainfall or irrigation water renew soil air. Dissolved oxygen also enters in soil when rainfall or irrigation water enters in soil. This water removes air from the soil pores. Evaporation, percolation and plant uptake remove water from pores. Therefore, new atmospheric air enters in soil.

Air Exchange Between Soil and Atmosphere

Mass flow and diffusion are two main processes for air exchange or movement between soil and air.

Mass flow: Mass flow of air is due to pressure difference i.e., movement of air from high pressure zone to low pressure zone. Extremely higher wind velocity creates pressure difference between soil and atmosphere at the surface of highly porous soil. Therefore, exchange of air takes place to equalize the pressure difference. Changes caused by presence of water in soil, soil temperature, wind velocity and atmospheric pressure variation influence the mass flow exchange of air in soil. In the total exchange of soil air part of the mass flow is negligible.

Diffusion: Most of the soil air is renewed by gaseous diffusion which is mainly due to the concentration gradient. However, temperature and pressure gradients also influence it. Gaseous diffusion is because of difference in concentrations of gases at two places, length of diffusion path and diffusion coefficient of gases and shape of air-filled pores.

Gaseous diffusion diffuses oxygen to plants roots from the atmosphere and diffuse carbon di oxide from there to atmosphere because concentration of oxygen is more in atmospheric air and carbon di oxide concentration is more in soil air.

Partial pressure of oxygen is more in atmosphere as compared to soil air. Therefore, oxygen diffuses from atmosphere to soil. Diffusion also takes place in the absence of pressure gradient as soil air contains more CO_2 and less O_2 as compared to atmosphere. Thus, carbon di oxide move from atmosphere to soil and oxygen from atmosphere to soil. This helps in the establishment of equilibrium between the gases of soil and atmosphere.

Length of diffusion path also governs diffusion of gases from soil to atmosphere and vice- versa. There will be lesser diffusion with the increase of length of diffusion path.

Diffusion coefficient of gases is another factor to govern movement of gases. Diffusion coefficient of CO_2 and O_2 is different in water and air. Diffusion coefficient of O_2 in water is $2.6x10^{5}$ and in air 2.3x 10^{-1} g/ cm^2/ second, whereas for CO_2 it is 2.0x 10^{-5} and 1.8x 10^{-1}/g/cm^2 second in water and air respectively. Hence if all the conditions are similar gaseous diffusion is 10000 faster in air filled pores as compound to water filled pores. Higher density of gases has lesser diffusion velocity. CO_2 having higher density diffuses slowly as compared to oxygen, nitrogen and water vapours.

Oxygen Diffusion Rate (ODR)

Oxygen is required for the respiration by the plant roots and soil organism. Rate of ODR for maximum growth of roots is about $40x10^{-8}$g/cm^2/minute. However, ODR 20x10-8g/cm^2/minute is essential for root growth. Therefore, oxygen diffusion rate is best parameter to know the aeration status of soil. Oxygen diffusion of soil is measured by platinum microelectrode method and calculated as

$$\text{ODR (ug/ cm}^2\text{/minute)} = \frac{it x 10^{-6} x 60 x 32 x 10^{6}}{\eta x f x A} \text{ where}$$

It = Current in time t

η = Electrons required for the reduction of one molecules of oxygen.

f = Faraday's constant (96500 coulomb)

A= Surface area of platinum electrode

In this technique diffusing oxygen is allowed to reduce at the surface of platinum electrode at a given electric current. The current flows between two electrodes are proportional to the rate of oxygen reduction.

Oxidation- Reduction Potential (Redox potential)

Redox potential is related with aeration condition of soil. Oxygen diffusion rate of submerged soils of paddy, water logged, water saturated soils or poorly drained soils are very less. In such soils, soil aeration is related with the oxidized or reduced forms of elements present in them. These soils have reduced forms of iron (Ferrous Fe^{2+}), Manganese (Manganous Mn^{2+}), Nitrogen (Ammonium NH_4^+), sulphur (Sulphide S^{2-}, H_2S), Oxygen (H_2O) and carbon (CH_4). But well drained and well aerated soil where oxidized forms of these elements are Ferric (Fe^{3+}), Manganic (Mn^{3+}), Nitrate (NO_3^-), Sulphate (SO_4^{2-}), O_2 and CO_2.

Besides these, many other chemical reactions are influenced by soil aeration such as production of methane, hydrogen sulphide, aldehydes, non-conversion of ammonia to nitrate etc. Redox reactions are the chemical reactions in which oxidation of one part took place reduction of another part simultaneously and there is transfer of electrons from one atom to another atom. The part or species which gain the electrons is called as reduced part and which loose the electron is termed as oxidized part of species.

Well aerated soils show redox potential (Eh) in the range of +0.3 to + 0.8 volt. This is decreased from +0.3 to - 0.4 volt in the poorly aerated soils. Redox reactions do not take place at a fixed redox potential for a particular element. But these take place in a range. When the oxidized forms of an element decreases, the redox potential starts to lower down and attain lowest values in the reduced forms.

Redox potential can be measured by Platinum micro electrode method. Do this method platinum electrode along with hydrogen reference electrode are inserted in the wet soil. The electrical potential difference between these two electrodes is measured by a voltmeter. The formula for the calculation of redox potential (Eh) is given here,

$$\text{Eh} = \text{EO} + \frac{RT}{nf} \ln \frac{(ox)}{(red)} \text{ where}$$

Eh = Redox potential (volt)

EO = Potential of reference electrode (volt)

R = Gas constant

T = Absolute temperature

h = Number of electrons transferred from reducer to oxidator.

f = Faraday's constant.

ln = natural log

ox = Concentration of oxidant

red = Concentration of reductant

4. Soil Density

The density is the mass of unit volume of any material. This is also the ratio between the mass of any material and its volume. Relative density is obtained by dividing density of any material to the density of water (1.0 g/cc). Thus, relative density is the ratio between the densities of any material to the density of water. Similarly, the soil density is the mass of unit volume of soil i.e., the ratio between mass of soil and its volume. Soil density is expressed as bulk density and particle density of soil.

Bulk Density: Mass of unit volume of air-dried soil is called as bulk density. This is the ratio between the mass of oven-dried soil and its volume. The bulk density is estimated after oven drying the soil at 105°C till constant weight is obtained. At this temperature the water or moisture content of soil is evaporated and dried soil particles and pores are left. Thus, volume of oven-dried soil has volume of soil particles i.e., solid phase and volume of pore space. This will be dry bulk density of soil, if dry weight of soil is divided by its volume.

Volume of pore space directly influences the mass of oven-dried soil. If volume of pore space is more the mass of soil will be less. Thus, bulk density will also less and this is the reason that loose and friable soils having more pore space show lower bulk density as compared to hard and compact soil. Sometimes weight of field soil sample (wet) is taken and its volume is calculated, the ratio between weight of field soil sample divided by its volume gives wet bulk density.

The bulk density of soils may vary from 1.00 - 1.80 mega gram per cubic meter (Mg m^{-3}). Bulk density of different soils or soil fractions are as: Organic matter = 0.5, organic matter rich soil = 1.0, aggregated clay soil = 1.1, aggregated loam soil and clay = 1.4, sands = 1.6 – 1.7 and highly compacted sub soil = 2.0 Mg m^{-3}. For calculations bulk density of sols is considered as 1.33 Mg m^{-3}. Soils with lower bulk density possess good physical characteristics. Their porosity pore space, pore velocity, oxygen diffusion rate, hydraulic conductivity, infiltration depth and rate, root development and yield will be more.

In general bulk densities of fine textured soils vary from 1.00 – 1.60 Mg m^{-3} and coarse textured soils vary from 1.20 – 1.80 Mg m^{-3}. This is because in fine textured soils, soil particles form very porous granules after mixing them with the organic matter. Therefore, mass per unit volume is less. But in case of coarse textured soils mass per unit volume increases due to less organic matter content and closely packed sand particles.

Generally, the fine textured soils (clays) are called as heavy soils but they are light in weight and coarse textured (sandy) soils are called as light soils but they are heavy in weight. This is mainly because of easiness in performing tillage operations. The fine textured soils often more resistance in tillage operations. Thus, they are heavy soils.

Calculation of Soil Weight

The weight of soil can also be calculated with the help of bulk density of soil. If the bulk density of a soil is 1.34g /cc, then weight of this in one hectare furrow slice will be 2010000 kilogram. This can be calculated as:

Depth of furrow slice = 15 cm.

Area of one hectare = 10000 x 10000 sq.cm.

Volume = 15 x 10000 x 10000 cubic cm.

$$\text{Density} = \frac{\text{Mass}}{\text{Volume}}$$

Mass (weight) = Density x Volume

= 1.34 x 15 x 10000 x 10000 gram.

$$= \frac{1.34 \times 15 \times 10000 \times 10000}{1000} \text{ kilogram}$$

$$= \frac{134 \times 15 \times 10000 \times 10000}{100000} = \text{kilogram}$$

= 134 x 15 x 1000 kilogram

= 20,10,000 Kilogram

Factors Affecting Bulk Density of Soils

Factors affecting bulk density of soils are Soil texture, Soil structure, Organic matter status of soil, Soil compaction, Soil depth and Nature of vegetation.

Soil texture: Fine textured soils (clay, clay loam and silt loam) have low bulk density as compared to coarse textured soils (sandy, sandy loam) which have high bulk density.

Soil structure: Crumb soil structure has low bulk density as compared to platy soil structure. Crumb structure is more porous in which fine particles make a larger granule.

Organic matter status of soil: Organic matter itself has low bulk density. Thus, soils rich in organic matter possess low bulk density. Also, the interaction of organic matter with soil mineral matter enhances the formation of porous soil aggregates, which reduces the weight of soil per unit volume.

Soil compaction: Highly compacted, puddle and water-logged soils show high bulk density. All these reduce the pore space and volume of the soil, which ultimately increase the bulk density.

Soil depth: Bulk density of surface soil is lower than the sub surface soil. This is mainly because of more amount of organic matter present in surface soil as well as tillage operations (ploughing and harrowing etc), done on soil surface.

Nature of vegetation: Soils grown with grasses have lower bulk density because grass roots bind the soil particles to form porous aggregates. Grass roots also provide more organic matter to the surface soil.

Estimation of Bulk Density

The bulk density of soil is estimated with the help of core sampler (metallic cylinder). The core sampler of known diameter and desired height is inserted into the soil with an iron plate and hammer. Then sampler with soil inside is taken out with the help of khurpi without disturbing its natural state. Excess soil from the both the sides and outer surface of sampler is removed with spatula so the volume soil sample should be exactly equal to the volume of sampler. The soil sample is put in a weighed moisture box and kept in an oven at 105°C temperature for constant weight. Finally, the weight of moisture box containing oven dry soil is taken and bulk density is calculated as:

$$\text{Bulk density (gm/cc)} = \frac{\text{Oven dries weight of soil}}{\text{Volume of soil}}$$

Particle Density

Particle density is the mass of unit volume of soil solids or soil particles. Soil mineral matter and soil organic matter are soil solids or soil particles. Thus, particle density is the cumulative density of mineral matter and organic matter of the soil. The particle density can be calculated with the help of relative proportion of mineral matter and organic matter contents of soil and their densities by using following formula:

Dp = (Dm x Cm) + (Do X Co)

Where

Dp = Particle density of soil

Dm = Density of mineral matters

Cm = Content of mineral matter

Do = Density of organic matter and

Co = Content of organic matter

Suppose any soil has 95% mineral matter and 5% organic matter and the density of these are 2.6 and 1.1 Mg m^{-3}, respectively. Then the particle density of this soil will be

$$Dp = (2.6 \text{ x } \frac{95}{100}) + (1.1 \text{ x } \frac{5}{100})$$

$$=2.470 + 0.055$$

$$= 2.525 \text{ Mg m}^{-3}$$

Liquid phase, gaseous phase and pore space of soils are not included in particle density. Therefore, particle density is the ratio between soil particles and their volume. The particle density of soil organic matter may vary from 1.1-1.4 Mg m^{-3}. Density of soil mineral matter ranges from 2.6-2.75 Mg m^{-3}, which is very close to the density of quarts and feldspars minerals present abundantly in sandy soil.

Amount of organic matter and amount of mineral matter are two main factors affecting the particle density of soil. Soils having organic matter more than 15-20% show particle density below 2.4 Mg m^{-3}. Similarly surface soil records less particles density as compared to subsurface soil due to higher organic matter content in surface soil. The particle density of soil may be more than 2.7 Mg m^{-3}, if soil content heavy mineral like magnetite, epidotic, garnet, zircon, tourmaline, limonite, hematite and hornblende. In general, 2.65 Mg m^{-3} particle density of arable soils is used for calculations.

5. Soil Pore Space and Porosity

Pore space is that part of the soil volume, which is not occupied by soil solids or pore space is that part of the soil volume which is occupied by soil water and soil air. Porosity is that percent volume of total soil bulk, which is not in the possession of soil solids. The pore space is equal to porosity when soil volume occupied by pore space is expressed in percentage. Porosity Generally, varies between 30 to 60 percent. The ratio of volume of soil pores to the volume of soil solid is called as void ratio.

The diameter of soil pores may vary from less than 0.002 mm to more than 0.2 mm. However, micro pores or macro pores have diameter less than 0.002 mm or more than 0.2 mm respectively. Water moves through micro pores due to capillarity. So micro pores are also called as water pores or capillary pores. Water moves through macro pores due to gravity and filled with air in the absence of water. Thus, macro pores are also known as aeration pores and non-capillary pores. Generally, macro pores occur between the peds and micro pores occur within the peds.

The sandy soils having more macro pores possess good aeration but poor water holding capacity. They require frequent and more irrigations for good crop growth and yield. The clayey soils having more micro pores of 0.1 mm diameter retain water for longer periods due to high water holding capacity. The aeration of these soils is poor. The excess of water and lack of air are not good for most of the plants. The medium textured loam soils are good for plants as their macro pores are filled with air and micro pores contain water.

Calculation of Percent Pore Space

Bulk density of soil has volume of soil pores plus volume of soil solids (soil particles) whereas particle density of soil has volume of soil solids only. Thus, bulk and particle densities of soil can be used for calculating percent pore space of soil by following formulae:

$$\text{Percent pore space} = 100 - \left\{\frac{\text{Bulk density}}{\text{Particle density}}\right\} \times 100$$

$$\text{Percent pore space} = \left\{1 = \frac{\text{Bulk density}}{\text{Particle density}}\right\} \times 100$$

Q.1. Suppose bulk density of a soil is 1.30 g/cc and particle density is 2.60 g/cc, then the percent pore space will be 50. This can be calculated as:

$$\text{Percent pore space} = 100 - \left\{\frac{1.30}{2.60}\right\} \times 100$$

$$= 100 - (0.5) \times 100$$

$$= 100 - 50$$

$$= 50$$

$$\text{2. Percent pore space} = \left\{1 - \frac{1.30}{2.60}\right\} \times 100$$

$$= \frac{2.60 - 1.30}{2.60} \times 100$$

$$= 0.5 \times 100$$

$$= 50$$

Q.2. The porosity of soil sample is 50% and its bulk density is 1.5 gm/cc. Calculate particle density of this soil.

$$\text{Percent pore space} = \left\{1 - \frac{\text{Bulk density}}{\text{Particle density}}\right\} \times 100$$

Or Porosity

$$50 = \{1 - \frac{1.5}{\text{PD}}\} \times 100$$

$$\frac{50}{100} = \{1- \frac{1.5}{PD})$$

$$0.5 = \{1- \frac{1.5}{PD}\}$$

$$\frac{1.5}{PD} = 1 - 0.5$$

$$\frac{1.5}{PD} = 0.5$$

$$PD = \frac{1.5}{0.5}$$

$$PD = 3.0 \text{ gm/cc}$$

Or

$$\text{Percent pore space} = 100 - \{\frac{BD}{PD}\} \text{x } 100$$

$$50 = 100 - \{\frac{1.50}{PD}\} \text{x } 100$$

$$50 = 100 - \frac{150}{PD}$$

$$-50 = -100 + \frac{150}{PD}$$

$$PD = \frac{150}{50} = 3.0 \text{ gm/cc}$$

Q.3. Find out the BD of the soil if porosity is 50% and particle density is 3.0.

$$50 = \{1- \frac{BD}{PD}\} \text{ x } 100$$

$$50 = \{1- \frac{BD}{3.0}\} \text{ x } 100$$

$$\frac{50}{100} = \left\{1 - \frac{BD}{3.0}\right\}$$

$$0.5 = \{1- \frac{BD}{3.0}\}$$

$$\frac{BD}{3.0} = 1 - 0.5$$

$$\frac{BD}{3.0} = 0.5$$

$$BD = 3.0 \text{ x } 0.5$$

$$= 1.5$$

5. Soil Temperature

Soil temperature plays an important role in many processes taking place in soil. Aeration, moisture, structure, nutrient availability, weathering and micro-organisms of soil directly or in directly depend upon the soil temperature. Various species of plants have different thermal requirements. There is no germination of seeds and growth of roots at extremely low soil temperature. In general soil temperature controls physical, chemical and biological processes of soil. Thermal properties of soil are related with pressure and transfer of heat. Some terms related with heat and its transfer is described here:

Thermal Capacity

Thermal capacity or heat capacity of any material is equal to its specific heat of its components multiplied by their masses. Heating of any material increases its temperature. Quantity of heat is given (θ) to the material is proportional to the mass (m) of the material and increase in temperature (ΔT). This can be written as:

$\theta = m \times \Delta T$ or

$\theta = c \times m \times \Delta T$

Here c is a constant which depends upon the mass and properties of the material.

Heat capacity is the ratio between quantity of heat given to a substance and increase in its temperature. This can be written as:

$$c = \frac{\theta}{mx\Delta T}$$

If m=1 and T=1 than c= θ. Therefore, specific heat is the quantity of heat required to increase one-degree centigrade temperature of one gram mass of any substance. The unit of specific heat is cal/g/c.

Volumetric Heat Capacity of Soil

The unit of specified heat for soil is cal/g/ soil. This is volumetric unit. Therefore, the definition of specified heat of soil is the quantity of heat required to increase one degree centigrade temperature of one cubic centimetre soil. The specific heat of water is 1.0 but normally specific heat of other substances and soil is less than one.

The volumetric heat capacity of soils can be calculated from volumetric heat capacities and respective volumes of the soil components as:

$Cs(\nu) = (Cm \times \nu m) + (Co + \nu o) + (Cw + \nu w) + (Ca + \nu a)$, where

$Cs(\nu)$ = volumetric heat capacity of soil

Cm, Co, Cw, Ca = volumetric heat capacities of soil mineral matter (0.46), organic matter (0.60), water (1.00) and air (0.0029), respectively, and ϖ m, ϖ 0, ϖ w and ϖ a = volumes of soil mineral matter, organic matter, water and air respectively.

Normally soil contains 45% mineral matter, 5% organic matter, 25% water and 25% air. The volumetric heat capacity of this soil: - $(0.45\times0.46) + (0.05\times0.60) + (0.25\times1.0) + (0.25\times0.0029) = 0.487725$ cal/g/cc/

Gravimetric Heat Capacity of Soil: The gravimetric heat capacity of soil and soil minerals are 0.16 and 0.23 cal/g/, and on an average they are considered as 0.10 and 0.20 cal/g/, respectively. Gravimetric heat capacity of soil can be calculated from gravimetric heat capacity and respective volumes of the soil components as:

Cs = (Cm × Mm) + (Co × Mo) + (Cw + Mw) + (Ca + Ma), where

Cs = Gravimetric heat capacity of soil

Cm, Co, Cw and Ca are specific heats of soil mineral matter, organic matter, water and air respectively and Mm, Mo, Mw and Ma are masses of soil mineral matter, organic matter, water and air respectively.

In the weight of soil, masses of mineral matter and water are more responsible as compared to masses of organic matter and air. Therefore, the gravimetric heat capacity of soil may be written as:

$$Cs = (Cm \times Mm) + (Cw \times Mw)$$

Heat capacity of soil increases with the increase in soil moisture content and show linear relationship. Therefore, highly moist soil can absorb or release more amount of heat per unit change in temperature. Such soils slowly warm up in summer and slowly become cold in winter.

Thermal Transmission in Soil

Transfer of heat from one substance to other substance or movement of heat from one place to other place is called thermal transmission. Conduction, convection and radiation are three processes of heat transmission. Conduction is main process of heat transmission in soils and convection and radiation has no influence. Therefore, heat conductivity, thermal conductivity is described here:

Thermal conductivity: Thermal conductivity is the quantity of heat that perpendicularly flows through long and uniform cross section area, thermal gradient and time as:

$$\theta \propto -A\frac{\Delta Q}{\Delta X}t \ or \ \theta = -KA\frac{\Delta Q}{\Delta X}t,$$ Where

θ = Quantity of heat

A= Area

$\frac{\Delta Q}{\Delta X}$ = Thermal gradient

t = time and

K= Constant which depends upon the material of section. Negative sign is the indication of flow of heat from higher temperature to lower temperature.

If the area of each end of this section is one square meter, distance between both end is one meter, thermal gradient is 1 and time is one second, then quantity of heat (θ) will be equal to thermal conductivity coefficient.

Thermal conductivity coefficient is defined as the quantity of heat under steady state perpendicularly flows through a body of unit length, unit square meter area in unit time under a unit degree centigrade temperature gradient. Its unit is kilo calorie/meter/second/°C or cal/cm/second/ °C,

Factors Affecting Thermal Conductivity of Soil

In general, the thermal conductivity of soil is in the order of sandy > loam > clay > peat. However, many factors affect thermal conductivity of soil.

Volume of solid, liquid and gas in soil: The thermal conductivity depends upon the proportion of solid, liquid and gaseous phases of the soil. Conduction of heat is more in solid medium as compared to liquid and gases. If solids are more in soil, thermal conductivity will also more

Soil temperature: Coarse textured soil is good conductor of heat as compared fine textured soils. In the summer, sand becomes hot very quickly. Reduction in size of soil particles also reduces the thermal conductivity. This is due to reduction in the surface contact among heat conducting particles.

Bulk density and porosity of soil: Porosity decreases with the increase in bulk density of soil. This creates good thermal contact among soil particles. Therefore, thermal conductivity increases.

Water content in soil: Thermal conductivity increases with the increase in moisture content in soil. When air present in pore space is replaced by water, then water transmits heat more rapidly as compared to air. This is because thermal conductivity of water (1.42 cal/cm/) is about 21 times higher than thermal conductivity of air (0.062 cal/cm/). Heat transfer is more in compact and wet soil as compared to loose and dry soils. But wet soils take more time to warm up because of higher specific heat of water.

Organic matter status of soil: Thermal conductivity is organic matter is remarkably less than many soil minerals. Increase in organic matter status reduces the thermal conductivity of soil. However, organic matter rich soils, due to their dark colour and low reflection, become warmer during day time.

Thermal Diffusivity

The rate of warming of soil against the thermal gradient is called thermal diffusivity. It is the change in temperature in one second when the thermal gradient changes by 1/cc. This is proportional to the thermal conductivity and inversely proportional to volumetric heat capacity.

$$\text{Thermal diffusivity} = \frac{\textit{Thermal conductivity}}{\textit{volumetric heat capacity}}$$

Thermal diffusivity increases up to certain maximum with the increase in soil moisture content and then decrease.

Factors Affecting Temperature of Soil

The variation in soil temperature is much more pronounced due to varying compositions and properties of soil. Numerous factors directly or indirectly affect the soil temperature are:

Amount of solar radiation: Quantity of solar radiation at any place depends upon the climate of that place. Soil receives the heat from sun through radiation. However, amount of solar energy received at soil surface is much less due to absorption by atmosphere and other atmospheric physical and chemical processes. The amount of solar radiation at polar region is much less than equatorial region. Therefore, soil temperature is more at equator than poles.

Exposure: The diurnal soil temperature changes with the position of the sun. More heat reaches on soil surface during afternoon when sun is at lower position. In the Himalayan region northern aspect is covered with snow for longer periods as compared to southern aspects.

Slope of the lands and angle of incidence: Land surface at the right angle of the sun's rays receives more heat because sun rays' approach to perpendicular. There will be more absorption of heat. However, if some rays make wider angle with the land surface, the heat is spread over larger area to cause low soil temperature. Southern slopes are warmest as compared to northern slopes. Steep slopes of Himalayas are cooler than mild slopes of south India. Angle of incidence depends upon latitude and slope of soil surface.

Latitude: In coming short-wave radiation and albedo are higher in lower latitudes. Albedo is lowest in tropical region. Therefore, thermal properties of soils of these areas are quite variable.

Shades: Bare soils become hot quickly after absorbing heat and cool down quickly after releasing it. But soils covered with good vegetation or shade trees do not have such rapid changes in temperature. In the summer vegetation cover intercepts solar radiation to prevent warming of the soil. However, in winter, vegetation acts as an insulator to reduce heat loss from the soil. Similarly vegetative cover reduces the diurnal variations of the soil temperature as compared to bare soil.

Vegetative cover: Vegetative cover checks the loss of heat from soil to atmosphere. Therefore, vegetation acts as an insulator against temperature change. The capacity of insulation depends type of vegetable cover. If soil surface has the dense canopy of grasses, then heat loss from soil surface through convection will be less. This is due to that dense canopy of grasses has very less air. But in the forests of larger trees, much air is present among the trees. After cooled down this air transmits heat from soil surface to atmosphere through convection.

Soil colour: Dark coloured soils absorb more solar heat than light coloured soils. Thus, maintain higher temperature during day time but good heat absorbers are also good radiators. Hence dark coloured soils emit more heat during night. This causes greater daily variation in soil temperature. Humus rich black coloured soils do not absorb much heat due to the less heat capacity of humus but absorb much amount of water. Energy is required for warming and evaporation of water. Hence these soils are cooler.

Heat of vaporization: Vaporization of soil water requires expenditure of energy (540 kilo calories/Kg soil). This causes cooling surface of the wet soil where evaporation of water takes place. Also, high specific heat of water presents to increase soil temperature. Thus, wet soil maintains lower temperature.

Heat Balance on Earth and Soil

Heat balance in soil depends upon incidence solar radiation, the heat exchange between air and atmosphere, soil surface and thermal properties of soil profile. Sun's energy comes at earth's atmosphere at the rate of 1.94 cal//minute or 1370 wm^2. Only 20% of this comes on soil surface. If solar energy is coming at the rate of 0.4 cal/sq.cm/minute on soil surface and considering heat capacity of soil (0.2 cal/gm/°C) the soil should be heated at the rate of 2 °C /minute. But it is not true because the energy is lost due to the radiation of energy.

An assumption has calculated that sum of all incoming and outgoing radiation on the surface is zero. About 31% incoming radiation is immediately reflected back to atmosphere as 21% by clouds, 6% by air, dust and water vapours and 4% by earth's surface. The remaining 69% heat up the earth and its atmosphere as about 18% is directly absorbed by ozone and water vapor, 3% by clouds and 48% is absorbed by earth's surface. Heat balance of a soil in one bright day of the March was as:

Table 3.5

Heat Balance of Soil (cal/gm/sq.cm)			
Total Gain	1616.00	Total Loss	1610.00
Solar radiation absorb by soil	740.13	Radiation to atmosphere during day time	680.00
Radiation from atmosphere	654.00	Radiation to atmosphere in during night time	480.00
Heat coming from lower layers of soil	221.40	Heat convection to atmosphere Heat conduction to lower horizon of soil	210.00 240.00

Soil Heat Flux

Soil heat flux represents the amount of radiant energy absorbed or released at the soil surface during a given time period. Many field studies in topics such as microclimatology and hydrology require reliable measurements of the soil surface flux

The soil heat flux is the energy receives by the soil to heat it per unit of surface and time. The Soil heat flux is positive when the soil receives energy (warms) and negative when the soil loses energy (cools).

Soil heat flux is important in micrometeorology because it effectively couples energy transfer processes at the surface (surface energy balance) with energy transfer processes in the soil (soil thermal regime).

Soil heat flux density: Soil heat flux density is the movement of thermal energy per unit area of soil in a unit time in response to a temperature gradient. Climatic conditions on the earth's surface are in part a function of varying physical position (elevation, latitude, and aspect) and the influence of large-scale meteorological forces such as air and ocean currents. The density and architecture of plant canopies in natural systems are directly influenced by these climatic factors. By contrast, for agricultural systems, it is the crop canopies that often influence local microclimate. In both instances, the soil plays an important role in affecting climate near the surface. Properties of the surface soil layer including colour, water content, texture, and density affect the partitioning of incident radiation and how much energy is used to evaporate water, warm the air above the ground, or warm the soil.

The ability of a soil to conduct heat determines how fast its temperature changes during a day or between seasons. Soil temperature is a key factor affecting the rate of chemical and biological processes in the soil essential to plant growth. Soil heat flux is important in micrometeorology because it effectively couples energy transfer processes at the surface (surface energy balance) with energy transfer processes in the soil (soil thermal regime). This interaction between surface and subsurface energy transfer processes has led to detailed investigations of soil heat flux for a wide variety of agricultural systems.

Soil Temperature Fluctuations

Soil temperature fluctuations indicate the change or increase or decrease in soil temperature. This fluctuation is distinct or clearer at soil surface and clarity decreases with depth of soil. At a specific soil depth there is no much change in soil temperature and daily, monthly, seasonal or annual temperature remain almost same.

Diurnal change: The diurnal or daily change in temperature depends upon nature of soil, soil surface cover, incidence of solar radiation and loss of heat. Temperature changes are clearer and deeply effective during bright sunny day as compared to cloudy day. Tripathi and Ghildyal (1987) recorded the temperature change from surface to 50 cm depth of a silty clay loam soil at Pantnagar during 23rd and 24th October. They indicated that soil surface started warming up at 6.00 AM and got highest temperature at about 2.00 PM, after it started to decrease.

There were 16.5 °C differences between maximum and minimum temperature of soil surface. Two centimetres below the soil surface, temperature change was almost similar to surface temperature change. But difference between maximum and minimum temperature was 10.5. At 5 cm depth of soil maximum temperature reached at 4.00 PM i.e., 2 hours after surface and difference between maximum and minimum temperature was 4.9. Highest temperature was recorded at 6.00 PM at 10 cm depth of soil and difference between maximum and minimum temperature was 3.80.

Temperature fluctuations, at 2.5 cm depth, the difference between the maximum and minimum temperature was only 0.6, at 5.0 cm depth the difference was only 0.1 and below 10 cm soil depth difference was very less and unremarkable. These temperature fluctuations indicated that at soil surface the solar heat is effective with the sunrise. Hence soil got maximum temperature in the midday but at the lower depth the solar heat is less effective, less intense and penetrates slowly. Therefore, lower depth gets maximum temperature only after mid day i.e., after some period in the night when surface cooled after release the heat then soil below 10cm depth show higher temperature. This is because slow entry of heat waves in the lower depths of soil. In night when surface is cool and lower layers are hot there is upward movement of heat. Therefore, there is alternate cooling and heating or rise and fall in soil temperature during sunrise and sunset (dusk).

Monthly or seasonal change: Temperature variations in Pant Nagar soil (Tripathi and Ghildiyal 1987) were 310- 27°C from third week of June to end of October and 15°-24°C from November and onward. In the month of January, upper layers of soil record lowest temperature but lower layers remain relatively warm. In February month upper layer warm up slowly and slowly due to higher air temperature and gain of heat from lower layers and lower layers become some cool. March and April months are warm up the upper layers and cease the cooling of lower layers. In the month of May whole profile gets highest temperature. In the end of June and on setting of Monsoon soil temperature starts decreasing.

During rainfall and excess of water in soil no appreciable changes are recorded in soil temperature up to September end. However, heat conductance remains downward. Commencement of winter from mid-October the cooling starts from soil surface and upper layers and lower layer of profile slow almost same temperature. In the end of October to the January profile starts cooling and record lowest temperature.

In general, the soil profile remains hot in summer, middle in raining and cold in winter season. Similarly, differences in temperature fluctuations are greatest in hot season and least during rainfall and monsoon.

Annual change: Annual temperature changes extend much deeper in soil or land as compared to diurnal changes. The diurnal waves of heat are effective only up to 10 cm depth in soil.

Control and Management of Soil Temperature

Soil temperature can be controlled up to some extent. However, the climate conditions can destroy the efforts of temperature control. Use of mulches, inter culture operations, vegetations and irrigation and drainage can control and manage the soil temperature.

1. Mulches: Mulches are the materials used to protect soil and plants from rain drops, prevent crust formation, check evaporation and freezing and increasing infiltration. Straw, saw dust, leaves, plastic sheets and loose soils etc. are some examples of mulches.

Mulches of plants residues like straw etc. decrease the maximum temperature of soil. During day prevent the soil from overheating by absorbing solar radiation and in night provide heat to soil to prevent cooling.

Plastic sheets may be black, transparent or non transparent. Plastic mulch increases the temperature of soil and decrease the evaporation. Black plastic sheets maintain a constant temperature of soil. They absorb more solar radiation. But due to their shading effect and air between plastic sheets and soil, there is less conduction of heat towards soil. Therefore, soil maintains a constant temperature.

Transparent plastic sheets are permeable for both incident solar radiations as well as outgoing radiation. Even though, they increase the soil temperature. During day time they permit the entry of incoming solar radiation. This increases the soil temperature and water evaporation. In night lower atmospheric temperature condenses water vapours inside the plastic sheets. These water droplets prevent the exit of heat from the soil. Therefore, plastic sheets mulch always maintains higher temperature of soil in all the depths as compared to bare soil.

Inter culture operations: Compaction of soil increases heat conductivity of soil and tillage acts as mulch. Tillage increases the porosity of surface soil which checks the movement of heat. Therefore, upper tilled soil has more temperature as compared to lower untilled soil.

Bunding also increases the soil temperature. Flat and leveled fields have lower temperature as bunded fields. Bunds decrease the albedo of soil. Albedo is the fraction of net radiation reflected by soil or any other material. Higher albedo of soils indicates lesser changes in temperature. In summer covering of soils with chalk powder decrease the temperature and similarly covering of soil with black powder in winter will increase the soil temperature. The albedo of light-coloured soil is more than black coloured soils. Albedo also increases when moist soils dry up.

Vegetation: Vegetation acts as organic mulches helps in soil temperature control. Solar radiations directly come on bare soils. Hence bare soils warm quickly and in cold or night rapidly release the heat to atmosphere. A good vegetative cover absorbs a significant amount of solar heat and prevents the warming of soil in summer. In winter it acts as insulator and prevents the heat loss from soil. The major effect of vegetation is due to their shade. Shade is proportional to temperature control i.e., more shade, more temperature control. Due to greater temperature fluctuations at coco and coffee growing areas, the roots of these crops cannot absorb nutrients. In these areas coffee and coco plants are grown with shade providing plants.

Irrigation and drainage: Irrigation increases the heat capacity and heat conductivity of soil, moisture in the air above soil and decreases the temperature of air above soil. All these decrease the temperature of soil. In the summer season irrigation decreases the maximum temperature and, in the winter, it increases the minimum temperature of soil because of high heat capacity of soil. Drainage of wet soil reduces the heat capacity of soil to increase temperature.

6. Soil Water

Water is natural resource. It is present in the nature in the forms of solid (ice), liquid (water) or gas (water vapour). Water is essential for all living beings including plants, where it is required to perform various processes including photosynthesis, conversion of starch to sugar nutrient, movement and turgidity. Plant protoplasm contains 85 – 95% water. Water is also an important component of soil. Soil stores water for plant use. But plants blow up soil water in to atmosphere through evapotranspiration. This creates a water deficit in the soil, which can be fulfilled by irrigation or rain full otherwise plants will wilt or die. Plants stomata also can absorb some amount of water but roots absorb major part of water required by plants.

Therefore, plants depend upon water present nearby their roots. Thus, storage of plant available water by the soil is important. Water is a good solvent. It dissolves added fertilizers and salts present in soil to help their movement from soil to roots. Therefore, sufficient irrigation facilities are essential to keep the nutrients in moist condition near active roots for good crop growth. Soil development, soil swelling and shrinkage, temperature and aeration and other chemical and biological properties of soils are governed by water present in the soil

Water is a simple compound of one oxygen and two hydrogen atoms. Small hydrogen atoms are covalently attached with large oxygen atom at an angle of 104.5° to form a water molecule. But individual water molecule does not occur as positive valency of hydrogen ions does not fully satisfy by negative valency of oxygen ion and vice-versa. This creates charge difference between two sides of water molecule i.e., weak negative charge at oxygen side and weak positive charge at hydrogen side. Thus, water is dipolar, where negative pole of one water molecule is attached with positive pole of other water molecule and so on by hydrogen bonding to form hexagonal lattice structure. In the same the water may be attached to the oxygen of soil mineral by hydrogen bonding to hold the water molecule tightly with the soil.

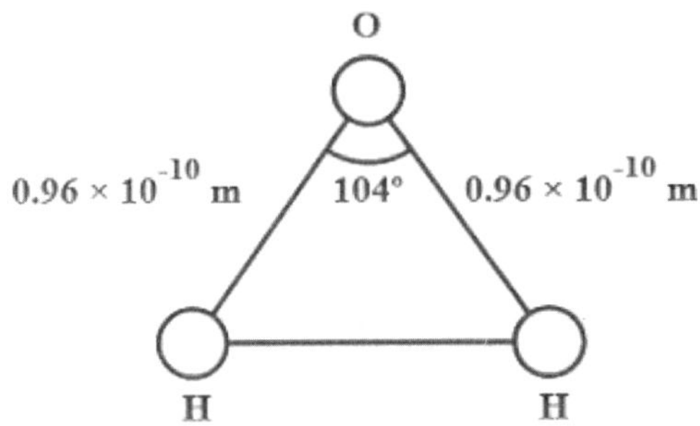

Fig. 3.8: Water molecule

Energy Concept of Soil Water

Energy concept of soil water is an important concept. The movement and retention of water in soil, evaporation of water from soil, uptake of water by plants and translocation of water with in plants, are related with energy state of water. The energy of water is kinetic, potential or electrical type. The kinetic energy of water is very much less or negligible as water moves very slowly in soils. Therefore, electrical and potential energy govern the condition and movement of water in soil. The water molecules have a symmetrical distribution of charge. Hence, they attract each other due to cohesion and attached to charged soil particles due to adhesion.

Thus, a film of water is formed around the soil particle and as much as thin is the film, water molecules will be held more tightly to that extent by soil particles. More adhesion of water takes place when surface area of soil particles is increased. All these reduce the movement and energy of water. Water remains in the micro pores of soil due to cohesion and adhesion forces which are much more than the gravitational force. Therefore, plants have to work, exert suction or expend energy to take water present in micro pores. Water present in macro pores move toward lower horizons due to gravitational force.

Here potential energy of water (½ m g h, m = mass of water, g = acceleration due to gravity and h = height or distance to be moved by water) come in picture. Potential energy of water increases when any work is done on it. But when soil water does any work its potential energy decreases. Lift of soil water upon some height will increase its potential energy and absorption of soil water by soil minerals will decrease its potential energy. Potential energy of water is Generally, referred as potential.

Term 'Free energy' is also used to denote energy state of soil water. It is sum total of all the energies working on soil water. Water present in a pond, water logged soil, submerged soil, wet paddy soil just after rainfall or just after irrigation has highest free energy as compared to dry or moist soil. Soil water moves from higher free energy to lower free energy. Water will move in soil from the place where it is in more amounts to the place where it is in less amounts i.e., in soil. Water will move from wetter places to drier places. Some other generalizations of free energy are; in a multiphase system the free energy of water in all the phases will be equal, if all the phases are in equilibrium. At a constant pressure free energy decreases with the increasing temperature. Increase in osmotic pressure decreases the free energy and increase in hydrostatic pressure increases the free energy.

Soil Water Potential

Many forces like cohesion, adhesion, gravitational etc. work on soil water and change its free energy. Pressure, tension, suction and potential terms are Generally, used to denote these forces. But now a day potential is mostly used. Soil water potential is the difference between energy state or potential of free water to that of energy state or potential of soil water at reference point in soil. The energy state or potential of free water is considered zero and energy state or potential of soil water, which is under various forces, is less than zero i.e., negative.

Total Soil Water Potential

Soil water potential (i.e., work to be done) is also called as total soil water potential which is the total of all the potentials generated by the action of each force acting on the soil water. Following equation can indicate this:

$\Psi T = \psi M + \psi g + \psi P + \psi X$

Where

ψT = Total soil water potential

ΨM = Matric potential

Ψg = Gravitational potential

ΨP = Pressure potential

ΨX = Osmotic potential

Matric Potential: Metric potential is due to adsorption and capillary forces working on soil water. Soil particles and exchangeable ions attract water molecules and adsorb them on surface of particles as a water film. Similarly micro pores hold water as capillary water. Both adsorption and capillarity reduce the free energy or potential of soil water to that of bulk water. Therefore, metric potential is less than zero or negative. Metric potential affects the movement of water from soil to seeds, plants' roots and microbes.

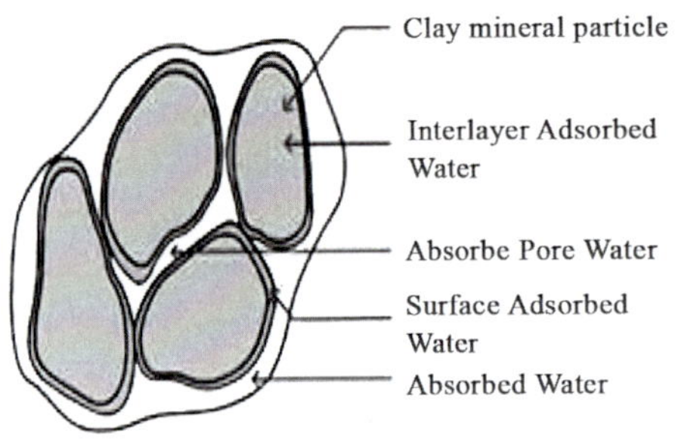

Fig. 3.9: Matric potential

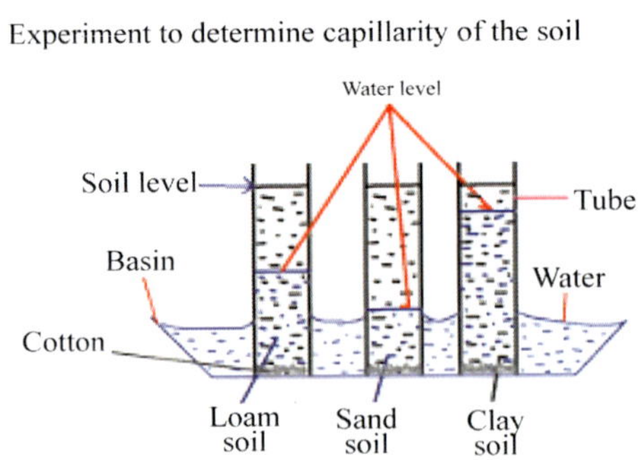

Fig. 3.10: Capillarity of the soil

Gravitational Potential: Gravitational potential is due to the force offered by earth's gravity which gravitate (moves) water vertically downward to the centre of earth. Gravitational potential is more important for submerged or water saturated soils where it drains excess water from soil by moving water from higher elevation to lower elevation i.e., from upper horizon to lower horizon.

Gravitational potential is indicated as $\psi g = \psi h$ where g is acceleration due to gravity and h is arbitrary reference point which may be at soil surface or at any other position in the soil profile. The gravitational potential is positive above the reference point and negative below it. Soil properties and soil water do not affect the gravitational potential. Gravitational and metric potential in combination is called as hydraulic potential.

Pressure Potential: Pressure potential is originated when more pressure than the atmospheric pressure is exerted on the soil water.

Osmotic Potential: Osmotic potential is due to soil solutes. These soil solutes (ions and organic molecules) reduce the free energy of soil water. Therefore, osmotic potential is negative. Osmotic potential is the hydrostatic pressure required to check inflow of pure water separated by semi permeable membrane from the liquid. Numerically osmotic potential is equal to osmotic pressure (Fig. 3.11). Importance of osmotic potential is negligible in normal soils. But in saline soils osmotic potential has great importance where it governs the movement of water from soil to roots and microorganisms up to some extent due to their higher salt contents.

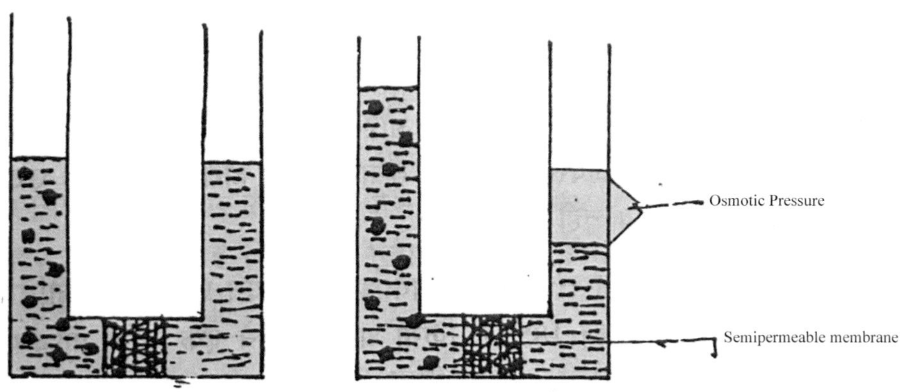

Fig. 3.11: Osmotic Potential

Movement or Flow of Water in Soil

When there is difference in potentials of water present at two points in any soil, water starts moving from one point to other to equalize the potential difference. This potential difference is Generally, produced more after rainfall or irrigation.

When much more amount of water is present on soil surface, the free energy of water is also greater. But at lower horizons of soil, due to lesser amount of water in them, negative potentials of water develop. Under such conditions due to greater potential differences of water at soil surface and lower horizons, water quickly enters in to soil from soil surface to establish the equilibrium. The rate of entry of water reduces gradually as the equilibrium is attaining.

Water entering in to the soil filled up into the soil pores after replacing their air and saturation condition comes when all the pores, macro and micro, are completely filled with water. The water excess or more than the saturation moves towards lower soil layers or horizons due to gravitational force. When irrigation or rainfall stops, the supply of water to soil also stops or reduces. But evaporation of water from soil surface and absorption of water by plant roots cause rapid loss of water from soil. Thus, pore space becomes empty again and partially filled pores create an unsaturated condition of water. Potential of water in soil tended to reduce. Therefore, in soil profile water moves from the zone of higher water content to zone of low water content. It indicates that water moves through soil under saturated and unsaturated conditions due to pressure of water potential differences, gravity and combination of all these.

Water Movement Under Saturated Condition

All the pores, macro as well as micro, are completely filled up with water during water saturation condition of soil. This condition is commonly occurring just after rainfall or irrigation, at lower horizons of poorly drained soils and at lower zone of ground water table. At this condition water is tension free, without any negative potential and possesses highest free energy.

In the year 1856 a French engineer, Darcy studied saturation flow of water through sand filters and proposed the equation for flow of water or any liquid through a porous medium. The soil is also porous. So, this equation can be used for soil also.

$$V = \frac{Q}{ta} = k\,\frac{p1-p2}{l}$$

where,

V = average velocity of flow of water

Q = quantity of flow obtained in time t

a = cross section area of porous body

k = permeability or hydraulic conductivity

p1-p2 = pressures at water inlet and outlet points of the porous body.

l = length of porous body

This rule says that flow of water through a sand filter in unit time depends upon the cross-section area of filter, length of filter, height of water on filter and proportionality constant 'K'. Proportionality constant 'K' depends upon nature of sand.

Darcy's law requires some conditions like no change in pore space geometry in compressible fluid, saturation of medium, steady state and laminar flow and constant temperature. However, Dracy's law can be applied for all soils except very sandy soils.

The simplified form of Darcy's law may be as V = Kf where V = movement of water in unit time, K = hydraulic conductivity of soil and f = potential difference or hydraulic head gradient or driving force of water. The difference in the height of water at top and bottom i.e., inflow and outflow points of soil column is the driving force of water which produces the gradient. Water movement from inflow point to out flow point in the soil column depends upon the gradient and hydraulic conductivity.

Hydraulic conductivity means the quickness by which water flows in soil in the response of potential gradient. Hydraulic conductivity of water saturated soils is less in soils having fine texture, unstable structure, low porosity and low water content as compared to soils with coarse texture, granular structure, more porous and more water content.

Vertical and horizontal saturated movement or flow of water

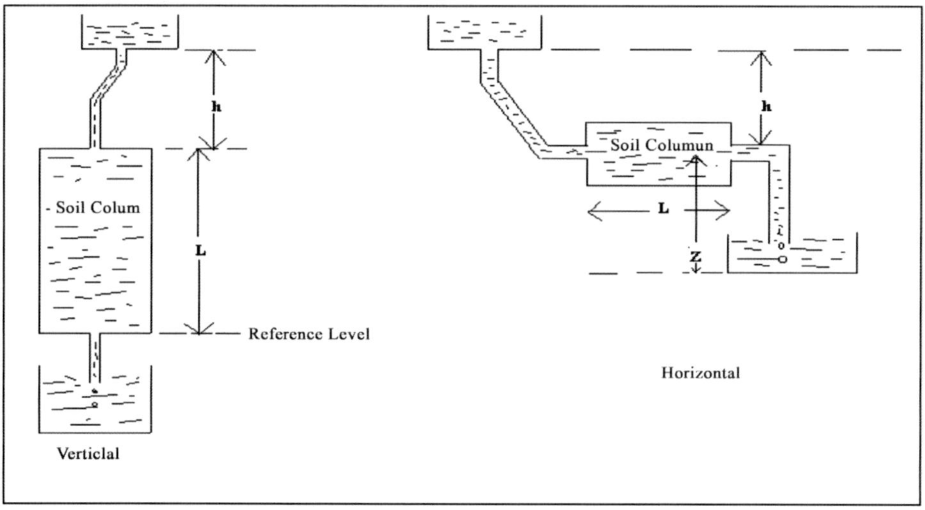

Fig. 3.12: Vertical and horizontal movement or flow of water

Vertical saturated flow of water is somewhat like to flow of water from overhead tank of a building to the tap at bottom through pipe (Figure 3.12). If the diameter of pipe is more, more water will come at tap. Similarly, diameter of soil pores is more; more water will flow through them and due to this reason, saturated flow of water is more in coarse textured soils as compared to fine textured soils.

The total head at inflow point will be equal to the water pressure head plus gravity head whereas total head at outflow point will be zero as water pressure and gravity heads are zero and water is freely flowing in the atmosphere. If the water reservoir is lowered down up to the inflow point, then water pressure head will be zero and total head at the inflow point will be equal to length of soil column. In this situation the water will flow only due to the gravity head. Therefore, vertical saturated flow of water in soil is greatly influenced by gravity.

Horizontal Saturated Movement or Flow of Water

In the horizontal saturated flow of water, the situation is quite different. In this case the total head at inflow point will be equal to pressure head plus height (h) of outflow point from the reference level. The total head at out flow point will be equal to pressure head (which is zero) plus height of outflow point from the reference level. The pressure head difference between in flow point and outflow point will be due to pressure head only. If the water reservoir is lowered down up to the inflow point, there will be no flow of water because pressure head is zero. This also indicates no influence of gravity head in the horizontal saturation flow of water. The water flows under the hydraulic gradient, h / L.

Water Movement or Flow under Unsaturated Condition

All soils have about 25 percent air and 25 percent water i.e., soils are not always saturated with water. Some soils pores are filled with air and some with water. This is unsaturated condition of water in soil where all the pores are not completely filled with water and they do not take part in water flow. Under highly unsaturated condition air also comes in between the water. This forms an air water interface, which breaks the continuity of water in pores.

Discontinuity of water increases the tension of water in soil. Increase in the tension reduces the flow of water, which may be stopped finally. Also saturated hydraulic conductivity (K) of water under unsaturated condition reduces due to development of resistance in flow caused by either decrease in water content, water pressure and or continuity of water film around the soil particles. This ultimately stops the flow of water in soil. Matric potential gradient or hydraulic gradient is driving force of water under unsaturated condition.

Unsaturated hydraulic conductivity is reduced very rapidly along with the reduction in matric potential gradient. Under the moist condition when matric potential is quite low, the unsaturated hydraulic conductivity of sandy soils is quite higher than the clayey soils. But in dry conditions unsaturated hydraulic conductivity of clayey soils is greater than sandy soils. It is because of quantity of water is most important for unsaturated flow of water in soil.

Therefore, Darcy's law may be modified after including water quantity in the simplified form as,

V=K (Q) f where V is unsaturated flow of water in unit time, K is hydraulic conductivity, Q is water quantity and f is hydraulic gradient. It indicates that under unsaturated condition hydraulic conductivity depends upon quantity of water present in soil.

Water Movement or Flow under Stratified Soils

Sometimes the soil layers are not homogenous. They differ very widely in texture, structure and other properties forming different strata. Therefore, the flow of water in stratified soils is quite different than the homogenous soils.

Numerous types of strata are found in soils. These may be impedance and impervious layers, hard pans, clay pans etc. Sometimes plant residues are also buried in soil by ploughing and holes and burrows are made by burrowing animals and insects. All these changes normal flow of water in soil. The down ward flow or movement of water in soil is restricted. When a coarse textured soil material like gravel or sand and plant residues with larger pores and minimum hydraulic conductivity come across the moving water front, the horizontal flow of water increases till all the pores are filled up with water and after their saturation down ward flow of water again starts.

Fine pores of hard pans and clay pans also remarkably restrict water flow. Initially due to high water absorbing capacity of finer particles of these pans, water saturation of pores is attained rapidly. But when the distance for the movement of water in fine pores is increased the rate of down ward flow of water is decreased. This may be due to capillarity, which causes upward flow of water. Therefore, downward flow of water in these pans may be so slow or minimum causing water stagnation upon them.

Movement or Flow of Water Vapour in Soil

The amount of water in soil as water vapours is quite small as compared to liquid water. Therefore, gaseous (vapour) flow of water in soil is also too much low. But instead of low vapour flow, energy required for the vaporization or condensation of water (about 585 cal/g at 20°C) is more important. In soil vapour flow of water is mainly due to vapour pressure. Vapour pressure depends upon the amount of moisture, amount of soluble salts present in soil and prevailing temperature. Excess moisture and higher temperature increase the vapour pressure but higher amount of soluble salts decrease it.

In the soil vapour flow is mostly done by diffusion process in which vapour pressure gradient acts as driving force. Therefore, vapour diffuses from moist soil where vapour pressure is more to the nearby dry soil where vapour pressure is low. Similarly, the water vapours generated from a warm soil diffuse into the nearby cool soil. However, in soil horizons the situation may be somewhat different. In case of moist and cool upper horizon and dry and warm lower horizon the flow of water vapour will be negligible or minimum because flow of vapour from moist to dry and warm to cool will nullify each other, However, in case of dry and cool upper horizon and moist and warm lower horizon the vapour flow from lower horizon to upper horizon will be considerable. Water moved through vapour flow fulfils the water requirement, up to some extent, of some desert plants to save their lives at extremely low soil moisture conditions.

Retention of Water in Soil

Soil becomes saturated with water after heavy rainfall or irrigation. At this stage all the pores of soils are filled with water and water can be seen even on the soil surface. But after some time, water disappears from soil surface. It indicates that all the amount of water, which saturated the soil up to the surface, is not retained by it. The water may be drained out from the soil, percolated to deeper layer of soil, evaporated from the soil surface and or retained or hold by the soil under various energy states.

The energy states of water in soil can also be used to classify or typify soil water as gravitational water, hygroscopic water, capillary water etc. The maximum water holding capacity of the soil is the saturation of water by rainfall or irrigation where water is free to drain out and metric suction is completely zero. But when unsaturated condition develops water is retained by soil under negative soil water pressure. The negative soil water pressure is also known as metric suction, tension or soil suction. This is related to radius (r) of soil pores i.e., height of water column in a capillary or capillary rise (0.15/r)

The water retention capacity of soil can be determined by applying known suction to a saturated soil and calculating the corresponding water content by gravimetric method. In the laboratory water retention can be measured by porous ceramic plate with water column assembly, mercury manometer and pressure plate apparatus and in the field by tensiometer. The figures of these are given here.

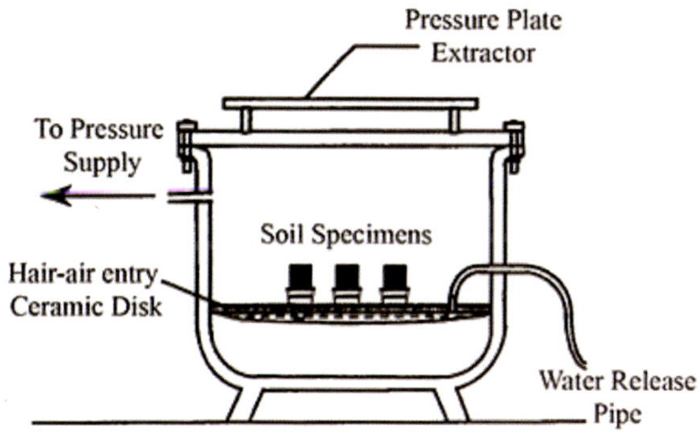

Fig. 3.13: Figure of pressure plate apparatus

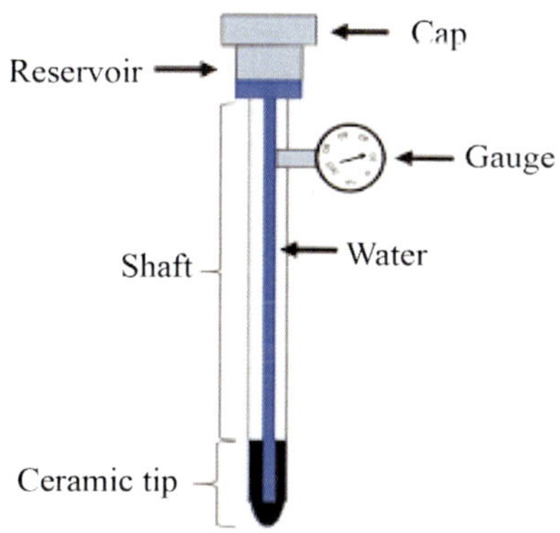

Fig. 3.14: Tensiometer

Gravitational Water

The water held in soil below 0.33 bars is not useful for plants. This water is filled in the macro pores of soil. Gravity removes gravitational water from soil pores. Retention of gravitational water in soil for longer period is harmful for plants because it reduces soil aeration. If gravitational water remains in soil for more than two or three days then drainage should be provided as per need of crop.

Ground Water

Ground water is present below soil surface and it is tension free. Depth of ground water is mainly depending upon hydrology and soil permeability of particular place. Upper surface of ground water is called as water table which increase or decrease according to weather conditions. The pores of lower portion of soil which are in contact of water table remain saturated with water. Water in such pores moves upward due to capillary rise. This water saturated zone above the water table is called as capillary fringe.

The height of capillary fringe may be up to 2 meters from water table. This height depends upon size of soil pores. The height of capillary fringe is less in clayey soil because of lesser speed of capillary rise and in sandy soils due to lesser capillary rise. The height of capillary fringe is maximum in silty soils. If roots of plants may reach up to the capillary fringe, then they can utilize ground water also.

Constants and Units to Denote Energy State of Water

Energy state of water present in soil can be denoted by many constants. The first constant is pressure. Pressure means the weight exerted by water column at a particular place. This can be understood by a cylinder filled with water. Suppose a measuring cylinder whose radius is 3cm is filled with water up to the height of 50 cm. The volume of water in this cylinder will be $3.14 x 3^2 x\ 50 = 1413\ cm^3$. If the density of water (1gm/ cc) is multiplied with this volume of water then this will give the weight of water, which will be 1413 gm.

On the top of this cylinder the pressure of water will be zero, as the weight of water is zero. But in the bottom of weight of water will be 50 gm/cm^2 by $\left(\frac{1413}{3.14X9}\right)$ These 50 gm/ cm^2 pressures will be equal to 50 cm high water column. At sea level the pressure of 1023 cm high water column is one atmosphere. Therefore, the value of 50 cm high water column will be about =0.049 atmosphere.

If in this cylinder a capillary of radius, r is dipped, then the capillary rise will be approximately 50 cm by the formula of capillary rise. The pressure inside the capillary will be negative. Potential of water present in capillary is negative as it reduces free energy of water. Therefore, as soon as the height of water column in capillary will increase the negativity of water pressure will also increase. At the height of 50 cm the pressure will be -0.50 gm / cc. This will be equal to – 0.049 atmospheres.

Another constant or unit to denote energy state of water is potentials and negative potentials of water are expressed as tension (Just like water retention in capillary) and suction (similar to absorption of water by plant roots). Negative potentials expressed as tension or suction give them positive sign.

pF is also one constant or unit of denoting the energy state of water present in soil. Logarithm of height of water column gives the pF values. About 3.009 will be the pF value of 1023 cm height water column. Likewise, pF 2 will be for 100 cm high water column and pF zero will be for 1 cm high water column. One digit change in the pF value will ten times change in the height of water column.

Now internationally energy state of water is denoted by megapascal (MPa) or kilopascal (KPa) instead of height of the water column, atmosphere, and bar or pF. Different soil water constant or units are given in Table 3.6.

Table 3.6: Soil water constants

Height of water column (cm)	Potential			pF	Soil water condition
	Atmosphere	Bar	Mpa		
1,000,0000	10,000	10130	1013	7	Oven dry condition
				4.5- 7.0	Hygroscopic water
1,000,000	1000	1013	101.3	6	Air dry
15,340	15	15.195	1.52	4.186	Wilting point
				2.5-4.5	Capillary water
1,023		1.013	0.101	3.0	
341	0.33	0.334	0.033	2.53	Field capacity
15340-340	15- 0.33	15.195-0.334	1.52-0.033	4.186-2.53	Available water
			0.33	2-7	Moisture equivalent
				2.5-0	Gravitational water
1	Negligible	Negligible	Negligible	0	Saturated soil/ Maximum water holding capacity

This whole system may be compared with a soil profile whose 10 cm of the lower portion is sunken or dipped in ground water and there is continuity of capillaries at upper portion. These capillaries raise the water up to 50 cm height. The water in sunken portion will be at saturated condition and at upper portion at unsaturated condition. The portion of the profiles touching the ground water will has negligible or zero pressure. But below this pressure of water will increase positively and at bottom (10cm) it will be 10 gram /cc. This will be equal to 0.01 atmospheres. Similarly at upper portion of the profile the pressure of water will increase negatively. This will be -50 gram/ cc or 0.05 atmospheres at 50 cm height of capillary rise.

Soil Colour

Soil colour is also a basis of soil classification since ancient time. Black soil, red soils, yellow soils, brown soils, dark yellow brown soil etc. are some examples of it. Soil colour can indicate some other characteristics of soil like soil genesis and formation, organic matter, mineral matter and other constituents, drainage conditions etc.

Factor affecting soil colour: Many factors like amount of humus and organic matter, mineral matter, stage of soil development, climate and drainage conditions influence the soil colour.

Humus and organic matter- Amount of organic matter and humus provide brown to black colour of the soils. Peat and humus provide brown and black colour respectively.

Mineral matter- Presence of montmorillonite, titaniferous and magnetite minerals give black colour to the soil. Oxides of iron provide yellow and red colour to the soil. Yellow colour is given by hydrated iron oxide (Limonite) and red by hematite. Mixture of iron oxide and organic matter provide red brown colour. Marl and quartz sand rich soils have grey colour. Soil surface may be whitish in arid areas due deposition of soluble salts as a result of higher evaporation.

Stage of soil development- Parent material influences the colour of young soils as red sandstone may develop red soils. Sometimes soil forming processes develop red soil from granite, genesis or schist. Soil colours tend to become more brownish or reddish than greyish or yellowish under increased temperature, rainfall and adequate drainage conditions due to increasing stages of soil development.

Climate: In cold humid temperate climate under the influence of nature and amount of organic matter, iron hydroxide and drainage conditions, greyish soil colour become brown and black. But in warm humid intertropical climate soil are red due to intensive weathering and release of iron oxides and hydroxides under well drained oxidizing condition. This gives red colour to soil

Drainage: Drainage conditions of an area are reflected in soil colours. Red or reddish-brown soil colours are indication of well oxidized conditions and under reducing conditions bluish grey soil colour develop. Periodic water logged and well drained conditions create both ferrous and ferric states of iron in soil. These accumulate as mottles having one colour in the background of another colour.

Identification of soil colour: Soil colour can be identified by Munsell's soil colour chart. This chart is a note book having 7 cards with 175 chips of different colours and a hole below each chip. These chips are systematically arranged according to colour variables i.e., hue, value and chroma

Hue: Hue is the dominant spectral colour and indicate dominant of wavelength light. Munsell's colour chart has three basic hues, Red (R), Yellow Red (YR) and Yellow (Y). These are indicated from 0-10 numbers. The hue becomes more yellow and less red with the increase in number. Hue is written on top right-hand corner of colour chart like 10 YR or 7.5 YR etc.

Value: Value indicates relative darkness or lightness of dominant spectral colour. Value is square root of total reflected light. It varies from 0 (absolute black) to 10 (Absolute white). Value 5 is between absolute black and absolute white and value 6 is relatively less dark than 5. Value is written on left hand side of the card and increase vertically.

Chroma: Chroma indicates relative purity or strength the dominant colour. Chroma has 0-8 number and increases with the decrease of greyness. Chroma is written on the bottom of card and increase horizontally.

10YR 6/3 notation of the soil colour indicates that Hue is 10YR, value is 6 and chroma is 3 and soil colour is pale brown. Similarly, 7.5YR 5/2 means hue 7.5YR, value 5 and chroma 2 and soil colour is brown.

8. Soil Consistence

Soil consistence is also an important physical and dynamic property, which varies with the changes in soil moisture and applied stress. Soil consistence is mainly related with tillage operations, constructions and transport. One cannot walk on sand due to its low cohesion and adhesion and it is very difficult to plough wet clayey soil because of higher stickiness. Soil consistency is the strength with which soil materials are held together or the resistance of soils to deformation and rupture. Soil consistency is measured for wet, moist and dry soils. Soil consistency may be estimated in the field using simple tests or may be measured more accurately in the laboratory.

Degree of resistance to deformation and rupture, cohesion of soil particles with themselves and attachment with other materials (adhesion) is called soil consistence and precisely defined as the expression of the physical forces of cohesion and adhesion acting within the soil at various moisture levels. It also indicates proper moisture content in soils for their puddling.

Soil textures, soil structure, amount of organic and inorganic colloids and specifically amount or condition of moisture in soil directly affect the soil consistency. The soils lose their stickiness and plasticity, become friable and soft and finally turn into hard and coherent with the decreasing moisture content. Therefore, soil consistency is classified according to three moisture conditions in soil as wet, moist or dry. Soil consistence is also expressed on the basis of degree of cementation.

Consistency in wet condition: Consistency in wet condition is tested when soil moisture is at field capacity or little more and indicated as Stickiness or Plasticity. First stickiness is determined, that is, the ability of soil materials to adhere to other objects. Then, determine plasticity, that is, the ability of soil materials to change shape, but not volume, continuously under the influence of a constant pressure and to retain the impressed shape when the pressure is removed.

Stickiness: It is the capacity of soil to adhere with other objects. It is estimated by pressing the soil between thumb and fore finger at soil moisture content that displays maximum adherence. Press a small amount of wet soil between your thumb and fore finger. Then slowly open thumb and fingers and see whether it sticks to your fingers or not. Classify the stickiness as Non sticky, slightly sticky, sticky and very sticky as per increasing stickiness.

1. Non-sticky- No soil or practically no soil sticks or adheres to thumb and finger after removal of pressure.
2. Slightly sticky- If the soil begins to stick to fingers but comes off one or the other cleanly and does not stretch when the fingers are opened.
3. Sticky- Soil adheres to both thumb and finger and tends to stretch and pull without free pulling from thumb and finger.
4. Very sticky- Soil strongly adheres to both thumb and finger and is definitely stretched when thumb and finger are separated.

Plasticity: Plasticity is the ability to change shape continuously due to the influence of applied stress or pressure and retain the changed shape on removal of the stress. It is a degree or capacity to which a soil can be moulded in a cast or change the form and remain in changed form or deformation without rupturing. Plasticity is identified by placing soil in between thumb and finger and pallets or wires are made by repeated rounding of soil. Plasticity classes are Non plastic, slightly plastic, moderately plastic and very plastic.

Non-plastic- No wire of 6 mm diameter, and 4 cm long is formed and if it is formed, it cannot retain the shape.

Slightly plastic- A wire of 6 mm diameter and 4 cm long is formed but can easily be broken and returned to its former state. Smaller wire of 4 mm diameter and 4cm length can also not support itself.

Moderately plastic- The soil is able to form a 4 mm diameter and 4 cm long wire which supports itself. If wire of 2 mm diameter and 4 cm length is formed, it cannot support itself.

Very plastic- The soil is able to form a 2 mm diameter and 4 cm long wire which supports itself.

Consistency in moist condition: Moist consistence is determined when the soil is moist but not wet, i.e., 24 hours after good rainfall or irrigation or when moisture content is approximately in between air dry and field capacity. The consistence of moist condition is most important to decide the condition of soils for tilling in the field. A small amount of moist soil is crushed by pressing it between thumb and fore finger or by squeezing it in the palm of hand to determine the consistence of moist soil. Moist soil consistency is described in six classes as Loose, Very friable, Friable, Firm, Very firm and extremely firm. These are described here:

Loose: Non coherent i.e., single-grain structure

Very friable: Soil can be crushed easily under very gentle pressure. But soil coheres or sticks together when pressed again.

Friable: Soil can be crushes easily under gentle to moderate pressure.

Firm: Soil can be crushed under moderate pressure between thumb and forefinger. The resistance is distinctly noticeable.

Very firm: Soil can difficultly be crushed between thumb and forefinger. But soil can be crushed under strong pressure,

Extremely firm: Soil cannot be crushed between thumb and forefinger. the soil crushes only under very strong pressure, but must be broken apart bit by bit.

Consistency in dry condition: Dry consistency of soil is characterized by rigidity, brittleness, maximum resistance to pressure, tendency to crush as powder and inability of crushed soil to cohere again when pressed together. Consistency in dry condition is done when the soil has been air-dried. A small amount of air-dried soil is tried to broke by pressing in between thumb and forefinger or by squeezing it in the palm of hand. Accordingly, classes of dry soil consistency are described as Loose, Soft, Slightly hard, Hard, Very hard and extremely hard.

Loose: Soil is non-coherent with single-grain structure.

Soft: Soil can be crushed under very low pressure. The soil is very weakly coherent and friable. It can be crushed to powder or individual grains under very slight pressure.

Slightly hard: Soil slightly resists pressuring, but can easily be crushed or broken between thumb and forefinger under low pressure.

Hard: Soil resists moderate pressure. It can be crushed between thumb and forefinger under high pressure but can be broken in the hands without difficulty.

Very hard: Soil resists great pressure. It cannot be broken or crushed between thumb and forefinger but can be broken in the hands with difficulty.

Extremely hard: Soil resists extreme pressure and cannot be broken in the hands but can be crushed after applying fully body pressure by foot.

Cementation: Independent to soil moisture content some soil horizons exhibit cementation consistency due to the presence of cementing agents like oxides of iron and aluminium, calcium carbonate and silica. The consistence of soil horizons due to cementation is given here as, weakly cemented, strongly cemented and indurated with their description.

Weakly cemented: Cemented horizon can be broken by hands.

Strongly cemented: Cemented horizon cannot be broken by hands but can be broken easily with a hammer.

Indurated: Cemented horizon can be broken only with sharp blows of a hammer.

Soil Consistency and Atterberg Limits: The consistency of soils, as studied earlier, changes with the moisture present in them. Such changes in soil consistency may be accurately measured in the laboratory following standard procedures by determining the Atterberg Limits. These limits may be used for judging the suitability of the soil for various uses by specialized technicians of civil works, aquaculture ponds including soil scientists.

Significance of Atterberg Limits

1. The different Atterberg limits represent the moisture content of the change from the friable to the plastic consistency.
2. Atterberg Limits represent the minimum moisture percentage at which the soil can be puddle. The moisture content of Atterberg limits depends upon the amount and nature of soil colloids present.
3. The plasticity number or the index of plasticity is an indirect measure of force required to mould the soil.
4. Atterberg Limit is the function of the number of the films and represents the amount of water that must be added to the soil system to increase the distance between the particles of maximum tension and the tension at which flow is produced.

Atterberg Limits: An Atterberg Limit corresponds to the moisture content at which a soil sample changes from one consistency to another. Atterberg limits are also the limits of water content used to define soil behaviour. Atterberg Limits, as suggested by a Swedish scientist, Atterberg, are three namely upper plastic limit, lower plastic limit and plasticity index.

Upper plastic limit: It is also called liquid limit. It represents the moisture content of soil at a point where the soil and water mass just flow under an applied force and does not retain its shape. In the upper plastic limit, the film of oriented water molecules becomes so thick that cohesion is decreased and the entire soil- water mass flows freely under an applied force. The tension of water at upper plastic limit is about pF 0.5.

Lower plastic limit: Lower plastic limit refers to the moisture content of a soil at a point where its consistency from plastic to friable. Soil and water mass is unable to change shape continuously under the influence of an applied force, and ultimately the mass breaks into fragments. The tensions of the water at the upper plastic limit are equivalent to pF values 2.8- 3.3.

Plasticity number (Index): If refers to the difference between moisture contents of a soil at its upper plastic limit and lower plastic limit. Different soils are characterized by a specific 'plastic number or index of plasticity'.

Plastic Index (PL) = Liquid limit (LL) – plastic limit (PL)

The liquid limit (LL): The percentage moisture content at which a soil changes with decreasing wetness from the liquid to the plastic consistency or with increasing wetness from the plastic to the liquid consistency.

The plastic limit (PL): The percentage moisture content at which a soil changes with decreasing wetness from the plastic to the semi- solid consistency or with increasing wetness from the semi-solid to the plastic consistency. The plastic limit is the lower limit of the plastic state. A small increase in moisture above the plastic limit will destroy the cohesion of the soil.

Factors affecting the Atterberg limit: Factors particularly affecting Atterberg limit are clay content, type of clay mineral, exchangeable cations and organic matter status.

Clay content: Amount of clay greatly influences the plasticity. The clay content indicates the surface area available for water absorption. An increase in the percentage of clay causes plastic limits to be higher with the moisture content and increases the plasticity number or index. With the decrease in clay content in the soil, the upper plastic limit decreases and thereby, decreases the plasticity number or the index of plasticity. Atterberg limits are raised as the surface is increased due to higher amount of clay present in the soil.

Type of clay mineral: The size of the clay particles present in soil directly influences the amount of absorbed water required at the plastic limit. Plastic limits of calcium saturated and sodium saturated clays are given in Table . The type of clay mineral has a tremendous influence on the absorption of water by the colloidal system. The plasticity of the montmorillonite group is high because of greater surface hydration, which results from more orientation of water and more of interlattice swelling as the thickness of the water films increases.

Table 3.7: Plastic limits of Ca and Na saturated clays

Types of clay	Calcium saturated (%)		Sodium saturated (%)	
	Lower Limit	Upper Limit	Lower Limit	Upper Limit
Kaolinite	36	73	26	52
Montmorillonite	63	177	97	700
Illite	40	90	34	61

Exchangeable cations: The exchangeable cations remarkably influence soil plasticity. The Na- saturated montmorillonite has the highest plastic limit as compared to the K, Ca and Mg saturated montmorillonite. The effect of these ions on non- expanding clays is quite different from the effect on the montmorillonite. The higher hydration energies of the divalent cations cause a rising of the Atterberg limits.

Organic matter: It has been observed that the plasticity limits decreased due to oxidation of organic matter with hydrogen peroxide. Organic matter has a high adsorptive capacity for water. Hydration of organic matter must be complete before sufficient water is available for film formation around the mineral particles. Consequently, the plastic limit occurs at relatively high moisture contents.

Chemical Properties of Soil

Presence of elements, chemical compounds, complexes and mixtures, their structures, properties and reactions in soils are studied under soil chemistry. Although soils contain many elements and relatively new techniques are providing rapid and simultaneous determination of the concentrations of many elements listed in Mendeleev's periodic table with minimal sample preparation time. But we are more concerned with elements, found in soil minerals, required for plant nutrition and or polluting our environment. Thus, knowledge of chemical composition of soil is necessary.

Chemical Composition of Soils

Soil formation and development directly influence the chemical composition of soils. Soils formed and developed from the underlying bed rock show a relationship with properties of their parent material and local climate. But soils formed and developed from the transported parent material may have no relationship with the under lying minerals and rocks. Here climate decide the chemical composition of soil. The chemical composition of numerous benchmarks soils of India (Das and Chatterjee, 1982 and Sohan Lal et. Al. 1994) and Bangladesh (Khan et. al., 1997) has been reported. The compositions reported by these authors or others anywhere also are the medium of actual chemical composition of soils.

In fact, the chemical composition of soils is quite variable due to inherent heterogeneity in physiography, climate, vegetation etc. and similar soils show different composition. For example, old alluvial soil of IARI (Delhi and Red sandy soil of Tiruchirappalli (Tamil Nādu) are taxonomically same i.e., Haplustalfs but their composition is quite different. Similarly, Ghior series, Gangachara series and Ghatail series of Bangladesh are same. All these are Aquic Heplaquepts, but their chemical composition shows remarkable differences.

Among Indian soils, Histosols (Peaty and Kari soils) show lower pH, whereas saline-alkali soils record pH too much more than 7.0. The other soils are Generally, neutral in reaction. Black soils and desertic soil show higher cation exchange capacities. Black soils are much fertile, but desertic soils are less productive due to lack of water.

Generally, Indian soils are poor in organic carbon and nitrogen content. Organic matter is greater in peat, tarai and black soils and lower in alluvial and lateritic soils. Total nitrogen in Indian soils is normally less them I percent. This is mainly due to high temperature which hinders nitrogen accumulation. Soils are low to medium in Phosphorus status and medium to high in potassium contents. Al_2O_3, Fe_2O_3 and SiO_2 status of soils help in assessing the degree of weathering and mineralogical composition.

Soils with more exchangeable Ca, Mg and K are good as compared to those having higher amounts of Al and Na. Calcareous soils contain more Ca and Mg and alkaline soils are rich in Na and B. Although the Fe status in the form of Fe_2O_3 is in the higher range but few alluvial, red and desert soils are deficient in iron. Manganese status is in the optimum range but some peaty, alluvial and poorly drained calcareous soils show its deficiency. Total Mn is more is the soils of arid and semiarid regions as compared to humid and sub humid region. But the reveres of this are true with available (DTPA extractable Mn).

Coarse textured alluvial soils and higher weathered red and laterites contain less zinc as compared to fine textured soils. Soils situated at higher elevations and receiving more rainfall are rich in copper contents. Deep black soils are rich in molybdenum status and red sandy, brown hilly and saline-alkali soils show less than 1 ppm Mo. However, the concentration of Mo in soil is Generally, very low. Generally, coarse textured, irrigated and saline soils contain more boron is compared to fine textured, unirrigated and normal soils.

Element Concentrations in Soils and Other Surficial Materials of the Conterminous United States was reported By Hansford T. Shacklette and Josephine G. Boerngen through U.S. Geological Survey Professional Paper 1270 (1984). This paper gave an account of the concentrations of 50 chemical elements in samples of soils and other regoliths. These elements were Aluminium, Antimony, Arsenic, Barium, Beryllium, Boron, Bromine, Calcium, total Carbon, Chromium, Cobalt, Copper, Fluorine, Gallium, Germanium, Iodine, Iron, Lead, Manganese, Mercury, Molybdenum, Neodymium, Nickel, Niobium, Phosphorus, Potassium, Rubidium, Scandium, Selenium, Silicon, Sodium, Strontium, Sulphur, Thorium, Tin, Titanium, Uranium, Vanadium, Zinc, Zirconium etc.

Bangladesh soils, as reported by Khan et. Al. (1997), have nitrogen content almost equal to that of similar soils. Phosphorus, potassium, calcium and magnesium content are slightly higher than average values of Bangladesh soils. The similar was true with micronutrient status. These soils are seemed to be illitic, containing other minerals also. The nature of clay minerals suggested that soil fertility is favourable for crop cultivation. Furthermore, every year seasonal flooding brings large amount of silt which is rich in plant nutrients.

The explorations of the relationships between element composition data and site factors using Random Forest regression demonstrated that soil-forming factors have important influence on total elemental composition in the soil. The fact that the soil-forming factors are related to the concentration of naturally occurring elements in the soil gives rise to the notion that they might be predicted from the soils' element composition. Results implied that > 70% of variation in soil element composition patterns can be predicted using information in existing databases or readily observable features.

Successful use of TXRF technique would open up possibilities for using total soil elemental composition fingerprints as a useful basis for characterizing soils in a way that relates to soil-forming factors and inherent soil functional properties.

Chemical Potentials in Soils

Mineral matter, organic matter, water and air are four components of soil. Each component in the soil has a chemical potential. Chemical potential is a fundamental property which decides the criterion for equilibrium about the shift and transformation of the matter from on phase to another. Solid, liquid and gas are three phases of soil. This matter may be shifted from solid to liquid and or gaseous phases. The reverse of this is also possible. The matter may also be transformed in to a new chemical compound in the soil. The shifts of matter from one phase or component to another one is just like the transfer of thermal energy from high temperature zone to low temperature zone. Similarly matter also transfer or shift from phase or component of high chemical potential to phases or components of low chemical potential.

Chemical potential (μ) depends upon the temperature (T) pressure (P) and amount of matter (mols, m) present in the system. The chemical potential of an element in its mass or protons and electrons in the aqueous system in the standard state (T=298.15 K and P=1 atmosphere) is considered to zero. The generalized equation of chemical potential may be written as:

$\mu = \mu 0 + RT \ln m$

μ = Chemical potential of an ion or electrolyte

$\mu 0$ = Chemical potential of that ion or electrolyte at standard state

m = Numbers of molecules of that ion or electrolyte in the system

R = Gas constant

T = Absolute temperature (K)

Further this equation may be converted in logarithmic form to the base 10 as:

$\mu = \mu 0 + 2.3$ RT log m

Chemical potential is also called as partial molar free energy. The free energy of any system, besides the temperature and pressure, depends upon many factors like gravity, osmosis, absorption, adsorption etc. The partial molar free energy is an important thermodynamic property in relation to electrolytes or ions dissolved in any solvent. In such situation the free energy of the solution as a whole will be the sum of free energies of all the constituents forming the solution. This may be represented as:

$G = \bar{G}$ a ma + $\bar{G}$ b mb + $\bar{G}$ c mc + _ _ _ _ where

G = Partial molar free energy or Gibbs energy

$\bar{G}$ a, $\bar{G}$ b and $\bar{G}$ c are the partial molar free energies of constituents a, b and c, respectively and ma, mb and mc are the number of moles of constituents a, b and c, respectively.

In soils numerous inorganic acids and bases are dissolved in soil water to form the soil solution. Thus, many chemical potentials are present in soil system. However, some important chemical potential mostly related with soils are described here.

1. Oxidation – Reduction Potential

Oxidation-reduction potential (redox potential) is related with the redox reaction. Redox reaction is chemical reaction in which one component is oxidized and another is reduced with the transfer of electrons from one atom to another. The simple examples are $2Mg + O_2 \rightarrow 2MgO$ and $2Fe + 3Cl_2 \rightarrow 2\ FeCl_3$. In these reactions there are transfer of four electrons from 2 Mg to O_2 and six electrons from 2Fe to $3Cl_2$. Oxidation-reduction reactions are just like electrolytic cells, made by dissolving a salt, acid or base in water. This situation is very true for soils in which numerous salts, inorganic and organic acids and bases are dissolved in soil water to make soil solution. Thus, at a time many oxidation-reduction reactions may take place in the soil.

Redox potential may be defined as the potential of an inert electrode (Platinum or Gold), inserted in the solution containing oxidizing and reducing agents, in relation to standard hydrogen electrode. Here the electrode potential will be reduced form oxidized form + n electrons and this can be represented by a general equation.

$$E = E^0 - \frac{RT}{nF} In \frac{\text{Oxidized form}}{\text{reduced form}} \quad \text{where}$$

E = Redox potential (volts)

E^0 = Standard electrode potential (volts)

R = Gas constant

T = Absolute Temperature

N = Number of electrons transferred

F = Faraday constant

An example of redox reaction may be Mg ⇌ Mg2++ 2e, and the potential of this reaction is – 2.37.

Redox potential indicates the oxidation-reduction or aeration states of any soil. In soils, oxidized as well as reduced state of an element may present depending upon water condition of soil. Under oxidized (well aerated) condition, plant nutrients are in oxidized form like NO_3^-, SO_4^{2-}, Fe^{3+}, Mn^{4+} etc. whereas under reduced conditions (water submergence) reduced forms of these nutrients such as NH_4^+, S^{2-} , Fe^{2+} and Mn^{2+} are found.

The redox potential of well aerated and well drained soils varies between +0.3 to +0.8 volts but in submersed soils it tends to decrease from +0.3 to -0.4 volts. The redox potentials of some plant nutrients are given in Table 3.8. These are the chemical reactions in which oxidation of one part took place, simultaneously reduction of another part and there is transfer of electrons from one atom to another atom.

Table 3.8: Redox potential of plant nutrients

Element	**Reaction**	**Potential (Volts)**
Hydrogen	$H^+ + e \rightleftharpoons {}^1/_2\, H_2$	0.000
Oxygen	$O_2 + 4H^+ + 4e \rightleftharpoons 2H_2O$	+1.229
Nitrogen	$3N_2 + 2H^+ + 2e \rightleftharpoons 2NH_3$	-3.100
Potassium	$K^+ + e \rightleftharpoons K$	-2.925
Calcium	$Ca^{2+} + 2e \rightleftharpoons Ca$	-2.87
Magnesium	$Mg^{2+} + 2e \rightleftharpoons Mg$	-2.375
Iron	$Fe^{2+} + 2e \rightleftharpoons Fe$ $Fe^{3+} + 3e \rightleftharpoons Fe$	-0.409 -0.036
Manganese	$Mn^{2+} + 2e \rightleftharpoons Mn$ $MnO_2 + 4H^+ + 2e \rightleftharpoons Mn^{2+} + 2H_2O$ $MnO_4^- + 8H^+ + 5e \rightleftharpoons Mn^{2+} + 4H_2O$	-1.029 +1.208 +1.510
Copper	$Cu^{2+} + e \rightleftharpoons Cu^+$ $Cu^{2+} + 2e \rightleftharpoons Cu$	+0.158 +0.340
Zinc	$Zn^{2+} + 2e \rightleftharpoons Zn$ $ZnO^{2-} + 2H_2O + 2e \rightleftharpoons Zn + 4OH^-$	-0.763 -1.216
Chloride	$AgCl + e \rightleftharpoons Ag + Cl^-$ (Silver chloride electrode) $½\, Cl_2 + e \rightleftharpoons Cl^-$ (Platinum electrode)	+0.222 +1.395

Redox potential can be measured by Platinum microelectrode method. In this method platinum electrode along with hydrogen reference electrode are inserted in the wet soil. The electrical potential difference between these two electrodes is measured by a voltmeter. The formula for the calculation of redox potential (Eh) is given earlier.

2. Donan or Membrane Potential

Two solutions, MCl and M proteinate are separated by an ion selective semi permeable membrane. The membrane is permeable only for M and Cl ions. The M and Cl ions will pass through membrane and after certain period equilibrium will be established. At equilibrium, the chemical potentials must be equal at both the sides of membrane i.e.,

μ M Cl (inner say i) = μ M Cl (outer say o). Chemical potential of these will be

$\mu_i + nF\ \Psi_i = \mu_o + nF\ \Psi_o$. But the chemical potential stated earlier was, $\mu = \mu_o + RT \ln m_o$. Thus, $\mu_i + RT \ln m_i + nF\ \Psi_i = \mu_o + RT \ln m_o + nF\ \Psi_o$ or $am_i \times aCl_i = am_o \times aCl_o$ and the membrane or Donnan potential ($\Psi 0 - \Psi_i$ say ΨD) will be

$$\Psi D = \frac{RT}{\eta F} \ln \frac{a_i}{a_o}$$

At this stage there will be potential difference between inner side and outer side of membrane and if solutions are dilute, activities (a) may be replaced by concentrations (c). Hence Donnan potential will be.

$$\Psi D = \frac{RT}{\eta F} \ln \frac{c_i}{c_o}$$

Diffused double layer covering the clay particles also acts as Donnan or membrane system. In this case attractive electric force between particle surface and counter ions ununiformly distribute the counter ions in the micellar (outer) solution.

3. Zeta Potential

Zeta potential or electrokinetic potential, indicated by Greek letter (ζ), is mainly related with the diffused double layer. It is the potential across the diffuse double layer and defined as the potential difference developed at the interface between solid – liquid phase. Zeta potential is responsible for the behaviour of the solid particles in the presence of applied electric field.

A diffuse double layer at the interface of a solid and a liquid may be considered as an electrical on condenser in which two plates are placed at a distance of d cm and each plate has a charge σ per sq. cm. The dielectric constant of the medium is D. Then the electro kinetic potential i.e., ζ potential will be the difference of potential between the plates. Electrostatically this can be represented as:

$$\zeta = \frac{4\pi\sigma d}{D}$$

Zeta potential is influenced by thickness of the diffuse double layer and electrolyte concentration in the solution. The thickness of diffuse double layer decreases with the increase in electrolyte concentration. Divalent cations decrease the thickness more as compared to monovalent cation. Thus, zeta potential will decease with the increase in electrolyte concentration as well as valency of cations.

Zeta potential can also be reduced by lowering the pH, i.e., removal of monovalent ions with H^+ ions. The point at which zeta potential becomes zero is called as isoelectric point. At isoelectric point diffuse double layer is very thin and repulsive forces are lowest. This situation enhances the flocculation of colloidal particles. The reduction in zeta potential will increase the flocculation.

Under an applied or external electric field, the electrically charged colloidal particles move in the liquid or gaseous medium. This is called as electrophoresis. In case of electro osmosis only solvent moves. Both electrophoresis and electro osmosis are known as electro kinetics.

The electrophoretic mobility can be calculated by assuming, a colloidal particle charge X under the strength E of an electrical field. The force F on the colloidal particle will be equal to XE i.e., F=XE. The frictional drag of the solvent medium will be 6 ηrv according to Stokes' law. Thus, XE = σπnrv and

$$V = \frac{XE}{\sigma\pi\eta r}$$

The zeta potential may also be indicated as:

$\zeta = \frac{X}{Dr}$ and by placing the value of ζ in to ionic mobility (μ)

$\mu = \frac{X}{\sigma\pi\eta r}$ we will get the formula as:

$$\mu = \frac{\zeta D}{\sigma\pi\eta}$$

Where D is dielectric constant, ŋ is viscosity of the medium and r is the radius of the particle.

4. Lime Potential

Lime potential is related with the pH, base saturation and lime requirement of the soil. It is derived from ratio law. Many factors like soil: water ratio, soluble salts, suspension effect and CO_2 gas remarkably influence soil pH. Consequently, different values of pH are obtained. Schofield and Taylor (1955) suggested using 0.1M $CaCl_2$ solution instead of soil water to make soil suspension because [H]/√ [Ca] provides the constant values of pH.

The lime potential is defined as the negative logarithm of the ratio of the hydrogen ion activity to the square root of sum of the activities of calcium and magnesium in the soil solution. Lime potential is Generally, expressed as pH – ½ pCa or pH – ½ p (Ca+Mg). In terms of activities of Calcium and Magnesium Hydroxides in soil solution, lime potential may be represented as, ½ log [(Ca + Mg) x $(OH)_2$] + pkw, where kw is the ionic product of water.

Corrected lime potential is another term which is used now a day. This is defined as the ratio of cube of the sum of the activities of Calcium and Magnesium Hydroxides to the square of the activity of Aluminium Hydroxide in the soil solution. This can be represented as pH-½ (pCa+Mg) – ½ (pkg – py) in which kg is the solubility product of gibbsite and y is the value of ionic products, (Al) $(OH)_3$ in the soil solution. Corrected lime potential has a relationship to the degree of base saturation based on the effective cation exchange capacity.

5. Phosphate Potential

Phosphate potential is related with available soil phosphorus. Schofield (1955) indicated that only a small quantity of soil phosphorus is available to plants, which can be estimated by using suitable chemical potential. This chemical potential of phosphorus can be measured by estimating pH and total phosphate concentration in 0.01 M $CaCl_2$ solution which has come to equilibrium with soil.

Phosphate potential is defined as the ratio of phosphate ion activity to the activity of calcium ion. The negative potential of monocalcium phosphate in soil is obtained by determining pH and concentrations of calcium and phosphorus in the equilibrium solution. Phosphate potential can be represented as ½ pCa+pH_2PO_4. Phosphate potential is supposed as an index of available phosphorus. This is like pF which is supposed as an index of available water.

6. Potassium Potential

Potassium potential is related with the available soil potassium. The labile potassium contains exchangeable potassium and solution potassium. Exchangeable potassium is present in the inner most part of the double layer and this potassium relates with the activity of potassium in soil solution. Release of potassium from exchangeable form to soil solution form depends upon concentration of potassium and composition of cations on exchange surface. Thus, estimation of potassium potential is necessary.

The term potassium potential was first introduced by Backett (1964) considering the activities of potassium and Ca+Mg. This is represented as:

$$\frac{aK}{\sqrt{(aCa + Mg)}}$$

The chemical potential of potassium in solid phase indicates the free energy per gram equivalent of potassium. If soil is in equilibrium with its soil solution, then RT Ln aK / (aCa) is proportional to the difference between the chemical potential of labile K and Ca held in the solid phase. This may be used as a measure of potassium availability and potassium potential among soils having somewhat equal amount of calcium.

Adsorption by Soil

Adsorption has a notable importance in the use of substances by cells, enzymatic reactions and processes occurring in soil. Adsorption is a surface phenomenon by which the concentration of ions, atoms or molecules is increased on the surface of a solid or liquid by physical and chemical forces. These ions, atoms or molecules are retained there. The ions, atoms or molecules concentrated on the surface of solid or liquid are called as adsorbed phase or adsorbate and a solid or a liquid on whom surface adsorption takes place is known as adsorbent.

Silica gel and activated carbon are adsorbents. The groups of organic and inorganic substances in soil which are capable to adsorb ions, atoms or molecules are called as adsorption complex. The colloidal substances like humus and clays are more capable of adsorption as compared to non-colloidal substances like sand and silt. The desorption is the opposite process of adsorption where adsorbed ions, atoms or molecules are removed from the surface of an adsorbent.

Adsorption is quite different than absorption. Absorption occurs in the bulk of material that absorbs. The ions, atoms or molecules cross the surface and distributed through out the body of absorbent. Absorption of water by root cells or by a sponge are the examples. Terms sorption is used where it cannot decide that adsorption or absorption is taking place.

Types of Adsorptions

Adsorption is classified as physical, chemical, negative and positive adsorptions. Physical adsorption or van der walls adsorption has low heat of adsorption (less than 10 kcal or about 5 kcal per mole of adsorbate) and rapid establishment of reversible adsorption equilibrium. Rapid establishment of equilibrium at low temperature indicates low activation energy.

Chemical adsorption or activated adsorption shows higher heat of adsorption (20-100 kcal per mole). There is establishment of irreversible adsorption equilibrium and stronger attachment of adsorbate on the surface of adsorbent at higher temperature. In the chemical adsorption activation energies are quite higher i.e., 5-20 Kcal or more. Therefore, it is also called as activated adsorption.

When solute increases the surface tension, then its concentration on the surface will be less than the bulk of the solution. This is negative adsorption. Negative adsorption is particularly related with the electrolytes. Positive adsorption has more concentration of the solute on the surface.

Adsorption Isotherms

Adsorption isotherm is the graph showing the amount of substance adsorbed by an adsorbent and the equilibrium solution concentration at constant temperature for example the graph showing the adsorption of phosphorus by soil from the solution of phosphorus at 25°C temperature. Five different types of isotherms have been noticed for physical and chemical adsorption and scientists developed equations for them. Some important equations are described here.

1. Langmuir Adsorption Isotherm

Langmuir considered unimolecular adsorption of gases on the plane surfaces. This is possible at low pressure or at moderately high temperatures. The Langmuir's adsorption isotherm can be indicated as:

$$\frac{P}{x/m} = \frac{1}{K_1 k_2} + \frac{P}{k_2} \text{ in which}$$

P = gas pressure

x = amount of adsorbent

K_1 = proportionality constant

K_2 = rate of evaporation of molecules, and the plotting of $\frac{P}{x/m}$ against P gives a straight line.

2. Freundlich Adsorption Isotherm

The pressure and temperature have direct influence on the adsorption and change in adsorption due to varied pressure was indicated by Freundlich (1926) by following equation:

$x / m = kp^{1/n}$, where

x = amount of gas adsorbed

m = amount of adsorbent

p = equilibrium pressure

k and n = constants depending upon nature of adsorbent and gas and temperature.

The graph obtained according to this equation show slight curvature. The log form of this equation can be written as:

$\text{Log } x/m = \log k + 1/n \log p$

The curve by resulted by plotting log x/m against log p is a straight line with slope (1/n) and intercept (log k). However, at higher pressures the plot may deviate from the straight line.

3. Brunauer, Emmett and Teller Adsorption Isotherm

The Langmuir adsorption isotherm deals with the unimolecular adsorption but BET isotherm is related with the multimolecular adsorption i.e., attachment of numerous layers of molecules on the single layer of molecules. It Generally, takes place at higher pressure. Brunauer, Emment and Teller (1938) expressed the following relationship:

$$\frac{P}{v(P_o - P)} = \frac{1}{vmc} + \left(\frac{c-1}{vmc}\right)\frac{P}{P_o} \text{ where}$$

v = volume of gas adsorbed at equilibrium pressure P and temperature T

P_0 = saturated vapor pressure at temperature T

vm = volume of gas adsorbed where surface is covered with an unimolecular layer.

c = constant related to heat of adsorption or BET constant.

Further it may be written as:

$$c = \exp\left(\frac{E1 - EL}{RT}\right) \text{ where}$$

E1= heat of adsorption for the first layer

EL=heat of adsorption for second and higher layers. This is also equal to heat of liquefaction or heat of vapourization.

R= rate of adsorption

The straight-line curve is obtained by plotting $P/v(P_o\text{-}P)$ against P/P_o in which slope is (c-1)/ vmc and intercept is 1/vmc.

4. Gibbs Adsorption Isotherm

Gibbs (1878) indicated the relationship between adsorption and surface tension. Positive surface activity decreases the surface tension and increases concentration while negative surface activity increases surface tension. Positive surface activity is adsorption of solute from the solution by the surface and negative surface activity is dispossession of solute from the surface. The equation of Gibbs adsorption isotherm is

$r = -\frac{a}{RT} \cdot \frac{d\gamma}{da}$ where

a = activity of solute

R = gas constant

T = absolute temperature

γ = surface tension and

$\frac{d\gamma}{da}$ = rate of change in surface tension of solution in relation to concentration

The activity (a) may be replaced by concentration (c) for dilute solutions and the equation will be

$$r = -\frac{C}{RT} \cdot \frac{dy}{de}$$

Negative - $\frac{d\gamma}{da}$ indicates positive γ and the adsorption of solute on the surface than the bulk solution. Positive $\frac{d\gamma}{da}$ shows negative γ, then solute will be more concentrated in the bulk solution than the surface. This is also called as negative adsorption.

Adsorption of Adsorbates in Soil

Plant nutrients, plant protection chemicals, industrial wastes and effluents, water, microorganisms and many more substances are adsorbates. These are adsorbed by soil components through various adsorption reactions and adsorption may be of physical or chemical type.

1. Adsorption of Plant Nutrients

Ionic forms of nutrients are taken up by plants. These may be cations or anions. Soil particles attract them to maintain electroneutrality. Silicate clays and humus are negatively charged particles, they attract cations and positively charged clays

(Fe and Al oxides in acid soils) attract anions. These ions are electrostatically held on the surfaces and called as adsorbed ions.

Aluminium, hydrogen, calcium, magnesium, potassium, sodium, ammonium, hydroxyl, phosphate, molybdate, borate, sulphate and chloride ions are mostly adsorbed by the soils. The adsorption of cations depends upon surface of adsorbents and valence and radius of ions. Potassium, Magnesium and Calcium are more strongly adsorbed than sodium and ammonium. Calcium is strongly adsorbed as compared to magnesium and so as potassium over sodium. The hydrogen ions are adsorbed by negatively charged clay particles as the counter ions or by acidic groups or surfaces of humus particles (Bolt, 1976).

Both negative and positive adsorptions of anions take place in soil. The negatively charged soil particles, repel anions. Thus concentration of anions remains more in bulk solution. This is negative adsorption and due to the same anion exchange capacity of silicate clays are quite low. Negative adsorption enhances the leaching losses of anionic plant nutrients as it increases their concentration in soil solution. Positively charged soil particles adsorb anions. This is positive adsorption. Silicate (SiO_4^{4-}) and phosphate (PO_4^{3-}) ions are adsorbed more strongly as compound to sulphate (SO_4^{2-}) and NO_3^-. Adsorption of Cl^- is very weak and even negative adsorption of anions may be indicated as $Cl^- \approx NO_3^- < SO_4^{2-} < PO_4^{3-} < SiO_4^{4-}$. It shows that adsorption of anions increases with the increase in the negative valence.

2. Adsorption of Plant Protection Chemicals

Plant protection chemicals are also adsorbed by soil and adsorption depend upon nature of plant protection chemicals and moisture content, pH and temperature of the soils. Fine textured soils adsorb more plant protection chemicals as compared to coarse textured soils. Soils rich in organic matter and clay adsorb more. Hydrogen bonding and protonation reactions mainly influence the pesticide adsorption. Pesticide adsorption Generally, obeys the Freundlich adsorption isotherm.

Industries, automobiles, sewage, sludge etc. increase the heavy metals and other pollutants in soils. These are adsorbed by soil particles. The inorganic compounds of mercury are easily adsorbed by soil. Positively charged clay particles adsorb negatively charged Arsenic tetra oxide (AsO_4^{3-}) ions. Soils also adsorb copper, manganese, iron and zinc. The immobile forms of these are mostly adsorbed by soil particles.

3. Adsorption of Water

Water is adsorbed by neutral as well as charged particles of soils and it takes place at low and high-water contents in soil. Hydration is somewhat similar to adsorption of water. Adsorption of water at uncharged surfaces is mainly due to Van der Waals forces whereas electrical forms are responsible for adsorption of water at charged surfaces. These electrical forms reorient the adsorbed water molecules. Thus, free energy of water is reduced.

At higher water content in the soil, there is interaction among water molecules. Therefore, many layers of water are attached over initially adsorbed one or two layers. This is mainly because of hydrogen bonding. Clays adsorb water in the inter micellar spaces also. This increases basal spacing of clay minerals and they swell. Inter layer water may attach between ions and charged surfaces of clays. Organic matter adsorbs water through various functional groups especially by ionized carboxylic group.

4. Adsorption of Microorganisms

Adsorption of micro-organisms by soil is quite heterogeneous. The nature of micro-organisms and various soil properties like particle size, surface charge, pH, amount of salts, organic matter and temperature directly influence the adsorption of micro-organisms. Generally, negatively charged (carboxyl group) micro-organisms are adsorbed by positively charged iron and aluminium hydroxide. Positively charged (amino group) microorganisms are adsorbed by silicate clays.

Finer fraction of soil adsorbs more micro-organism as compared to coarse fraction. This is mainly due to the surface area available for adsorption. In sandy soils more micro-organisms are adsorbed by cations around sand particles. Highly acidic pH of the soil shows stronger adsorption of microorganisms. The multivalent cations (from mono to di and tri) in the soil solution increase the adsorption. But increase in desorption take place after adsorption due to increase in salt concentration. Most of the bacteria are adsorbed by organic fraction of the soil. Increase in soil temperature also increase the microbial adsorption up to some extent.

5 Adsorption of Organic Substances

Many organic substances like proteins, amino acids, enzymes, antibiotic etc. are also adsorbed by soil. These substances are adsorbed on soil particles according to their charges. Negatively charged substances are adsorbed on positively charged surface and vice-versa. However, adsorption of uncharged organic substances also takes place. Physical adsorption of organic substances is restricted to the external surfaces of soil particles. Adsorption through hydrogen or other bonding form linkages among organic compounds and soil particles. The organic substances adsorbed in the intermicellar spaces of the clay minerals expand the basal spacing. Adsorption reduces the rate of decomposition of organic substances. This is mainly true with proteins. Adsorption also reduces the activity of antibiotics formed or added into the soil.

Soil Reaction

Soil reaction indicates the extent of acidity or alkalinity of any soil in terms of pH. Any soil contains more amounts of hydrogen ions than hydroxyl ions, the soil is acidic in reaction. Soil contains more hydroxyl ions, than hydrogen ions, that soil is alkaline and if soil contains equal amounts of hydrogen as well as hydroxyl ions, the soil is termed as neutral in reaction.

Effect of pH on Crops, Nutrient's Availability, Plant's Diseases etc.

Soil reaction is an important property of soil because it directly or indirectly influences the soil formation and development, plant nutrients availability, microbial population and the plant growth and development. Acid medium helps in the weathering of minerals. Soil fertility remains good between slightly acidic to slightly alkaline soil reaction and similar is the case with microbial population and plant growth.

Effect of pH on crops: Optimum pH range is essential for better yield and growth of crop plants. Generally, plants can grow very well in the pH range from 6.1 to 7.8 without any remarkable change in their growth and yield. However, specific pH ranges for different crops are summarized here,

1. Paddy and Tea 4.0-6.0
2. Jowar, Bajra, Millets, Cotton and Potato 5.0-6.5
3. Berseem and Oats 5.0-7.5
4. Ground nut 5.3-6.5
5. Soybean, French bean, Peas, Lentil and Gram 5.5-7.0
6. Tobacco 5.5-7.5
7. Wheat, Barley and Sugarcane 6.0-7.5

Effect of pH on nutrient's availability: Nutrient availability is influenced strongly by soil pH. Generally, most of the plant nutrients are available between pH 6.5 and 7.5. However, specific pH range for different plant nutrients is given here,

1. Nitrogen 6.0- 8.0
2. Phosphorus 6.5- 7.5 and 8.7- 10.0
3. Potassium and Sulphur 6.0-10.0
4. Calcium and Magnesium 7.0- 8.5
5. Iron 4.0-6/0
6. Manganese 5.0- 6.5

7. Zinc and Copper 5.0- 7.0
8. Molybdenum 7.0- 10.0
9. Boron 5.0- 7.0 and 8.7- 10,0
10. Chlorine around 7.0

Effect of pH on plant's diseases etc.: Scab disease of potato, black root rot disease of tobacco and club root disease of crucifers are remarkably influenced by soil pH. Composition of the root exudates may be changed by soil pH. This change may attract soil microorganisms and influence their behaviour.

Normal Ranges of Soil pH

Normally the soil pH ranges from 3.5 to 10. The peat soils show pH about 3.5 whereas pH of alkaline soils may rise upto 10.5. Soils of arid regions record higher pH (7-9) i.e., alkaline in nature. But humid region soils are low in pH (6-7). Based on pH, the acidity and alkalinity of soil may be classified in to less than 4.5, extremely acid; 4.5-5.0, very strongly acid; 5.1-5.5, strongly acid; 5.6-6.0 moderately acid; 6.1-6.5, 6.6 to 7.3 neutral and 7.4-7.8 slightly alkaline; 7.9-8.4, moderately alkaline; 8.5-9.0, strongly alkaline and greater than 9.0 very strongly alkaline.

What is pH?

It is the negative logarithm of the hydrogen ion activity or logarithm of reciprocal of the hydrogen ion activity in gram mole litre. It indicates the amount of free hydrogen ions (H^+) i.e., the amount of hydrogen ions present in the soil solution. This is active acidity.

It may be indicated as

$$pH = \log [H^+] \text{ or}$$

$$pH = \log \frac{1}{[H^+]}$$

The term pH is derived from French word 'pouvior Hydrogine' means power of hydrogen. Basic principle of pH depends upon the amphoteric nature of water, in which equal amount of H^+ and OH^- ions are present with the undissociated water.

However, there is dissociation of only one molecule of water out of about 554 million molecules. The dissociation of water may be as:

$$H_2O + H_2O \rightleftharpoons H^+ + OH^- + H_2O \text{ or}$$

$$H_2O + H_2O \rightleftharpoons H_3O^+ + OH^- \text{ or}$$

$$H_2O \rightleftharpoons H^+ + OH^- \text{ and at Equilibrium}$$

$$K = \frac{[H^+][OH^-]}{[H_2O]} = 1.00 \text{ X } 10^{-14} \text{ at } 25^\circ C$$

constant of water.

Thus, Kw = $[H^+]$ $[OH^-]$ = 1.00×10^{-14}, and this equation can give the H^+ and OH^- concentration in water as:

$[H^+]$ $[OH^-]$ = 1.00×10^{-14}

[X] [X] = $1.00 \times 10^{+14}$

$X^2 = 1.00 \times 10^{-14}$

$X = 1.00 \times 10^{-7}$ i.e.

$X = H^+ = 1.00 \times 10^{-7}$ and

$X = OH^- = 1.00 \times 10^{-7}$

It indicates that amount of dissociated H^+ and OH^- will be equal. If H^+ ions are increased in the system the concentration of hydroxyl (OH^-) ions will decrease to maintain the dissociation constant of water and vice-versa. Thus, sum of H^+ and OH^- will always 14. In pure water pH = pOH = 7. Therefore, pure water is neutral in reaction.

All acidic solutions have pH below 7 and alkaline solutions have pH more than 7. Notation of 1×10^{-7} in decimal form (1x0.0000001) or in fraction form (1x1/10000000) was quite difficult and erroneous. Therefore, Sorensen (1909) devised the pH scale (0-14) using the logarithm of surd small values of H^+ and OH^- ions. The relationships between pH and pOH and concentration of free H^+ and OH^- ions i.e. active H^+ and OH^- ions all presented here:

H+	OH	pH	pOH	Reaction
1	10-14	0	14	Acidic
10-1	10-13	1	13	
10-2	10-12	2	12	
10-3	10-11	3	11	
10-4	10-10	4	10	
10-5	10-9	5	9	
10-6	10-8	6	8	
10-7	10-7	7	7	Neutral
10-8	10-6	8	6	
10-9	10-5	9	5	
10-10	10-4	10	4	
10-11	10-3	11	3	
10-12	10-2	12	2	
10-13	10-1	13	1	
10-14	1	14	0	Alkaline

Numerical on pH

The term pH is from French word 'pouvior Hydrogine' or hydrogen power. Soil pH is the negative logarithm of hydrogen ion (H^+) concentration in solution. Where water (HOH) ionizes to H^+ and OH^- (a neutral solution, both H^+ and OH^- and in concentration 10^7 mole per liter (a mole is one molecular weight of ion or the molecule).

Q–1– Find out the pH of the following solutions.

0.01 N HCl, 0.01 N H_2SO_4, 0.05 M H_2SO_4, 0.05 M HCl, 0.01 M NaOH.

Solution

(a) 0.01 N HCl or 0.01 M HCl

$pH = -\log [H^+]$

$= -\log [0.01]$

$= -\log [10^{-2}]$

$= -(-2) \log 10$

$= 2 \log 10 \ (\log 10 = 1)$

$= 2$

$pH = 2$

Or

$$= -\log \left(\frac{1}{100}\right)$$

$= -(\log 1 - \log 100)$

$= -(0-2)$

$=2$

pH=2

(B) $0.01\ NH_2SO_4 = \frac{0.01\ N}{2} = 0.005\ M$

$= 0.005\ M\ H_2SO_4$

$pH = -\log [H^+]$

$= -\log (0.005)$

$= -\log \left(\frac{5}{1000}\right)$

$= -(\log 5 - \log 1000)$

$= -(0.699 - 3)$

$= 3 - 0.699$

$= 2.301$

$= 2.3$

pH = 2.3

(C) $0.05\ M\ H_2SO_4$

$pH = -\log [H^+]$

$= -\log (0.05)$

$= -\log \left(\frac{5}{100}\right)$

$= -(\log 5 - \log 100)$

$= -(0.699 - 2)$

$= 2- 0.699$

$= 1.301$

pH = 1.301

(D) 0.05 M HCL

$pH = -\log [H^+]$

$= -\log (0.05)$

$$= - \log \left(\frac{5}{100} \right)$$

$$= - (\log 5 - \log 100)$$

$$= - (0.699 - 2)$$

$$= - 0.699 + 2$$

$$= 1.301$$

(E) 0.01 M NaOH

$$KW = [H^+] \times [OH^{-1}] = 10^{-14}$$

$$[H^+] = \frac{KW}{[OH^-]}$$

$$= \frac{1X10^{-14}}{1x10^{-2}}$$

$$[H^+] = 1 X 10^{-12}$$

$$pH = - \log [H^+]$$

$$= - \log [1 X 10^{-12}]$$

$$= 12$$

$$pH = 12$$

or

$$pOH = - \log [OH^-]$$

$$= - \log \left(\frac{1}{100} \right)$$

$$= - (\log 1 - 100)$$

$$= - (0-2)$$

$$= - (-2)$$

$$pH = 2$$

$$pH + pOH = 14$$

$$pH = 14 - pOH$$

$$= 14 - 2$$

$$=12$$

$$pH = 12$$

Q-2- What is the concentration of H^+ and OH^- in 0.0001 M HCl

Solution – 0.0001 M HCl

$$[H^+] = 1 \text{ X } 10^{-4} \text{ mole/ L}$$

$$\text{Therefore, } [OH^-] = \frac{1.00 \text{ X}10^{-14}}{1 \text{ X}10^{-4}}$$

$$= 1 \text{ X } 10^{-10} \text{ mole/ L}$$

Q-3- Find out the pH of the solutions with

$H^+ = 1 \text{ x } 10^{-6}$ mole /L

$[H^+] = 1 \text{ x } 10^{-6}$

$pH = - \log [H]$

$= - \log (10^{-6})$

$pH = 6$

and

$5 \text{ x } 10^{-3}$ mole/L.

$[H^+]\ 5 \text{ x } 10^{-3}$ mole/L.

$pH = - \log [H^+]$

$= - \log [H^+]$

$= - \log \left[\frac{5}{1000}\right]$

$= - (\log 5 - \log 1000)$

$= - 0.6999+3$

$pH = 2.301$

Q-4 - What is the concentration of H^+ and OH^- in 0.001 M NaOH?

Solution-

$pOH = - \log (OH^-)$

$pOH = 0.001 \text{ M} = 1 \text{ x } 10^{-3}$ mole/L.

OH^- ion concentration = $1 \text{ x } 10^{-3}$ mole/L

$$OH = \frac{KW}{\left[H^+\right]}$$

$$1 \text{ X } 10^{-3} = \frac{1.0 \text{ X}10^{-14}}{\left[H^+\right]}$$

$$H^+ = \frac{1\ X10^{-4}}{1\ X10^{-3}}$$

$H^+ = 1\ X\ 10^{-11}$ mole/L

Q-5- What is the concentration if H^+ and OH^+ in 0.001 N HCl.

Solution

0.001 N HCl = 0.001 M HCL

$[H^+] = 0.001$ M

$= 1 \times 10^{-3}$ mol/L

$$[OH^-] = \frac{KW}{\left[H^+\right]}$$

$$= \frac{1.00\ X10^{-14}}{1.00\ X10^{-3}}$$

$OH^- = 1.0\ X\ 10^{-11}$

Q-6- What will be the resultant pH of 0.1 mL of 1.0 N HCl added to pure water to make 1 liter solution.

Solution –

$N_1V_1 = N_2V_2$

$1.0 \times 0.1 = N_2\ X\ 1000$

$$N_2 = \frac{1\ X.1}{1000} = \frac{.1}{1000}$$

$$= \frac{1}{10000}$$

So, the normality of 1 liter pure water is = 0.0001 N

for HCl N = M

$pH = -\log [H^+]$

$$= -\log\left(\frac{1}{10000}\right)$$

$= -(\log 1 - \log 10000)$

$= -(0\text{-}4)$

pH = 4

Q-7- What is the $[H^+]$ concentration of a solution having pH = 5.4

$$pH = -\log [H^+]$$

$$H^+ = -\log pH$$

$$= -\log^{5.4}$$

Determination of pH

pH has direct and indirect effect on the plant growth, nutrients availability, plant diseases and microbial activities. There is a favourable pH range for each plant to grow well, each nutrient to be available in optimum quantity and each microbe to show the better activities. Therefore, determination of pH is very essential.

Methods of pH Determination

Colourimetry method and Potentiometric method are two main methods for the determination of pH of any solution or soil suspension. Both are described here in short:

1. Colourimetry method: In this method indicator strip or universal indicator solution is used for the determination of pH. Indicators are weak organic acids or bases. They dissociate to give one colour in acid medium and another in alkaline medium. This property of indicators is used to determine the pH.

Indicator strips or universal indicators are prepared by careful mixing of many acid - base indicators in correct proportion. Colour of strip or indicator changes with the change in pH. These colour changes are pre standardized for different pH values. Indicator strip is dipped or few drops of universal indicator are added in to the soil: water suspension and pH is recorded.

The main indicators or dyes in the universal indicator are thymol blue, methyl red, bromothymol blue and phenolphthalein. However, Himedia Laboratories, Technical data of Universal indicator, 1013 (himedialabs.com 3, 2021) indicated composition of universal indicator as; Thymol blue 0.026 gm, Methyl red 0.060 gm, Bromothymol blue 0.30 gm, Phenolphthalein 0.50 gm and Ethanol 500 ml. Dilute NaOH is added until solution turns green i.e., neutral pH. This method simply gives approximate pH. However, universal indicators of standard companies should be used for more correct pH.

2. Potentiometric method: This method gives accurate pH of any system. pH meter is used for the determination of pH of soil: water suspension. Commonly used pH meters have a glass electrode (indicator electrode) and a calomel electrode (reference electrode). Now pH meters have both electrodes in one electrode called as combined electrodes.

The glass electrode is a thin glass bulb (about 1/20 mm) filled with 0.1 M HCl and attached with silver wire coated with AgCl. Electrode potential of glass electrode is 0.0591 volt. A potential (E) is developed when this glass electrode is dipped to the soil: water suspension. This potential difference (E) is due to the variation in H^+ ion activities in the bulb and in soil: water suspension and can be calculated as:

$$E = \frac{RT}{nF} \log \frac{1}{H^+}$$

But for measuring E, a reference electrode is also required. Because the emf of a half cell (glass electrode) cannot be measured alone. Calomel electrode is Generally, used as a reference electrode. Calomel electrode contains mercury in contact with saturated solution of Mercurous Chloride (Hg_2Cl_2) and saturated solution of Potassium Chloride. Electrode potential of calomel electrode is 0.246 volt.

Now both glass and calomel electrodes are dipped in a solution of known pH to standardize them. The potential of these two electrodes (ET) will be equal to E glass – E calomel.

The coupling of two half cells (glass and calomel electrodes) and soil: water suspension will be as:

Ag	Ag Cl O.1 N HCl	Soil water suspension (unknown H^+)	Hg_2Cl_2 Saturated KCl

The potential (Eu) of this system will be

$$E_u = E \text{ calomel} - \log a H^+$$

$$= E \text{ calomel} + 0.0591 \text{ X pH}$$

It is given E calomel = 0.246 and E glass i.e., hydrogen electrode $\frac{E\mu - 0.246}{0.0591}$ = 0.0591 at 25 ^{0}C.

Therefore,

$$Eu = 0.246 + 0.0591 \text{ x pH}$$

$$Eu = 0.246 + 0.0591 \text{ pH}$$

$$pH = \frac{Eu - 0.246}{0.0591}$$

Buffering in Soils

Maintenance of constant pH is very essential for many biological and chemical systems operating in soil. Sudden and remarkable change in soil pH will adversely affect the plant nutrients' availability, microbial activities and ultimately plant growth. Thus, stability of soil pH is essential. The buffering also determines the quantity of amendments required to reclaim the acidic or alkaline soils. Various chemicals, fertilizers, manures are added to the soil and activities of soil micro-organisms are capable to create enormous changes in soil pH. But there is very little change in soil pH. It indicates that any process is going on in the soil which maintain the soil pH near constancy or which is resisting the abrupt change in pH.

This resistance to change in pH due to the addition of acids or bases is called as buffering or buffer action. Buffering capacity of the soil is the resistance offered by soil against the change in pH. It is the amount of strong base needed to cause unit increase in soil pH. Buffering capacity of water is very low or negligible whereas the buffering capacity of soil solution is quite high. If 100 ml solution of N/10 HCl is added to one litre of pure distilled water (pH 7.0), the resultant pH will be about 2.0, but the same amount of N/10 HCl is added to one litre neutral soil solution the resultant pH will be about 6.8. Thus, soil solution resists changing in pH and maintaining somewhat constant pH.

Buffering Agents

Chemically the buffer agents are weak bases and their salt or weak acids and their salts and buffer solutions can be prepared by dissolving required quantities of these in water. Ammonium hydroxide and ammonium chloride and acetic acid and sodium acetate are the example of buffer agents. Sometimes buffer solutions are also prepared by dissolving two salts, in which one acts as an acid and another as the salt. Buffer agents keep a constant concentration of hydrogen ions.

However, in case of soils, buffer action is provided by soil colloids, organic matter, organic acids and carbonates, bicarbonates, phosphates and borates of various cations. So these may be called as buffer agents of soils. Higher amounts of these in soils enhance the buffering capacity of soils. The buffering capacity also increases with the increase in cation exchange capacity (CEC) of soil. Montmrillonitic clays (2:1 type) possess more CEC as compared to kaolinitic (1:1) type, Thus, buffering capacity of 2:1 type clay is higher than 1:1 type clay. It is because high CEC can exchange more H^+ from the soil solution against the cations present in soil colloids.

Mechanism of Buffering

Cation exchange capacity and neutralization perform the buffering in soil. Soil pH is the indication of amount of free hydrogen ions present in soil solution. These hydrogen ions are released from the exchange surfaces of soil colloids and their amount is thousands time less than the hydrogen ions remained on the soil colloids. The unreleased hydrogen ions or hydrogen ions remained on the soil colloids are called as reserved H^+ ions. These are also known as potential H^+ ions. In terms of acidity both may be named as reserve acidity or potential acidity, respectively. Both reserved and active hydrogen ions take part in the buffering.

Suppose any base is added to the soil, it will increase the concentration of hydroxyl ions in soil solution. These increased hydroxyl ions will react with the H^+ ions present in soil solution (active H^+ ions) and form water. The formation of water will decrease the H^+ ions concentration of soil solution and to overcome this deficiency, H^+ ions will be released from the colloidal surfaces (reserved H^+ ions). This process will continue till all the OH^- are neutralized. The simple depiction of whole may be as,

(Soil colloid)2H^+ + H^+ (soil solution) + NH_4OH (added base) $\rightarrow$ (Soil colloid) H^+ +H_2O

Constancy of soil pH is also maintained when H^+ ions are added to this soil through acidic materials. The acidic material will increase the H^+ concentration in soil solution. In this situation cation exchange phenomena will work and due to the same some H^+ ions move toward soil colloid where they release equal number of cations already present at exchange sites. By this way remaining few H^+ ions will not change the soil pH remarkably. It can be understood as:

(Soil colloid)2H^+ +H^+ (soil solution) +HCl (added acid) $\rightarrow$ (Soil colloid) +NH_4

Thus, reserve acidity of soil increases and cations released from the exchange sites are utilized by the plants.

Buffer Curves

Buffer curves are titration curve showing the change in soil pH due to the addition of base or acid in the soil: water suspension. The slope of the curve is the inverse to the buffering capacity of soil. Buffer curves are given in figure 3.15.

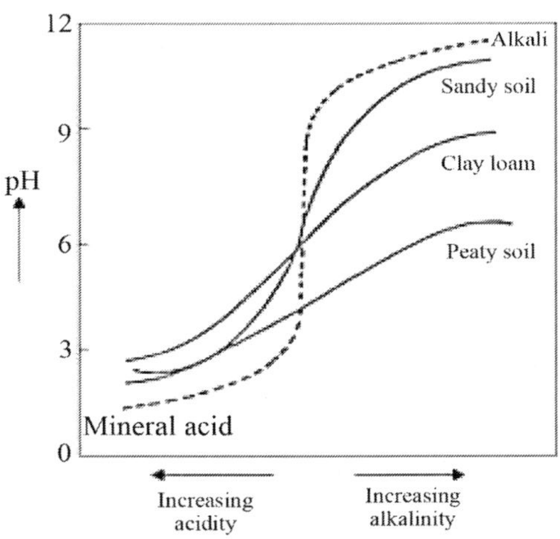

Fig. 3.15: Buffer curves

Ion Exchange in Soils

Ion exchange phenomenon has been studied by various disciplines like soil science, plant physiology, physical chemistry, nuclear chemistry, chemical engineering etc. In soil science its main importance is related with swelling and shrinkage of clays, weathering of minerals and rocks, leaching of salts, reclamation of salt affected soils, absorption of nutrients by plants, flow, exchange and adsorption of ions by soil materials. In agriculture ion exchange has importance next to photosynthesis.

Historically Tomas Way discovered it in the form of Base Exchange while identifying the materials to check ammonia losses from the manure heaps. Since than zeolites, permutit, sulphonated coals and synthetic resins have shown ion exchange properties.

Ion exchange is a reversible process by which ions are exchanged between two solids or between a liquid and a solid. The important characteristics of ions exchange are,

It is a reversible process. Ion can be replaced by another ion and that also replaceable. Ca ion may be replaced by NH4 ions and NH4 can be replaced by Ca or K in the same system. Ion exchange is a stoichiometric process. A single ion of Ca will be replaced by a single ion of Mg or two ions of K or Na. It is due the fact that exchange sites in given mass of exchanger are fixed. Ion exchange is instantaneous in which equilibrium is rapidly attained. The equilibrium constants are neither too low nor too high, Thus, the exchange of ions does not take place completely in a single treatment. This reduces the leaching losses of nutrients. In ion exchange, each ion behaves independently. Although complimentary ions influence the quantity of ions exchanged. In agriculture ion exchange has importance next to the photosynthesis.

Ion Exchange Equations

Ion exchange is supposed to be the redistribution of ions present on soil solids and in soil solution. Knowledge of this redistribution is essential for plant nutrition, use of poor quality ground water in agriculture, reclamation of salt affected soils and interaction of pollutants and heavy metals with soils. This distribution on ions is mainly governed by amount and type of clay minerals, amount and quality of organic matter, charge density of adsorbing surface, pH of the soil and ionic strength and ionic composition of soil solution.

Many equations to know the distribution of ions in soil solid and solution phases has been developed and tested by various workers. Basically, these equations are related with the distribution of cations. Rajpal (1992) and Poonia (1998) grouped these equations in to Semi-empirical exchange equations, equations based on thermodynamics and equations based on diffused double layer theory. The short description of these equations is given here:

Semi empirical equations

These equations are based upon the theories put forward by Vanselow (1932), Gapon (1933) and Krishnamoorthy, Davis and Overstreet (1948). These equations may be presented as:

$$KV = \frac{\left(M^{+}\right)^{2} a^{++}}{M^{++}\left(a^{+}\right)^{2}}$$

$$KG = \frac{N^{+}\left(a^{++}\right)^{1/2}}{N^{++}a^{+}}$$

$$KKDO = \frac{\left(M^{+}\right)^{2} a^{++}}{M^{++}(1.5^{++} + M^{+}(\left(a^{+}\right)^{2}}, \text{ where}$$

N and M are equivalent and mole fractions of cations, respectively,

a is the activity of cations (mol L^{-1}) in the equilibrium solution and

KV, KG and KKDO are the selectivity coefficients for Vanselow, Gapon and Krishnamoorthy Davis-Overstreet equations, respectively.

Thermodynamic equations

The thermodynamic equation can be presented as:

ZABx + ZBA = ZBAx + ZAB

$$KN = \frac{(NA)ZB\ (AB)ZA}{(NB)ZA\ (NA)ZB}$$

where

x = soil solid phase

Z = valence

A and B = cations

a = activity of ions in the solution

KN = equilibrium constant

In this equation the free energy of the exchange reaction (ΔGr) may be written as:

ΔGr = ΔGr2 + RT ln K

where R is gas constant, T is temperature and K is thermodynamic exchange constant.

Diffused Double Layer Equations

Eriksson (1952) applied the diffused double layer for heterovalent cation exchange is soils and presented equation:

$$\frac{\left[Na^{+}\right]^{2}\left(Ca^{2+}\right)(c)}{\left(Na^{+}\right)^{2}\left[Ca^{2+}\right]\left[Na^{+}+Ca^{2+}\right]}=k$$

where

[ ] and ( ) indicated adsorbed and free ions respectively, c is exchange capacity of soil solid and K is equilibrium constant.

Bolt (1955) modified the Eriksson's equation as:

$$\frac{\Gamma+}{\Gamma}=\frac{\Gamma}{\Gamma\beta ½}\sin h^{-1}\frac{\Gamma\beta ½}{\Gamma+4VC_{2}½}\text{ Where}$$

Γ = Total surface charge on the exchanger (mol m^{-2}), Γ + = Surface charge neutralized by monovalent cation (mol m^{-2}), Γ = C1.C2-½ (mol m^{-2}), C1 and C2 = Concentrations of monovalent and divalent cation in solution (mol L^{-1}), V = Cos h Y. Y is the potential at the mid-way plane between two adjacent clay particles and is 1.08 x 10.3 (m mol) at 298^{0} K.

Later, Lagerwerff and Bolt (1959) proposed a very complex ion exchange equation which can be reduced to Gapon equation as

$$\frac{Na\ x}{Ca\ x}\cdot\frac{\sqrt{Ca}}{Na}=K\text{ Where}$$

x is the exchange complex and K is constant.

Types of Ion Exchange

There are two types of ion exchange, Cation exchange and Anion exchange.

Cation exchange: Cation exchange is the exchange or interchange of cation present in the solution and on the surface of the surface-active clay or organic colloid. Negatively charged clay and humus take part in the cation exchange. This is dependent upon the valence of cations. One divalent cation will be replaced by one divalent cation or two monovalent cations and one monovalent cation will be replaced by only half of the divalent cation.

The cation exchange may take place between two solids or between plant roots and soil solution or solid. The effective sites of cation exchange are edges and corners of soil colloids as compared to plane surfaces.

Importance of Cation Exchange

Cation exchange directly influences the characterizations of soil, plant nutrition and reclamation of sodic or acidic soils. Higher exchangeable calcium enhances the aggregation of finer particles in to large and stable aggregates. Thus, soil structure becomes granular and porous which shows good aeration, water holding capacity, drainage etc.

More than 15 percent exchangeable sodium imparts poor physical condition to the soil. Excess sodium causes toxicity to plants and disperses the soil. Dispersed soil becomes impervious to water movement. Thus, aeration and drainage become poor. Higher stickiness of Na rich soil adversely affects the tilth and land preparation for crops. Ultimately all these reduce the plant growth and yield.

Presence of excess hydrogen ion lowers the pH of the soil, ultimately soil becomes acidic. Excess soil acidity converts Al ions into exchangeable aluminium. Excess quantities of Al and H ions are harmful for plant growth because they imbalance the nutrients availability. Excess of sodium and H hydrogen ions can be replaced by the addition of calcium i.e., gypsum and lime, respectively in to the soil.

Numerous cations are found in soils. Among these Ca, Mg, K, NH_4, Fe, Mn, Zn and Cu are utilized by plants for their nutrition. Na, Al, and H ions adversely affect the soil properties and many others like, Ni, Cr, Li etc. pollute the soils.

Cations are released in to the soil by weathering of rocks and minerals and decomposition of organic matter. Excess cations from the retention capacity of soil solution is adsorbed by the soil colloids. If any cation is added in to the soil, the quantity of that cation will increase in the soil solution. Therefore, that cation will be adsorbed by the soil colloids and on the basis of valency equal amount of the other cations will come in to the soil solution.

Similarly, plants roots are absorbing cations from soil solution. The quantity of these cations will reduce in the soil solution and to maintain equilibrium equal amounts of cations will be released from soil colloids in to the soil solution.

Adsorption of cations by soil colloids also reduces the leaching losses, otherwise a large portion of ammonical and potassic fertilizers will be drained out from the soil. These adsorbed cations are released slowly and slowly as per the need of crop plants. Adsorption of cations mainly depends upon their valency and ionic radii and moisture content in soil. Cations of higher valency and lesser hydrated radii are adsorbed with more strength as compared to lower valency and higher hydrated radii.

The divalent cations are adsorbed more with the increase in moisture content of soil. Simultaneously equal number of monovalent cations is released in the soil solution. Opposites of this take place when moisture content in soil is reduced.

Cation Exchange Capacity of Soils

Cation exchange capacity (CEC) is the total quantity of cations adsorbed by a soil. It also indicates the total amount of cations that can be exchanged by a soil. Effective cation exchange capacity (ECEC) and pH dependent cation exchange capacity are other terms used in this context. Effective cation exchange capacity is the total amount of cations adsorbed by a soil in its natural condition. Generally, ECEC is the total of exchangeable bases (Ca, Mg, Na and K) and exchangeable Aluminium and hydrogen ions. The term ECEC is used for taxonomic classification of acidic soils. pH dependent cation exchange capacity is the difference between effective cation exchange capacity and cation exchange capacity measured at the pH greater than the actual pH of the soil.

Cation exchange capacity directly influences the fertility of soil. Soils having higher CEC are more fertile and if there is no constraint, they will be more productive also. Soils, soil minerals and humus show remarkable variations in their CEC (Table 3.9)

Table 3.9: Cation exchange capacity (C mol Kg^{-1})

Soils	CEC	Soil minerals and humus	CEC
Black soils	28.6 – 61.3	Kaolinite	3-15
Alluvial soils	3.2 – 38.0	Halloysite ($2H_2O$)	5-10
Red and Laterite soils	2.9 – 11.1	Halloysite ($4H_2O$) Muscovite Illite	10 – 140 11 30 – 40
Desertic soils	51.0	Nontronite Saponite Montmorillonite Vermiculite Chlorite Humus	57 – 64 69 – 81 80 – 150 100 – 150 10 – 40 200 – 400

Black soils due to their higher CEC are more fertile and productive. Alluvial soils are in the middle and fertility status of red and laterite soils is low because of their low CEC. CEC of desertic soils is quite high, their fertility is also high but the productivity is low because of inadequate supply of water etc.

Factors Affecting CEC

Soil pH, soil texture, amount and type of clay mineral and amount of organic matter or humus are some factors which directly or indirectly influence the CEC of the soil.

Soil pH: Negative charge of the clays minerals particularly kaolinite increases with increase in pH and with the increase in negative charge also increases cation exchange capacity. pH dependent charge increases with the increase in pH due to the dissociation of H^+ ions from the OH^- ions present on soil colloids. Thus, remaining O^{2-} ions increase the negative charge.

Soil texture: Fine textured soils show high CEC as compared the coarse textured soils. Fine textured soil, have more amount of clays. Similarly, soils having more amount of montmrillonitic clays will record more CEC as compared to soils having more amounts of kaolinitic clays.

Amount of organic matter: Soils having more amount of organic matter also have higher CEC. Humus which is a negatively charged organic colloid of soil can rctain and exchange tremendous number of cations and if any soil contains more amount of humus, the CEC of that soil will also high.

CEC Determination

Cation exchange capacity is determined by replacing the adsorbed cations by the solution of a cation like K, NH_4 or Ba say Ammonium acetate and estimating individually the quantities of replaced cations in the filtrate. This can be represented as:

(Soil colloid)Ca Mg Na NH_4 K H +8NH_4 (added) $\rightarrow$ (Soil colloid)8NH_4 + Ca Mg Na NH_4 K H(in filtrate)

The quantities of Ca Mg Na NH_4 K H in filtrate can be estimated by suitable methods. This ammonium soil (as all the exchange sites are occupied by NH_4 ions, Thus, called as ammonium soil) is again leached with the solution of cation other than NH_4 (Say Na or Ca) to replace ammonium ions. The amount of replaced NH_4 ions is determined in the filtrate, which will be equal to the CEC of the soil. This can be represented as:

(Soil colloid)8NH_4 + 8 Na $\rightarrow$ (Soil colloid)8Na + 8 NH_4 (in filtrate)

This method gives more accurate results of CEC besides the number of individual cations.

Numerous electrolytes are used to replace adsorbed cations in the soil. But ammonium acetate is good enough among them because it is highly buffered, so there is no change in pH up to the last part of the solution and it is more effective in the displacement of cations from the soil used. While expressing the CEC, the name of cation used for displacing adsorbed cations should invariable be quoted. Thus, 61.3 CEC of soil will be expressed as 61.3 (mol (Na^+) Kg^{-1}. It indicates that all the adsorbed cations are displaced by Na ions.

Units of Cation Exchange Capacity

Centimoles of positive charge per kilogram of soil (Cmol (p+) kg^{-1}) and milli equivalents per 100 grams soil (meq/100g) are two units of CEC. Now a day according to International System of Units (SI), Cmol (p+) kg^{-1} soil or soil colloids is used for expressing CEC. However, meq/100g and Cmol (p+) kg^{-1} both are same.

Molarity

Molarity also called as molar concentration is the number of moles of a solute or molecular weight in gram per litre of solution or per kilogram of soil or other material. Thus, equal volumes of equimolar solution contain equal number of molecules. If concentration of a solution in terms of grams of dissolved solute is given, the molarity can be calculated.

Molarity = Number of moles/litre solution or per kilogram material.

$$\text{Number of moles} = \frac{\text{Weight in grams}}{\text{Molecular weight}}$$

Suppose 500 mL solution contains 10g KOH. The molarity (M) of this solution will be 0.356 as:

$$\text{Number of moles} = \frac{\text{Weight of KOH in grams}}{\text{Moleculae weight of KOH}}$$

$$= \frac{10}{56.108}$$

$$= 0.178$$

$$\text{Molarity (M)} = \text{number of moles/ litres}$$

$$= \frac{0.178}{0.5}$$

$$= 0.356$$

Similarly, weight of KOH in 6 litres solution of 0.356 M KOH will be 119.846 g. This can be calculated as:

: Weight of KOH in 1 litre solution of 1 M KOH = 56.108 g

: Weight of KOH in 6-litre solution of 1 M KOH = 56.108 x 6 g

: Weight of KOH in 6-litre solution of 0.356 M KOH = 56.108 x 6 x 0.356 g

=119.846 g

Suppose CEC of a soil is 15 Cmol/kg. It means 15 Cmol of hydrogen ions, or other cations are adsorbed on one kilogram of soil. These hydrogen ions will be exchanged by equal (15 c mol) of monovalent cations (Na^+ or K^+ etc.), half (7.5 c mol) of divalent cations (Mg^{2+} or Ca^{2+} etc.) and 1/3rd (5 c mol) of trivalent cation (Al^{3+} or Fe^{3+} etc.). Thus, 15 c mol is chemically equal to 15 c mol of negative charge by which soil can adsorb equal positive charge i.e., 15 c mol cations. One centimol is equal to 0.01mol, 1/100th part of a mole or one mole divided by one hundred.

Conversion of Cmol kg^{-1} to kg ha^{-1}

Cmol kg^{-1}quantity of any ion can also be converted in kg ha^{-1} by taking into consideration the weight of soil in one hectare (ha) area of land. Suppose one kg soil contains 15 cmol Ca. The weight of Ca will be calculated as:

$$1 \text{ c mol Ca} = \frac{1 \text{ x atomic weight of Ca in gram}}{\text{valency of Ca x100}}$$

$$= \frac{1 \text{ x}40.08}{2 \text{ x}100} \text{ gram}$$

$$= 0.2004 \text{ gram}$$

$$= 200.4 \text{ milligram/ kilogram soil}$$

$$= 200.4 \text{ mg kg}^{-1} \text{ soil}$$

$$15 \text{ cmol Ca} = \frac{15 \text{ x}40.08}{2 \text{ x}100} \text{ gram}$$

$$= \frac{601.2}{200} \text{ gram}$$

$$= 3.006 \text{ gram or } 3006 \text{ mg kg}^{-1} \text{ soil}$$

The mg/kg is also known as parts per million (ppm) as one Kg has 1000 gram and 1 gram has 1000 mg. Thus, 1 kg has 1000 x 1000 = 1000000 mg. By this way 3006 mg/kg may be called as 3006 ppm Ca mg/kg or ppm can be converted into kg/ha.

The weight of 15 cm deep (furrow slice) soil in one hectare area is 2.2 million kg. Thus, 3006 mg/kg or 3006 ppm Ca can be converted in to kg/ha as:

: 1 kg soil has Ca = 3006 mg/kg

: 2200000 kg soil has Ca =3006 x 2200000 mg

$$= \frac{3006 \text{ x}2200000}{1000} \text{ g}$$

$$= \frac{3006 \text{ x } 2200000}{1000 \text{ x } 1000} \text{ kg}$$

$$= \frac{3006 \text{ x}22}{10} \text{ kg}$$

$$= 3006 \text{ x } 2.2 \text{ kg}$$

$$= 6613.2 \text{ kg}$$

Thus, multiplication of mg/kg or ppm with 2.2 will give kg/ha quantity of an ion. ppm can be converted into lbs/acre by multiplying with two i.e., ppm x 2 = lbs/ acre and a multiplication of 1.1 to the lbs/acre will give kg/ha i.e., lbs/acre x 1.1 = kg/ha. Thus, 3006 ppm Ca will be 6012 lbs/acre Ca or 6613.2 kg/ha Ca.

Anion Exchange in Soils

Anion exchange is another type of ion exchange. Lesser emphasis is given on anion exchange as compared to cation exchange because anionic plant nutrients are lesser in number as compared to cationic plant nutrients. Nitrate, phosphate, sulphate, molybdate, borate and chloride are anionic plant nutrients. Anion exchange capacity is the total amount of exchangeable anions that can be adsorbed by a soil. AEC is expressed as cmol per kg soil. Anion exchange capacity is lower than the cation exchange capacity.

Anions in soil are dependent upon anion repulsion and anion binding. Anion repulsion is called as negative adsorption. A series of negative adsorption of anions (Bolt, 1976) is as:

Anions in soil are dependent upon anion repulsion and anion binding. Anion repulsion is called as negative adsorption. A series of negative adsorption of anions (Bolt, 1976) is as:

$Cl^- = NO_3^- < SO_4^{2-} << PO_4^{3-} < SiO_4^{4-}$

Thus, Cl^- and NO_3^- show highest anion repulsion or lowest adsorption and due to the same leaching losses of Cl^- and NO_3^- are more as compared to other anions.

Anion binding may be weak, moderate and stronger. Chloride and Nitrate show weak binding with soil solids and they can be easily leached. Sulphate ions show moderate binding. They don't form insoluble compounds and can be leached with water. The phosphate and silicates ions are tightly adsorbed due to strong binding.

Phosphorus (orthophosphate) reacts with all soils. Reactive surfaces in solid phase containing Fe, Al, Ca etc. change soluble phosphorus in to insoluble forms. The association of anions with the solid phase is more complex as compared to cations. Therefore, no general equations are possible for anions as in the case of cations. Thus, each anion is considered individually for repulsion and binding studies.

Anion exchange in soil is also shown as cation exchange

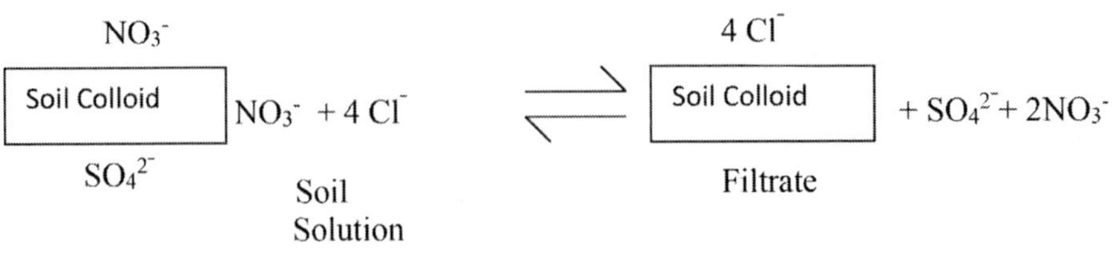

However, the exchange of sulphate and phosphate ions is not so simple. These ions are attached with the hydroxyl groups of clay.

Factors Affecting Anion Exchange Capacity

The pH of the soil, type of clay mineral, valency of ions and salt concentration are some factors which affect the AEC of the soil.

pH of the soil: Anion exchange capacity decreases with the increase in the pH of the soil. It is reverse of CEC which increase with the increase in pH.

Type of clay mineral: Anion exchange is the phenomenon of positively charged minerals. They attract anions and repel cations. Hydrous oxides of iron and aluminium, allophane and kaolinite have positively charged surface. They adsorb negatively charged ions. The anion exchange capacity of kaolinite (40 cmol kg^{-1}) is too much higher than montmrillonitic clays (about 5 cmol kg^{-1}) at acidic pH. The CEC/AEC ratios of kaolinite, illite and montmorillonite are 0.5, 2.3 and 6.7, respectively.

Valency of ions and salt concentration: Anion exchange capacity increases with the increase in the valency of anions but increasing valency of cation decreases it. However, the addition of cations like Zn (OH^+), Zn Cl^- or both increase the AEC. An increase in the salt concentration reduces the AEC of the soil.

Base Saturation of Soil

Numerous cations are adsorbed by the soil colloids. These may be basic or acidic in nature. Ca, Mg, Na and K are basic cations and H, Al and Mn are acidic cations. The amount of basic cations in a soil is also called as exchangeable bases and indicates its base saturation. Base saturation is related with cation exchange capacity. It may be defined as the magnitude of saturation of exchangeable cations other than Hydrogen and Aluminium on the adsorption complex of a soil. Generally, base saturation is expressed as the percent of the total cations exchange capacity and called as percent base saturation (PBS). Percent base saturation (PBS) can be represented as:

$$\text{PBS} = \frac{\text{Sum of exchangeable bases}\left(\text{cmol kg}^{-1}\right)}{\text{Cation exchange capacity}\left(\text{cmol kg}^{-1}\right)} \text{ x } 100$$

Suppose a deep black soil contains 42.7, 15.2, 1.4 and 1.04 c mol kg^{-1} of exchangeable Ca, Mg, Na, and K, respectively. If the CEC of this soil is 61.3 cmol kg^{-1} than PBS will be 98.43%., which can be calculated as:

$$\text{PBS} = \frac{42.7 + 15.2 + 1.4 + 1.04}{61.3} \text{ x } 100$$

$$= \frac{60.34}{61.3} \text{ x } 100$$

$$= 0.9843 \text{ x } 100$$

$$= 98.43$$

Soils show remarkable variation in their percent base saturation. The fertility status of a soil can be assumed by the PBS also. Higher PBS greater will be the fertility status. At higher PBS there will be more availability of exchangeable Ca, Mg and K. These are essential plant nutrients. Percent base saturation is remarkably influenced by pH of the soil. As shown in (below). At pH 5.5 most soils have 45–55% Base Saturation and at pH 7 Base Saturation is about 90%. This is mainly due to pH dependent CEC of the soil which increases with the increase in soil pH. Generally, pH of the arid region soils is higher as compared to humid regions. Thus, PBS of arid region soils is also higher than the humid region soils. Low PBS is associated with high Hydrogen ions or more acidity in soil which can be corrected by the application of lime after determining the lime requirement of the soil.

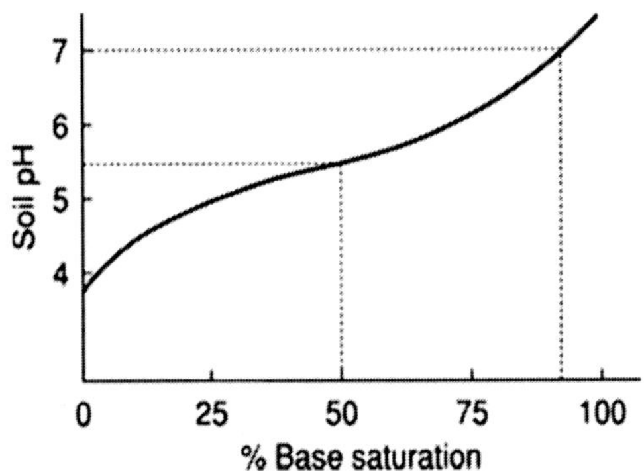

Fig. 3.16:

Source: Hillel et al. 2005: Encyclopedia of Soils in the Environment (Ref. J.L. Havlin, in Encyclopaedia of Soils in the Environment, 2005)

Biological and Biochemical Properties of Soil

Soil biology and Soil ecology, Organisms in soil, soil flora and fauna, macro as well as micro, Fixation of Nitrogen in soil, Biological Nitrogen Fixation, symbiotic and non-symbiotic.

Soil biology is defined as the branch of soil science related with study of living organisms of soil and soil ecology is the study of interactions among soil biology and biotic and abiotic aspects of soil environment. The soil environment includes various physical, biological and chemical properties and factors that affect the abundance and diversity of microorganisms found in the soil. Soils ecosystem comprises both living organisms and non-living matter with a multilevel interaction among them.

Food chain is the sequence of transfers of matter and energy in the form of food from one organism to another organism under one linear energy pathway through an ecosystem. Locally many food chains are found into a food web because any organisms consume many types of flora and fauna. Food web is defined as the transfer of energy between species in any ecosystem and food web is more complex and shows all the possible potential pathways. Energy to soil food web and all food webs is provided by primary producers.

Producer, consumer and decomposer all are the part of soil ecosystem. Higher and chlorophytic plants in the presence of solar energy, CO_2 from air and water and nutrients from soil manufacture their own food by photosynthesis for growth and development of themselves. Thus, higher and chlorophytic plants are producer. Many herbivores depend upon these producers and numerous carnivore and omnivores depend upon herbivorous for their food. Therefore, these are consumer. Residues and wastes of producer and consumer are changed into simple compounds and molecules by the microorganisms through decomposition. Therefore, these microorganisms are come in the category of decomposer.

A simple example is that plant leaves are eaten by locust or grasshopper, these insects are eaten by frog or bird and frog or bind are used as food by snake. All the residues of plants and residues, wastes, excreta and dead bodies of animals are invaded and decomposed by microorganisms. In this example plant is producer, animals are consumer and microorganisms are decomposer. By this way many simple and complex food chains act simultaneously in the soil food web.

Soil biology is mainly related with organisms residing in soil, their functions, reactions, and activities in context to microbial ecology, microbiology and biochemistry of carbon and other nutrients and their cycling and bioremediation. It also studies the biological interfacial systems at the molecule to field or landscape level. Soil microbiology is also part soil biology. Antony von Leeuwenhoek discovered microorganisms but Winogradsky, Hillreigel, Wilfarth and Beijerinck were founder of soil microbiology by their work on aerobic and anaerobic bacteria in soil, organic matter decomposition, biological nitrogen fixation and nitrification.

Organisms in Soil

Numerous organisms also reside in soil during part or whole of their life besides soil mineral matter, organic matter, soil water and soil air. These organisms are less than one percent of total soil volume but they make the soil a living entity. A spoonful of soil contains millions of different macro and microorganisms to bring numerous functions or role in soil. In the Vedic period it was very clearly indicated that when normal animals roam on the surface of land, worms and bacteria roam on the surface of land as well as inside the land also. Nature of some of these is harmful but some of these make more vital to soil and its products (Atharv Veda, 12.1.46). Soil organisms directly or indirectly play very significant and remarkable role in soil which may be summarized as:

Living organisms are one component of soil ecosystem and come in the category of producer (seed or plants), consumer (nematodes, slug) and decomposer (bacteria, fungi etc). The plant nutrients in the wastes, residues, excreta etc. are in the organic forms. There are unavailable to plants. But soil microorganisms decompose these and convert into inorganic forms and available forms to be utilized by plants and other microorganisms. Soil organisms take part in soil formation and development through weathering of rocks and minerals, development of soil structure and soil aggregates and formation of humus.

Soil microorganisms increase the fertility status of soil. They fix atmospheric nitrogen, change insoluble forms of plant nutrients into soluble forms and actively take part in the mineralization and immobilization of nutrient elements. They also influence soil and plant health and sustainability of agriculture. Soil organisms prevent leaching losses of plant nutrients by retaining them in soil. Soil organisms promote plant growth and control crop pests and diseases.

Classification of Soil Organisms

There are several types of organisms present in soil. These may be classified according to their size, feeding, trophic nature, kingdom etc. These are briefly described.

1. Size: Soil organisms are classified according to their size into three classes as;

Soil Micro-organisms: The width is less than 0.1 mm and length is less than 0.2 mm

Soil Meso-organisms: The width varies 0.1 to 2 mm and length 0.2-10 mm

Soil Macro-organisms: The width is more than 2 mm and length is more than 10mm

2. Feeding: This classification is based on the feeding of soil organisms on flora, fauna etc. for their survival and placed into seven classes as:

Herbivores: Those feed on living plants

Detritivores: Those feed on dead plant debris

Carnivores: Those feed on animal tissues

Bacterivores: Those feed on bacteria

Fungivores: Those feed on fungi

Parasites: Those feed on or in host organism and get the food from or at the expense of host organisms

Predators: Those survive by killing and eating another organism i.e., its prey animals

3. Trophic nature: Soil organisms are grouped into two because of food relationships with other organisms as:

Autotrophs: Those manufacture their own food from simple inorganic sources; obtain Carbon from CO_2 and energy from photosynthesis or oxidation of various elements.

Heterotrophs: Those do not manufacture their own food and depend on organic compounds for their Carbon and energy obtained from other plants and animals.

4. Kingdom: Kingdom is also a basis of Soil organisms' classification. Earlier Plantae (flora) and Animalia (fauna) were only two kingdoms. Now these two are divided into many kingdoms. Among them seven main kingdoms (Ruggiero, et.al. 2015) are:

Kingdom: Bacteria - Prokaryota

Kingdom: Archaea - Prokaryota

Kingdom: Protozoa – Eukaryota - Amoeba

Kingdom: Chromista – Eukaryota - Brown Algae, Diatoms etc.

Kingdom: Plantae – Eukaryota- Red and green Algae, Land Plants

Kingdom: Fungi- Eukaryota

Kingdom: Animalia- Eukaryota

Michael A. Ruggiero, Dennis P. Gordon, Thomas M. Orrell, Nicolas Bailly, Thierry Bourgoin, Richard C. Brusca Thomas Cavalier-Smith, Michael D. Guiry, Paul M. Kirk Published: April 29, 2015 A Higher Level Classification of All Living Organisms https://doi.org/10.1371/journal.pone.0119248

Even though soil organisms are mostly classified into broad classes, 1. Soil flora and 2. Soil fauna and these are described here:

1. Soil flora: Soil flora is divided into soil microflora and soil macroflora.

Soil microflora: Bacteria, actinomycetes, cyanobacteria, fungi, algae and virus are six major groups of soil microflora. These are described here:

Bacteria: Bacteria itself is Kingdom of Prokaryotic organisms. Bacteria are soil microflora and most active among all the micro-organisms present in soil. They play important role in the formation of micro aggregates, decomposition of organic matter, transformation of organic compounds into humin, fixation of atmospheric nitrogen and conversion of fixed nitrogen into plants' available forms, conversion of oxidized and reduced forms of elements or minerals, and vice versa, dissolution and precipitation of minerals, conversion of unavailable forms of nutrients to plants available forms and transformation of soil parent material.

Length of bacteria is less than 0.005 nm and width is even less than length. Bacterial population in Pant Nagar soil is about 1.2x106 to 2.7x106 counts /gram dry soil. Bacteria are classified based on their morphology, shape, size, temperature tolerance, requirement of molecular oxygen, source of energy, biochemical processes etc.

Morphologically bacteria have the shape of rods, spheres or spiral. Some bacteria move by flagella and some by gliding.

Since source of energy bacteria may be heterotrophs (Xanthomonas and Pseudomonas) and autotrophs (Chlorobium). Heterotrophs produce energy for their life cycle by the oxidation of organic compounds whereas autotrophs produce energy by the oxidation of inorganic compounds or mineral matter.

Some bacteria live in the presence of molecular oxygen or air are called as aerobic (Aerobacter) and other in the absence of oxygen or are known as anaerobic (Clostridium)

Actinomycetes: Actinomycetes come under Kingdom Monera and next to bacteria in population. Actinomycetes are heterotrophs, aerobic soil microflora. They are single-celled, prokaryotic, filamentous and often have profoundly branched mycelium to look like them as fungi. When mycelium breaks in small pieces then actinomycetes look like bacteria and also called as branched Eubacteria.

Actinomycetes grow very slowly in soil. Because first of all bacteria decompose organic matter and make intermediary substances on which fungi feed. Lastly actinomycetes act upon residual and hardy substances to decompose and mineralized them in to plant available forms. Actinomycetes population in Pant Nagar soil is about 3.8x106 to 12.1x106 counts /gram dry soil.

Production of more than 500 antibiotics and decomposition of hardest part of organic matter are two important functions of actinomycetes. Streptomyces is famous to produce Streptomycin and Nocardia has big role in the decomposition of chitin, cellulose and fats.

Cyanobacteria: Cyanobacteria are classified within the Kingdom Monera (Photosynthetic Prokaryota). In prokaryotic structure, protoplasm is covered by semi permeable cell membrane. Earlier they are also known as blue-green algae. Cyanobacteria are soil microflora, prokaryotic and most ancient organisms. Cyanobacteria are either autotrophic or heterotrophic. Many genera of Cyanobacteria can fix atmospheric nitrogen (e.g., Anabaena sp.) and are especially plentiful in rice paddies and other wetland soils, where they fix large quantities of nitrogen.

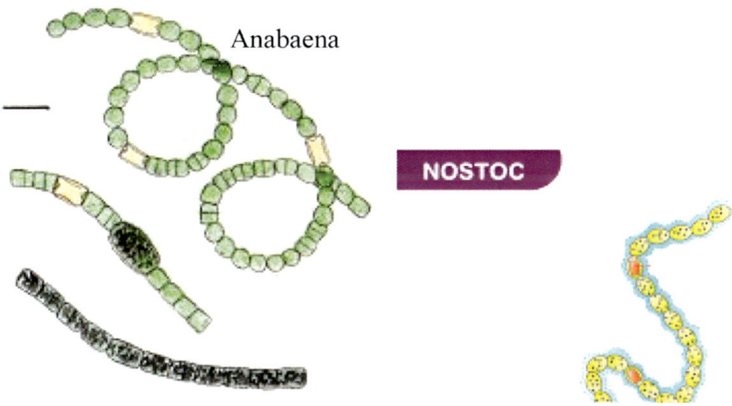

Fig. 3.17: Blue green algae

Fungi: Soil fungi are microflora and come under Kingdom Fungi. Soil fungi are a very heterogeneous group of heterotrophic, aerobic and eukaryotic organisms. Fungi may be unicellular (yeast) to multicellular filamentous (mildew, rusts and smuts). The filaments (hyphae) may be septate or aseptate mycelium. Molds and mushroom fungi have more importance in soil. Soil fungi accounts many thousands in number and in Pantnagar soils fungi were 47.0 x 10^5 -98.9 x 10^5 counts/gram dry soil.

Fungi mostly depend upon dead or alive organisms according to their functions and activities and act as circulatory system in soil. In soil fungi play vital role from soil development to plant nutrition. Their mycelia increase cohesion of soil particles to form stable aggregates and transport water and nutrients.

Soil fungi are divided in the three main groups: 1. Saprophytic or Decomposers Fungi 2. Mutualistic or symbiotic Fungi and 3. Pathogenic or parasitic Fungi. These are described here briefly.

1. Saprophytic or Decomposers Fungi: As heterotrophs and saprophyte or decomposers fungi depend on living or dead organic materials for both their C and energy. They decompose various organic substances, cellulose, hemicellulose, and lignin, present in soil. The metabolic by-products of decomposition are mostly CO_2, organic acids etc. Mucor is a saprophyte or decomposer fungi. Mushrooms grow well in diversified locations and at different substrates to derive their nutrients.

2. Mutualistic or symbiotic fungi: Mutualistic or symbiotic fungi are also called as mycorrhizal fungi. These make symbiotic relationships with plant by residing or colonizing on host plant roots. This symbiotic association is very widespread and beneficial for natural as well as agricultural ecosystems. These soil fungi are most important for plant nutrition.

The thin fungal hyphae and their mycelia can extend several centimeters into surrounding soil particles from the root and increase the surface area of the plant roots. This enhances extraction of more nutrients and water from the soil. These fungi take carbohydrates and other organic compounds from host plants and provide acquired nutrients specially phosphorus including nitrogen, micronutrients, and water to them. So, this is a mutualistic or symbiotic relation in which both partners are benefitted.

Mycorrhizal fungi are divided in two major groups according to their position on host plants. These are Endomycorrhiza and Ectomycorrhiza. These are described here briefly.

1. Endomycorrhiza: The hyphae of endomycorrhiza enter in to the plant root cells and grow there. The Phycomycetous fungi of Endogenaceae family have symbiotic association with various agricultural and industrial crops. Other crops having this association in their roots are maize, wheat, sugarcane, rice, potato, beans, vegetables, shrubs, apple, coffee, poplar etc. Phoma, Pythium and Rhizoctonia are examples of such fungi. Spores of these fungi are present in soil. Spores germinate when suitable host and favourable conditions are available and develop their hyphae. Endomycorrhiza are obligate symbionts.

Endomycorrhiza are also called as vesicular arbuscular mycorrhizae (VAM) as vesicle and arbuscle are two main morphological structures which name fungi vesicular arbuscular mycorrhizae. Vesicles are present with in cells or in intercellular spaces. These are storage bodies for lipids and also important for reproduction.

Arbuscle are intracellular which provide nutrients to fungi from roots of host plants. Arbuscules are club-like, branched filaments within root cells where the transfer of energy and nutrients takes place.

2. Ectomycorrhiza: Another mycorrhiza is the ectomycorrhiza, which grow on the surface of woody plant roots and are frequently seen on the alder, beech, birch, eucalyptus, oak, pines etc. trees of Fagaceae and Pinaceae families. The Amanita, Boletus etc. fungi have symbiotic ectomycorrhizal associations with above and other trees.

Fungi do not enter in root cells. But between epidermal cells fungi form a mantle around the root surface. Fungal hyphae also enter into the soil and increase nutrient and water absorption area. Hyphae in this association also enhance nutrient uptake, decompose organic matter, and improve soil structure. Many ectomycorrhizal fungi can be cultured without the help of host plant.

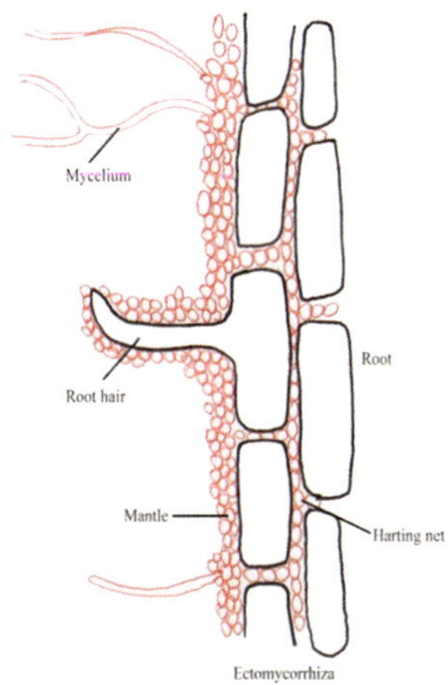

Fig. 3.18: Ectomycorrhiza

3. Pathogenic or parasitic Fungi: Pathogenic or parasitic fungi are reverse of mycorrhizae in action. Their colonization may decrease plant growth and yield even death of plants. Blights, rusts, smuts and wilts are pathogenic or parasitic fungi which cause remarkable loss in agricultural production. However, many fungi can be used for biological control of nematodes, worms, insects etc.

Algae: Algae come under soil microflora. Some algae come under kingdom Chromista (Brown Algae, Diatoms etc.) and some under Kingdom Plantae (Red and green Algae) of unicellular or unicellular colonies of invisible Eukaryotes. Soil algae are main component of soil microorganisms. They help in the nutrient cycling of the soil. Algae are autotrophic due to the presence of chlorophyll for photosynthesis and eukaryotes. Soil Algae are found in nearly all cultivated soils but most active and abundant in wet soils. In dry soil, the water film that surrounds soil particles becomes too thin for their free movement. Algae remain close or very near to the soil surface due to their need of light for photosynthesis. The most common soil algae are: green, yellow-green and golden- brown or diatoms. Green algae also have symbiosis with fungi, lichens, moss and ferns and fix atmospheric nitrogen. In paddy fields algae can fix atmospheric nitrogen from 15 to 40 kg/year. Algal counts vary from hundred to millions per gram soil.

Algae play important role in rock weathering and soil formation, soil fertility, saline and alkaline soil reclamation and aggregate stability. Saline and alkaline soils have more pH and salts and less organic matter. Autotrophic algae use soil salts for their growth and development. Thus, reduce salt content in soil and their living or dead cells increase soil organic matter. Also, algae use atmospheric Carbon di oxide for photosynthesis which makes carbonic acid with water. This acid reduces soil pH and increase nutrients availability. Algae also release polysaccharides which increase soil aggregation and aggregate stability.

Virus: Soil viruses are also considered as soil microflora. Soil virus is too small and visible under electron microscope much abundant in number. Soil viruses are reservoir of genetic diversity and play important role in geochemical processes. They always require host for multiplication due to their obligate parasitism. These are also called as bacteriophage when parasites on bacteria. Soil colloids help in the retention and spread of virus as they are adsorbed on soil colloids. Soil viruses may influence the soil organisms by transferring genes from host to host or other and causing their death.

Various viruses can affect plant growth in different manners. Soil-borne pathogenic viruses have harmful effect on plants. These viruses may remain freely in soil or associated with nematodes and other microorganisms already present in soil. Virus can infect soil microorganisms and change their functioning. The changes may be positive or negative on plants depending on the nature of infected microorganism. Virus can recycle nutrients within soil microbes also.

Soil Macroflora

Soil macroflora are the part of Kingdom Plantae i.e., kingdom of multicellular Eukaryotic producers. Mostly roots of higher plants and plants' bulbs, corm, rhizome, tubers etc. are considered as soil macro flora. Higher plants provide a large quantity of organic residue to the soil. Tap roots and adventitious roots are two types of roots of dicot and monocot plants respectively. Tap roots are long, thick and penetrate deeply in soil. Many small roots are attached with tap root. Adventitious roots are small, thin, do not go deeper in soil and many roots emerge from same place.

The roots and organic residue after decomposition increase organic matter status of soil and provide energy and nutrients to soil organisms. Roots assist in rock and mineral weathering, organic matter accumulation, cation exchange capacity, water holding capacity, soil structure and aggregation.

Roots consume oxygen, water and nutrients and release CO_2 and exudates. Root exudates are mixture of organic acids, sugars and other soluble plant compounds that escape from roots. Root exudates provide food and energy to soil flora and fauna. Roots influence microbial activities especially in the rhizosphere. Rhizosphere means the soil zone under the influence of roots where living bodies and microorganisms interact effectively. In this zone different organic acid, proteins and mucilage are released from new and smaller roots in to the soil. Therefore, there is tremendous increase in the population of micro-organisms near roots (1-2 cm from root surface) as compared to other places. It was observed that rhizospheric zone may have 13 times more fungi, 83 times more actinomycetes and 271 times more bacteria than non rhizospheric zone.

2. Soil fauna: Soil fauna come under Kingdom Animalia. This kingdom has multicellular Eukaryotic consumers. Soil animals provide very less organic matter to soil. However, their biting, cutting and chewing activities divide soil organic matter into smaller particles. Smaller particles have more exposed surface area for microbial attack. Soil fauna are also divided in to soil micro fauna, soil meso fauna, soil macro fauna and soil magna faunas well as in to soil invertebrates and soil vertebrates. Soil vertebrates are rat, mice, gopher, porcupine etc. These are described here:

Soil micro fauna: Soil micro fauna are small animals with less than 0.2mm length and less than 0.1 mm width. Protozoa and nematodes come under soil micro fauna.

Protozoa: Protozoa are soil micro fauna and come under kingdom Protozoa. Protozoa are eukaryotic, motile, single-celled organisms. Protozoa form cysts due to soil drying and depletion of food. Protozoa are heterotrophic and ingest their food. Euglypha also live autotrophically due to the presence of chlorophyll. Bacteria, fungi, actinomycetes, algae, nematodes and even protozoa are their food. Therefore, protozoa make a balance among the population of all soil microorganisms.

Protozoa are divided in to three groups on basis of their motion. These are amoeba when move by pseudopodia or cytoplasmic waves, flagellate when move by flagellum and ciliate when move by cilia.

Generally, Protozoa are found near roots and places of bacterial accumulation and thrive on them. This may affect plant nutrients cycle in soil. But their lesser number does not remarkably affect nutrients availability or organic matter decomposition. Population of protozoa increases with the increase in bacterial population. Nitrogen requirement of protozoa is lesser than many bacteria. Hence protozoa after their death and decomposition provide more nitrogen to soil as compared to other nutrients. Many diseases of animals and human being are caused by protozoa.

Nematodes: Nematode is also called as threadworm (Nema means thread), roundworm or eelworm and belongs to soil micro fauna under Phylum Nematoda. Nematodes are microscopic to few millimetres in length. They are found in almost all soils. They thrive well in the water-filled pores of soil and most active at rhizosphere.

Nematodes show three types of feeding behaviour. Mostly they are saprophytic and feed on decaying organic matter. The predator nematodes feed on other nematodes or bacteria, fungi, and protozoa. Parasitic nematodes are of two types, ectoparasite and endoparasite. Ectoparasites remain in the soil and feed on the root surface. Endoparasites enter in roots by puncturing and cutting root cells and tissues with their needle like mouthpart and live and feed within the root. They infest roots.

Root infestation increases the possibility of other diseases in plants. Summer ploughing, use of organic manures, cultivation of resistant crops etc. controls parasitic nematodes. Also soil compaction reduces population of nematodes as it reduces space among soil aggregates for their movement.

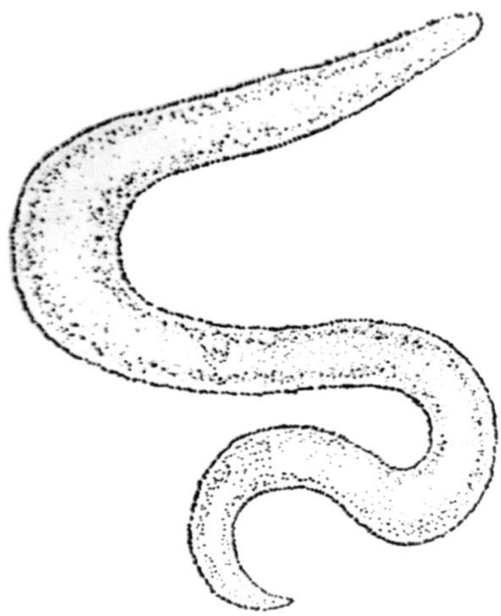

Fig. 3.19: Soil Nematode

Soil Meso Fauna

Organisms of soil mesofauna group are larger than soil microfauna. The size varies from 0.2 to 10 mm in length and 0.1 to 2 mm in width. Soil mesofauna are mostly arthropods including beetle, caterpillar, insects, wood louse, mites, myriapods, spiders, springtails etc. They regularly used to feed on roots and stems of plants growing on soil. They also help to improve soil structure, root elongation, water infiltration, drainage, and aeration. Their importance in minimal disturbed soil, forests, rangelands, perennial croplands etc. is well known. Here they also help in decomposition of organic matter by shredding plant residues, mixing them with soil, feeding on them and stimulating decomposition within their guts. Few soil meso fauna are described here:

Springtails feed on plant roots, decaying vegetation, fungi and nematodes also. They can rapidly grow in aggregations with favourable conditions. Springtails are major food source for mites, beetles and ants.

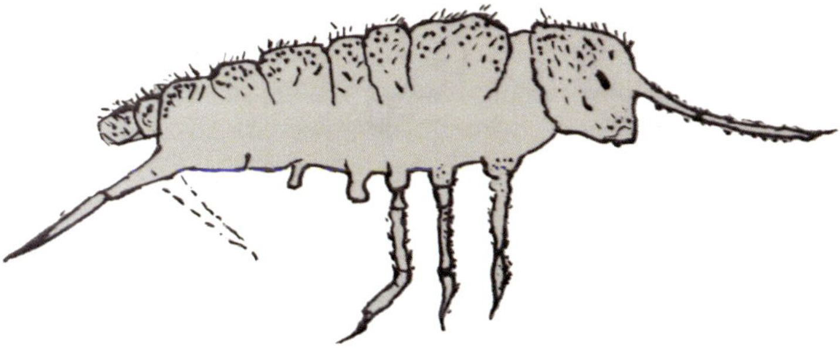

Fig. 3.20: Spring Tail

Termites are social insects and famous for invasion on stems and branches of trees and completely destroy them. Their habitat is called as termitarium is made up by moving large quantities of soil from lower depths to soil surface to form 3-4 meters high mounds. Termitaria are so strong to bear excessive rainfall due to addition and mixing of self-secretions with soil and organic residues.

Termites are most important for decomposing woods in grasslands and forests due to their ability to digest cellulose. This is because of colonization of cellulose digesting microorganisms in their intestine. Besides, wood termites also feed on plant residues, humus, and fungi.

Fig. 3.21: Termitarium

Ants are social insects found in soils. They feed on dead organisms, organic residues, seeds, bio secretions etc. and collect these in their burrows and nests. They form small surface mounds by bringing soil from lower depths (some time up to 3 meters deep) on the soil surface. Mites are commonly found in many soils including moist soils rich in organic matter. Spiders present in soil and surface litter layers are major predators of soil insects. They also feed on fungi, algae, nematodes etc.

Fig. 3.22: Mite

Beetles are the most diversified group of mesofauna present in soil. Beetles of the soil can be divided into six as ground beetles, plant feeding beetles, predators, rove beetles, saprophytic beetles and tiger beetles. Tiger beetles are predacious for other organisms. Wireworms are larvae that feed on roots. Dung beetles Generally, feed on dung and excreta of large animals. Dung beetles with the help of their mouth and legs move dung or excreta by rolling and collect it into soil. This increases organic matter status of soil.

Fig. 3.23: Photo of dung beetle (*Courtesy* Uttarakhand Darshan)

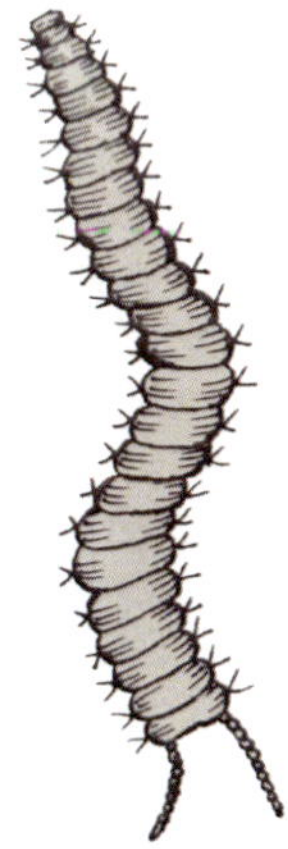

Fig. 3.24: Millipedes

Soil Macrofauna

Soil macrofauna are large organisms and also called as megafauna. Their length and width are more than 10 mm and 2 mm respectively. Soil macrofauna include myriapods, annelids, millipedes, centipedes, earthworms etc. and vertebrates like mice, mole etc. Soil macrofauna mix soil through their burrowing activities and help in soil formation.

Millipedes and centipedes are myriapods. Millipedes are herbivorous and Generally, feed on plant roots, bulb, corm etc. and calcium rich decaying organic residues. Soil microorganisms grown on excreta of millipedes help in organic matter decomposition. Centipedes are carnivorous and feed on various insects and micro arthropods present in soil and surface litter.

Annelids are soil macrofauna. They are segmented worms. They are detritivores. Neeris and earthworms are in this group. Earthworms are the most famous soil macrofauna. Farmers know importance of earthworms in agriculture. They are natural tillers of soil and consume only soil and soil organic matter. The length is 10-15 cm and colour vary brownish grey and bright blue in Philippines also. There are hundreds of species of earthworms but most common are *Phertima posthuma, Lumbricus terrestris* and *Allolobophora caliginosa*. Medium textured, well aerated, moist, calcium and organic matter rich and moderately acidic to slight alkaline soils are good for earthworms.

Earthworms eat particles of silt, clay and organic matter of soil during their entry in soil. These particles are ground and mixed well in their alimentary canal. This material is deposited on soil surface in the form of casts. In these casts nutrients present in organic and mineral matters of soil are converted in to plants available forms due to the action of digestive secretions of earthworms. These casts are also rich in organic matter, microbial populations and nutrients. Earthworms improve nutrient cycling, root penetration and elongation due to presence of small burrows, aeration, aggregation, infiltration and water holding capacity of upper layer of soils. Burrows of earthworms are Generally, localized in upper layer of soils.

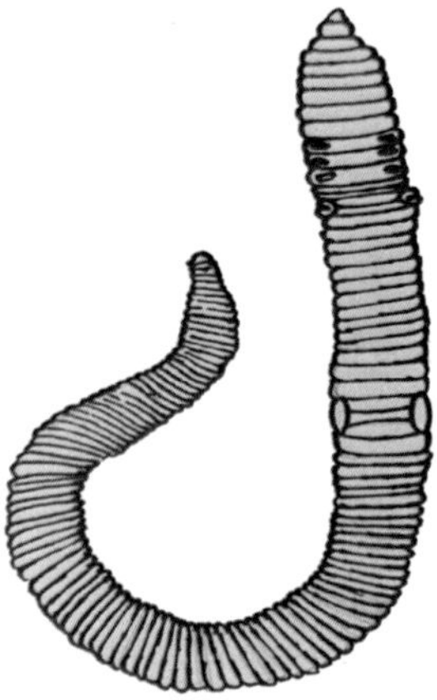

Fig. 3.25: Earthworm

Soil Magna fauna

Soil magna fauna are come under Kingdom Animalia and divided in to vertebrate and invertebrate.

Vertebrate: Most common vertebrates found in soils are rats, mice, gopher and porcupine. These are burrowing animals and redistribute soil materials by making burrows of different size and shape. They also store significant amount of plant materials and seeds in the subsurface soil. Excreta and dead bodies of these animals contribute to organic matter and plant nutrients to subsurface soil. This enhances weathering of bedrock and formation of new soil to support growth of vegetation.

Invertebrate: Molluscs are invertebrate and slugs and snails are examples. Molluscs are herbivore, fungivore, predator and detritus. But mostly feed on surface vegetation including soft roots and enhance organic matter of surface soil.

Fig. 3.26: Mollusc

Biologically Active Chemicals in Soil

Numerous biologically active chemicals are also found in soil. These may be extracellular or produced by soil microorganisms. Antibiotics, humic substances, plant growth regulators, natural insecticides, herbicides and pesticides, soil organisms' stimulants, vitamins and soil enzymes are some examples of biologically active chemicals in soil. Among these soil enzymes are most important which are described here,

Soil enzymes play a major role in agriculture by transferring energy through organic matter decomposition and cycling of nutrients in soil. Soil microorganisms are the primary source of soil enzymes besides plants and animals. Soil enzymes are adsorbed by soil organic matter and clays. Therefore, remain active and viable for longer periods. The cropping systems, crop management, nutrient management practices, elevated carbon dioxide, temperature, drought and factors affecting soil microorganisms also affect production and activity of soil enzymes.

Amylase, cellulase, chitinase, dehydrogenase, hydrolase, glucosidase, lipase, phosphatase, protease, phytase, sulphatase, urease etc. are some examples of soil enzymes. Urease is a main and permanent enzyme present in soil. It converts urea in to ammonium carbonate by hydrolysis process which changed to ammonium hydroxide and ultimately to plant available ammonium and nitrate.

Factors Affecting Soil Organisms

There are many factors which directly or indirectly affect the growth, development and activity of the soil organisms in general and soil microorganisms in particular are:

1. **Soil Moisture:** Water is very essential for the growth and survival of all living organisms. Soil moisture affects the soil organisms in many ways. Very dry conditions are not good as plants may not be able to uptake

sufficient water from the soil and may wilt permanently. Moisture little more than field capacity (1/3 bar) is better for soil microorganisms. Most of the soil microorganisms are aerobic and they perform best at this soil moisture status.

Under excess moisture i.e., wet conditions water logging will arise. This will destroy all aerobes due to lack of oxygen and air. In such anaerobic conditions anaerobic organisms will develop profusely. More biomass and biodiversity are found under moist climatic conditions as compared to dry climatic conditions.

2. **Soil Temperature:** Temperature directly affects the number and activity of the soil organisms as too much cold or hot condition is not good for them. Generally, soil organisms may be active in the temperature ranging from less than 5^0C to more than 60 ^{0}C. Very hot springs and vents record remarkably higher temperature. Highest temperature that could be tolerated by microorganisms is around 120 ^{0}C.

   But most of the organisms are mesophiles and they thrive well in the temperature range between 15 to 45 ^{0}C. Temperature from 25 to 37 ^{0}C is best suitable for mesophiles. This range is Generally, obtained during rainy season along with sufficient moisture enhance growth. Temperature above 45^0 C kill mesophiles but it is good for thermophiles whose counts and activities increase tremendously. Similarly, psychrophiles can complete their lifecycle at the temperature below 10 ^{0}C and can survive temperature below 5 ^{0}C.

3. **Soil pH:** Most of the soil organisms grow and develop well with their full activities at neutral pH as pH 7.0 is proper internal cellular pH. However, pH ranging from 5.5 to 8.0 is also good for them. Some species of organisms can tolerate very extreme acidic or alkaline conditions which are not found in agricultural soils. Larger soil organisms are more sensitive to pH than microorganisms.

   Earthworms are more active in agricultural soils as compared to forest or peat soils due to their susceptibility to pH less than 5. In soils bacteria in pH 6.5 to 8.0, fungi 4.5 to 5.5 and actinomycetes in 6.8 to 8.8 are more active.

4. **Soil aeration:** Soil organisms are mostly aerobic. They require oxygen for their respiration and release Carbon di oxide in soil. This will increase CO_2 status of soil which will kill aerobic organisms. Therefore, good drainage and renewal of soil from atmospheric air is essential.

5. **Soil mineralogy:** Microorganisms are also influenced by the soil mineralogy. There are many studies indicating that rocks phosphate, mica, basalt etc. with different composition influence growth and activities of

different soil organisms including bacteria and fungi. Chemoautotrophs and fungi weather soil minerals foe direct source of energy.

6. **Availability of food and nutrients in soil:** Major part of soil organisms is heterotrophic in natures which get their food from organic matter decomposition. Therefore, enough organic matter as manure, biochar and bio solid, compost, humus etc. should always be present in soil. These substances are direct source of carbon for soil organisms. Similarly, many soil organisms take food and energy from simple inorganic compounds. Therefore, nitrogen, phosphorus, sulphur and micronutrients should be present in sufficient quantities in soil. Deciduous forests and grasslands are good for soil organisms as they increase organic matter and nutrient status of soil.

7. **Presence of other organisms:** Struggle for Existence rule is also applicable on soil organisms as they require food, nutrition, space, air and light. This creates interspecies competition among plants and other organisms in soils. This competition is stronger in the rhizosphere and develops many antagonistic and symbiotic relationships. These relationships influence the number and activities of organisms up to some extent. Protozoa feed on other soil organisms and springtail compete with arbuscular mycorrhiza are example of antagonism. Similarly symbiotic relationship is nitrogen fixation by bacteria and nutrient solubilisation by mycorrhiza.

8. **Sun light:** Direct sunlight is harmful for dark loving organisms present on the soil surface. Soil solarisation (soil heating by sun) is done to kill or decrease many biotas of soil. Some parts of the light spectrum are more harmful to organisms than others. Ultraviolet radiation can damage DNA which induces mutations or biochemical changes in the organisms to kill them. Photoautotrophs (plants, algae and cyanobacteria) use photosynthetically active radiation of sunlight to synthesize their food as carbohydrates.

Nitrogen Fixation in Soil

Nitrogen is essential for life on earth. Sufficient nitrogen is present in atmosphere, soil, rocks and sea, but plants or animals cannot use it directly as it does not exist in plants or animals utilizable forms. Plants use inorganic forms (nitrate and ammonium ions) of nitrogen to build their cells and ultimately proteins. Animals use proteins synthesized by plants as essential nitrogenous substance for their bodies.

Inorganic forms of nitrogen used by plants remain sufficiently in soil in organic forms as plant and animal residues. These residues are converted in to ammonium salts and nitrates by microbial activities. Simultaneously nitrogen is continuously lost from soil due to denitrification, leaching and plant uptake. So, in agricultural lands output of nitrogen is often more than its input creating its depletion in soil. However, equilibrium is established after some time at a lower level.

This level of equilibrium depends on nature of soil, crop, interculture operations and climate. After this new equilibrium, for obtaining good yields of crops, supply of nitrogen from external sources is necessary. These external sources may be nitrogenous fertilizers, organic manures and biological fixation of nitrogen. Production of nitrogenous fertilizers requires lots of money, energy, space, specific industrial infrastructure, processes etc. Therefore, cost of production of these is more and prices are high due wider gap between demand and supply. Also nitrate is highly leachable and pollutes the water bodies. Therefore, use of nitrogenous fertilizers is not economically beneficial. This automatically increases the importance of biological nitrogen fixation.

Biological Nitrogen Fixation

In context of economy and fertility of nitrogen little attention was paid to nitrogen fixers as they consume more energy and fix less nitrogen. Heterotrophic nitrogen fixers consume at least one kilogram of carbohydrates and fix merely 25 gm nitrogen. There is always shortage of readily oxidizable carbohydrates in most of the soils. Also, in soils the microbes have hard competition with each other for organic substance. More availability of nitrogen in soil reduces their efficiency besides the lesser numbers of nitrogen fixers. All these factors remarkably reduce nitrogen fixation. However, increasing prices and lesser availability of nitrogenous fertilizers show the importance of biological nitrogen fixation.

Biological nitrogen fixation was discovered by German agronomist Hermann Hellriegel and Dutch microbiologist Martinus Beijerinck. Triple covalent bond, $N \equiv N$ of inert nitrogen gas (N_2) has bond energy of 942 KJ/mol, is highly stable and resistant that cannot be broken in normal conditions. Haber-Bosch chemical process to break and fix atmospheric nitrogen requires a catalyst or container made up of iron or ruthenium (Ru) inside temperature more than 426 °C (450-500 °C) and pressure about 200 atmospheres, intense reducing agent H_2 and exclusion of oxygen. These force nitrogen and hydrogen to react together and convert them into liquid ammonia. The liquid ammonia is then used to manufacture nitrogenous fertilizers.

But many microorganisms break this $N \equiv N$ bond in very much normal and natural condition and fix nitrogen in their cells. Organic combinations of molecular nitrogen or its biological conversion in plant usable forms is called biological nitrogen fixation.

Necessities or Requirements of Biological Nitrogen Fixation

Biological nitrogen fixation is done by various micro-organisms. These micro-organisms reduce nitrogen into ammonia (NH_3) by their activities. Some main requirements for this reduction are enzyme, reductant, energy and partial pressure. These are described here in short:

This reduction is enzymatic and in this process enzyme nitrogenase is used. Nitrogenase contains water soluble and easily separable two protein complexes namely dinitrogenase and dinitrogenase reductase. Both are active in the presence of each other. Their sizes vary with the microbial source.

These complexes are molybdenum iron protein (MoFe protein) and iron protein (Fe protein). Molecular weight of MoFe protein is more (about18000 to 22000 Dalton). There are 28 to 32 atoms of Fe and 28 atoms of acid labile sulphur with two atoms of Mo. It is also called FeMo cofactor P1or dinitrogenase. This part of nitrogenase enzyme takes active part in nitrogen fixation, because substrate sites are present on it.

Molecular weight of Fe protein is less (nearly 51000 to 65000 Dalton). This is nonheme Fe complex which is also called P - cluster, P2 or dinitro reductase. There are four atoms of acid labile sulphur with four atoms of Fe. This part of enzyme gives essential electrons for reduction and adenosine triphosphate (ATP) is attached with it.

Second necessity of biological fixation is reductant or reducing agent with less redox potential, Eh $\leq$ 4000mv. This reduction is also electron transporting agent. Ferredoxin, FeS protein, Flavodoxin and Azoflavin are included in this group.

There is also necessity of energy for biological nitrogen fixation. This energy is obtained from Mg salts of ATP. About ten atoms of ATP are required to manufacture one molecule of ammonia.

Fourth necessity of biological nitrogen fixation is low partial pressure of oxygen on substrate site of nitrogenase. Nitrogenase enzyme is very much sensitive for oxygen. Its activity is reduced in the excess of oxygen. So, it is essential to protect nitrogenase from oxygen. But most of the nitrogen fixing microorganisms are aerobic. Therefore, to fulfil their oxygen requirements the nature has provided them oxygen buffer, respiration protection and change in internal morphology. These provisions stop contact of enzyme with oxygen.

Biochemistry of Biological Nitrogen Fixation (BNF)

In the biological nitrogen fixation (BNF) process atmospheric nitrogen gas is reduced to ammonia by the enzyme nitrogenase. This conversion of nitrogen to ammonia is an exergonic reduction reaction. The BNF produces two moles of ammonia from one mole of atmospheric nitrogen gas by expensing 16 moles of ATP and the supply of electrons and hydrogen ions (protons).

Ammonia is very first stable product of nitrogen fixation. Only nitrogenase enzyme has the capacity to perform this activity in the nitrogen cycle. But after so many years of research the actual mechanism of nitrogen fixation is still not very clearly known because of the complexity of the FeMo-cofactor of nitrogenase. There are four steps in the process of biological nitrogen fixation which add hydrogen at each step from nitrogen to ammonia and then ammonium. These steps are:

1. **Conversion of atmospheric nitrogen into diimide or diazene**
2. **Conversion of diimide into hydrazine**
3. **Conversion hydrazine into ammonia and**

$$N \equiv N \xrightarrow{2^{e-}\ 2H^+} HN = NH \xrightarrow{2^{e-}\ 2H^+} H_2N = NH_2 \xrightarrow{2^{e-}\ 2H^+} 2NH_3$$

(Dinitrogen) (Hydrazine) (Diamide) (Ammonia)

The intermediates of N_2 fixation.

However, over all reaction in the form of equation can be expressed as:

$N_2 + 8H^+ + 8e^- + 16ATP \rightarrow 2NH_3 + H_2 + 16ADP + 16P_i$

4. Conversion of ammonia into ammonium: This ammonia is immediately protonated to form ammonium (NH_4^+) ion. But ammonium is toxic to plants except rice. This ammonium quickly reacts with various organic acids present in plants and synthesises amino acids and proteins. Amino acids synthesis takes place by three processes namely

1. Reductive amination
2. Catalytic amination and
3. Transamination

1. Reductive amination: In this process ammonium in the presence of the glutamate dehydrogenase enzyme reacts with α-ketoglutaric acid and form glutamic acid as:

α-ketoglutaric acid + NH_4^++ NADPH → glutamic acid + H_2O + NADP

2. Catalytic amination: It is a two steps process catalysed by glutamine synthetase (GS) and glutamate synthetase (glutamine – 2-oxyglutarate aminotransferase, or GOGAT).

Glutamate + NH_4^++ATP → Glutamine +ADP + Pi

Glutamate + 2- oxyglutarate + NADH + H^+ → Glutamate +NAD where

ATP Adenosine triphosphate, ADP Adenosine diphosphate, P Phosphorus, NADH Nicotinamide Adenine Dinucleotide + Hydrogen (H), NADP Nicotinamide Adenine Dinucleotide + Phosphorus (P) and NAD Nicotinamide Adenine Dinucleotide.

3. Transamination: Transamination produces other amino acids from Glutamic acid or Glutamate. The enzyme transaminase catalyses all such reactions. Transamination involves transfer of amino group from amino acid to oxaloacetic acid and α-ketoglutaric acid to form aspartic acid and glutaric acid, respectively. Both Aspartic acid and Glutaric acid produce many amino acids and amides which are transported to the different parts of the plants.

Organisms Involved in Biological Nitrogen Fixation

Many organisms mostly microorganisms independently or in association with plants fix atmospheric nitrogen are called as nitrogen fixers. The nitrogen fixers can be classified as:

1. Nitrogen fixers: These are divided into two 1. Asymbiotic / Free living / non-symbiotic fixers and 2. Symbiotic fixers

Asymbiotic fixers: These are divided into two 1. Heterotrophic bacteria and 2. Autotrophic Cyanobacteria

Heterotrophic bacteria: These are divided into four 1. Aerobic 2. Facultative aerobic 3. Obligate aerobic and 4. Anaerobic

Symbiotic fixers: These are divided into two 1. Leguminous and 2. Non leguminous

Leguminous: These are divided into two 1. Nodules on roots and 2. Nodules on stems

Types of Biological Nitrogen Fixation

Based upon the classification of nitrogen fixers the biological nitrogen fixation may be of two types 1. Asymbiotic Nitrogen fixation and 2. Symbiotic Nitrogen fixation. Here both are briefly described.

Asymbiotic Nitrogen Fixation

In asymbiotic fixation of nitrogen micro-organisms does not require any association with plants. They live their lives independently and fix the atmospheric nitrogen. Although in 1862 Beijerinck, knew about nitrogen fixing micro-organisms. But in 1894 Winogradsky also searched and identified nitrogen fixation.

Nitrogen Fixation by Free-Living Bacteria

Azotobacter, Azospirillum, Beijerinckia and *Derxia* are aerobic nitrogen fixers. Azotobacter is most important among asymbiotic nitrogen fixing bacteria. There are many species of Azotobacter. Among them *Azotobacter chroococcum* is most superior as it fixes about 20 kg nitrogen/ha/year. Azotobacter requires good aeration because its respiration rate is 10 to 15 times more than other aerobes.

Excess oxygen reduces activity of nitrogenase which is overcome by covering oxygen sensitive sites by another protein. This protein is not a part of nitrogenase enzyme. Thus, activity of enzyme is reduced temporarily due to excess of oxygen. When oxygen is utilized by respiration then enzymes again restore its highest activity. Presence of acetic and butyric acids, 25-30°C temperature, neutral soil (pH 7-8) and phosphorus, calcium and molybdenum in appropriate quantities are beneficial for its good growth.

Application of Azotobacter with cellulose decomposing microbes in soils can be more beneficial because cellulose decomposers will decompose substances with cellulose into simple compounds. Azotobacter will utilize these compounds for their source of energy.

Azospirillum has several species. Its application in cereal crops is also beneficial. *Beijerinckia* and *Derxia* grow well in acidic soils. In acidic soils, when Azotobacter is not able to fix nitrogen then *Beijerinckia* successfully perform this work even at very low pH. It can also develop at pH 3. Generally, it occurs in temperate soils but present in Indian soils etc.

Bacillus, Klebsiella and *Enterobacter* are facultative nitrogen fixers and grow well in both aerobic and anaerobic conditions. They fix nitrogen in absence of oxygen. *Klebsiella* does not manufacture nitrogenase enzyme in the presence of oxygen.

Rhodospirillum, Chlorobium and *Disulphovibrio* are obligate aerobic nitrogen fixers of lesser importance.

Clostridium plural *Clostridia* are anaerobic nitrogen fixating bacteria. First of all, in the year 1894 Winogradsky searched nitrogen fixing property in *Clostridium pasteurianum. Clostridium butyrium* and *Clostridium acetobutylicum* also fix nitrogen. They grow very well in acidic pH, but their growths have been observed in alkaline pH also. They grow very rapidly when organic matter is added in soil. Clostridiums are also found in excess near root of plants. They often develop along with saprophytic bacteria, which use oxygen and create anaerobic condition for Clostridiums.

Some photosynthetic bacteria also have property of nitrogen fixation. These bacteria are non-sulphur violet bacteria, *Rhodomicrobium, Rhodopseudomonas* and *Rhodospirillum*, sulphur violet bacteria, *Chronaticum* and *Ectinorhodospira* and green sulphur bacteria *Chlorobium*. Light is beneficial and presence of oxygen is very much harmful for them. Although aforesaid all asymbiotic bacteria fix very less amount of nitrogen but their nitrogen fixing capacity may be increased in future by biotechnology.

Autotrophic Cyanobacteria

Autotrophic cyanobacteria are also commonly known as Blue Green Algae. Green algae also have symbiosis with fungi, lichens, moss and ferns and fix atmospheric nitrogen. Everyone can see these algae in paddy fields. Because these make their food by photosynthesis, so there is no shortage of energy. By this reason these fix nitrogen in more quantity than asymbiotic bacteria. In Indian condition quantity of fixed nitrogen by these is 14 to 50 kg/ha. Besides of nitrogen they also provide oxygen and organic matter in more quantities. In water logged conditions excess of oxygen promotes nitrification but in normal conditions excess of oxygen and organic matter are beneficial for Azotobacter.

These algae also secret root growth promoting hormones. More growth has been seen in roots of seedlings treated by extracellular products of *Aulosira fertilissima* algae. Some blue- green algae emit ascorbic acid. It is supposed that ascorbic acid is directly beneficial for paddy crop or indirectly by taking part in nitrogen fixation and nitrate reduction processes.

Cyanobacteria or blue green algae are found at every place where sun light is available for photosynthesis. But excess sun light is also harmful for these. Algae can grow successfully on rocky surfaces, waste lands and virgin soils because of their cell formation ability with the help of light, carbon dioxide, nitrogen, water and other nutrients. Neutral pH is good for algae. Dry conditions destroy algal cells. But availability of moisture in low rainfall area rapidly enhances their growth.

Generally, counts of *Anabaena, Nostoc* and *Calothrix* algae are more in our country. Anabaena and Nostoc like filamentous blue green algae also have some special cells. These cells are called as heterocyst. These are present in algal filaments at the regular intervals. Colour of heterocyst is some lighter than normal cells. These cells fix atmospheric nitrogen. The heterocysts have some specific arrangement to protect oxygen sensitive nitrogenase from oxygen produced by photosynthesis. Nitrogenase of Anabaena is iron combined metallic protein. Filamentous *Lyngbya* and unicellular *Gleocapsa* can fix nitrogen without heterocysts.

Factors Affecting Asymbiotic Nitrogen Fixation

Asymbiotic nitrogen fixation is performed by various groups of microbes. These microbes are very much influenced by nearby edaphic, physical and chemical conditions. Their own specific requirements and diversified environmental conditions significantly affect nitrogen fixation. However, major factors affecting asymbiotic fixation of nitrogen are, quantity of nitrogenous compounds in soil, quantity of nutrients in soil, soil reaction, quantity of organic matter in soil, quantity of moisture in soil and soil temperature. These factors are briefly described here:

Quantity of Nitrogenous Compounds in soil: It is well established fact that inorganic forms of nitrogen present in soil can inhibit N_2 fixation by asymbiotic or nonsymbiotic microorganisms. Medium or more concentration of ammonium or nitrate compounds reduces nitrogen fixation. More quantity of urea is harmful, but effect of amino acids is less harmful than urea.

Quantity of Nutrients in Soil: Asymbiotic nitrogen fixers need many nutrients for their growth and development to fix nitrogen effectively. More amounts of Cu, Ni, Cd, Cr, Pb, Hg, As etc. can reduce their population and Nitrogen fixation. In addition to phosphorous, potassium and calcium, molybdenum, cobalt, boron, zinc, copper and iron (Fe) are also required by nonsymbiotic nitrogen fixers.

Deficiency of any nutrient negatively influences nitrogen fixation. Molybdenum is required for metabolism of nitrogen. Microbes cannot use of nitrate if molybdate is in short supply. Azotobacter need iron (Fe) for the metabolism of nitrogen and function well in soils with higher available phosphorus status. Also, Azotobacter, Beijerinckia and Clostridium require cobalt to use nitrogen. Blue green algae like excess of calcium in soil.

Quantity of Organic Matter in Soil: Availability of carbon as an energy source is essential for asymbiotic N2 fixation. Organic matter gives energy to heterotrophic microbes. There is more fixation of nitrogen in aerobic, microaerophilic and anaerobic conditions if excess of organic matter with broad carbon: nitrogen ratios as sugar, cellulose and straw and plant residues are added in the soil. Mostly Nitrogen fixing heterotrophic bacteria can use the products of cellulose and hemicellulose decomposition.

Rates of Nitrogen fixation are proportional to the amount of crop residue available and to rates of decomposition. Generally, asymbiotic N_2 fixing microbes depend on decomposing organic matter of crop etc. present on surface or in sub surface of soil and carbon containing compounds of root exudates of plants. Azospirillum spp. can use straw directly for nitrogen fixation.

Soil Reaction: Soil reaction is very much effective on microbes. Increase in pH of soil reduces number of Azotobacter. At pH less than 6 number of Azotobacter is almost negligible and they cannot fix nitrogen also. Contrary to this Beijerinckia can effectively fix nitrogen in acidic soils with pH ranging from 3 to 9. Acid tolerance of Clostridium is in between Azotobacter and Beijerinckia. Blue green algae very well work at normal pH and if pH is less than 6, their number reduces remarkably. However, nitrogenase activities of free-living bacteria is best at pH 7-7.5

Amount of Moisture in Soil: Increase in amount of moisture in soil also increases nitrogen fixation. Higher soil water contents enhance N_2 fixation by reducing O_2 at the nitrogen fixing sites. In general. there is more nitrogen fixation at water holding capacity (field capacity). Asymbiotic microbes also fix more nitrogen in waterlogged soils. In this condition organic compounds are decomposed into simpler compounds. Nitrogen fixing microbes utilize these simpler compounds.

Soil Temperature: The most favourable temperature for Nitrogen fixation varies between 25 to 35°C in the range of 4 to 45° C. The best temperature range may vary due to the organisms present and the climatic conditions.

Biochemical Process of Asymbiotic Nitrogen Fixation

Reduction of atmospheric nitrogen in ammonia is done by enzyme dinitrogenase. Continuous research is being done to know intermediary products of this process. However, it is supposed that diamide (HN=NH) and hydrazine (N_2H_4) are intermediary compounds between atmospheric nitrogen and ammonia. Ammonia makes glutamic acid after reacting with α ketoglutaric acid. Later, other amino acids, glutamine and proteins are formed.

Symbiotic Nitrogen Fixation

Symbiotic Nitrogen Fixation is divided into two 1. Symbiotic Nitrogen Fixation by Leguminous crops 2. Symbiotic Nitrogen Fixation by Non-leguminous crops. Again, Symbiotic Nitrogen Fixation by Leguminous crops is divided into two 1. Nodules on roots 2. Nodules on stems. These are briefly described here:

All the living organisms require numerous nutrients for the growth and development during their whole life span. Nitrogen is one of them. Nitrogen is used by microbes as ammonia, nitrate or amino acids, by plants as ammonium, nitrate and urea and by animals and human beings as proteins. But these forms of nitrogen are always in shortage particularly for animals and humans. A remedy to eliminate this shortage is fixation of atmospheric nitrogen by the microorganisms and symbiotic nitrogen fixation is one of them.

Symbiosis is the living of two dissimilar organisms in association for the benefit of each other. The symbiosis may be between legume crops and bacteria, non legume crops and actinomycetes, lichens and blue green algae and blue green algae and aquatic ferns. In these associations nutrition mainly carbohydrates and favourable environment to bacteria and blue green algae are provided by host plants and in return microbes provide nitrogen fixed by them. It is presumed that in the world 20 to 30 of total amount of fixed nitrogen is contributed by legume crops. Therefore, importance of legume crops cannot be denied.

Symbiotic Nitrogen Fixation by Legume Crops

Centuries ago, benefits of symbiotic association between legume crops and nitrogen fixing bacteria were very well known and about 100 years ago microbiologists also knew it. First of all Beijerinck discovered the nitrogen fixing bacteria and named as *Bacillus radicicola*. Later on its genus was identified as *Rhizobium*. When these bacteria infect suitable legume crop, form nodules on their roots and start fixing nitrogen. The Leguminosae or Fabaceae family is also called as legume, pea, or bean family. This family includes about 751 genus and about 19,000 studied species. About 90% of legumes can develop nodules.

Rhizobia and Rhizobium Bacteria: Along with Rhizobium. Sinorhizobium, Mesorhizobium, Azorhizobium, and Allorhizobium are soil Rhizobia. Rhizobia mean a group of bacteria which can fix atmospheric nitrogen. The word Rhizobium is made up from two Greek words Rhiza and Bios referring root and life, respectively. Rhizobium is a group of gram-negative soil bacteria. These are aerobic, non-endospore former and have symbiotic relationship with various leguminous crops. These are medium sized ranging from 1.2 to 3.0 μ m long and 0.5 to 0.9 μ m wide. These bacteria are heterotrophic and feed on the remains of dead organisms.

In the soil the bacteria are free living and motile. Free living rhizobia cannot Independently fix nitrogen. They are regular in shape as straight rods. But in root nodules they are present as irregular club and Y-shaped cells called Bacteroides. Their mobility is by a single polar flagellum or two to six tails like flagella all over body. The relationship between leguminous plants or legume crops and Rhizobium bacteria is a symbiotic relationship because the bacteria and the legumes benefit each other. The rhizosphere of plants protects and provides nutrients and energy to the bacteria, whereas bacteria convert atmospheric nitrogen to ammonia which is useful for plants.

Many species of genus *Rhizobium* are identified. However, few prominent species are *Rhizobium meliloti, Rhizobium trifoli, Rhizobium legumimosarum*, *Rhizobium phaseoli, Rhizobium lupini, Rhizobium japonicum, Rhizobium alamii, Rhizobium lentis, Rhizobium metallidurans* and *Rhizobium smilacinae.*

But later on, genus Rhizobium was divided in to two genus on the basis of their growth rate in the culture medium. First genus is Rhizobium whose bacteria grow very rapidly in the culture medium. Bacteria of this genus Generally, produce nodules in legumes of temperate region. Second genus is Bradyrhizobium. Bacteria of this genus grow slowly in medium and often develop nodules in legumes of tropical region.

Cross Inoculation Group

It was noticed in various studies on different legumes and Rhizobium species that not all species are capable to produce nodules in all groups of the legumes. This emphasised to group Rhizobium into appropriate groups of legumes (cross inoculation groups). Crop inoculation group signify those legumes, in which nodules are developed on inoculation of bacteria obtained from the nodules of any member of the specific legume group i.e., species of Rhizobium produce nodules only on the roots of any legume of that group, even other legumes are growing in the field, for example *Rhizobium trifoli* will form nodules on the roots of Trifolium only even peas are growing in same field. Various cross inoculation groups are given here in Table 3.10

Table 3.10: Cross inoculation group and bacteria legume association

CrossInoculation group	Bacteria	Host Genus	Leguminous Crop
Lucerne group	*Rhizobium meliloti*	Medicago Melilotus Trigonella	Alfalfa Sweet Trifolium Fenugreek
Trifolium group	*Rhizobium trifoli*	Trifolium	Trifolium
Pea group	*Rhizobium legumimosarum*	Pisum Lens	Pea Lentil
Bean group	*Rhizobium phaseoli*	Lathyrus Vicia Phaseolus	Sweet pea Vetch Bean
Lupine group	*Bradyrhizobium Species* *Rhizobium loti*	Lupinus Lotus	Lupine Lotus
Soybean group	*Bradyrhizobium japonicum*	Glycine	Soybean
Cowpea group and miscellaneous	*Bradyrhizobium species*	Vigna Lespedeza Crotalaria Pueraria Arachis Phaseolus	Cowpea Lespedeza Crotalaria Kudzu Groundnut Lima bean

It is clear from the table that the bacteria present in nodules of Lathyrus can inoculate Lathyrus, Vicia and Phaseolus. Thus, bacteria *Rhizobium phaseoli* obtained from Vicia can form nodules in other legumes of its group.

Although the principle of cross inoculation group is not completely correct as Rhizobium bacteria also inoculate legumes of other groups and form the nodules. In Indian conditions Gaur et al. (1979) reported nodule formation in groundnut by Rhizobium bacteria obtained from 51 other species of legumes. Simultaneously they also found that Rhizobium bacteria obtained from nodules of groundnuts were invasive on 25 species of legumes. But it is not necessary that such nodule formation other than the cross-inoculation group can fix nitrogen. Mostly this type of inoculation is ineffective.

Process of Nitrogen Fixation by Symbiotic Bacteria in Legumes

The whole process of nitrogen fixation by symbiotic bacteria in legume crops can be divided in four parts as:

1. Infection of Legumes by Bacteria: Symbiotic nitrogen fixating bacteria remain in soil independently. Bacteria grow in rhizosphere of leguminous host plants before forming infection thread after attack or invasion on root hairs.

The root exudates of various leguminous plants contain dicarboxylic acids, flavonoids etc. which attract Rhizobium bacteria. When the bacteria detect these chemicals, they actively move towards and attach to the legume root. Attraction to the legume roots is followed by transcription of Nod genes in preparation of the symbiotic relationship. The flavonoids also activate the genes involved Nod factors production. Sets of genes in the bacteria control different aspects of the nodulation process. One Rhizobium strain can infect particular species of legume but not others.

These bacteria infect roots by two ways or methods. In first method Rhizobium bacteria attack or invade on roots' hairs and form infection thread. This method is applicable to most of the legumes. In second method bacteria directly attack on cortical cell of roots. This method is applicable to some legumes mainly groundnut. In this method bacteria attack at the spaces between root hair emerging sites near root hair wall, cracks where these roots emerge and adjacent epidermal and cortical cells.

2. Formation of infection thread: After attachment with the root hair, the bacteria excrete nod factors (NF). Nod factors (Nodulation factors or NF) are signalling molecules produced by soil rhizobia during the initiation of nodules on the root of legumes. These NF stimulate the hair to curl. Rhizobium then invades the root through the hair tip where they induce the formation of an infection thread (IT). Later, due to very close attachment between root hair and bacteria, curling or deformation of root hairs starts.

Sometimes curling may be more than 360 which form a loop. The attachment is done in two-steps. The bacteria first attach with the help of Rhicadhesin Ca^{2+} - binding protein. In the second step bacteria accumulate and stay at the root hair surface and firmly attach there due to lectins and/or cellulose produced by the bacteria and host plant. It is also supposed that nucleic acid and a protein or polysaccharides are beneficial to produce curling in root hairs. Also, tryptophan secreted by roots of legumes and B- indoleacetic acid secreted by bacteria produce deformation in root hairs.

After curling and deformation bacteria enter into root hair and invagination of root hair start to form a cavity. This invagination is continued, until a tubular structure is formed. This particular tubular structure is called infection thread. Wall of this hyphae like tube is formed by cellulose obtained from host plant. Wall of this tubular structure is in continuum with wall of root hair. In this infection thread number of bacteria are lesser in number and they remain in single row. This thread is constructed by the root cells and not by the bacteria and is formed only in response to infection. The infection threads grow through the root hair cells and penetrate other nearby root cells.

Infection thread containing bacteria continuously expands and reaches up to cortical cells of the root. Infection thread after reaching to the cortical cells divided in branches and release bacteria there. Cortical cells in which bacteria are released multiply very rapidly. Simultaneously bacteria begin to increase their number in cytoplasm of host cells. Thus, host cells are completely filled up with bacteria. Finally, such a structure is formed whose central part is filled with bacteria and nearby there are cortical cells with vascular system.

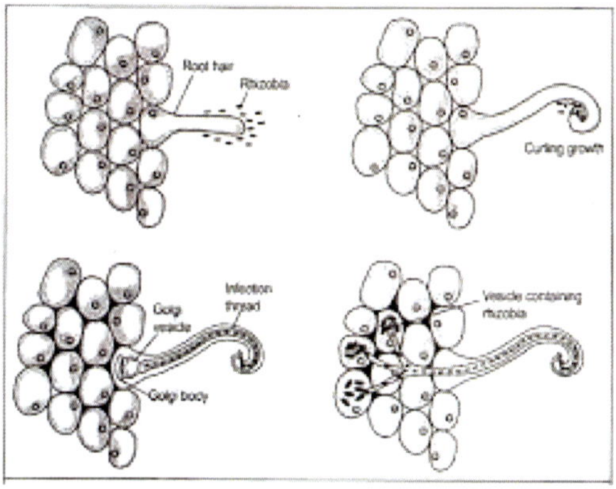

Fig. 3.27: Process or steps of nodule formation

3. Nodule formation and development: A structure which has bacteria in its central part and nearby cortical cells with vascular system is the start of nodule formation. Before reaching up to base of root hairs about 70 to 95 infections are destroyed. Therefore, extremely lesser number of curled or deformed root hairs is developed in nodules. Flavonoids, Nod factors and exchange of signal molecules are important in nodule development. In the process of nodulation number of cortical cells with bacteria of host plants increases rapidly. But these cells remain in endodermis of root only and make nodules. Within a week of infection small nodules are visible by the naked eyes.

Fig. 3.28: Nodules on one month old Ground nut and cowpea plant's roots respectively (Photo: Dr. Venkatesh Bharadwaj)

Bacteria get a specific shape or misshaped after coming in cytoplasm of host plants. Each root nodule is filled with mostly branched star or club shaped living Rhizobium bacteria which are called as Bacteroides. Difference in Bacteroides of one legume species can easily be distinguished with another species. In the nodule Bacteroides are covered with the membrane obtained from host plant cells forming a capsule like structure. These structures are known as symbiosomes and defined as specific capsule or section in host plant cells that consists an endosymbiont in a symbiotic relationship. Endosymbiont is any organism living within the body or cells of another organism that may or may not form a symbiotic relationship.

The symbiosomes contain several Bacteroides or just one depending on the host plant varying from single bacteroid in clover, peas and ground nut, four to six Bacteroides in soybean, more in cowpea and crossing the number up to fifty thousand. In the symbiosome, Bacteroides fix atmospheric nitrogen which is transported to the host plant. These Bacteroides within the capsule are drowned in the liquid containing leghaemoglobin. This makes them to work at M concentration of oxygen.

Morphology of nodules: Morphology of nodules is too much diversified. Strain of soil bacteria and variety of legume plants influence nodule morphology. In clover club shaped and lobe shaped nodules has been observed. Nodules of cowpea, groundnut and Lima bean are spherical and nodules of alfalfa are branched and long.

Size of nodules may vary from tip of needle to one centimetre. But in velvet bean size of nodules also may be as big as a ball. Numbers of nodules are less on well-developed tap root and more on fibrous root. Similarly, on roots large nodules is less in number as compared to small nodules which is more in number. Generally, large nodules are not spherical so that they have more surface area than volume. Therefore, they acquire more active and expanded area to get more atmospheric nitrogen.

Just before flowering there is maximum development of nodules and they contain maximum quantity of leghaemoglobin. Leghaemoglobin provides red tinge to nodules. Quantity of leghaemoglobin gradually decreases when nodules become older and older. This results nodule colour begins to become brown. Ultimately Bacteroides are changed in to bacteria and they use whole liquid present in cells including cell walls for their nutrition. At this stage destruction of nodules starts and bacteria again come in the soil.

Types of nodules: Different types of nodules are found in nature. But legumes show two types namely determinate nodules and indeterminate nodules and these are described here:

1. Determinate Nodules: Determinate nodules are found on soybean, bean, ground nut and lotus. The Site of initial cell division is outer cortex. The nodule growth is due to cell expansion. The meristematic activity is ceased very early after initiation. There is no active apical meristem. Thus, developed nodules are spherical in shape. Figure of Determinate nodule

2. Indeterminate Nodules: Indeterminate nodules are found in pea, alfalfa, clover and vetch. The Site of initial cell division is inner cortex. The nodule growth is due to cell division. The meristematic activity is persistent after initiation. The active apical meristem produces new cells at the tip of nodule providing cylindrical shape.

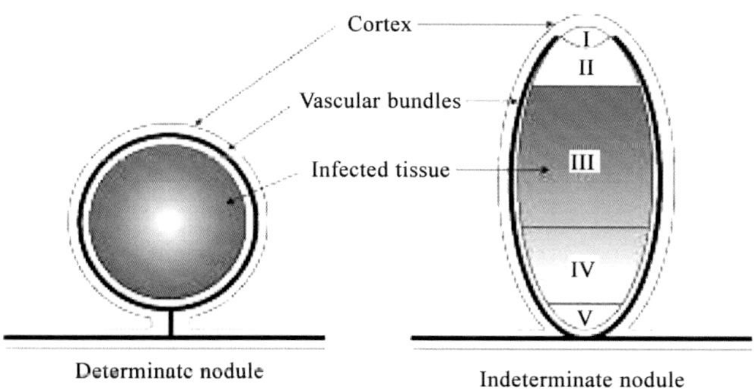

Fig. 3.29: Determinate and Indeterminate nodule

3. Nitrogen Fixation: In symbiotic nitrogen fixation infectiveness and efficiency of Rhizobium is most important. Infectiveness is the capacity of any Rhizobium strain to infect legume and form nodules. All bacteria invading legume are not infective. Thus, nodules are not formed on roots of plants. But some strains can make nodules but inefficient to fix nitrogen or fix it in very small quantity. Rhizobium which are unable to fix nitrogen are ineffective or inefficient and which fix nitrogen in small quantity are partial effective or inefficient. Effective or efficient Rhizobium can fix maximum possible quantity of nitrogen or as much as required by host plant.

Factors Affecting Symbiotic Nitrogen Fixation

In addition to infectiveness and efficiency of Rhizobium and type of legumes other factors also influence fixation of nitrogen. These factors are quantity of various nutrients in the soil, supply of carbohydrates to roots, carbohydrate: nitrogen ratio, soil reaction, soil organic matter, presence of other micro-organism in the soil, temperature, moisture, light and soil texture.

1. Type of legume crops: When all the essential requirements and conditions are optimum and favourable still various leguminous crops fix nitrogen according to their capacity. However, naturally legumes can fix nitrogen approximately from 28–90 kg per hectare per year which may go beyond hundreds of kilograms. Generally, Berseem fixes maximum nitrogen as compared to other legumes. Quantity of nitrogen fixed by various legume crops is given in the table. 3.11.

Table 3.11: Approximate quantities of nitrogen fixation (kilogram/hector) by legume crops (Please set table)

Legume Crop	Nitrogen fixation	Legume Crop	Nitrogen fixation	Legume Crop	Nitrogen fixation
Alfalfa	194	Moth	80	Guar	56
Sweet clover	119	Pea	72	Groundnut	42
Kudzu	107	Cowpea	70	Bean	40
Lespedeza	85	Soybean	58	Green gram	35

2. Quantity of Various Nutrients in Soil: Excess of nitrogen in soil as ammonical and nitrates forms reduces fixation of nitrogen. Therefore, inorganic nitrogen should be in lesser quantity in soils. Although in the initial stage of growth of legumes little application of nitrogenous fertilizers is beneficial. The presence of appropriate quantities of phosphorous and potassium in the soil does not directly influence nitrogen fixation, but indirectly these elements develop healthy and nutritive host plant for bacteria.

Calcium deficiency restricts nitrogen fixation due less growth of leguminous host plant. Calcium present in the soil helps bacteria to tolerate low pH value. Its deficiency reduces weight of nodules. Various strains of Rhizobium have different calcium requirements. In nitrogen fixation magnesium has remarkable importance. Magnesium makes easy supply of ATP to nitrogenase enzyme. Magnesium salt of ATP provides energy required for nitrogen fixation. Rhizobium of groundnut and lucerne require more sulphur for nitrogen fixation. So there would be good availability of sulphur in soil during growth period of these legumes.

Fe (iron) is very much useful in nitrogen fixation. Iron is a major constituent of ferredoxin. Ferredoxin is electrons transporter for nitrogen fixation. Molybdenum is required by legumes only for the nitrogen fixation. Molybdenum deficient soils can develop root nodules in legume crop but they are unable to fix nitrogen. Molybdenum is the part of high molecular weight constituent of nitrogenase enzyme. Boron is also essential for nitrogen fixation. A very small quantity of cobalt is sufficient for nitrogen fixation. Cobalt is a constituent of vitamin B12 present in nodules and essential for the manufacture of leghaemoglobin.

Absence of reddishness or increase of brownness in nodules is the indication of their inactivity. It is known that these nodules are not fixing nitrogen. Leghaemoglobin is found in the nodules of leguminous plants and supplies oxygen to the nitrogen fixing bacteria. Also, it protects nitrogenase enzyme to be inactivated by oxygen by maintaining low level of oxygen at substrate site. Leghaemoglobin is similar to haemoglobin of blood.

3. Supply of Carbohydrate to roots: A low carbohydrate reserve in nodules reduces nitrogen fixation. Therefore, Rhizobium bacteria require continuous supply of carbohydrate from host plant. It is supposed that nearly 18-gram carbohydrate is required to fix one gram nitrogen. Asymbiotic bacteria require about 50-gram carbohydrates for fixing one gram nitrogen because they continuously need carbohydrates for their growth and development.

4. Carbohydrate: nitrogen ratio: Excess nitrogen in soil reduces formation of nodules due to decrease in carbohydrate: nitrogen ratio. Therefore, supply of carbohydrate from roots of host plant decreases resulting reduction in process of nodulation. Wide carbohydrate: nitrogen ratio enhances nitrogen fixation due to sufficient supply of carbohydrates.

5. Soil Reaction: Soil pH remarkably influence symbiotic nitrogen fixation. *Rhizobium* can grow well between pH 5.00 to pH 8.00 and no growth was observed with pH 3.5. Generally, pH 5.5 to 7.0 is also suitable for growth of *Rhizobium*. Optimum pH range for the growth of Rhizobia is considered 6.5 to 7.5. *Rhizobium legumimosarum* grow well between pH 6.5 to 8.0. Very much acidic soils should be treated by lime and alkaline soils by gypsum or calcium carbonate before sowing for good nitrogen fixation. Pelleting of seeds by lime or gypsum will reduce ill effects of acidity or alkalinity.

Rhizobium meliloti and *Rhizobium trifoli* can tolerate more salinity. But salinity more than 3 dS m^{-1} is harmful. It is also seen that more salinity causes more harm to host plant than Rhizobium. Comparison of Rhizobia obtained from normal soils and saline soils shows that there is also localized adaptation in Rhizobium because Rhizobium present in saline soils can tolerate more salts than Rhizobium present in normal soils.

6. Soil organic matter: More nitrogen is fixed in the soil with the low soil organic matter content and the amount of nitrogen fixation increases with the decrease in soil organic matter status. Addition of different types of organic materials in the acidic and saline soils improved the shoot growth and nodule formation due to the improvement of soil physical conditions, soil moisture and nutrient uptake.

7. Presence of other Micro-organisms in Soils: Activities of other organisms present in soil can be harmful or beneficial for Rhizobium and nitrogen fixation. Just after inoculation number of Rhizobium decreases rapidly. This decrease is because of eating or consumption of the Rhizobium colonies by protozoa. But the Rhizobia which are escaped being prey of protozoa rapidly multiply their number beyond expectations. Root knot nematodes (*Meloidogyne incognita* and *Heterodera cajian*) making knots on the roots of cowpea adversely influence nodulation, growth and nitrogen fixation. *Meloidogyne incognita* is more harmful.

Rhizobium inoculation of pea, lentil and gram along with Azotobacter is more beneficial for yield as compared to only Rhizobium inoculation. Rhizobium eating bacteriophages develop in soils, if same legume crop is grown continuously in same field for many seasons. This reduces Rhizobia population and ultimately nitrogen fixation is decreased. Generally, this is true with continuous cultivation of berseem and lucerne. Various toxins are produced by bacteriophages cause harm to Rhizobium. Various agricultural chemicals used in fields have various adverse effects on legume crop, related Rhizobia and nitrogen fixation. Therefore, agricultural chemicals should be used judiciously in recommended quantities to exert negligible effect on symbiotic nitrogen fixation.

8. Temperature: Temperature is important for survival and persistence of symbiotic nitrogen fixing bacteria in soils. Higher or lower temperatures cause delayed nodule formation, development, nodule structure, functions of legumes and nitrogen fixation. However, both the legumes and their rhizobia can successfully adapt to adverse temperature conditions.

9. Moisture: There is a high degree of correlation between soil water content and N-fixing activity. Maximum N fixation occurs at about field capacity of soil and above this, the N-fixing activity may be reduced due to water logging. Slow natural drying of soil over a 6 week period may result in a progressive reduction of N-fixing activity, which can be restored by irrigation. This indicates that soil water availability is the major environmental factor affecting nitrogen fixation. It is plausible that water stress directly affects nodule activity. This effect may further be aggravated by reduced supplies of photosynthate due to wilted plant leaves.

10. Light: Light is responsible for photosynthesis as well as it is a primary regulator of symbiotic nitrogen fixation. It is understood that at a constant temperature more day time light does not reduce nitrogen fixation. But significant decrease in nitrogen fixation was recorded at short day time light.

11. Soil texture: Soil texture directly influences the symbiotic nitrogen fixation in different soil textural classes in the order: medium texture soil > heavy texture soil > light textured soil. It indicated that medium texture soils are good for nitrogen fixation.

Biochemistry of Symbiotic Nitrogen Fixation

Isotopic nitrogen is used to know some probable intermediary compounds formed during symbiotic fixation of atmospheric nitrogen. Also, it is to understand that fixed nitrogen is converted into organic molecules by bacteria or by legumes itself. Symbiotic nitrogen fixation starts after the fulfilment all the essential requirements in the presence of nitrogenase enzyme. Reduction of atmospheric nitrogen takes place gradually in diamide, hydrazine and finally to ammonia. The ammonia combines with oxalacetic acid obtained from metabolism of carbohydrate and makes amino acid like α-aspartic acid. α- aspartic acid reacts with other amino acids to form various proteins. The universal equation for nitrogen fixation may be written as:

$$N_2 + 8H^+ + 8e^- + 16ATP \rightarrow 2NH_3 + H_2 + 16ADP$$

Some quantity of fixed nitrogen directly comes in the soil from nodules amino acids. Some quantity of nitrogen transfers very rapidly to upper parts of plant. In new plants the transfer takes place in few hours. Later on, slowly and slowly most part of fixed nitrogen reaches to aerial parts of plants.

Nodulation on Stems of Legume Crops

The soil bacteria of the Rhizobiaceae family form atmospheric nitrogen fixing nodules on the roots. Similar to the formation of nodules on roots, nodulation also occurs on stems of some legumes. Few legume species form nodules both on their roots and stems. Bacteria present in these nodules also fix atmospheric nitrogen. Ladha et.al. (1990) observed nodules on the stems of about 26 species of *Sesbania*, *Aschynomen* and *Neptunia*. These plants can be grown successfully in waterlogged soils and used for green manuring in paddy fields. The nodules or primordia are found over the whole stem or only on lower portion of the stems. They are arranged in straight vertical rows or in spiral rows around the stem.

The stems nodules do not suffer from shortage of energy required for nitrogen fixation, as stems also produce energy by photosynthesis. In plants as quantity of carbohydrate increases number of active nodules on stem also increases.

Bacteria, which produce nodules or primordia on stems, are included in genus *Azorhizobium* and *Photorhizobium*. These can fix nitrogen independently. Bacteriods of these bacteria produce bacteriochlorophyll. This chlorophyll helps to utilize that wavelength of sun light which cannot be used by host plants. Thus, these can produce energy alone. Therefore, competition for energy between these and host plants decreases.

Generally, in root nodulating plants bacterial inoculation is to be done every time before sowing. But in stem nodulating plants symbiosis continues to generations due to seed born bacterial spores. So, there is no need of repeated inoculation of bacteria. Provision of same type is not in other nitrogen fixation systems. Stem nodulating rhizobia are also soil bacteria and they can live saprophytically on soil organic matter. They are highly mobile in the water due to their unipolar flagellum. The phyllosphere of the host plant is also a place for the growth of stem nodulating bacteria. The phyllosphere term is used in microbiology indicating the above ground parts or aerial region of plants. This is a habitat for microorganisms like bacteria, fungi, algae, yeasts, nematodes etc. The aerial inhabitants are called epiphytes.

Soil and the phyllosphere provide inoculum for stem infection. Rain has major role in stem inoculation. Soil splash due to rain brings bacteria to the lower portion of stem. Insects and wind also help in stem inoculation by transporting bacteria laden soil up to the stem.

Many plants like *Sesbania rostrata*, *Aschynomen afraspera* and *Aschynomen nilotica* may be utilized to fix nitrogen by these bacteria. Later, these plants are turned into soil as green manure for the benefit of next crop.

Symbiotic Nitrogen Fixation in Non-Legume Plants

The symbiotic nitrogen fixation also occurs in non-legume plants. Actinorhizal plants can develop an endosymbiosis with the nitrogen fixing soil actinomycete *Frankia*. The establishment of the symbiotic process results in the formation of root nodules in which *Frankia* provides fixed nitrogen to the host plant in exchange for reduced carbon. Few examples of non-leguminous nitrogen fixing plants are Cycads, Ceanothus, Cercocarpus, Purshia, Shepherdia, Allocasuarina etc. Now *Gluconacetobacter diazotrophicus* a non-rhizobial, non-nodulating and atmospheric nitrogen-fixing bacterium is the ray of light to know about atmospheric nitrogen fixation theories. This bacterium was isolated from the sugarcane sap. Strains of *Gluconacetobacter diazotrophicus* may intracellularly colonize the roots and shoots of the wheat, maize, rice, potato, tomato, tea, oilseed rape, grass etc.

Dicotyledon non-legume actinorhizal host plants are mainly divided into three angiospermic orders, eight families, 24 genera and more than hundred species. Plants of these eight families are mostly found in temperate regions. But several genera of these also grow very well in hilly areas of tropical and sub-tropical regions. Plants of genus *Hippophae* of Elaeagnaceae family also are found in Himalayan region. *Casuarina* of Casuarinaceae family is generally found in Indian Sub continental regions. *Discaria* of Rhamanceae family is found in tropical and sub-tropical regions. These generas are suitable for low nitrogen, sandy and acidic soils areas as they complete their life cycle successfully here. *Alnus glutinosa* and *Myrica gale* grow and develop well at pH 3.5 to 5.0 and 4.5 to 5.5, respectively.

Parasponia is only non-legume host plant nodulated by *Rhizobia* with atmospheric nitrogen fixation mutualism. *Parasponia* is a genus of the Cannabaceae family having five tropical tree species. All species develop nodules with *Rhizobia*. The trees are medium in height and good for growing on nitrogen deficient and problematic soils.

This symbiosis is also influenced by various factors likewise other nitrogen fixating systems. Soil, climate and plant species are major factors. *Alnus* fixes nitrogen up to 300 Kg/ha/year. Nitrogen fixating capacities of other genera are lower than this like *Hippophae, Coraria, Ceanothus,, Casuarina* and *Myrica* as 179, 150, 60, 59 and 10 Kg/ha/year, respectively.

Biochemistry of nitrogen fixation by non-legume crops

Biochemistry of nitrogen fixation by non-legume crops is like other systems but not clearly understood. In this process, also nitogenase enzyme converts atmospheric nitrogen to ammonia as $N_2 + 8\ H^+ + 8\ e^- \rightarrow 2\ NH_3 + H_2$. This is a general reduction reaction in which nitrogen is gaining eight electrons as:

$$N_2 + 8H^+ + 8e^- + 16ATP \rightarrow 2NH_3 + H_2 + 16ADP$$

Plant and Algae Symbiosis

Plant and algae symbiosis is also found in nature just like legume and *Rhizobium* and non-legume and *Frankia* symbiosis. Some examples of plant algae symbiosis to fix atmospheric nitrogen are Lichens and Cycas and Anabaena and Nostoc, Anthoceros and Gunnera and Nostoc, Azolla and Anabaena. Here Lichens, Cycas, Anthoceros, Gunnera and Azolla are different plants. Among all these Azolla and Anabaena symbiosis fixes remarkable quantities of atmospheric nitrogen to show its importance in agriculture and major role in environmental protection. Under good to very good conditions from 400 to 2400 Kg nitrogen per hectare per year may be fixed by this symbiosis. Azolla is a good biofertilizer and green manure for the rice crop and used as poultry and livestock feed. Many centuries ago, farmers of China know about this particular symbiosis.

Azolla

Azolla is an aquatic fern and generally found in ponds, ditches and rice fields of warm-temperate and tropical regions of the world. Its common names are mosquito fern, duckweed fern, fairy moss, water fern etc. There are six species of Azolla in two families Azollaceae and Salviniaceae. These species are *Azolla caroliniana, Azolla filiculoides, Azolla microphylla, Azolla Mexican, Azolla nilotica* and *Azolla pinnata*.

Azolla is an ideal longtime partner for *Anabaena* because the two organisms are living together continuously without separation from centuries. *Azolla* provides *Anabaena* to pass down characters together during reproduction from one generation of *Azolla* to the next generation. Azolla is a small plant. It floats on water surface in the paddy fields and water bodies. Its diameter ranges from 1-2.5 cm for small species like *Azolla pinnata* to 15 cm or more for *Azolla nilotica*. The diameter of mature plant ranges from 0.50-0.7 cm with simple individual roots. It has too small and lobed leaves overlapped on each other. The upper lobes of the leaf are green due to the presence of chlorophyll and perform photosynthetic activities. The lower lobe is colourless due to absence of chlorophyll, transparent and remains in contact with water. *Azolla's* leaf cavities are ideal habitat for strains of *Anthrobacter, Corynebacterium, Agrobacterium* bacteria and *Anabaena azollae* alga. Anabaena alga resides in cavities of upper lobes of leaves of Azolla as Anabaena alone cannot grow very well. Anabaena also has heterocysts like other blue green algae.

In normal conditions nitrogen fixation rate of *Azolla* may be up to 1 to 2.5 Kg/ha/day and under ideal conditions, this rate may be up to 3.5 Kg/ha/day. On the dry weight basis *Azolla* contains 4-5% nitrogen. Azolla grown with paddy also uses nutrients present in layer of water and sunlight for photosynthesis and nitrogen fixation. *Azolla* also enhances the activity of other nitrogen fixing micro-organisms present in layer of water and soil surface. *Azolla* remarkably improves fertility and other characteristics of soils of paddy. All these increase yield of paddy.

Azolla pinnata

Azolla pinnata is also known as mosquito fern, feathered mosquito fern and water velvet. It is native to tropical Africa, South Africa, South East Asia, Australia and New Zealand and prominently found and used in India, Bangladesh, Pakistan, Sri Lanka, Korea, China, Japan, Malaysia, Philippines, Vietnam, New Guinea etc. *Azolla pinnata* is a small aquatic fern which generally found and grow in lakes, slow moving streams and rivers, marshes, ponds, wetlands, paddy fields, ditches. Swift currents and waves break up the plant. It grows very rapidly and its biomass doubles in about two days.

Azolla pinnata has been used for centuries in Southeast Asia as a biofertilizer in rice crop because *Azolla-Anabaena* can fix almost three times more atmospheric nitrogen than legumes. Typical rates for legumes are 400 kg of nitrogen per hectare per year whereas for *Azolla-Anabaena* is 1100 kg of nitrogen per hectare per year. However, it dies in the hot summer and the decomposing plants release nitrogen into the soil. This is also a weed because it quickly spreads water bodies and covers them. Its thick surface mats impede water flow, clog irrigation channels and reduce availability of oxygen and sun light essential for other aquatic organisms. But this is also good animal feed.

Azolla pinnata is a small 1.5-2.5 cm long plant with a straight main axis and pinnately arranged side branches. Progressively longer branches towards the base provide it a triangular shape with a triangular frond measuring up to 2.5 centimetres in length which floats on the water. The pinnate basal branches fragment themselves. The main axis decomposes to form new plants. Leaves are very small 1-2 mm long and overlap in two rows or lobes. Upper lobe is green, brownish green or reddish and often has cylindrical unicellular hairs which repell water from the top of leaf surface. Thus, the fern float upon water. Lower lobe is translucent brown. The reproduction is vegetative by broken branches or sexual by sporocarps on the leaves which release spores in the water. Roots have fine lateral rootlets, giving a feathery appearance in the water.

Fig. 3.30: *Azolla pinnata*

Anabaena azollae

Anabaena azollae is blue green alga live in symbiotic relation with Azolla and fixes atmospheric nitrogen. This is a small filamentous phototrophic cyanobacterium generally look like as multicellular organism. It has with two distinct and interdependent cell types. The first cell type is small, circular, photoautotrophic and blue green coloured vegetative cell. It performs photosynthesis. The second cell type is a heterocyst. This is a larger, more homogenous and pale coloured cell. This fixes atmospheric nitrogen.

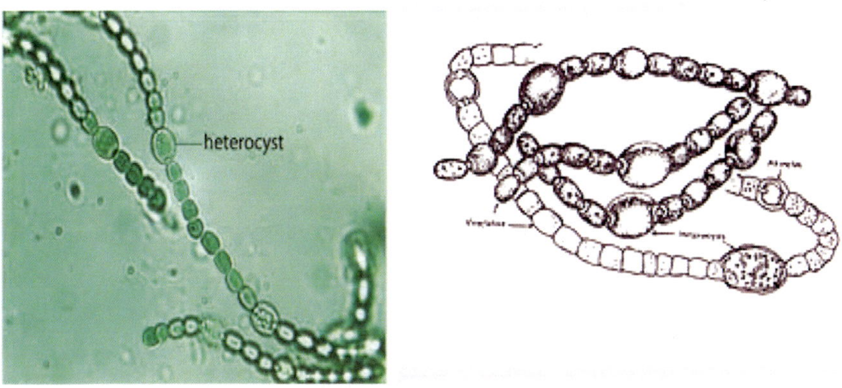

Fig. 3.31: *Anabaena azollae*

Biochemistry of nitrogen fixation by *Azolla-Anabaena* symbiosis

This symbiosis can fix the atmospheric nitrogen in heterocysts through the nitrogenase enzyme and convert it into the ammonium ion. *Anabaena azollae* fixes N_2 on the leaf of Azolla. The fixed nitrogen is supplied to the host plant for its colonization, reproduction and development in nitrogen deficient aquatic environments. Azolla turns ammoniumious into various amino acids and finally to proteins. In the return *Anabaena azollae* gets continuous supply of nutrients and protection against enemies and desiccation. A general reduction reaction in which nitrogen is gaining eight electrons as:

$$N_2 + 8\ H+ + 8e- +16\ ATP + 16\ H_2O \rightarrow 2\ NH_3 + H_2 + 16\ ADP +16\ Pi$$

Factors Affecting Growth and Nitrogen Fixation by Azolla

Chlorophyll containing Azolla sufficiently provides organic matter and nitrogen to the soil. The nitrogen fixation and organic matter both are influenced by many environmental factors. The main factors are availability of plant nutrients, pH, salinity, temperature, quantity of water in the field, relative humidity and sunlight.

Availability of plant nutrients: Azolla requires all macro and micro nutrients which are essential for normal growth and development in optimum quantities. Nitrogen, phosphorus, potassium, calcium and magnesium are important macronutrients. Among micronutrients iron, molybdenum, cobalt and zinc are required for growth and nitrogen fixation.

In the paddy fields nitrogen is sufficiently dissolved in the water. Azolla uses this nitrogen for its growth and development. Azolla simultaneously fixes nitrogen during initial 10-15 days of growth and development while absorbing dissolved nitrogen. Later on when fronds and roots of old Azolla begin to die and fall off, then fixed nitrogen comes in soil and water and utilized by growing paddy or other crops. In the presence of nitrogen, atmospheric nitrogen fixation is a main characteristic of Azolla-Anabaena symbiosis. However, this system does not work when the quantity of ammonium nitrogen reaches up to 60 ppm in the field.

Phosphorus is a limiting element for growth and nitrogen fixed by Azolla. Generally, phosphorus contents in soil solution and in water of paddy fields are low. Good quantity of phosphorous in soil is beneficial for Azolla and nitrogen fixation. Various experiments indicated that split application of phosphorus at the rate of six kg/ha is good for Azolla and nitrogen fixation even 0.2 % phosphorous is present in the soil.

In the initial stages more availability of potassium is good for Azolla. Later on becoming excess of potassium reduces the growth of Azolla.

pH: Azolla can survive in the soil pH range of 3.5-10.0 but pH range from 4.5 to 7.0 is good for optimum growth. Highly acidic and alkaline soil pH is not good for Azolla.The pH 6.5 to 7.0 of soil and water layer is better for Azolla. Azolla performs properly between 5.0 to 8.0 pH. Generally, the pH on which paddy can grow and develop nicely is also suitable for Azolla.

Salinity: Salt tolerance of various species of Azolla is quite different. Quantity of salt in the aqueous layer should be lower than 0.01% for proper growth of Azolla and nitrogen fixation. Higher salinity reduces the growth and nitrogenase activity in *Azolla-Anabaena* association.

Temperature: Temperature tolerance of various species of Azolla is quite different. These species can effectively grow and fix nitrogen in the range between 5°C to 45°C. But temperature between 25°C to 35°C is most suitable for growth and development and nitrogen fixation. Higher temperature tolerant varieties of *Azolla* should be grown in the areas where the temperature of field water reaches up to 40°C to 45°C in the paddy season. In our country, *Azolla imbricata* for tropical areas, *Azolla rubra* for cold areas and *Azolla pinnata* for all areas are suitable.

Quantity of Water in the field: The Azolla contains 90-95% water. This is required for the maintenance of its structural turgidity and many physiological and biochemical processes. Azolla floats on the surface of water. Violent air accumulates Azolla at a place if more water is present in the field. This condition adversely affects the absorption of nutrients from water and nitrogen fixation by Azolla. Therefore, water should be maintained in the field in such a way that roots of Azolla remain in contact with the surface of soil. Azolla also grows very well in wet soils and performs very excellently. However, 3-5 cm water depth is recommended for optimum growth of Azolla.

Relative humidity: Azolla attains normal growth if relative humidity is in between 85-90% in rows of paddy.

Sun light: Sun light has direct effect on growth and development of *Azolla* and *Anabaena* as both are photosynthetic organisms. *Azolla* fixes nitrogen from sunrise to sunset i.e., morning to evening and acquires energy during night for next day nitrogen fixation. *Azolla* fixes maximum nitrogen in the midday and noon time as sunlight is highest during this period. Fixation of nitrogen is relatively more when average intensity of light is 5-15 Lux.

Geological and Mineralogical Properties of Soil

Geology and Soil geology, Mineralogy and Soil mineralogy, Rocks and their classification, Minerals, their classification and identification, Occurrence of minerals in soils, Weathering of rocks and minerals, Weathering sequence and products, Soil colloids, Inorganic and organic colloids of soils, Characteristics of clay minerals and humus.

Geology: Geology is the science which studies the dynamics and physical history, the rocks composed it, the physical, chemical, and biological changes that has undergone and still undergoing and organisms residing on earth. Geology also includes geodesy, mineralogy, and stratigraphy. Generally, the geology is associated with the study of rocks as rocks are the basic record of the major of geological history of the Earth.

Soil geology: Generally, any loose material on the surface of earth is referred as soil. But the geologists first of all studied the origin and development of soils in field. They indicated that the mantle of loose-weathered material on the earth's surface is called as regolith and upper 100-200 cm part of regolith can be termed as soil which includes organic matter and is essential for sustainable plant growth. Soil is developed by the gradual physical, chemical and biological weathering of rocks and minerals and other processes.

Mineralogy: Mineralogy is a part of geology dealing with study of the chemistry, crystal structure, physical and optical properties of minerals and mineralized artifacts including their origin, occurrence, formation, classification, geographical distribution, uses etc.

Soil mineralogy: Soil mineralogy is the study of the formation, occurrence, properties, composition, transformation and classification of the minerals present in the soil. The sub branches of soil mineralogy are sand mineralogy, silt mineralogy and clay mineralogy according to particles of soils. It aims to know the soil mineral matter which makes up about 90% of the solid volume of mineral soils. Soil mineralogy is a key determinant of many soil functional properties like nutrient quantities and intensities, pH and buffering, anion and cation exchange capacities, aggregate stability, soil carbon protection, dispersion and resistance to erosion.

The Soil geology and Soil mineralogy are described here. This scientific discipline encompasses a diverse range of topics including: i) the occurrence and distribution of soil mineral materials, ii) their chemical composition, and crystallographic properties, iii) the stability, transformations, and interactions of soil minerals in natural environments, and iv) their consequent influence on soil physical and chemical properties and regulating air and water chemistry (Dixon and Schultze, 2002).

The knowledge of rocks began with the use of stones as tools and arms by primitive man. They were able to select and differentiate the stones after picking them from the ground. They knew about the extraction of clay for pottery and stones of different colours, shape and brightness which can be used as ornaments. Archaeology indicates that by the end of prehistoric period man knew about the practical use of sandstone, quartzite, quartz, silicon, nephrite, rock salt and other rocks and minerals.

Gradually all these developed a science called as Petrology. Petrology is the branch of geology which deals with the scientific study of the origin, composition, structure, changes and classification of rocks. Petrology also has two branches, Petrogenesis which studies the origin and genesis of rocks and Petrography which deals with the description and classification of rocks. Soil Petrology is the scientific study of such rocks which are the sources of soil parent material. Although we can study this under Soil geology also. Fossils are prehistoric hard-rock remains or traces of plants or animals preserved in sedimentary rocks.

Fig. 3.32: A Fossil rock depicting skull of any animal (may be elephant) kept at Garjia Mata Mandir near Ramnagar, Nainital (Photo Dr. Venkatesh Bharadwaj)

Rock

Rock is a hard and consolidated mass of one, two or more minerals whose characters depend mainly upon the minerals present in them. Rocks are generally formed due to cooling and consolidation of molten magma, sedimentation and transportation of fragmented rock materials and alterations of existing rocks. Thus, according to genesis or formation, rocks can be classified in to igneous, sedimentary and metamorphic rocks.

Fig. 3.33: A hanging rock structure at Tirumala hills near Shri Tirupati Balaji temple (Photo by Dr. Venkatesh Bharadwaj)

Igneous Rocks

Igneous rocks are also called as primary rocks or magmatic rocks. These are formed due to the cooling and crystallization of magma. These are non laminar and unstratified massive bodies which constitute 95% of the earth's crust. Magma is a silicate melt containing water vapour, gases and elements with a temperature range from 700-1400^{0}C. Magma may be solidified at various depths in the earth or on the earth's surface.

Place of the solidification of magma provides the natural classification of igneous rocks. The solidification of magma upon the surface earth produces extrusive or volcanic rocks and solidification underneath the surface of the earth creates plutonic or intrusive rocks. At deeper layers of earth, cooling and solidification of magma take much time and time taken in crystallization is quite long. Hence holocrystalline coarse-grained rocks are formed. But in case of extrusive conditions magma solidifies quickly due to the rapid fall in temperature. Therefore, there is incomplete crystallization of magma and fine-grained vitreous rocks are formed.

Percent weight of silica (SiO_2) in the igneous rocks is another classification system in which igneous rocks are grouped as ultra basic rocks (less than 45% silica), basic (44-55% silica), intermediate (56-65% silica), acid or oversaturated (more than 66% silica) groups and alkaline igneous rocks having 5–15% alkali (K_2O + Na_2O) content. Extrusive alkaline rock is Phonolite. Table 3.12 contains examples of igneous rocks according to these classification systems.

Table 3.12: Classification of igneous rocks

Silica Depth	Acid	Intermediate	Basic	Ultrabasic
Volcanic	Rhyolite Dacite	Trachyte Andesite Rhonolite	Basalt Diabase	Limburgite
Plutonic	Granite	Syenite Diorite	Gabbro	Peridotite Dunite Pyroxenite Pickrite

Gabro, Basalt, Diabase, Andesite and Granite rocks are related with soil petrology as these convert into soil parent material after weathering. The igneous rocks contain feldspars, quartz, mica, augite, hornblende, olivine and Lucite minerals. Commonly found igneous rocks in India are described here:

Granite: Granite is a plutonic, acidic, coarse grained, massive and hard igneous rock. It contains mainly 50-90% feldspars (Orthoclase), 10-50% quartz and mica with or without hornblende and ferromagnesian minerals. In secondary minerals epidote, kaolinite, actinolite, apetite etc. are also found in granite. The different coloured granite can also be polished for shining and used in buildings. Porphyritic textured granite rocks are also found. The density of granite varies from 2.52 to 2.81 g/cc

Basalt: More than 90% of all volcanic rock on Earth is basalt. Basalt is a basic, fine grained and compact igneous rock. It contains mainly plagioclase feldspar (anorthite) along with augite and magnetite mineral. This is dark coloured (dark grey to coal black) rock showing 2.8 to 3.0 g/cc density.

Syenite: Syenite is a plutonic, intermediate, and coarse to medium grained and very hard igneous rock. General composition of syenite is similar to granite except deficient in quartz (< 5%). It contains mainly orthoclase with lesser amounts of free quartz, hornblende and augite. The colour of syenite varies from blackish pink to blackish grey and density ranges between 2.63 to 2.90 g/cc.

Diorite: Diorite is a plutonic, intermediate, coarse grained and hard igneous rock which contains mainly plagioclase feldspar (andesine), augite, hornblende, biotite and/or pyroxene. The chemical composition of diorite is in between mafic gabbro and felsic granite. Black coloured rock shows density from 2.72 to 2.96 g/cc.

Sedimentary Rocks

Sedimentary rocks are formed due to the weathering, transportation, deposition and compaction of pre-existing igneous and metamorphic rocks. Blowing wind, running and percolating water, moving glaciers etc. cause continuous wear and tear of rocks to produce loose sediments. These sediments may remain at their original place or may be transported to other places by various agents like water, wind, ice, organisms etc.

Also, these sediments may remain as loose mass or change into compact and cemented bodies due to the action of cementing agents (silica, lime or iron oxides) or pressure exerted by overlying materials or water. Thus, the rocks developed from the sediments are called as sedimentary or secondary rocks. Many layers of sediments deposit one by one to develop numerous strata. These sedimentary rocks are also called as aqueous or stratified rocks. These strata can be separated by breaking. Sandstone, limestone, laterite, shale, coal and dolomite are common sedimentary rocks. Fragmental, chemical and organogenic rocks are three main groups of sedimentary rocks.

Fragmental Rocks

These are also called as clastic rocks. These rocks are originated from the mechanical sediments like pebble, gravel, silt and sand. The sediments may be residual or transported by wind, water or ice. Laterite and bauxite are the rocks of residual sediments. Mechanical disintegration of rocks does not produce new rock but different grain size materials are obtained like clay size in shale and claystone, silt in siltstone, sand in sandstone etc. Movement of fragmented material suspended in runoff water and then sedimentation produce rocks of varied texture, structure, colour and thickness. Sandstone, conglomerate, breccia, shale, grit and mudstone are few examples of fragmental rocks.

Chemical Rocks

These rocks are formed by the evaporation or precipitation of the true or colloidal solutions. The evaporation is greatly influenced by the temperature and amount of dissolved salts in water. In the arid region higher evaporation causes super saturation of salts in the water of lakes etc. Later on crystallization produces salt deposition. Gypsum, anhydride, halite (rock salt), sylvite and carnallite are the examples of such depositions which produce chemical rocks. Limestone, dolomite, bauxite, siliceous and bog iron ore are also chemical rocks and ooilitic limestone or ooilitic hematite in shallow marine system.

Organogenic Rocks

Organogenic rocks are also called as organic rocks. These formed due to the action of organisms on chemical sediments. These rocks are the products of accumulation of organic remains of plants or animals as organic matter and changed remarkably in the course of time. The largest organogenic rocks consist of plant remains which may be preserved for longer periods due to devoid from oxygen, buried under sediments, oxygen-free water, younger vegetative cover, anaerobic conditions etc. Various organisms worked on chemical sediments to form organogenic rocks. Chalk, dolomite, diatomite, marl, fossiliferous limestones, peat, coal, lignite and native sulphur are common examples of organogenic rocks.

The limestone, dolomite, sandstone, shale, laterite and chalk are some abundantly found sedimentary rocks of our country. These are described here:

Limestone: It is a medium hard sedimentary rock. It mainly contains calcite mineral. It is non-laminar, light coloured rock which effervesces rapidly with dilute HCl. It most commonly forms in clear, warm, shallow marine waters due to the accumulation of shell, coral, algal, and fecal debris. Hardness is one and density of lime stone is 2.7 g/cc.

Dolomite: Dolomite is also called as dolostone and dolomite rock and found in sedimentary basins of the world. It is a medium hard sedimentary rock which contains appreciable quantities of magnesium and calcium carbonates, $CaMg(CO_3)_2$. This is non-laminar, white to light brown in colour, hardness ranges from 3.5 to 4.0 and density is 2.9 g/cc. It shows weak effervescence with dilute hydrochloric acid.

Sandstone: Sandstone is one of the most common types of sedimentary rock and is found in sedimentary basins worldwide. Sandstone is made up of sand size particles of mineral or rock. Sandstones are formed by the cementing of sand particles. Sand may be monomineralic (single mineral), oligomictic (two minerals) or polymictic (many minerals). Mica, glauconite, calcite, gypsum, limonite, hematite and feldspars minerals are also found in sandstones. Cementing with silicates or carbonates produce white or light coloured sandstones but cementation with hematite provides yellowish, brownish or reddish colour to the rocks. The hardness is 7 and density ranges from 2 to 2.65 g/cc.

Shale: Shale is a most common soft and fine-grained sedimentary rock which is about 70 percent of the rocks found in the Earth's crust. It is made up of compacted mud consisting clay and smaller particles of quartz, calcite, mica, pyrite, other mineral, kaolinite, silt, iron oxides and organic materials. Its hardness is 3 and density is 2.65 g/cc.

Laterite: Laterites are soft to medium hard sedimentary rocks which mostly contain alumina and hydroxides of iron along with kaolinite, goethite, hematite and gibbsite. The colour of laterite is red-brown due to the presence of iron oxides and iron hydroxide but these may have yellow spots. 'Later' is a Latin word which means 'brick'. Its Density varies from 1.019 to 1.778 g/cc and hardness is 2 to 6.

Chalk: Chalk is a soft, fine grained and porous sedimentary carbonate rock which is a form of limestone. It is composed of the mineral calcite. Calcite is an ionic salt called calcium carbonate or $CaCO_3$. Mixture of various shales also produces chalk. It effervesces with dilute HCl. Its hardness is 1 and density is 2.49 g/cc.

Metamorphic Rocks

Metamorphic rocks are the product of metamorphism of pre-existing igneous, sedimentary and even metamorphic rocks. Metamorphism is the change in texture, mineral composition or some times in the chemical composition of rocks. Temperature, pressure, gases and fluids are the main agents of metamorphism. Metamorphosis causes dehydration and recrystallization of pre-existing rocks i.e., protolith. Chemical composition may remain same or change due to the addition of other materials during recrystallization.

Thermal metamorphism is related with the change of rocks under the influence of temperature. Higher temperature may be provided by radioactive processes and geothermal gradient. The molten magma also provides much heat due to the interpenetration of magmatic masses to the rocks. This is called as contact metamorphism.

Pressure or load metamorphism is due to the weight of overlying rocks and mountains. Dynamo thermal metamorphism is the result of joint activity of pressure and temperature upon rocks. At each 100-meter depth there is increase in pressure and temperature by 27 KP/cm^2 and 3^0C, respectively. Dynamo thermal metamorphosis is most powerful force causing to more or less complete recrystallization of rocks with new characteristics.

Based on structure and texture, metamorphic rocks are divided into, foliated, and unfoliated metamorphic rocks.

Foliated Rocks

These have parallel structure as their constituent minerals are flattened and oriented in parallel layers. These rocks break along parallel planes. Gneiss, phyllite, schist and slate are common examples.

Unfoliated Rocks

Unfoliated rocks are also called as non-foliated rocks. These rocks have massive structure and do not show foliation because of large grain size. These rocks break down irregularly. Common examples are marble, quartzite and novaculite.

Minerals present in the metamorphic rocks are quartz, feldspars, pyroxenes, amphiboles, micas, calcite, dolomite, hornblende and chlorite. Table contains the examples of some metamorphic rocks.

Table 3.13: Metamorphic rocks

Unconsolidated sediment	Original rock	Metamorphic rock	Nature of metamorphic rock
Calcareous mud	Lime stone	Marble	Fine grained Unfoliated
Quartz sand	Sand stone	Quartzite	Coarse grained foliated
	Conglomerate	Gneiss	Coarse grained foliated
	Granite	Gneiss	Coarse grained foliated
	Gabbros	Schist	Coarse grained foliated
	Shale	Slate	Fine grained Unfoliated

Description of Some Common Metamorphic Rocks is given here

Marble: It is an unfoliated fine grained metamorphic rock which has mainly calcite and sometimes dolomite also. Pure marble is white in colour but impurities give rise yellow, red, green and grey colours. Its hardness is 3 to 4 and density is 2.6 to 2.7 g/cc. It effervesces with dilute HCl.

Quartzite: It is a coarse grained unfoliated massive metamorphic rock. It is made up of quartz. Sometimes mica and feldspars may also be present. Light in colour but may be yellow, red, grey and black due to impurities. Hardness is 6 to 7 and density is 2.6 to 2.8g/cc.

Gneiss: Gneiss is a coarse grained foliated metamorphic rock with alternate light and dark bands or streaks. Lighter bands are due to mica, quartz and feldspars and darker bands are because of ferromagnesian minerals (biotite, amphiboles and or pyroxenes). Hardness is 6 to 7 and density is 2.6 to 2.8g/cc.

Schist: It is a coarse grained foliated metamorphic rock with variable colour and density. Minerals present are micas and chlorite with garnet, quartz and staurolite in some varieties. Hardness is 4 to 5 and density is 2.6 to 2.7g/cc.

Slate: Fine grained foliated metamorphic rock made from shale or mudstone. Colour is dark grey to black. It is used for roofing, flooring, and flagging due to its durability and attractiveness. Minerals present are mica, quartz and chlorite. Hardness is 2.5 to 4 and density is 2.6 to 2.7g/cc.

Minerals

Minerals are naturally occurring homogenous inorganic substances which have fixed chemical composition and distinctive crystalline and internal atomic structures. Minerals are integral part of rocks. Many minerals may have variations in their chemical composition and internal structure.Thus, various mineral groups are formed. Mineral species have similar chemical composition but dissimilar structure.

Minerals are formed by the natural physical and chemical processes, mostly inorganic chemical processes. However, some chemical compounds produced due to the vital and organic activities may be placed under minerals. The examples are coal, amber, pearls, elemental sulphur, apatite (bones), fluorite (teeth) etc.

Mineral Forming Elements

Oxygen, Silicon, Aluminum, Iron, Calcium, Sodium, Potassium, Magnesium and Titanium are abundantly distributed mineral forming elements. After these the distribution of Phosphorus, Hydrogen, Manganese, Fluorine, Chlorine and Sulphur are in middle. Carbon, Barium, Chromium, Zirconium, Vanadium, Nickel, Strontium, Lithium, Copper, Cerium, Beryllium, Cobalt, Boron, Zinc, Lead, Arsenic, Cadmium, Tin, Mercury, Antimony, Molybdenum, Silver, Tungsten, Bismuth, Selenium, Gold, Bromine, Tellurium, Platinum etc. are least distributed. The least distributed elements form rare minerals,

Identification of Minerals

The minerals are identified by knowing their crystal system, physical, optical and chemical properties.

1. Crystal System

Most of the minerals are crystalline in nature. Amorphous minerals are also found. Crystals are solid geometric bodies or forms and have ordered internal arrangements of molecules, atoms or ions. The perfect crystals have these both characters, where as crystalline has only internal arrangements of molecules, atoms or ions without developing external geometrical body or form. Amorphous minerals have distribution of atoms with in irregular or ungeometrical bodies. The quartz is the example of crystal or crystalline body and ordinary glass is amorphous body.

The definite proportional, systematic or regular arrangements of the faces (smooth surface confining a crystal), edges (straight lines joining the faces) and angles (joining point of three or more faces) is called as the symmetry. Symmetry is described on the basis of elements of symmetry. Plane, axis and points are the elements of symmetry. Symmetry of a crystal is identified by rotating the crystal toward elements of symmetry. Only thirty-two classes of crystal symmetry are possible. Every crystal class is a member of one of the six crystal systems. These systems include the cubic or isometric, tetragonal, hexagonal, orthorhombic, monoclinic and triclinic crystal systems. The hexagonal crystal system is further divided down into hexagonal and trigonal or rhombohedral sub systems. These crystal systems are also shown in figure 3.21. Crystal system has definite crystallographic axes along with the one or more crystal elements. Crystal systems are again grouped in higher, intermediate and lower categories.

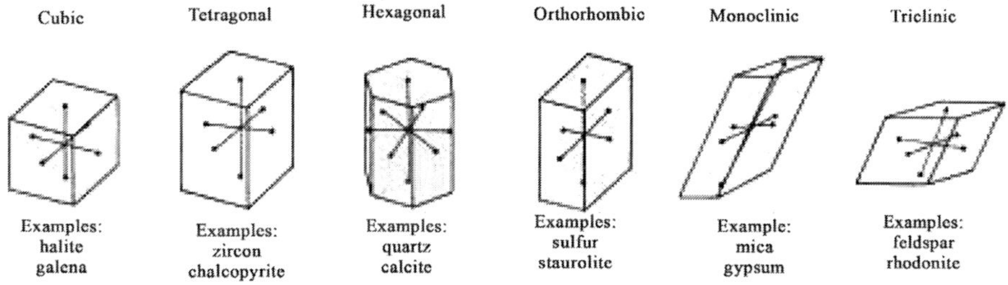

Fig. 3.34: Seven Crystal system

2. Physical Properties of Minerals

Identification of minerals can be made by their physical properties. Colour, streak, hardness, specific gravity, cleavage, fracture, form and structure and magnetism are some important physical properties of minerals.

1. **Colour:** Only few minerals show distinctive colours. But colour is still most appealing and unreliable property to identify them. Malachite, Azurite, Hematite, Limonite etc. have specific colours. Dichromatic minerals show specific colour due to their composition. The specific colour of minerals is changed due to inclusion of impurities. Thus, a minerals can show different colours which is called as allochromatic.

2. **Streak:** The colour of fine powder of a mineral is called as streak. Streak is produced by rubbing the mineral on the unglazed porcelain plate. The colour of streak may be same or different than the colour of original mineral. Colour of malachite is green and streak is also green. Colour of azurite is blue and streak is light blue. Colour of haematite is black or steel grey whereas colour of streak is cherry red. Colour and streak of limonite are yellow and yellowish brown, respectively.

3. **Hardness:** Hardness is an important property of minerals to identify them. This indicates the resistance against the abrasion or scratching. Mohs described it in the scale of 1-10 in the increasing order of hardness. Softest mineral talc has a hardness of 1 whereas hardest diamond shows the hardness 10. No mineral can scratch the diamond. Hardness is identified by scratching the mineral with the objects having specific hardness. Vickers scale, Brinell scale, Rockwell scale, Meyer scale and Knoop scale are other hardness tests. Mohs test indicates hardness of minerals because of scratch test. Mohs scale is given in Table 3.14.

   It is not necessary that calcite is having fifty percent hardness than orthoclase. The Mohs scale gives only relative hardness. Thus, other suitable method for the estimation of harness should be investigated. Hardness depends upon the type of bonds. Minerals having covalent bond between their atoms are very hard. Ionic type bonding provides moderate and hydrogen and Van der Walls bonds give low hardness to minerals.

4. **Density:** Density is a basic property of minerals. It indicates relative weight (heavy or light) of minerals. Minerals showing close packing have more density than the open packed. Similarly atomic weight of atom present in mineral is more, greater will be the density. Specific gravity is the ratio between weights of mineral to the weight of equal volume of water at 4^0C.

Table 3.14: Mohs hardness scale

Hardness category	Mineral	Hardness of specific object
1	Talc	
2	Gypsum	
2-2.5	-	Finger nails
3	Calcite	Copper coin
4	Fluorite	
5	Apatite	Window glass
5.5-6.0 5.5-6.5	-	Blade of steel knife Most of hard glasses
6	Orthoclase	
6.5-7.5	-	Steel file
7 7.5	Quartz Garnet	
8	Topaz	
9	Corundum	
10	Diamond	

5. **Cleavage:** Cleavage is the break down of minerals along a particular direction to provide plane, smooth and lustrous surfaces. Cracks developed due to cleavage are called as cleavage cracks. Minerals may or may not have cleavage but crystalline minerals always show cleavage where as amorphous material does not have this property. Cleavage may be perfect, good, distinct, imperfect and none.

6. **Fracture:** Fracture is the uneven breakage of minerals having none or imperfect cleavage.

7. **Lustre:** Lustre is the shining of minerals due to the reflection of light from their surface. Lustre may be metallic and non-metallic.

8. **Magnetism:** Some minerals like magnetite and pyrrhotite have magnetic property. Minerals showing moderate or slight magnetism are called as paramagnetic and diamagnetic mineral. Paramagnetism is weak attraction

to *magnetic* fields. Some minerals show negative magnetism i.e., they do not attract with magnet. Ferromagnetism indicates strong attraction to *magnetic* fields.

3. Optical Properties of Minerals

Optical properties play an important role in mineral identification. Thin sections of minerals are used to study the optical properties. The ordinary, plane polarized and polarized lights are used for this purpose and the reflection, refraction, dispersion and absorption of light are determined.

Minerals also reflect and the incident rays as other transparent, translucent and opaque objects. Some minerals also show double refraction, where incident rays broken down in to two refracted rays. These broken refracted rays move with different velocities. Thus, different refractive indices are obtained for a mineral. The phenomena of double refraction or birefringence are common with mineral systems other than cubic i.e., isometric system. Calcite mineral shows double refraction.

4. Chemical Properties of Minerals

Chemical properties of minerals show the presence and arrangement of atoms in minerals. There are different types of atoms and each type of atom is classified as an element. Out of 118 elements identified so far ninety-four occur naturally on earth. About twenty elements make more than 95 percent of all matter on earth and others are rare elements. Bonding of atoms together forms molecules. Using their chemical properties, minerals are identified by how they react to other substances. Minerals have some elements and chemical compounds in their composition.

Chemical composition indicates some properties of minerals. Minerals are classified according to their composition because it has prime importance in their identification. Minerals either have single element or elements are combined with some anion. Thus, minerals can be classified according to the presence of elements or anion. Since 1950s, the classification of minerals has been carried out by the International Mineralogical Association. The IMA placed minerals into three broad classes as:

1. Non-silicate minerals, 2. Silicate minerals and 3. Organic minerals.

1. **Non-Silicate Minerals:** Non silicates minerals don't have silicon, oxygen combination i.e. silicate tetrahedral as a part of their structure. They have a less complex structure. Oxides, sulfides, carbonates, sulfates, halides and phosphates are the six classes of non silicate minerals. These are about 8% in the earth crust. Non silicate minerals have importance as ore minerals.

2. **Silicate Minerals:** They are composed of silicon and oxygen atoms. Silicate has the chemical formula of SiO_4^{4-}. Silicate minerals are the most abundant minerals on the earth surface. Most of the silicate minerals are rock forming minerals. There are six groups of silicate minerals on the basis of number of oxygen atoms shared by one silicate tetrahedron as neosilicates (e.g., forsterite), sorosilicates (e.g., epidote), cyclosilicates (e.g., beryl), inosilicates (e.g., tremolite), phyllosilicates (e.g., talc), and tectosilicates (e.g., quartz).

3. **Organic mineral:** Organic minerals are carbon containing organic compounds in mineral forms. Organic minerals are formed by biological processes or due to the presence of biological materials. These are not considered as true minerals. There are three classes of organic minerals as hydrocarbons, containing hydrogen and carbon, salts of organic acids, and miscellaneous. Organic minerals are rare.

The minerals are also grouped in five or more broad classes as: 1. Primary minerals 2. Secondary minerals 3. Sialic minerals 4. Mafic minerals and 5. Accessory minerals.

1. **Primary minerals:** Any mineral formed during the crystallization and solidification of the molten magma and host igneous primary rock. Feldspars, quartz and micas are primary minerals.

2. **Secondary minerals:** Any mineral which is formed from the geochemical, chemical, hydrothermal alterations and weathering of primary minerals and solidification of precipitates of soluble substances. Aragonite, calcite, dolomite, gypsum and clay minerals are secondary minerals.

3. **Sialic minerals:** Minerals having more amounts of silicon and aluminium with low density and light in colour are called as Sialic minerals. The examples are feldspars, quartz and muscovite.

4. Mafic minerals: Minerals having more amounts of magnesium and iron and with more density and dark colour are called as mafic minerals.The examples are pyroxene, amphiboles, olivine and biotite.

5. **Accessory mineral:** Accessory mineral is a constiuent mineral present in small quantity in any rock which is not considered for identifcation and name the rock. It is also called as minor accessory when present in small amounts. Some accessory minerals are pyroxene, amphibles, topaz, zircon, corundum, fluorite, garnet, monazite, rutile, magnetite, ilmenite, allanite, tourmaline, storolite, biotite, muscovite, olivine etc. However, for the sake of simplification the minerals can be placed in nine classes as Table 3.15.

Table 3.15: Nine classes of minerals

Class	Class name	Description
1.	Native elements	Minerals of Cu, S, Pt, C, Au and Ag
2.	Sulphides and analogues	Chalcocite (Cu_2OS), Sphalerite (ZnS), Pyrite (FeS_2) Molybdenite (MoS_2) and Smaltite $(Co\ Ni)_4\ (AS_4)_3$
3.	Halides	Fluorite (CaF_2), Halite (NaCl), Sylvite (KCl) and Carnatlite ($K\ MgCl_{3.}6H_2O$)
4.	Oxides and Hydroxides	Cuprite (Cu_2O), Hematite (Fe_2O_3), Brucite ($MgOH_2$), Goethite (FeOOH) and Quartz (SiO_2)
5.	Borates, carbonates and Nitrate	Borax ($[Na_2BO_5(OH)_46H_2O]$, Metaborate (HBO_2), Calcite ($CaCO_3$), Rhodochrosite ($MnCO_3$), Magnesite ($MgCO_3$), Smithsonite ($ZnCO_3$), Dolomite (Ca Mg $(CO_3)_2$ and Salt peter ($NaNO_3$)
6.	Sulphate and analogues	Anhydrite ($CaSO_4$), Gypsum ($CaSO_{4.}2H_2O$), Powellite ($CaMoO_4$) and chromite ($FeCr_2O_4$)
7.	Phosphate and analogues	Apatite $Ca_5(PO_4)_3F$
8.	Silicates	Fayalite (Fe_2SiO_4), Zircon ($ZnSiO_4$), Hemimorphite ($Zn_4Si_2O_7OH_2$ 4O), Enstatite ($Mg_2Si_2O_8$), Talc ($Mg_3Si_4O_{10}\ (OH)_2$), Kaolinite ($Al_4Si_4O_{10}\ (OH_8)$ and Albite ($NaSi_3A10_8$)
9.	Organic Compounds	Pearl, kidney stones and salts of organic acids (whewellite)

Occurrence of Minerals

Red hot lava is the product of volcanic eruptions or discharge which consolidates into the rocks. Crystals of numerous minerals are present in these rocks. The rocks forming minerals as well as those of economic value all are found in igneous, sedimentary or metamorphic rocks. But their presence in rocks is under certain mineral associations. Therefore, quartz containing igneous rock also has orthoclase, muscovite, biotite and albite. Similarly other minerals in olivine bearing igneous rocks are pyroxenes and plagioclases. The other examples of Paragenetic associations are pyrite, chalcopyrite and goethite. Kyanite, Sillimanite, Chlorite, andalusite and garnet are found in metamorphic rocks. The sedimentary rocks are the product of other rocks. Thus,, various minerals or mineral association may also be found in these rocks. The natural mineral associations or coexistence of minerals in rocks is called as Paragenesis i.e. the regular and combined occurrence of minerals in the earth's crust that are related due to common conditions of their formation.

Presence of main minerals in Earth's Crust and Soils

Earth's crust contains more amounts of feldspars, augite, hornblende and quartz and soils contain more amounts of quartz, clay minerals and numerous other minerals. The approximate mineralogical composition of soil and earth's crust is given in table.

Table 3.16: Mineral composition of Earth's crust and soils (%) by weight)

Mineral/Mineral group	Earth's crust	Soil
Feldspars	58.0	2.5
Augite and hornblende	17.0	3.0
Quartz	12.0	44.5
Mica	4.0	0.5
Clay mineral	1.0	21.5
Others	8.0	28.0

Minerals Present in Soils

The abundantly present minerals in soils are quartz, mica, feldspars, pyroxenes, amphiboles, goethite, hematite, gibbsite, calcite, gypsum, tourmaline, garnet, zircon, apatite, kyanite, rutile, rheolite, epidote, magnetite, ilmenite, monazite, sillimanite, cyolite, barite, siderite, pyrite, illite, smectite, montmorillonite, beidellite, vermiculite, kaolinite, halloysite, chlorite etc. along with some mixed layer minerals. Soils contain numerous minerals which are found in their coarse (sand and silt) and fine (clay) fractions. Therefore, the presence of minerals may be summarised as: minerals present in coarse fraction of soils and minerals present in fine fraction of soil.

1. Minerals present in coarse fraction of soils: Primary minerals present in coarse fraction of soils: Quartz, Mica, Feldspars, Pyroxenes, Amphiboles and Olivine.

Secondary minerals present in coarse fraction of soils: Goethite, Hematite, Gibbsite, Calcite, Gypsum, Illite and Kaolinite.

Accessory minerals present in coarse fraction of soils: Tourmaline, Garnet, Zircon, Apatite, Kyanite, Rutile, Rheolite, Apidote, Magnetite, Ilmenite, Monazite, Sillimanite and Barite.

2. Minerals present in fine fraction of soils: Illite, Smectite, Montmorillonite, Beidellite, Vermiculite, Kaolinite, Halloysite, Chlorite, and partly mixed layered minerals, Mica, Quartz and Feldspars.

The short description of some minerals is given is given here:

1. Name and formula: Quartz SiO_2 2. System Trigonal 3. Colour and Streak Colourless white to grey and colourless 4. Lustre Vitreous 5. Hardness 7 6. Density 2.0-2.5 7. Cleavage & fracture none and conchoidal 8. Class Oxide 9. Occurrence and specific feature Igneous and sedimentary rock. It is a highly weathering resistance mineral.

2. Mica: Mica is the group of layered aluminosilicates. Muscovite, phlogopite and biotite are three types of mica which are described here:

A. **Name and formula:** Muscovite, $KAl_2AlSi_3O_{10}$ $(OH)_2$ 2. System Monoclinic 3. Colour and streak Colourless and colourless 4. Lustre Vitreous 5. Hardness 2.0-2.5 6. Density 2.78-2.88 7. Cleavage & fracture Eminent 8. Class Phyllosilicate 9. Occurrence and specific feature Igneous and metamorphic rocks. Also called potash Alumina mica. More in Indian soils.

B. **Name and formula:** Phlogopite (Mica) K Mg_3 $AlSi_3O_{10}$ $(F,OH)_2$ 2. System Monoclinic 3. Colour and streak Yellowish brown and colourless 4. Lustre Vitreous 5. Hardness 2.3-3.0 6. Density 2.7-2.9 7. Cleavage & fracture Eminent 8. Class Phyllosilicate 9. Occurrence and specific feature Metamorphic rocks, Also called as Magnesian mica.

C. **Name and formula:** Biotite (Mica) $K(Mg.Fe)_3$ Si_3AlO_{10} $(OH)_8$ 2. System Monoclinic 3. Colour and streak Brown and Black and colourless 4. Lustre Dull 5. Hardness 2.5-3.0 6. Density 3.0-3.2 7. Cleavage & fracture Eminent 8. Class Phyllosilicate 9. Occurrence and specific feature Metamorphic rocks, Also called as Magnesian mica.

3. Feldspars: Feldspars are potassium, sodium and calcium rich aluminosilicates. They also contain Barium and Strontium. Alkali feldspars and plagioclase feldspars are two groupes of feldspars. Alkali feldspars include albite, anorthoclase, sanidine, orthoclase and microcline. Plagioclase feldspars include albite, oligoclase, andesine, labradorite, bytownite and anorthite. These are described here:

Alkali Feldspars

a. **Name and formula:** Albite and $Na(AlSi_3O_8)$ 2. System Triclinic 3. Colour and streak White greyish 4. Lustre Vitreous 5. Hardness 6.0-6.5 6. Density 2.6 7. Cleavage & fracture perfect 8. Class Tectosilicate 9. Occurrence and specific feature Igneous and Metamorphic rocks.

b. **Name and formula:** Anorthoclase and $(Na\ K)AlSi_3O_8)$ 2. System Triclinic 3. Colour and streak Mostly White vitreous pearly 4. Lustre Vitreous 5. Hardness 6.0-6.5 6. Density 2.5 -2.8 7. Cleavage & fracture Perfect 8. Class aluminosilicate 9. Occurrence and specific feature Igneous and Metamorphic rocks.

c. **Name and formula:** Sanidine and $(K\ Na)AlSi_3O_8)$ 2. System monoclinic 3. Colour and streak Mostly White Vitreous pearly 4. Lustre Vitreous 5. Hardness 6.0-6.5 6. Density 2.53 -2.56. 7. Cleavage & fracture Perfect 8. Class aluminosilicate 9. Occurrence and specific feature Igneous rocks.

d. **Name and formula:** Orthoclase and ($KAlSi_3O_8$), 2. System monoclinic 3. Colour and streak Colourless reddish tints and colourless 4. Lustre Vitreous 5. Hardness 6.0-6.5 6. Density 2.54 -2.57. 7. Cleavage & fracture Perfect and angle 8. Class tectosilicate 9. Occurrence and specific feature Igneous rocks , schists and Gneisses.

e. **Name and formula:** Microcline and $K(AlSi_3O_8)$2. System triclinic 3. Colour and streak Reddish and brownish and colourless 4. Lustre Vitreous 5. Hardness 6.0-6.5 6. Density 2.54 -2.57. 7. Cleavage & fracture Perfect 8. Class tectosilicate 9. Occurrence and specific feature Igneous and metamorphic rocks.

Plagioclase Feldspars

a. **Name and formula:** Albite and $Na(AlSi_3O_8)$ 2. System triclinic 3. Colour and streak White greyish 4. Lustre Vitreous 5. Hardness 6.0-6.5 6. Density 2.62. 7.Cleavage & fracture Perfect and 8. Class tectosilicate 9. Occurrence and specific feature Igneous and metamorphic rocks.

b. **Name and formula:** Oligoclase and $Na_4/_5Ca_1/_5$ $(Al_3 Sio_8)$ 2. System triclinic 3. Colour and streak White greyish 4. Lustre Vitreous 5. Hardness 6.0-6.5 6. Density 2.6. 7.Cleavage & fracture Perfect Oligoclase $Na_4/_5Ca/_5$ $(Al_3 Sio_8)$ 8. Class tectosilicate 9. Occurrence and specific feature Igneous and metamorphic rocks.

c. **Name and formula:** Andesine and $Na_3/_5Ca_2/_5$ (Al_3Sio_8) 2. System triclinic 3. Colour and streak White greyish 4. Lustre Vitreous 5. Hardness 6.0-6.5 6. Density 2.68. 7.Cleavage & fracture Perfect 8. Class Aluminosilicate 9. Occurrence and specific feature Igneous and metamorphic rocks.

d. **Name and formula:** Labradorite and $Na_3/_5Ca_2/_5$ (Al_3Sio_8) 2. System triclinic 3. Colour and streak White greyish 4. Lustre Vitreous 5. Hardness 6.0-6.5 6. Density 2.70 Cleavage & fracture Perfect 8. Class Aluminosilicate 9. Occurrence and specific feature Igneous and metamorphic rocks.

e. **Name and formula:** Bytownite and $Na_1/_5/Ca_4/_5$ (Al_3SiO_8) 2. System triclinic 3. Colour and streak White greyish 4. Lustre Vitreous 5. Hardness 6.0-6.5 6. Density 2.75 7.Cleavage & fracture Perfect 8. Class Aluminosilicate 9. Occurrence and specific feature Igneous and metamorphic rocks.

f. **Name and formula:** Anorthite and Ca (Al_2SiO_8) 2. System triclinic 3. Colour and streak White greyish 4. Lustre Vitreous 5. Hardness 6.0-6.5 6. Density 2.76 7. Cleavage & fracture Perfect 8. Class Aluminosilicate 9. Occurrence and specific feature Igneous and metamorphic rocks.

4. Pyroxenes: Pyroxenes are the minerals of Mafic group and mainly contain Magnessium and Iron. Generally found in alkaline rocks. High temperature and/ or high pressure are favourable for the formation of pyroxenes. The pyroxenes are single chain silicates where the long chains of Silica tetrahedron are attached with Calcium, Aluminium, Lithium, Manganese and other cations. The higher weathering rates provide many plant nutrients. There are four types of Pyroxenes namely, Augite, Diopside, Hypersthene and Enstatite. These are described here:

a. **Name and formula:** Augite $(Ca,Na)(Mg,Fe,Al,Ti)(Si,Al)_2O_6$ 2. System Monoclinic 3. Colour and streak Black and Colourless 4. Lustre Vitreous to pearly 5. Hardness 5-6.0 6. Density 3.2-3.6 7. Cleavage & fracture Perfect 8. Class Inosilicate 9. Occurrence and specific feature Mafic Igneous rocks.

b. **Name and formula:** Diopside CaMg (Si_2O_8) 2. System Monoclinic 3. Colour and streak Dull green and Colourless 4. Lustre Vitreous to pearly 5. Hardness 5.5-6.0 6. Density 3.3-3.4 7. Cleavage & fracture Medium 8. Class Inosilicate 9. Occurrence and specific feature Metamorphic rocks, silicates, limestone and dolomite.

c. **Name and formula:** Hypersthene $(MgFe)_2$ (Si_2O_8) 2. System Rhombohedral 3. Colour and streak Greenish to brownish black and colourless 4. Lustre Vitreous to pearly 5. Hardness 5.5 6. Density 3.3-3.5 7. Cleavage & fracture Medium 8. Class Inosilicate 9. Occurrence and specific feature Metamorphic rocks, silicates, limestone and dolomite.

d. **Name and formula:** Enstatite $Mg^2(Si_2O_8)$ 2. System Rhombohedral 3. Colour and streak Colourless and colourless 4. Lustre Vitreous 5. Hardness 5.5 6. Density 3.1-3.3 7. Cleavage & fracture Medium at 85^0 C 8. Class Inosilicate 9. Occurrence and specific feature Metamorphic rocks.

5. Amphiboles: Amphiboles are Mafic minerals with two groups and many subgroups. Amphiboles are double chain silicates where the long chains of Silica tetrahedron are attached with Calcium, Magnesium, Iron, Manganese. Sodium, Aluminium and other cations. They are weathering products of pyroxenes and generally found in acidic rocks. The higher weathering rates provide many plant nutrients. High temperature and/or high pressure are favourable for the formation of Amphiboles. Amphiboles are also called as Asbestos and Hornblende. The four main subgroups of amphiboles are, 1. Iron-magnesium-manganese amphibole, 2. Calcic amphibole, 3. Sodic-calcic amphibole and 4. Sodic amphibole. Main Amphiboles are Anthophyllite, Tremolite, Actinolite and Hornblende. These are described here:

a. **Name and formula:** Anthophyllite $(Mg,Fe)_7Si8O_{22}(OH)_2$ 2. System Orthorhombic 3. Colour and streak White, greenish grey and White to greyish white 4. Lustre Vitreous, Pearly 5. Hardness 5.5-6 6. Density 2.9-

3.6 7. Cleavage and fracture Perfect and conchoidal 8. Class Inosilicate 9. Occurrence and specific feature Ultrabasic igneous rocks.

b. **Name and formula:** Tremolite $Ca_2Mg_5Si_8O_{22}(OH)_2$ 2. System Monoclinic 3. Colour and streak Many colours varying from White to purple and colourless 4. Lustre Vitreous, silky 5. Hardness 5.0-6 6. Density 2.9-3.2 7. Cleavage and fracture Prismatic and uneven 8. Class Inosilicate 9. Occurrence and specific feature Igneous rocks.

c. **Name and formula:** Actinolite $Ca_2(Mg,Fe)_5Si_8O_{22}(OH)_2$ 2. System Monoclinic 3. Colour and streak Bottle green grayish green to black and colourless.4. Lustre Vitreous, silky 5. Hardness 5.5-6 . Density 2.9-3.5 7. Cleavage and fracture Prismatic and uneven 8. Class Inosilicate 9. Occurrence and specific feature Igneous and metamorphic rocks.

d. **Name and formula:** Hornblende $Ca_2(Mg, Fe, Al)_5 (Al, Si)_8O_{22}(OH)_2$. System Monoclinic 3. Colour and streak Black, dark green,dark brown and white, colourless.4. Lustre Vitreous 5. Hardness 5-6 . Density 2.9-3.4 7. Cleavage and fracture Prismatic and uneven, splintry 8. Class Inosilicate 9. Occurrence and specific feature Igneous and metamorphic rocks.

6. Olivine: Olivine is one of the most common Mafic minerals in the earth. It is a major rock forming mineral which is not classified as an individual mineral and considered as a mineral group. It is easily weatherable mineral which helps in the build up of soil fertility. The olivine is $(Mg\ Fe)_2\ SiO_4$ which is somewhere in between Forsterite (Chrysolite) and Fayalite seires. However, the properties of Olivine, Forsterite and Fayalite are given here:

a. **Name and formula:** Olivine $(Mg\ Fe)_2\ SiO_4$. System Orthorhombic 3. Colour and streak Black, dark green, dark brown and white, colourless.4. Lustre Vitreous 5. Hardness 5-6 . Density 2.9-3.4 7. Cleavage and fracture Prismatic and uneven, splintry 8. Class Inosilicate 9. Occurrence and specific feature Igneous and metamorphic rocks.

b. **Name and formula :** 1. Forsterite $Mg_2\ SiO_4$. 2. System Monoclinic 3. Colour and streak Black, dark green, dark brown and white, colourless.4. Lustre Vitreous 5. Hardness 5-6 . Density 2.9-3.4 7. Cleavage and fracture Prismatic and uneven, splintry 8. Class Inosilicate 9. Occurrence and specific feature Igneous and metamorphic rocks.

c. **Name and formula :** 1.Fayalite $Fe_2\ SiO_4$. 2.System Orthorhombic 3. Colour and streak Greenish yellow, yellow-brown, brown; pale yellow to amber and white.4. Lustre Vitreous 5. Hardness 6.5- 7 . 6. Density 4.4 7. Cleavage and fracture Distinct in two directions at 90 degrees and conchoidal 8. Class Inosilicate 9. Occurrence and specific feature acidic and alkaline igneous rocks.

Secondary Minerals Present in Coarse Fraction Of Soils

1. **Name and formula :** 1. Goethite FeOOH. 2. System Orthorhombic 3. Colour and streak Dark Brown to yellow brown and brown.4. Lustre Adamantine 5. Hardness 5- 5.5 . 6. Density 4.4 7. Cleavage and fracture Perfect 8. Class Hydroxide 9. Occurrence and specific feature Sedimentary iron ore, laterite and limonite.

2. **Name and formula:** 1. Hematite Fe_2O_3. 2. System Trigonal 3. Colour and streak Iron black and bright red to cherry red.4. Lustre Submetallic 5. Hardness 6-6.5 . 6. Density 5-5.3 7. Cleavage and fracture None and irregular 8. Class Oxide 9. Occurrence and specific feature Metamorphic and sedimentary rocks.

3. **Name and formula :** 1. Gibbsite $Al(OH)_3$. 2. System Monoclinic 3. Colour and streak White to off white and white 4. Lustre Pearly 5. Hardness 2.5-3.5 . 6. Density 2.4 7. Cleavage and fracture Perfect and Tough 8. Class Hydroxide 9. Occurrence and specific feature Laterites, highly-weathered soils and clay deposits and Weathering of bauxite.

4. **Name and formula :** 1. Calcite $CaCO_3$. 2. System Trigonal 3. Colour and streak White various coloures due to impurities and white 4. Lustre Vitreous 5. Hardness 3. . 6. Density 2.6-2.8 7. Cleavage and fracture Perfect and conchoidal 8. Class Carbonates 9. Occurrence and specific feature Sedimentary rocks limestone, marbles.

5. **Name and formula :** 1. Gypsum $CaSO_4.2H_2O$. 2. System Monoclinic 3. Colour and streak White and white 4. Lustre Vitreous 5. Hardness 1.5-2.0 6. Density 2.3 7. Cleavage and fracture Perfect and conchoidal 8. Class Sulphates 9. Occurrence and specific feature Horizons of arid and semiarid soil and sedimentary rocks.

6. **Name and formula :** 1. Illite $(K,H_3O)(Al,Mg,Fe)_2(Si,Al)_4O_{10}[(OH)_2,(H_2O)]$ variable due to isomorphous substitution 2. System Monoclinic 3. Colour and streak Yellow, green, brown and white 4. Lustre Pearly to bronze 5. Hardness 1.5 6. Density 2.3 7. Cleavage and fracture Perfect and conchoidal 8. Class Phyllosilicates 9. Occurrence and specific feature Alteration of biotite.

7. **Name and formula :** 1. Kaolinite $Al_4 (Si_4O_{10}) (OH)_8$ 2. System Monoclinic 3. Colour and streak White with coloured tints and white 4. Lustre Pearly to dull earthy 5. Hardness 2- 2.5 6. Density 2.6 7. Cleavage and fracture Perfect and conchoidal 8. Class Phyllosilicates 9. Occurrence and specific feature Chemical weathering of rocks in hot, moist climates of tropical rainforest areas and Weathered Aluminosilicates.

Accessory Minerals Present in Coarse Fraction of Soils

1. **Name and formula:** 1.Tourmaline Na Mg_3Al_6 $(BO_3)_3$ $Si_60_{18}(OH)_4$ 2. System Trigonal 3. Colour and streak Redsish pink and colourless 4. Lustre Vitreous 5. Hardness 7- 7.5 6. Density 2.4-3.2 7. Cleavage and fracture None 8. Class Borosilicates (Cyclosilicates) 9. Occurrence and specific feature Metamorphic rocks.

2. **Name and formula:** 1. Garnet (CaMgFe,Mn) (FeAlTiCr) $(SiO_4)_3$ 2. System Isometric 3. Colour and streak Many except dark blue and colourless 4. Lustre Vitreous 5. Hardness 6.5- 7.5 6. Density 3.5-4.2 7. Cleavage and fracture Imperfect and conchoidal 8. Class Silicates 9. Occurrence and specific feature Metamorphic rocks.

3. **Name and formula:** 1. Zircon $ZrSiO_4$2. System Tetragonal 3. Colour and streak Yellow, orange, brown and colourless 4. Lustre Vitreous 5. Hardness 7.5 6. Density 4.7 7. Cleavage and fracture Imperfect and conchoidal 8. Class Silicates 9. Occurrence and specific feature Igneous and Metamorphic rocks.

4. **Name and formula:** 1. Apatites $(Ca_3(PO_4)_2)_3$ CaF_2,$CaCl_2$, $Ca(OH)_2$ $CaCO_3$ 2.. System Hexagonal 3. Colour and streak White, green, blue, violet and white, colourless 4. Lustre Vitreous 5. Hardness 5.0 6. Density 3.2 7. Cleavage and fracture ImPerfect and conchoidal 8. Class Phophates 9. Occurrence and specific feature Basaltic and sedimentary rocks.

5. **Name and formula:** 1. Kyanite Al_2SiO_5 2.. System Hexagonal 3. Colour and streak White, green, blue, violet and white, colourless 4. Lustre Vitreous, pearly 5. Hardness 4.5 – 5 along length and 6.5-7 across width 6. Density 3.6 7. Cleavage and fracture Perfect and conchoidal 8. Class Silicates 9. Occurrence and specific feature Metamorphic rocks.

6. **Name and formula:** 1. Rutile $TiO_2$2.. System Tetragonal 3. Colour and streak Dark yellow brown and white, light brown 4. Lustre Adamantine 5. Hardness 6 6. Density 4.2 7. Cleavage and fracture Perfect and conchoidal 8. Class Oxide 9. Occurrence and specific feature Metamorphic rocks and Granite.

7. **Name and formula:** 1. Epidote $Ca_2(Al_2.Fe)$ Si_2O_7 (SiO_4) O (OH) 2. System Monoclinic 3. Colour and streak Pistachio green and white 4. Lustre Vitreous 5. Hardness 6-7 6. Density 3. 4 7. Cleavage and fracture Perfect and Regular 8. Class Sorosilicates 9. Occurrence and specific feature Metamorphic rocks.

8. **Name and formula:**1. Magnetite Fe_3O_4 2. System Isometric 3. Colour and streak Black to silvery grey green and black 4. Lustre Submetallic 5. Hardness 5.5-6.5 6. Density 4.9-5.2 7. Cleavage and fracture None and

Uneven 8. Class Oxide 9. Occurrence and specific feature Igneous and Metamorphic rocks.

9. **Name and formula:** 1. Ilmenite $FeTiO_3$ 2. System Trogonal 3. Colour and streak Black and black 4. Lustre Metallic 5. Hardness 5-6 6. Density 4.7 7. Cleavage and fracture None and Conchoidal 8. Class Oxide 9. Occurrence and specific feature Igneous and Sedimentary rocks.

10. Name and formula:1. Monazite (Ce,La,Th) (PO_4) 2. System Monoclinic3. Colour and streak Brown Yellow and Colourless 4. Lustre Vitreous greasey 5. Hardness 5-5.5 6. Density 4.8-5.5 7. Cleavage and fracture Distinct and Conchoidal 8. Class Phosphate 9. Occurrence and specific feature Metamorphic rocks and ore of Thorium.

11. **Name and formula:** 1. Sillimanite Al_2SiO_5 2. System Orthorhombic 3. Colour and streak Colourless, white grey and white 4. Lustre Vitreous 5. Hardness 6.5-7.5 6. Density 3.3 7. Cleavage and fracture Perfect and Splintry 8. Class Orthosilicate 9. Occurrence and specific feature Metamorphic rocks.

12. **Name and formula:**1. Barite $BaSO_4$ 2. System Orthrhombic 3. Colour and streak White grey and white 4. Lustre Vitreous 5. Hardness 3-3.5 6. Density 4.4 7. Cleavage and fracture Perfect and uneven 8. Class Sulphates 9. Occurrence and specific feature Sedimentary rocks.

2. Minerals Present in Fine Fraction of Soils

1. Illite, Kaolinite Mica, Quartz and Feldspars are already described earlier.
2. **Name and formula:**1. Montmorillonite $(Al\ Mg)_2\ Si_4\ O_{10}\ OH_2.4H_2O$ 2. System Monoclinic 3. Colour and streak White with tints and colourless 4. Lustre Dull 5. Hardness 1 6. Density 2 7. Cleavage and fracture Perfect and uneven 8. Class Phyllosilicates 9. Occurrence and specific feature Weathering of soils and clays.
3. **Name and formula:**1. Beidellite $(Na,Ca_{0.5})_{0.3}Al_2((Si,Al)_4O_{10})(OH)_2 \cdot nH_2O$ 2. System Monoclinic 3. Colour and streak White, reddish, brownish-grey and white 4. Lustre Waxy 5. Hardness 1-2 6. Density 2 7. Cleavage and fracture Perfect and uneven 8. Class Phyllosilicates 9. Occurrence and specific feature Bentonitic clays, alteration product in hydrothermal mineral deposits and soils derived from mafic rocks.
4. **Name and formula:** 1. Vermiculite $Mg_{0.7}(Mg,Fe,Al)_6(Si,Al)_8O_{20}(OH)_4 \cdot 8H_2O$ 2. System Monoclinic 3. Colour and streak Yellow, green,brown and greenish white 4. Lustre Vitreous 5. Hardness 1.5-2 6. Density 2.3 7. Cleavage and fracture Perfect and uneven 8. Class Phyllosilicates 9. Occurrence and specific feature Volcanic rocks and alteration product of biotite and phlogopite.

5. **Name and formula:** 1. Halloysite $Al_2Si_2O_5(OH)_4$ 2. System Monoclinic 3. Colour and streak White grey, green, blue, yellow, red from included impurities.4. Lustre Pearly, waxy, or dull 5. Hardness 2-2.5 6. Density 2-2.6 7. Cleavage and fracture Probable and Conchoidal 8. Class Phyllosilicates 9. Occurrence and specific feature Basaltic rocks and Rhyolite.

6. **Name and formula:**1. Chlorite $(Mg,Fe)_3(Si,Al)_4O_{10}(OH)_2{\cdot}(Mg,Fe)_3(OH)_6$ 2. System Monoclinic 3. Colour and streak White; grey, green, blue, yellow, red and Pale green to grey 4. Lustre Pearly, waxy, or dull 5. Hardness 2-2.5 6. Density 2.6-3.3 7. Cleavage and fracture Perfect and Lamellar 8. Class Phyllosilicates 9. Occurrence and specific feature Igneous rocks and alteration product of mafic minerals.

7. **Mixed layered minerals:** In the fine fraction of soils many minerals, mostly clay minerals, are found partially or completely mixed as intermediate products of different clays. These are stacked together in various ways to make bundles or packets of new structures different from their primary components. This is known as interstratification and such minerals are termed as interstratified or mixed layered minerals. Interstratification may be regular or random type.These minerals cannot be separated by physical methods. Mostly mixed layer minerals have two components, but interstratifications with more minerals have also been noticed. Segregation, precipitation, solid state transformation, dissolution and crystallization are the main mechanisms. The weathering reactions with high-temperature are responsible to form mixed layer minerals.

Weathering of Rocks and Minerals

Soils are formed due to the action of various pedogenic processes on the soil parent material. The soil parent material may be unconsolidated and more or less wearthed inorganic mineral or organic matter. Weathered mineral matter is obtained from weathering of rocks. This may be present in situ on bed rock or deposited at newer places after transported by ice, water, air or gravity. Similarly, organic parent material is obtained from organic wastes and residues of plants and animals.

Physical and chemical changes in the rocks caused by environmental agents are called as weathering. This is the sum of all the processes which create mechanical disintegration and chemical decomposition of rocks present on or nearby the earth's crust. In the weathering process larger rocks break down in to smaller particles due to various physical factors which upon actions of chemical and biological agents convert into new minerals. Destruction and synthesis processes run simultaneously during weathering. Water, air, temperature, presence of Oxygen and CO_2 gases in air and organisms are main weathering agents.Weathering depends upon chemical composition of rocks, climate, relief and hydrochemical condition of the area.

Types of Weathering

Physical, chemical and biological are three types of weathering which consist many processes. These processes are interrelated to each other. But due to geoclimatic condition of the area, any one type of weathering may be more effectual.

Physical Weathering

This is also called as mechanical weathering where breaking and disintegration of rocks takes place. The rocks are converted into smaller fragments without any chemical change. Water, temperature and air are main agents of physical weathering whose effectiveness depends upon the texture of rocks.

Water performs physical weathering of rocks in many ways. Rain drops, flowing rivers, hail storms and moving glaciers cause abrasion of minerals and rocks due to their beating action. Flowing water in the rivers has significant scouring and load carrying capacities. Rapidly moving water can lift rocks from the river bottom for a short time. When these rocks fall to bottom, they collide with other rocks and break them in smaller pieces. Also, the transported rock pieces in flowing water generate particles of different sizes. When these particles meet already weathered surfaces, they remove them and newer surfaces for weathering are again exposed.

Ice wedging or frost wedging breaks many rocks. Ice wedging means repeated freezing and melting of water within small cracks in the rock surface. Water seeps into these cracks. There is increase in the volume of water when it freezes into ice. When the temperature drops below freezing, the water expands as ice in the cracks. Thus, freezed water exerts tremendous pressure in the cracks of the rocks. This extra pressure develops new cracks; widen old cracks and breaks large rocks into small pieces. Hitting rain drops and hailstorms are also abrase rocks and minerals. In the cold hilly areas glaciers melt due to the increase in temperature. The force of melted and moving ice along the slope is quite abrasive for rocks. This causes their fragmentation.

Plants also grow in cracks. Their growing roots produce substantial pressure for widening the cracks. Later, the increasing depth and width of roots further widen the cracks to break rocks apart.

Temperature plays an important role in the shattering and exfoliation of rocks and minerals. Minerals and rocks expand due to higher temperature and lower temperature contract them. The expansion and contraction is more pronounced in the heterogenous rocks having different minerals. Thus,, separate coefficient of expansion and contraction increase the shattering and abrasion of rocks and minerals.

The wind carrying fine sand particles show sand blasting which bruise the uncovered rocks and minerals. In the coastal areas oceanic waves produced by the joint action of air and water act as a significant weathering agent for rocks and minerals. Sometimes rock boulders etc. slide from higher places due to the gravity and break down into smaller pieces. It indicates the effect of gravity on weathering.

Chemical Weathering

Chemical decomposition of rocks and minerals takes place in chemical weathering. Thus, there is formation of new and stable minerals having different physico-chemical properties than the earlier rocks. The simple example of chemical weathering is the formation of clay minerals from the orthoclase. Generally water, oxygen and acids perform chemical weathering with the help of chemical weathering processes. Solution, Hydration, hydrolysis, oxidation, reduction, carbonation and restoration are chemical weathering processes. These are described here:

1. Solution: Water is an universal solvent. Thus, it can dissolve minerals transported by flowing or percolating water and deposit at newer places. The rocks having water soluble and insoluble minerals develop holes and sills due to the removal of water soluble minerals. Although solution is not an important weathering process but it significantly influence the weathering of halite/ rocksalt (NaCl) and Chile salt peter ($NaNO_3$). The dissolution power of water is enhanced by dissolved organic and inorganic acids.

2. Hydration: Chemical combination of water with the rocks and minerals in different quantities is called as hydration. There is rigid attachment of H^+ and OH^- ions of water to the molecules and atoms of the mineral. Hydration increases the volume and softness of minerals and they are weathered easily. Hydration forms the hydrated compounds of feldspars, amphiboles, micas, pyroxenes, hematite, bauxite, anhydrite etc. Hydration also removes the ferrous oxide (FeO) from the chrysolite $(MgFe)_2\ SiO_4$.

$2\,Fe_2O_3 + H_2O \rightarrow Fe_2O_3.\,3H_2O$

(Hematite, Red) (Limonite, Yellow)

$CaSO_4 + 2H_2O \rightarrow CaSO_4.\,2H_2O$

(Anhydrite, Grey bluish) (Gypsum, White)

$Al_2O_3 + 3H_2O \rightarrow Al_2O_3.3H_2O$

(Bauxite, Dark red) (Hydrated Aluminum Oxide, White)

3. Hydrolysis: Hydrolysis is an important process of chemical weathering and differs from solution and hydration. The term hydrolysis refers hydro means water and lysis means to loosen or decompose i.e. hydrolysis is a chemical reaction where water loosens or decomposes the chemical bonds within mineral and breakdown them. The hydrogen and hydroxyl ions of water react with rocks and minerals to form their hydroxides as well as new minerals and compounds. All these are more weatherable than the earlier ones. Here are some examples:

Potash feldspar (orthoclase) is a less soluble mineral which upon hydrolysis converts into acid silicate clay and Potassium hydroxide. Both are more soluble.

$KAlSi_3O_8 + HOH \rightarrow HAlSi_3O_8 + KOH$

Potash feldspar   Acid silicate clay+Potassium hydroxide

The hydrolysis of Orthoclase also produces Kaolinite as under

$4KAlSi_3O_8 + 4H^+ + 2H_2O \rightarrow 4K^+ + Al_4Si_4O_{10}(OH)_8 + 8SiO_2$

The Potassium ions may be leached, absorbed by plants or adsorbed by clay minerals. Kaolinite formation process is called as Kaolinization

4. Oxidation: In the chemical weathering oxidation means the addition and combination of oxygen to rocks and minerals or loss of electrons from elements. The ample amounts of air are present in atmosphere as well as in water. The oxidation is more active in the presence of moisture. The combination of oxygen with minerals increases their volume and softness. The change in electron number of elements alters their electro neutrality. Thus, minerals and elements become more susceptible for water and acids. The rusting of iron in the airy and aqueous medium is one of the examples of oxidation. The other examples are:

$4FeO$ (Ferrous oxide) $+ O_2 \rightarrow 2Fe_2O_3$(Ferric oxide)

Generally the oxidation of ferriferous minerals and di-trivalent iron takes place in the presence of moisture. Thus, oxidation and hydration processes take part in the formation of Goethite from Ferrous oxide. The reaction is given below:

$4FeO + O_2 + 2H_2O \rightarrow 4FeOOH$

(Ferrous oxide)   (Goethite)

Similarly FeS (Troilite) and $FeSiO_3$ (Ferrosilite) change to $Fe(OH)_3$

$4\,FeS + 10\,H_2O + 9\,O_2 \rightarrow 4\,Fe(OH)_3 + 4H_2SO_4$

$FeSiO_3 + 6H_2O + O_2 \rightarrow 4\,Fe(OH)_3 + 4SiO_2$

5. Reduction: Reduction is a chemical process in which electrons are gained, negative charge is increased and positive charge is decreased. There is also loss or removal of oxygen from the rocks and minerals. Reduction is somewhat reverse of oxidation. It generally takes place under excess water, water logged and oxygen deficient areas where Ferric Oxide (Fe_2O_3) is again changed to Ferrous Oxide (FeO). Changing oxidation-reductions conditions provide different colour to the soil as ferric iron is converted to ferrous iron. The equation is given below:

$2Fe_2O_3 - O_2 \rightarrow 4FeO$

6. Carbonation: Carbonation performs chemical weathering of rocks, minerals etc. Carbonation is a two-step reaction. First, carbon-dioxide reacts with water to form carbonic acid and then carbonic acid reacts with minerals and rocks to produce carbonate, bicarbonates or others. Carbonic acid is a weak acid. Metabolic activities of organisms, decomposition of organic matter, respiration and atmosphere are the sources of CO_2 gas to form the carbonic acid. Carbonation is an important process in the formation of many caves and sinkholes. Carbonic acid and other organic or inorganic acids dissolve rocks and minerals more easily and rapidly than the water alone. The few examples of carbonation are,

Insoluble Calcite mineral is converted into soluble Calcium Bicarbonate by the action of carbonic acid as follows:

$CO_2 + H_2O \rightarrow H_2CO_3$

$CaCO_3 + H_2CO_3 \rightarrow Ca\,(HCO_3)_2$

(Calcite) (Calcium Bicarbonate)

Carbon dioxide gas vigorously reacts with strong alkali to from their carbonates and bicarbonates:

$CO_2 + NaOH \rightarrow Na_2CO_3 + H_2O$

$Na_2CO_3 + H_2CO_3 \rightarrow 2\ NaHCO_3$

In iron-rich minerals such as olivine i.e., Fayalite (Fe_2SiO_4), the carbonation brings out iron of the mineral into solution in the divalent ionic form:

$Fe_2SiO_4 + 4H_2CO_3 \rightarrow 2Fe_2^{+} + 4HCO_3 + H_4SiO_4$

Iron and oxygen dissolved in water react in the presence of bicarbonate to produce hematite and carbonic acid:

$2Fe_2^{+} + ½\ O_2 + 2H_2O + 4HCO_3^{-} \rightarrow Fe_2O_3 + 4H_2CO_3$

7. Restoration: It is not an important chemical weathering process. However, activities of micro-organisms in soil horizons, water bodies with complete absence of free oxygen and buried organic materials induce the restoration process. This process is somewhat reverse of oxidation. Restoration, in the presence of organic carbon, changes $Fe_2O_{3.}\ 3H_2O$ in to Siderite.

Biological Weathering

Both flora and fauna take part in the biological weathering. Biological weathering is the process in which plants, animals and microorganisms break down rocks into smaller pieces. But considerable role is played by plants and microorganisms. They weather the rocks and minerals by physical and chemical processes. Biological weathering is also known as organic wethering and biochemical weathering.

The roots of growing plants enter in the joints, cracks and cervices of rocks. They exert pressure during their growth and development which widen the cracks and cervices and break the rocks in small pieces. In the hilly areas growth of plants and grasses can be in seen in rocks surfaces. These plants and grasses also provide tremendous amount of organic matter. Various organic acids are produced during the decomposition of organic matter which dissolves the chemical compounds present in rocks and minerals.

Micro-organisms weather rocks and minerals in many ways and even they change the redox conditions of weathering site. Conversion of bitumen and gypsum to the native sulphur by sulphur bacteria is one example.

Lichens have chelating agents. They trap the elements of decomposable rocks and convert them in organometallic complexes. Blue green algae and diatoms weather the minerals in to amorphous and clay minerals. Earthworms ingest soil minerals and change their properties. The casts of earthworms are helpful in physical and chemical weathering. Formations of burrows by large animals, cutting of rocks by human beings for civil structures and provision of organic acids by decaying bodies also weather the rocks and minerals. Normally plants and animals increase the solvent action of rain water by the production of CO_2 and organic acids during their growth, development, death and decomposition.

Factors Affecting the Rate of Weathering

Climate, relief and topography and properties of rocks and minerals and pollution are main factors which determine the rate of weathering.

1. Climate: Rainfall and temperature are two main parameter of climate. If there is more rainfall, there will be more physical and chemical weathering. Increase in temperature boost up the action of rainfall. Wet and hot climate increase the solubility and chemical decomposition of minerals. Water speeds up rate of weathering. Weathering occurs fastest in hot and wet, very slowly in hot and dry and little in very cold and dry climates. Ice wedging occurs with the changes in temperature. Under variable climatic regimes biotite weathers to vermiculite and to smectite (arid climate) which decomposes rapidly to kaolinite mineral.

2. Relief and topography: Relief and topography also influence the rate and type of weathering through runoff and exposure to the sun. At high relief position of hills and mountains most of the water is lost as runoff. Therefore, the intensity of physical weathering by water is greater at these areas. When the slope is less, slow movement and stagnation of water for longer period, increase the chances and intensity of chemical weathering. A steep slope enhances quick fall away of weathered rocks and exposes new rock surfaces for weathering. Exposed rocks are more affected by temperature of sun and wind at elevated areas as compared low level areas. Surface covered by the vegetation and thin and small roots plants are more protected from weathering as compared to bare surfaces.

3. Properties of rocks and minerals: Size of mineral particles, hardness, tightness of packing, composition of rocks and chemical structure of minerals are some important properties of rocks and mineral which influence the weathering.

The more surface area of a rock weather more quickly. When a large piece of rock or mineral is divided in to many small pieces or particles, they have more surface area. Thus, these fine particles weather more rapidly and easily as compared to coarse particles. Fine particles are more affected by weathering agents. Percolating water weather fine grained feldspars more rapidly as compared to coarse grained feldspars.

Hardness resists the minerals for mechanical disintegration. Thus, hard minerals are not easily weatherable as compared to soft minerals. Quartz is a weathering resistant mineral as its hardness is 7.0 where as gypsum or coarse lime stone having low hardness are quite susceptible for weathering.

Monomineralic rocks are less weatherable as compared to rocks having many minerals because rocks having many minerals also have separate coefficient of expansion for each mineral. Thus, change in temperature will cause different expansion and contraction. This will cause easy physical weathering of rocks. Rocks made up of feldspar, calcite and iron, weather more easily and quickly.

Chemical structure of minerals is important for their weathering. The minerals having close packing of atoms are resistant to weathering as compared to minerals having loose or open packing of minerals. Quartz (SiO_2) has close frame structure is quite resistant to weathering where as soapstone with three layered structure is easily weatherable.

4. Pollution: Pollution increases rate of weathering. Factories and automobiles release carbon dioxide and other gases into the air. These gases dissolve in the rainwater to cause acid rain. Acid rain containing nitric and sulfuric acid dissolve rocks and minerals to speed up weathering.

Weathering Sequence and Products

First of all, the volume of weatherable compounds increases during the weathering processes. Then main rock forming minerals of earth crust (silicates and aluminosilicates) are decomposed. Simultaneously the soluble salts of K, Na, Ca, Mg, and Fe^{2+} are removed from the weathering zone and carried away by different agencies to the longer distances. Soluble products of weathering also form new minerals. The least soluble or insoluble products like Al_2O_3, TiO_2, SiO_2 etc. remain at the place of weathering or move to an inconsiderable distance.

Weathering Sequence

Jackson (1968) indicated the weathering sequences for minerals from most easily weatherable to least easily weatherable minerals as :

Olivine – Pyroxene – Amphiboles – Biotite – Muscovite – Quartz---------1

Ca feldspars – Na feldspars – K feldspars – orthoclase – Muscovite – Quartz. -----2

According to these chains or sequences olivine and feldspars are most easily weatherable and muscovite and quartz are least easily weatherable.

Also the series proposed by Jackson (1968) indicate that gypsum and halite are most easily weatherable and anatase, rutile and zircon minerals are most hardy to weather.

Gypsum, halite – calcite, apatite – olivine, pyroxenes – Biotite, glauconite – Albite, anorthite – quartz, cristobalite – Muscovite, sericite, vermiculite – montmorillonite – Kaolinite, halloysite, allophone – Gibbsite, boehmite – Hematite, goethite – Anatase, rutile, zircon.---3

These chains indicate the power of crystallization of minerals. Accordingly quartz will crystallize in last and gypsum is first to crystallize.

Weathering Products

Very simply weathering products are change of rocks shapes in to spherical, a floor of loose and decayed rock regolith and ions in solution. But it is not so simple. The olivine changes to serpentine, limonite, hematite and quartz minerals, suspended silica and iron and magnesium carbonate. Quartz changes to sand grains and some suspended silica in solution.

Pyroxenes, amphiboles and mica change to calcite, hematite, limonite, quartz and clay minerals. The other products of these are suspended silica and calcium and Magnesium carbonates and iron and magnesium in solution. The weathering products of feldspars are calcite, quartz, clay minerals, suspended silica and calcium and potassium carbonates and potassium, sodium and calcium in solution.

Weathering products of Olivine are serpentine and iron and magnesium in solution. Weathering products of Pyrite are Iron oxide minerals, sulphuric acid and iron in solution. Calcite products are Calcium and carbonate in solution.

Soil Colloids

Soil colloids are extremely small particles of soil with particle sizes of 2 micrometers in diameter or smaller suspended in a soil with larger particles. Soil colloids are finer fractions i.e. clay or humus and impart electrical and chemical properties to a soil. These have high surface area and net negative charge. The net negative charge of the soil colloids attracts and holds positively charged ions.

Colloids are also known as colloidal solutions or colloidal systems. Colloid is any non-crystalline substance remarkably larger than atoms or ordinary molecules and invisible to naked eye. Colloids include gels, sols, and emulsions. The colloidal particles do not settle and cannot be separated by ordinary filtration or centrifugation. The common examples are: Solid in liquid (Dispersion of clay in water, Cheese, Butter, Jellies, Jam and Shoe Polish), Liquid in gas (Fog or clouds in atmosphere). Other examples are smoke and dust in air; milk, liquid aerosols, pigmented plastics, toothpaste etc.

Basic Properties of Colloid

The colloids have numerous basic physical, mechanical, optical, electrical properties. These properties help in the identification of colloids from true solution or crystalloid. Some of them are summarised here:

1. **Size:** The size of the suspended particles in a colloid can range from 1 to 1000 nanometres.
2. **Heterogeneity:** A colloid is a heterogeneous system in which one substance is dispersed (called dispersed phase) as very fine particles in another substance called (dispersion medium).
3. **Tyndall Effect:** This phenomenon was first observed by Tyndall in 1869. Tyndall effect is defined as the scattering of light by colloidal particles present in a colloidal solution. The Tyndall effect is due to the scattering of light by colloidal particles. The sizes of colloidal particles are comparable to the wavelength of ultraviolet and visible radiations. They scatter these and get illuminated. Tyndall effect can be used to differentiate a colloidal solution from a true solution.
4. **Brownian Movement:** This property was discovered by a botanist Robert Brown in 1827. The colloidal particles are seen continuously moving in a zigzag path when they are observed under a microscope. The continuous zigzag movement of the colloidal particles in the dispersion medium in a colloidal solution is called as Brownian movement. Brownian movement is due the moving molecules of the dispersion medium continuously attack on colloidal particles from all sides and provide movement to them.
5. **Presence of electrical charge on colloidal particles:** Colloidal particles have a definite electrical charge. In a particular colloidal solution, all the colloidal particles have the same type of charge; an equal opposite charge in dispersion medium and colloidal solution is electrically neutral. On the basis type of charge, the colloids may be positively charged or negatively charged.

6. **Electrophoresis:** The colloidal particles present in a colloidal dispersion move towards a particular electrode under the influence of an electric field. The nature of charge present on them decides the direction of movement of the colloidal particles.

7. **Coagulation:** Coagulation may be defined as the phenomenon involving the precipitation of a colloidal solution. Addition of an electrolyte can coagulate the colloidal solution.

8. **Flocculation:** The colloidal particles are also coagulated by adding an oppositely charged ion. Formation of flocs is known as flocculation. If the cations are held close to the negatively charged particles, the negative charge would be neutralized and the colloidal particles flocculate and settle down.

9. **Dialysis:** The process of separating the colloid particles from the crystalloid by diffusion process through a suitable membrane.

Importance of Soil Colloids

Soil colloids have much more importance or significance in plant nutrition and soil fertility, reclamation of problematic soils, soil pollution, construction control etc.

Soil colloids are the active fraction of soils. They regulate many physical and chemical properties of the soil. The small size of colloids results in a large surface area per unit weight. Because of their small size, all soil colloids have a larger external surface area per unit mass. The total surface area of soil clays ranges from 10 m^2/g for only external surfaces to more than 800 m^2/g for extensive internal surfaces.

Soils with high clay and humus contents could be highly productive for crops due to their high amount of charge that facilitates the retention of nutrients like Ca^{2+}, K^+, Mg^{2+} etc. Therefore, the clay and humus fractions are a storehouse of plant nutrients like potassium, calcium, and magnesium. The fertility of the soil-water solution for plants is based on the capability of the soil to hold and exchange cations.

The type, amount, and mineralogy of colloids remarkably influence many soil management practices. Reclamation of saline, alkali, sodic and acid soils mainly depends upon the presence of colloids in soils.

The soil colloids also serve the soil as a filter or treatment medium for land application of waste materials, soil pollutants etc. The organic and inorganic contaminants are often transported via colloidal particles. In the absence of soil colloids plants nutrients may leached away.

Types of Soil Colloids

Inorganic soil colloids and organic soil colloids are two types of soil colloids.The inorganic soil colloid (clay) and organic soil colloid (humus) are very intimately intermixed in soil. These are smaller than 2000 nanometres in size. But these two exhibit colloidal properties. Therefore, size limit of soil colloids is considered up to 2 micrometres i.e., 2000 nanometre.

Difference between Inorganic and Organic Colloids

Clay is made up of Si, Ai and O and Humus is made up of C, H and O. Clay is inorganic and crystalline and Humus is complex amorphous organic colloid. Clays are more stable but Humus is more dynamic and more rapidly formed and destroyed. Clays have definite and well-known structure but Humus has relatively complex structure which is not well known.

Inorganic Colloids of Soils

As stated earlier, there are two types of soil colloids and clays are the inorganic colloids of the soil. The term clay is used in various senses. Clays is a soil separate whose diameter is less than 0.002 mm or soil texture in which less than 45% sand, less than 40% silt and 40% or more clay are found. The clay is also a mineral having specific physical and chemical properties. Here clays will be described as a mineral.

Inorganic colloids of soils i.e. clays are of two types, silicate clays and sesquioxide clays. The sesquioxide clays are also called as iron and aluminium oxides clays. Most of thc clays are crystalline. A single clay particle or crystal is called as micro cell or micelle. Clays are neither lyophobic nor lyophilic i.e., they neither liquid or water loving nor liquid or water hating, respectively.

1. Silicate Clays

These have a definite number and arrangements of atoms along with numerous layers. Mineralogical composition and organization of layers depend upon the type clays which influences the characteristics of clays. Silicon tetrahedron and aluminium or Magnesium octahedron are two basic units of silicate clays.

Silicon Tetrahedron

Silicon tetrahedron has four faces or sides. An atom or ion of silicon is covered by four atoms or ions of oxygen at equal distances to form silicon tetrahedron. Formula of silicon tetrahedron is SiO_4 which has a net negative charge of 4. Net charge can be calculated as:

$SiO_4 = Si^{4+} = +4$

$O^{2-} x\ 4 = -8$

$= -4$

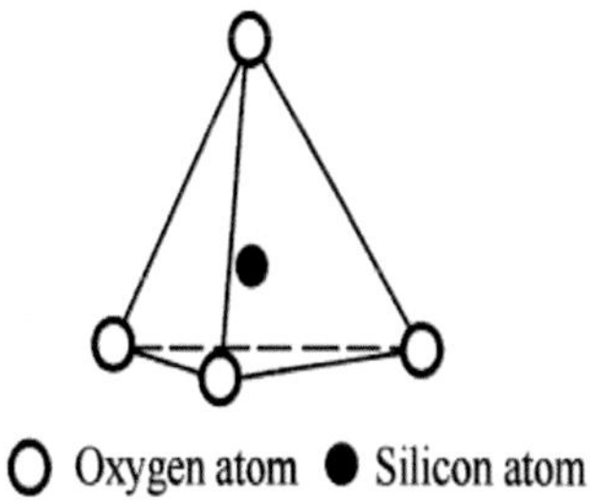

Fig. 3.35: Figure Silicon tetrahedron

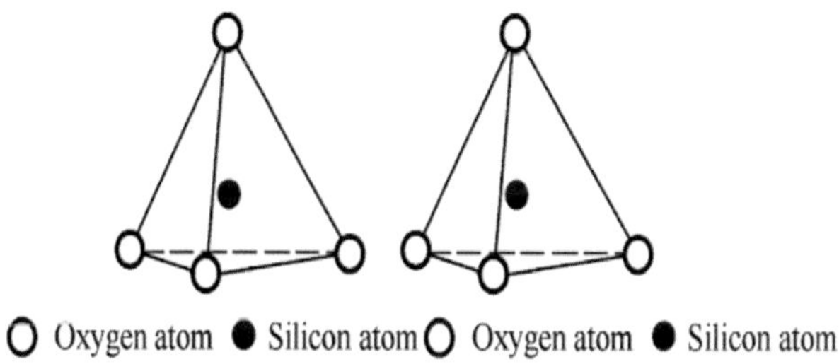

Fig. 3.36: Combination of two silica tetrahedrons

Combination of basal oxygen of two silica tetrahedrons together reduces the net negative charge proportionally. Suppose two silica tetrahedrons combine together, then net negative charge will be -6 as:

$2Si = Si^{4+} = +8$

$O^{2-} x\ 7 = -14$

$= -6$

The three out of four oxygens of a tetrahedron combine with oxygens of other tetrahedrons and form a sheet like structure. This arrangement is in a hexagonal pattern with the basal oxygens linked and the apical oxygens pointing up or down. This tetrahedral sheet looks like a net (sieve) having hexagonal holes if seen from upper side or apical oxygen side. A figure of tetrahedral sheet is given here:

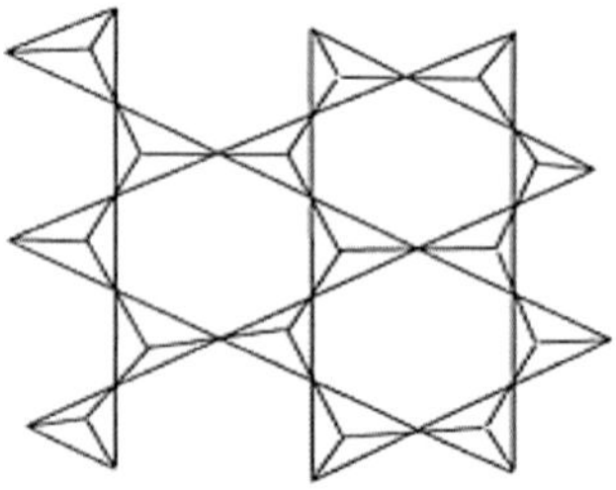

Fig. 3.37: Tetrahedral sheet

Aluminium Octahedron

Aluminium octahedron is second basic structure of clays. One atom or ion of Aluminium is surrounded by 6 atoms of oxygen or 6 hydroxyl ions at equal distances to form 8 faced or 8 sided structures. Sometimes Magnesium or iron is also surrounded by O^{2-} or OH^- to form octahedron. The aluminium based octahedron is called as dioctahedral where Al^{3+} occupies only two third portion of the octahedron. Gibbsite, $Al_2 (OH)_6$, is an example. Similarly, Magnesium or divalent iron based octahedron is known as trioctahedral in which all the ionic positions are occupied by Mg or Fe^{2+}. Brucite, $Mg_3 (OH)_6$, is an example.

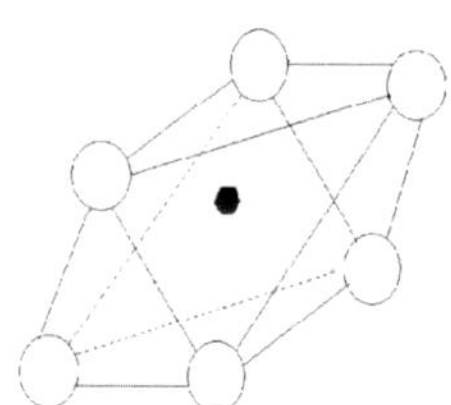

Fig. 3.38: Aluminium octahedron

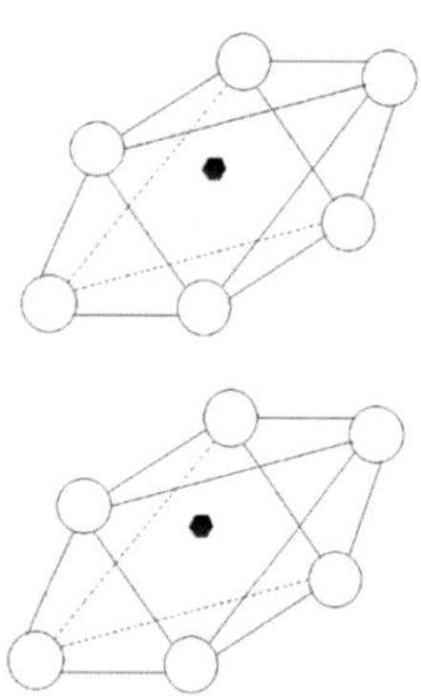

Fig. 3.39: Silica Tetrahedron

Numerous octahedrons join together by sharing nearby hydroxyl or oxygen anions and form horizontal octahedral sheet. The octahedral sheet is made up by the Al, Fe and Mg cations, in six-coordination with the O and OH anions. The octahedrons are also arranged in a hexagonal pattern.

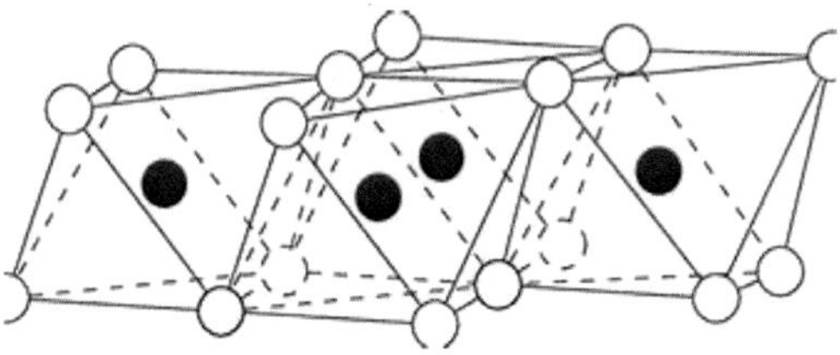

Fig. 3.40: Octahedral sheet

Formation of Layer Silicon Tetrahedron and Alumina Octahedron Sheets

Silicate and alumino silicate clays are basically layered structures. Thus, they are also called as Phyllosilicates. Phyllo is a Greek word meaning thin leaves. In the clays tetrahedral and octahedral sheets form new structures due to different types of configurations and combinations among them. Tetrahedral and octahedral sheets join or bond together by sharing the apical oxygen of silica tetrahedron. This bonding twists the bases of tetrahedrons in such a way that hexagonal holes of tetrahedral sheet look like triangles. The structure formed due to this joining or bonding of silica tetrahedral and Aluminium or Magnesium octahedral sheet is called as layer. Layer may be formed in three types.

1. 1:1 type (Kaolinite group) in which one tetrahedral sheet bonds with one octahedral sheet.
2. 2:1 type (Smectite group) in which two tetrahedral sheets join at both the sides of one octahedral sheet and
3. 2:1:1 type (chlorite) where two tetrahedral sheets share with one octahedral sheet alongwith one separate octahedral sheet i.e., Brucite sheet. This configuration is also called as 2:2 type. The basic figure of a layer is given here:

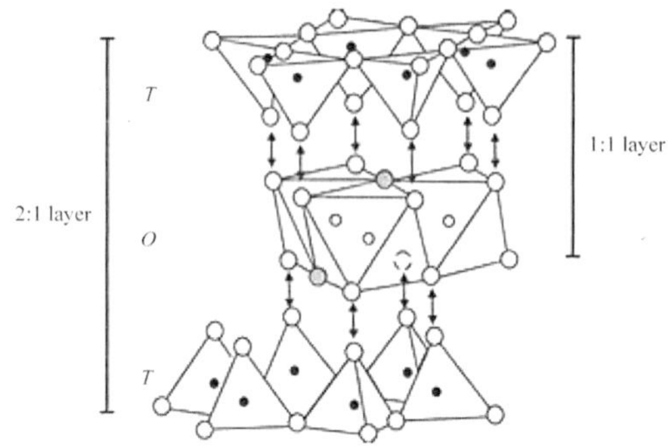

Fig. 3.41: Combination of 1:1 and 2:1 layers

At least two or more layers of 1:1, 2:1 or 2:1:1 combine together to make a crystal or micelle. The space between two layers is called as interlayer space, which has water or adsorbed cations.

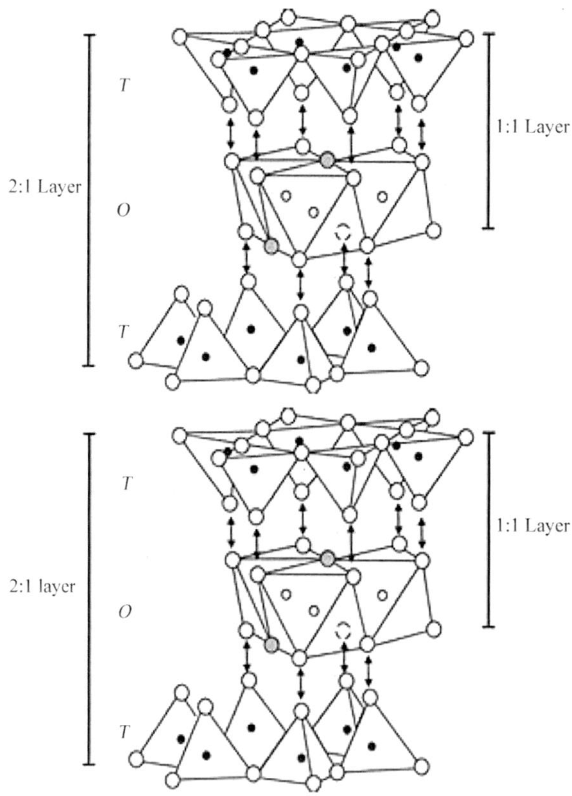

Fig. 3.42: Crystal or Micelle

Groups of Silicate Clays

Many groups of silicate clays are present in soil. Some of them are described here. Please c spacing to nm in all figures

1. Kaolinite group: This is also called as kaolin group. This is dimorphic i.e. two sheet structure. One sheet of silica tetrahedron and one sheet of alumina octahedron make a layer. These layers are strongly attached to each other by hydrogen bonding. The hydrogen bonds are so strong that cations and even water cannot enter in the layers. Therefore, the lattice is fixed showing 'C' spacing as 0.72 nm without any swelling and shrinkage.

Kaolinite, dickite, nacrite and halloysite are clay minerals of this group. Kaolinite is most abundantly found in soil. Halloysite contains interlayer water. Therefore, its 'C' spacing may increase up to 1.02 nm. Alterations in hydrogen bonding also provide tubular shape to halloysite. Dehydrated halloysite is called as metahalloysite.

Kaolinite is produced from the feldspars, mica and other aluminosilicates. Ca, Mg, Na and some silica are removed from aluminosilicates under medium to high acidic conditions. Remaining Al and Si ions recrystallize to form the kaolinite. Halloysite is supposed to be precursor of kaolinite in the weathering sequence of montmorillonite–halloysite- metahalloysite -Kaolinite.

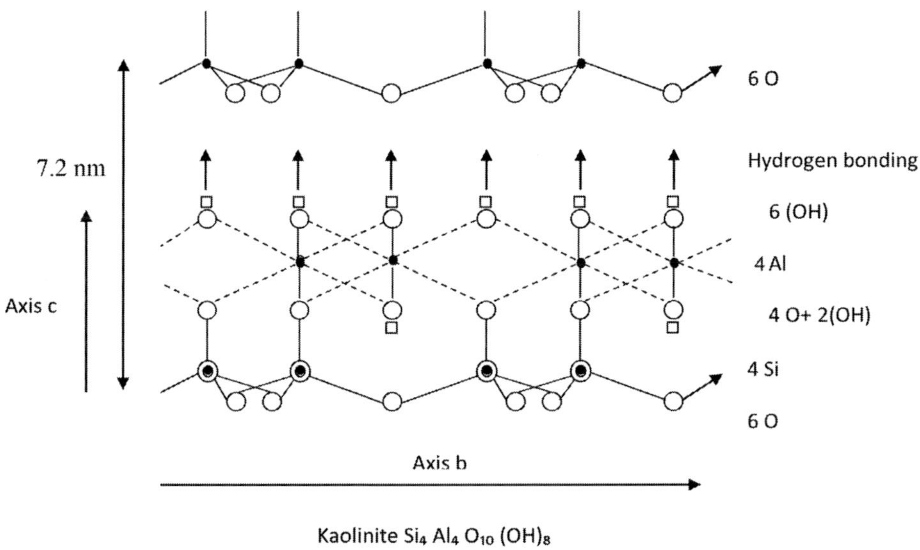

Fig. 3.43: Layer Structure of Kaolinite (1:1 layer silicate) mineral

2. Smectite group: This is expanding clay mineral with three sheet (trimorphic) structure. This group is also called as montmorillonite group. A unit layer of these minerals consists of one octahedral sheet in between of two tetrahedral sheets. Minerals of this group have variable composition because silica is partly replaced by aluminium from silica tetrahedral sheet and part of aluminium is replaced by magnesium and iron from alumina octahedral sheet.

These substitutions create electrical imbalance in the lattice which generates the negative charge. The negative charge attracts the H^+ Al^{3+}, Ca^{2+}, K^+, NH_4^+ and other cations besides the water molecules to satisfy or balance the charge.

The water and cations attached with the interlayer space make the mineral expandable. Thus, 'C' spacing of montmorillonite varies from 0.96 to 2.1 nm. The movement of cations and water molecules in the interlayer spaces provide openness in the mineral. Thus, internal surface of mineral is too much more than the external surface. The variable amount of water molecules also develops plasticity and swelling and shrinkage in montmorillonite.

Soils having more amount of montmorillonite are more plastic when wet and very hard when dry. Such soils also develop wide cracks upon drying. Structure of montmorillonite is shown in figure.

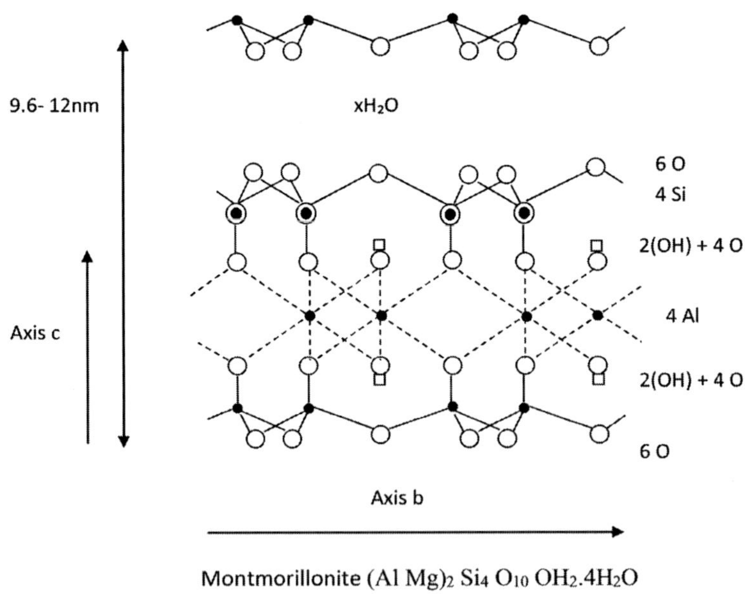

Fig. 3.44: Structure of Montmorillonite

Saponite, Nontronite and Beidellite are some common mineral of this group. These minerals show substitution from Al octahedral and Si tetrahedral layers. In saponite Mg substitutes the Al of octahedral layer and in nontronite Fe^{3+} substitutes the Al. But in case of beidellite Si is substituted by Al.

Weathering of effusive rock 'tuff' produces the montmorillonite. Basaltic parent material enriched with Mg, Fe and Ca bearing minerals and impeded drainage enhanced the formation of smectite. Alternations in vermiculite, illite or chlorite also produce smectite. In the alkaline medium biotite mica may also produce montmorillonite.

3. Illite: This is trimorphic non-expanding clay mineral. Illite is a fine grained micaceous mineral. Coarse grained illite is known as hydrous mica. Muscovite and biotite micas are of primary origin and found in sand and silt fractions of soil. Illite is also of secondary origin which found in clay fraction of soil. Thus, also known fine gained mica or soil mica.

The basic structure of illite contains one octahedral sheet covered by two tetrahedral sheets at both upper and lower surfaces. However, some of the silicon (about 15%) is replaced by the Al from tetrahedral sheet. Thus, negative charge is developed which is satisfied by the potassium ion. The thickness of interlayer space is somewhat equal to the ionic radii of hydrated potassium ion (0.133 nm). Thus, potassium ions very comfortably fit into the inter layer spaces. The Aluminium is also replaced by the Fe and Mg from the octahedral sheet.

The interlayer potassium binds or links two layers very tightly. Thus, interlayer spaces do not expand due to the addition of water and 'C' spacing does not go beyond 1.0 nm. Therefore, the swelling, shrinkage, plasticity and cations exchange capacity of illite remain in between kaolinite and montmorillonite. Fig. 3.45 represents the diagrammatic structure of illite.

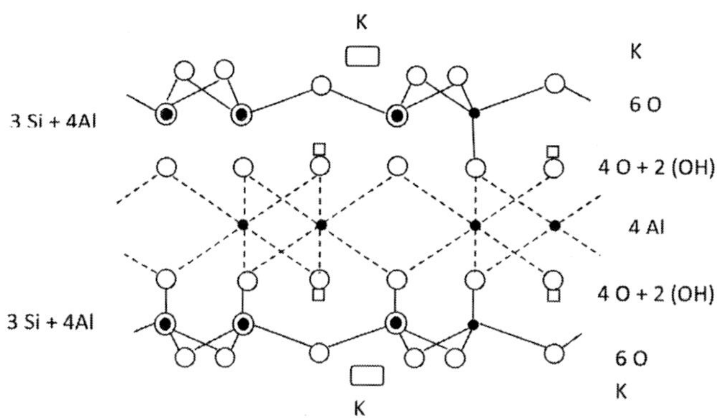

Dioctahedral illite $K_2 (Si_6 Al_2) (Al_4) O_{20} (OH)_4$

Fig. 3.45: Structure of dioctahedral illite

Weathering of muscovite and biotite micas produces illite. Removal of potassium from micas generates various clay minerals as: Mica –K illite –K vermiculite –K Montmorillonite. In nature reverse of this reaction may also takes place due the addition of potassium. The solid phase transformations of potash feldspars also produce the Illite. However, in soils illite generally comes from the sedimentary rocks.

4. Vermiculite: Vermiculite is somewhat like to illite or chlorite. In its crystalline structure hydrated Mg^{2+} ions are present instead of K^+ ions in the interlayer spaces. Hydrated Mg ion means one Mg ion is covered by six molecules of water. The thickness of the sheet due to the presence of hydrated Mg ions is about 0.5 nm. Thus, its structure also resembles to chlorite (2:1:1). The sheet of hydrated Mg^{2+} ions binds both the silicon tetrahedral sheets of the layer. Thus, swelling is not as much as montmorillonite.

Vermiculite may be of dioctahedral or trioctahedral type. Heating at 100^0C increases its volume 15-30 times more than the original and shape is changed to look like a worm. Thus, its name is derived from Latin word "Vermiculite" means worm like. But heating at 350^0C removes its water of hydration. Thus, original 'C' spacing of 1.4 nm reduces to 1.0 nm.

The vermiculite has highest cation exchange capacity (150 cmol Kg^{-1}). It also has high negative charge which is originated due to substitution of silicon by aluminium in the tetrahedral sheet. Vermiculite also fixes more amounts of K^+ and NH_4^+ ions as compared to the other clays minerals. This is due to the present of wedge zones. The structure of vermiculite is shown in Fig. 3.46.

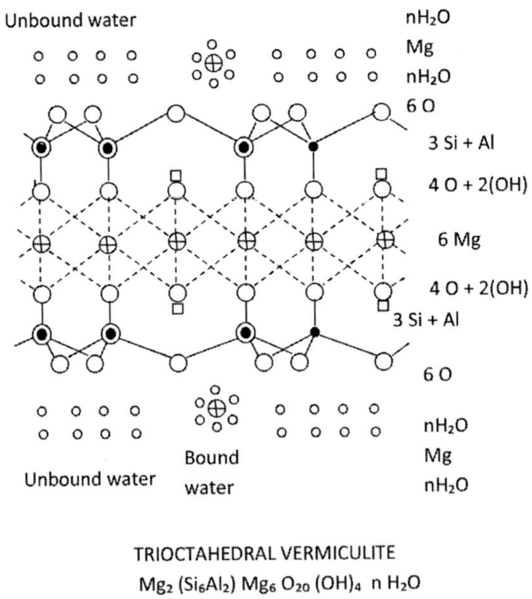

Fig. 3.46: Structure of trioctahedral vermiculite

Vermiculite is formed on the expanse of biotite and phlogopite micas and chlorite. Weathering or hydrothermal alterations of these develop the vermiculite. Acidic environment converts biotite into vermiculite. High rainfall conditions, oxidation of ferrous iron and loss of Fe and Mg from the brucite sheet also change chlorite into vermiculite.

5. Chlorite: This group has tetramorphic minerals in which a sheet of Brucite mineral, Mg $(OH)_2$, is attached between two tetrahedral sheets. Therefore, 2:1:1 or 2:2 type structural configurations as Si tetrahedron – Al octahedron – Si tetrahedron – Brucite is observed. Some times sheet of Gibbsite/Hydrargillite, $Al(OH)_3$ may also attach with Si-tetrahedral sheet instead of Brucite.

Basically, chlorites are hydrated Magnesium and Aluminium Silicates. Chlorites show extensive isomorphous substitution of main elements from tetrahedral, octahedral as well as brucite sheets. Magnesium is substituted by Fe^{2+}, Mn^{2+} or Ni^{2+}, Al^{2+} by Fe^{3+} or Cr^{3+} and Si is substituted by Mg^{2+} or Al^{3+}. Thus, chlorites have varied chemical composition. Pennatite, chamosite, thuringite etc. are some minerals of chlorites group. Fig. 3.47 shows the structure of chlorite.

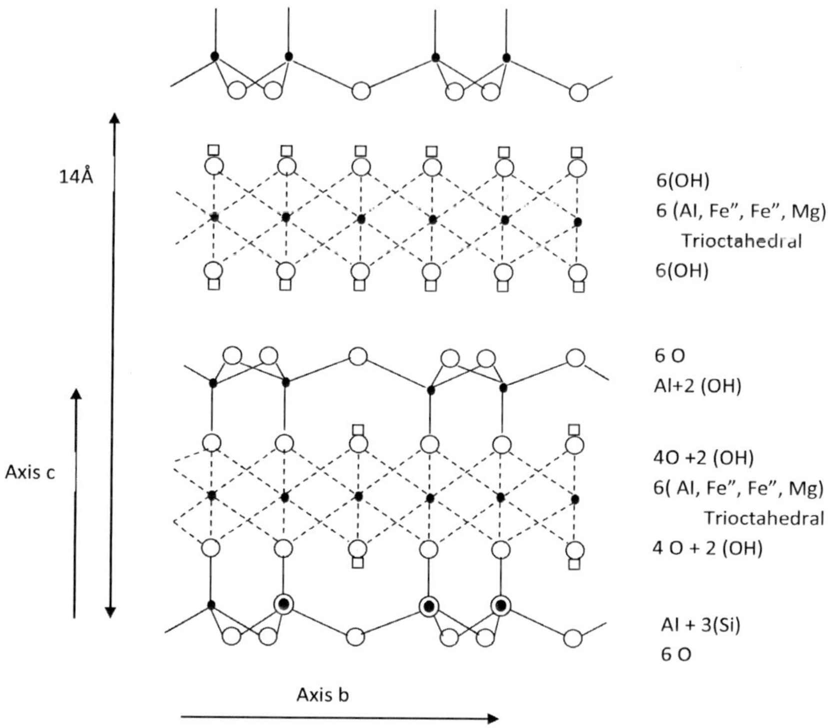

Fig. 3.47: Chlorite Structure of Chlorite (2:2 layer silicates) mineral

Chlorite for example $AlMg_5(HO)_{12}$ (AL_2Si_6) $AlMg_5$ $O_{20}(OH)_4$ Chlorites are generally formed due to the process of chloritization. This process is widespread in nature and changes biotite and hornblende into chlorite. The magnesium and

Iron sulphates are also changed into chlorites under low temperature hydrothermal conditions.

Mixed and Interstratified Minerals

Continuously running weathering and pedogenic processes in nature or in soil mix together numerous clay minerals. These minerals may be bundled together in regular or random fashion. The interstratified clay minerals may be defined as the aggregations composed of regular or random intergrowth of two or more clay minerals. Smectite –mica, smectite – kaolinite, vermiculite – illite, montmorillonite – chlorite-mica and mica-illite are some of the examples of stratification.

Various processes and conditions enhance the stratification of clay minerals. However precipitation, recrystallization and polymerization are three main processes. Amount of expanding clay minerals, mica, hydroxides of Fe and Al, oxidation or alternate oxidation and reduction and pH of the system are some important conditions to generate interstratified minerals.

Gibbsite sheet is developed in interlayer spaces after removal of exchangeable cations due to precipitation and recrystallisation. Silicate ion of biotite produce three layered mixed minerals and acidic environment helps in the inter stratification of montmorillonite -kaolinite. Inter stratification of biotite – vermiculite is generally observed in the presence of water rich in Ca and Mg ions. Identification of regular interstratified clay minerals is much easy than the random interstratified clay minerals.

2. Sesquioxide or Hydrous Oxide Clays

These clays are also called as iron and aluminium hydroxide clays. Tropical climate and intensive rainfall enhance leaching of silica and exchangeable bases from the soil. Thus, there is accumulation of oxides of iron and aluminium. These clay minerals may be crystalline to amorphous in nature. However, crystalline oxides are found more abundantly in soils. The crystalline iron bearing hydroxides and oxides are goethite and lepidocrocite and Hematite and maghemite, respectively. Ferrihydrite, $Fe_3(H_4O_3)_3$, an oxyhydroxide is also in this list. Gibbsite, diaspore and boehmite are the examples of crystalline Al bearing minerals.

These Fe and Al minerals have amphoteric characters. They develop weak electronegative charge in acidic condition and weak electropositive change in alkaline condition. Fe and Al oxides provide more surface area for the adsorption of phosphorus. Therefore there is less availability of phosphorus to the plant when these are rich in soils.

Many physical properties of soils are also related with Fe-Al oxides. Hematite provides bright red and goethite provides yellowish brown colour to the soil. Hydromorphic soils generally generate more amount of epidocrocite. Its presence indicates the process of 'gleying' in soils. Maghemite and gibbsite mostly found in the highly weathered soils of tropical regions.

Soils have these minerals in various shapes, sizes and strengths. These may be present as coatings on other minerals where they act as a cementing agent to develop aggregation. They may also present as localized streaks or stains. Size may vary from 2 cm to 7.5 cm in diameter. The strength may vary from soft to massive.

3. Amorphous or Non-crystalline Clays, Allophane

Some amorphous or non-crystalline clays are also found in the soils. These are in the form of phosphates, silicates and silica gel and sesquioxide gels and generally present as the coatings on other minerals. These are classified in a general group of 'Allophane'. Allophone is a group name of amorphous hydrated alumino-silicate which has variable composition. These are generally found in volcanic ash soils. High allophanic soils show low bulk density, high plasticity, non-sticky when wet, high water holding capacity, high CEC, low AEC, amphoteric and high phosphorus fixing capacity and less phosphorous availability. Japanese soils 'Andosols' are rich in allophane and other related minerals (imogolite, hisingerite, ferrihydrite etc). These minerals are associated with humic and fulvic acids of soil organic matter to form complexes or chelates.

Characteristics of Clay Minerals

Most of the soils are rich in kaolinite, montmorillonite and illite clay minerals. Therefore, characteristics of clay minerals in accordance to these are described here.

1. Size: Particle size of clay minerals is sub microscopic ranging from 0.1 – 5.0 micron (Kaolinite), 0.01 – 1.0 micron (montmorillonite) and 0.1-2.0 micron (illite). 1 micron = 1000 nanometre (nm).

2. Shape: Generally, particles are laminated having layers of plates, flakes or tubes. Their horizontal axis is too much more than the vertical axis. The shape of kaolinite is hexagonal plates whereas montmorillonite and illite have irregular flakes. Edges of kaolinite are clean cut, montmorillonite fluffy and edges of illite are in between.

3. Surface area: Surface area of clay minerals is too much higher because of their smaller size. Sheet like structure of clay minerals provide both external as well as internal surface area to them and sum of these two is the total surface area. Surface area per unit mass or volume is called as specific surface or specific surface area. The unit of this is square meter per gram (m^2g^{-1}).

Surface area increases with the decrease in particle size. External surface area is the outer surface of the mineral particles which is available for the adsorption ions etc. While internal surface area is the area between two crystal sheets or layers. Internal surface area is generally higher than the external surface area and also available for the adsorption of ions, gases, solutes etc. Generally, expanding type clay minerals have greater internal surface area as compared to external surface area. Specific surface areas of some clay minerals, as determined by Ethylene Glycol (EG) method, are given in Table 3.14.

Table 3.14: Specific surface area of clay minerals (m^2g^{-1}).

Clay mineral	External	Internal	Total
Kaolinite	30-35	7-10	37-45
Montmorillonite	80-150	500-600	580-750
Illite	50-70	70-100	120-170
Chlorite	70-100	60-80	130-180
Vermiculite	80-100	700-800	780-900

Source: Indian Society of Soil Science, Bulletin, 9:130, 1974.

4. Chemical composition: Most of the clay minerals contain silicon and Aluminium in higher amounts. Therefore, they are called as aluminosilicates minerals. Besides these, clay minerals also contain some quantities of Iron, Calcium, Magnesium, Potassium, Sodium and Titanium. However, the chemical composition of clay mineral is quite variable due to isomorphous substitution.

Generally, kaolinite contains 48 % SiO_2 and 39 % Al_2O_3. Montmorillonite contains 43 % SiO_2 and on the basis substitution 0-28 % Al_2O_3, 0-25 % MgO, 0-30 % Fe_2O_3. Besides these Titanium, Calcium, Sodium and Potassium are also found in Montmorillonite. Illite contains about 53 % SiO_2, 25 % Al_2O_3, 6 % K_2O, 3 % Fe_2O_3, 2 % MgO, 1 % CaO and 0.5 % Na_2O. Chlorite contains 32 % SiO_2, 19 % Al_2O_3 and 36 % MgO. Chlorite contains more Magnissium than aluminium.

5. Isomorphous substitution: Isomorphous substitution is the replacement or substitution of one atom or ion by another atom or ion with somewhat similar size in a crystal lattice without changing or disrupting the crystal structure of the mineral. In isomorphous substitution size of atoms or ions and geometry of interlayer space is more important than the valency of atoms. Isomorphous substitution takes place among the atoms or ions having size difference not more than 15 percent.

By this way Si^{4+} can replace Al^{3+} and vice-versa because the ionic radii of these two ions are 0.042 and 0.051 nm, respectively, (1 nm = 10^{-9}m). Other examples are Fe^{3+} (0.064 nm), Mg^{2+} (0.066 nm) and Fe^{2+} (0.074 nm) and Na^+ (0.098 nm) and Ca^{2+} (0.099 nm). In the replacement or substitution systems trivalent can replace trivalent or tetravalent ions, divalent can replace trivalent and monovalent can replace monovalent or divalent. This is also possible that ions of similar valence can replace each other. This replacement is called as isovalent isomorphous substitution (e.g., Zn^{2+} replaces the Fe^{2+} and vice versa).

When there is difference in the valence of ions, the system is called as heterovalent isomorphous substitution (e.g., Fe^{3+} replaces Mg^{2+} and vice-versa). However, heterovalent substitutions are more common as compared to isovalent substitutions. Isomorphous substitution takes place in both the silicon tetrahedral and aluminium octahedral sheets where part of Si^{4+} is replaced by Al^{3+} and Al^{3+} is replaced by Mg^{2+}, or Fe^{3+}

Electrical charge: All the clay minerals are electrically charged particles. This charge may be negative or positive. Generally, silicate clays are negatively charged and sesquioxide clays and allophanes are positively charged particles. This charge can be calculated by multiplying each element present in the chemical formula of the mineral by their respective valency, e.g., the net charge of simple Kaolinite, $Si_4 Al_4 O_{10} (OH)_8$, will be zero as,

$Si_4 = +16$

$Al_4 = +12$

$O_{10} = -20$

$(OH)_8 = -8$

+28-28= Zero

7. Types of electric charge: There are two types of charge on the clay minerals. These are permanent and pH dependent charges. Permanent charge is originated due to the isomorphous substitution. Suppose tetravalent silicon is replaced by trivalent aluminium, then one negative charge is left. Because one silicon was satisfying four negative charges but one aluminium is satisfying only three negative charges. Similarly, one negative charge will also left when one aluminium ion is replaced by one magnesium ion.

The pH dependent charge is also called as variable charge because it changes with the change in the pH. This change is produced due to the oxidation of protons (release of protons) or deprotonation of H^+ from OH^- at the edges or surfaces present on the clay surfaces. Besides pH presence of H^+ and OH^- in soil solution also regulates variable negative charge. Highly weathered soils of tropical areas show very low pH or more H^+ ions. These H^+ ions partially attached with the OH^- group and instead of forming water they positively charge the clays.

This pH dependent variable charge is most significant for kaolinitic clays, Fe and Al oxide clays and organic colloids. The pH dependent variable charge is positive at low pH, neutral at medium pH and negative at higher pH. Higher pH indicates presence of less H^+ ions. In alkaline medium OH attached with Al dissociates and make water. Therefore, only oxygen ions are left which are negatively charged.

$Al - OH + OH = Al\text{-}O^- + H_2O$
$Si - OH + OH^- = Si\text{-}O^- + H_2O$
Alkaline condition

But in case of acidic medium excess H^+ ions attach themselves on the exposed surfaces of OH^- ions to develop positive charge.

$Al - OH + H^+ = Al - OH^+_2$

Acidic condition

Origin or Source of Negative Charge

Silicate clays are generally negatively charge minerals. Isomorphous substitution and exposed surfaces or broken edges of clays are the sources of negative charge.

1. Isomorphous Substitution

The general chemical formula of montmorillonite showed net positive charge of +4. But after substitution in tetrahedral and octahedral sheets, the formula will be $Al_{1.7}\ Mg_{0.3}\ Si_{3.9}\ Al_{0.1}\ O_{10}\ (OH)_2.\ 4H_2O$ and net negative charge will be - 0.40.

$Al_{1.7} = + 5.1$

$Mg_{0.3} = + 0.6$

$Si_{3.9} = +15.6$

$Al_{0.1} = +0.3$

$O_{10} = -20$

$(OH)_2 = -2$

+ 21.6

-22

Net charge = -0.4

This tremendous change in the charge is due to the isomorphous substitution. The charge in the crystal will remain balanced when there is no substitution or two isovalent cations substitute each other. But when cations of higher valency are replaced by cations of lower valence, such as replacement of Si^{4+} by Al^{3+} and Al^{3+} by Mg^{2+}, the charge is imbalanced and unsatisfied negative charge originate with each such type of replacement or substitution.

Also, positive charges of silicon or aluminium are satisfied by equivalent negative charges of oxygen or hydroxyl surrounding them. Hence there is no charge on them. But after substitution of Si by Al or Al by Mg, one negative valence of oxygen or hydroxyl remains unsatisfied. These cause net negative charge in the crystal. This is very clear in silica tetrahedral and alumina octahedral sheets.

No net charge	Originated one negative (1^{--}) charge of oxygen
O^{-} Si^{4+} O^{--}	O^{-} Al^{3+} 1^{--}
O^{--} Al^{3+} Al^{3+} O^{--}	O^{--} Al^{3+} Mg^{2+} 1^{--}
O^{--} Al^{3+} Mg^{2+} 1^{--1}	Mg^{2+} Mg^{2+} 1^{--1} 1^{--1} (Here one more negative charge)

2. Exposed Surfaces or Broken Edges

Negative charge due to exposed surfaces or broken edges is mostly developed in the 1:1 type of clays. The O^{2-} or OH^{-} ions attached with Si or Al are also exposed towards the surfaces of crystal. Out of these ions some loosely attached OH^{-} ions dissociate at higher pH to produce H^{+} ions. Hence negatively charged oxygen ions remain there which provides temporary negative charge to the clay mineral.

Dissociation of H^{+} ions from OH^{-} ions of the alumina sheet gradually increases with the increase in soil pH from 5 to 9. Therefore, at pH 5-9 charge is positive, zero at pH between 6 to 7 and becomes negative at uprise of the pH to nine. The pH at which the charge of any system is zero is called as zero point of charge i.e., ZPC. Mechanism of development of negative charge at the exposed surfaces or broken edges is given in Fig. 3.48.

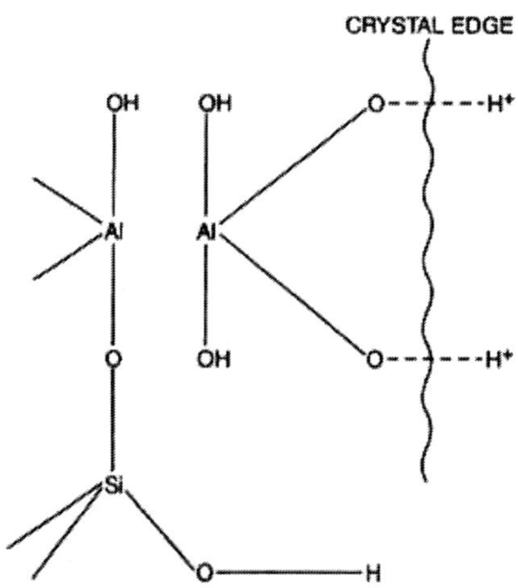

Fig. 3.48: Development of nagative charge at the exposed crystal edges (*Source*: A.E.: Lesson 8. Charges on Soil Colloids)

8. Adsorptions of Ions

The negative charge developed due to the isomorphous substitution or dissociation of H^+ from OH^- at higher pH is satisfied by the adsorption of cations. Similarly the positive charge of allophane, hydroxyl Fe-Al oxide clays and amorphous or positive charge developed due to the replacement of Mg^{2+} of trioctahedral sheet by Al^{3+} is satisfied by the adsorption of anions. Adsorption is the retention of gases or dissolved substances on the surface forms ionic double layer, diffuse double layer or electrical double layer.

1. Diffuse Double Layer

Diffuse double layer is the net electrical charge (usually negative) on the surface of solid or disperse phase along with the ionic assemblage (usually positive) in the solution. The micelle (Microcell) i.e., small colloidal size clay particles have negative charge. Thus, they attract positively charged ions i.e., cations. These ions are also called as counter ions. The concentration of these ions is more near the surface of clay particles and gradually reduces with the increase in the distance from the clay particles.

This forms an ionic double layer near the clay particles carrying negative charge and second layer or outer layer is the liquid near by the clays surface carrying the assemblage of counter ions. The counter ions are attracted by the inner layer, but they are free to move and distribute themselves in liquid. This attraction and distribution of ions developed various theories indicating the distribution of ions in the diffused double layer. About eight theories namely, Helmholtz, Gouy Chapman, Stern, Grahame, Bockris/Devanathan/Mullen, Trasatti/Buzzanca, Conway and Marcus were developed so far. However, first three are more considered and these are described here in detail.

Helmholtz (1853) proposed first theory. He stated that negative charge is uniformly present on the colloidal (clay) surface and in the second layer counter ions are concentrated in the plane parallel to the colloidal surface at a fixed distance. In this system the electrical potential is maximum at colloidal surface which decreases linearly with the increase in distance from the surface. This model was just a good foundation for the description of the interface without considering diffusion or mixing of ions in solution, adsorption on the surface and the interaction between solvent and the electrode (Fig. 3.49).

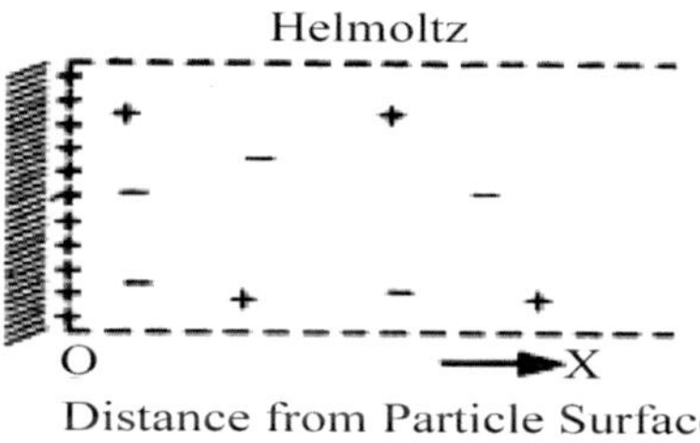

Fig. 3.49: Helmholtz diffused double **layer**

Later, Gouy (1910) and Chapman (1913) independently proposed the theories. Theories of both were same. Hence the name is Gouy-Chapman theory. In this theory the negative charge is again homogenously present on the colloidal surface but the counter ions, instead of concentrating on colloidal surface, are distributed in the liquid. Thus, electrical potential shows a curved decrease with the distance. In this layer the charge distribution of ions was a function of distance from the surface and the electric potential decreases exponentially away from the surface of the bulk of fluid. This theory holds good for positively charged colloids, but not good highly charged diffuse double layers. The Gouy-Chapman diffused double layer is depicted in Figure 3.50.

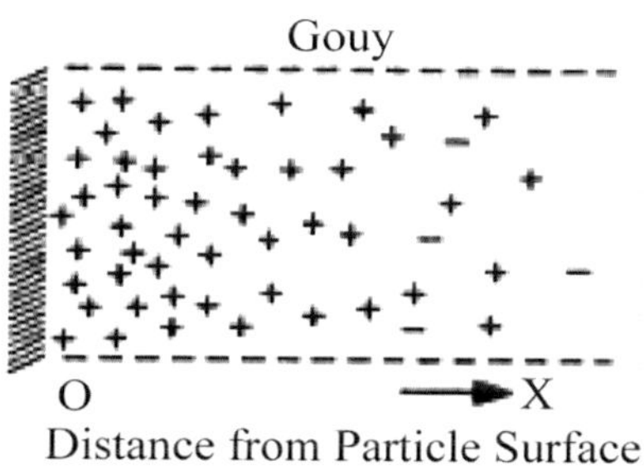

Fig. 3.50: Gouy-Chapman diffused double layer

Gouy-Chapman diffused double layer theory was not good for highly charged diffuse double layers. Therefore, Stern (1924) also considered ionic dimensions and combined both earlier theories. In Stern's diffuse double layer, some ions adhere to the electrode as indicated by Helmholtz, making an internal Stern layer and some ions form a Gouy-Chapman diffuse layer. The influence of ionic dimensions is more near the colloidal surface. The negative charge is homogenously present on the colloidal surface and counter ions are distributed in two parts according to Helmholtz and Gouy Chapman systems. Thus, the assemblage of counter ions is more at the immediate vicinity of colloidal surface along with the significant number of counter ions is evenly distributed in the liquid. Therefore, electrical potential show linear decrease in first part (Stern layer) and curved decrease in second part (diffuse layer), with the distance (Fig. 3.51).

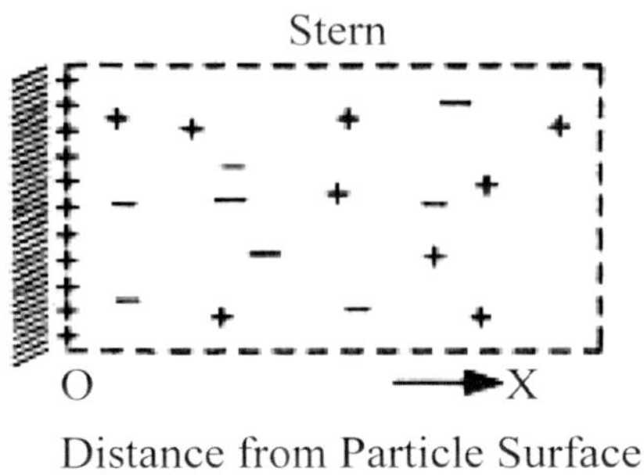

Fig. 3.51: Stern diffused double layer

Concentration of electrolyte and valence of cations directly influence the thickness of diffused double layer. Generally, thickness reduces with the increase in concentration and valence. Higher concentration of electrolytes increases the cation concentration in the liquid phase. Thus, cations do not move from the colloidal surface to the liquid phase. This reduces the thickness of double layer. Similarly monovalent ions (K and Na) construct thicker layer as compared to Ca and Mg and the double layer constructed by Al ions is thinnest. This is only due to power of attraction. Greater power of attraction will cause thinner diffused double layer.

9. Other properties: Some other important properties of clays are described here,

Dispersion: Dispersion is related with the fineness of particles. Fine clay particles repel each other. Negatively charged clay particles attract cations. But force of attraction is decreased with the increase in the distance between clay surface and cations. By this way sodium which produces thicker diffused double layer partly neutralize the negative charge. When clay particles having partly neutralized negative charge come in contact they repel each other. Thus, clays remain in dispersed phase in the presence of water. Dispersion is essential for the estimation of soil texture. Dispersion is somewhat similar to deflocculation. Dispersion is greater at low electrolyte concentration. At this condition the clay particles are covered by thicker diffused double layer.

Flocculation: The repulsion among the clay particles is reduced or attraction is increased by the addition of electrolytes like calcium and magnesium.These electrolytes reduce the thickness of diffused double layer as these cations come in close contact of negatively charged clay surface. This reduces the unneutralized negative charge on the clay. Thus, clay particles instead of repelling each other, they attract themselves and unite to form floccules. The process of floccules formation is beneficial for good soil structure. In flocculation Vander Waals forces (Forces of attraction) exceeds the diffused double layer forces (repulsive forces). Flocculation is somewhat similar to coagulation.

Swelling and shrinkage: Swelling and shrinkage are more effective in 2:1 type (montmorillonite) than 1:1 type (Kaolinite) clays. Montmrillonitic clays swell when wet and shrink upon drying. Swelling is due to the increase in volume of clays by absorbing water between the sheets. Shrinkage causes decrease in volume of clays. Thus, particles come in close contact to develop cracks of different width and depth. Swelling and shrinkage is more in sodium enriched clays as compared to calcium clays.

Cohesion and Adhesion: Clay particles have property to stick with each other (cohesion) and with other bodies (adhesion). Molecular attraction among clay particles i.e. cohesive forces help in the aggregation. Cohesion is greater in the oriented clays. Adhesion is mainly due to the Vander Waals forces. It forms metallic or ionic linkages. Thus, adhesion tends to cover sand particles with clay particles. They also stick with hands, implements and both. Cohesion and adhesion are related with soil consistence, water retention in soil pores and absorption of nutrients.

Plasticity: Moist clays can be moulded in any shape and they can retain the given shape. Plasticity is related with structure and tillage of the soil. Saturation of clays with ions show different plastic limits. Ca-saturated illite and kaolinite and Na-saturated montmorillonite are more plastic.

Clay Minerals in Different Soil Orders

Clay minerals are important for terrestrial biogeochemistry and other earth sciences. But globally their data are partially available. Akihiko Ito & Rota Wagai (2017) presented a global dataset of clay-size minerals in the topsoil and subsoil for the soil orders of Soil Taxonomy for each of the ten groups of clay minerals like gibbsite, kaolinite, illite/mica, smectite, vermiculite, chlorite, iron oxide, quartz, non-crystalline and others (Figure 3.52). The validity of the clay-size mineral distribution obtained was examined by comparing with other datasets. The global soil clay data provides spatially relations for terrestrial biogeochemical cycles, dust emission to the atmosphere and other interdisciplinary earth sciences. Parent material, climate and stage of profile development directly influence the type of clay mineral found in any soil. Different soil orders contain numerous clay minerals.

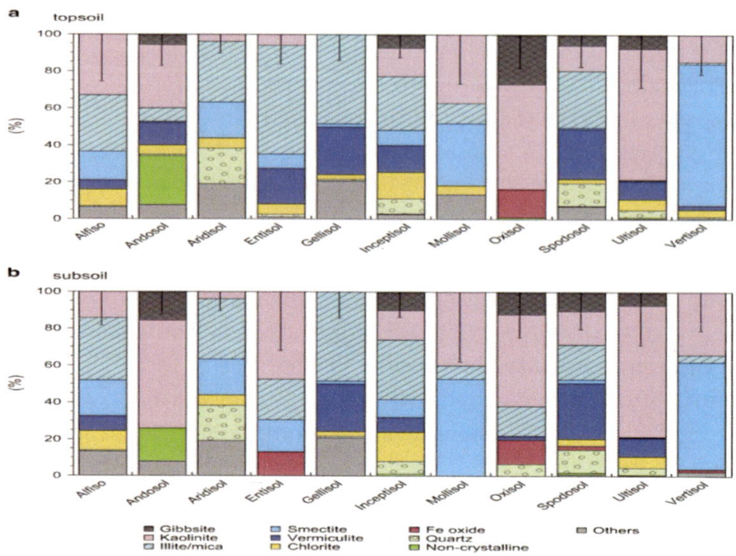

Fig. 3.52: Clay minerals in soil orders of soil taxonomy

Clay Minerals in Indian Soils

Similarly Indian soils also have many clay minerals. There is too much diversity in climate and soil parent material of India. Therefore, many types of minerals are found in Indian soils. Clay minerals in Indian soils may be in pure forms but intergrades of them and interstratified minerals are also found in significant quantifies. Illite is the main clay mineral of Indian soils. Due to the same Indian soils are rich in potassium.

Table 3.15: Main clay minerals in Indian soils

Soil types	Main clay minerals
Black soils	Montmorillonite
Alluvial soils	Illite
Red soils	Kaolinite, illite
Lateritic soils	Kaolinite
Desertic soils	Illite, chlorite
Hill soils	Illite, chlorite
Tarai soils	Chlorite
Saline – sodic soils	Illite, chlorite

Reference, Distribution of clay minerals in Indian soils

Ghosh, S.K.; Raychaudhuri, S.P. 1974. Bulletin, Indian Society of Soil Science 9: 143-164

Organic Colloids of Soils

Humus is an organic colloid of the soils. It is developed from the soil organic matter and defined as a brown to black organic substance present in soil which comprises of partially or wholly decomposed plants and animal residues and influence many soil properties including plant nutrition. Humus consists of three different humic substances, fulvic acid, humic acid and humin. The term acid is used to describe humic substances because humus behaves like weak acids.

Fulvic acids: This fraction of humus is soluble in water under all pH conditions. Their colour is commonly light yellow to yellow-brown.

Humic acids: This fraction of humus is soluble in water, except more acid than pH 2. Common colours are dark brown to black.

Humin: This fraction of humus is not soluble in water at any pH and that cannot be extracted with a strong base like sodium hydroxide (NaOH). Common colour is black.

Composition of humus directly depends upon organic matter added or present in the soil. The composition of added or already present organic matter is quite variable. Therefore, composition and characteristics of humus are also variable. Humus mainly contains 42.0-59.3% C, 4.2-5.5% H, 1.2-6.6% N, 34.5-46.8% O and 0.1-3.7% S. Humus of agricultural soils contains more oxygen as compared to forest soils which have more carbon, hydrogen and nitrogen. Among the fractions of humus, humic acid contains more, C, H and N less oxygen than fulvic acid.

Humus is high molecular weight compound. It may be about 800 varying from 996-1015 for humic acid and 661-679 for fulvic acid (Mishra and Srivastava, 1991). Humic acid is quite heavier than fulvic acid.

Carboxylic COOH, hydroxyl OH, phenolic-OH, alcoholic-OH, Ketonic C=O, quinoid -C=O, carboxyl-C=O, aliphatic C-H, aromatic C=C and amide N-H are many functional groups present in humus. Among these carboxylic COOH and phenolic- OH groups are more important. Humic acid contains less carboxylic and hydroxyl groups than fulvic acid.

Humus particles are negatively charged. Negative charge originates due to the dissociation of H^+ ions from – COOH and –OH groups. At pH 3.0 carboxylic group dissociates its protons (H^+) to provide some negative charge. But higher negative charge is achieved at pH 9.0 when phenolic group starts dissociation. The negative charge of humus is pH dependent or variable which increases with the increase in pH.

Cation exchange capacity (CEC) of humus is quite high than clay. It may vary from 392-1014 c mol (P+)/kg. CEC of the humus is pH dependent and varies with the change in pH cations like Ca, Mg, K, NH_4, etc. are attracted by the negative surfaces of humus. Thus, keep them available to plants after preventing from leaching losses.

Humus has affinity to interact with metallic ions, clay and plant protection chemicals to make the complexes. Humus metallic complexes (also called as organometallic complex) are of chelate type. Metallic ions are attached with the nearby functional groups of humus to form complexes having different stability constants. Normally stabilities of metallic complex are greater than other complexes. This influences the solubility, availability and losses of micronutrients of the soil. Nutrients are easily available from metallic- fulvic acid complexes as they are more soluble. There is less leaching of nutrients from metallic - humic acid complex as they are less soluble. Organometallic complexes also help in the soil formation. They tear off loose Al and Fe from various soil forming minerals and enhance their rate of decomposition.

Humus and clays both are negatively charged, even then there is formation of clay humus complexes. This is possible due to the origin of positive charge on the surfaces of clay or humus. Negatively charged clays adsorb cations on their surface. If there are excess cations, the unsatisfied positive charge attracts negatively charged humus to form clay-cation-humus complexes. Another possibility is formation of water molecules with the interaction of hydroxyl ions of clays and humus. The remainder oxygen atoms will not satisfy positive charges of Aluminium or silicon, which will attract the negatively charged humus. Although these are not simplest one and very complex chemical reactions take place to perform this.

Soil Fertility and Integrated Plant Nutrition

Management of soil fertility in relation to plant nutrition is an important aspect of soil science. The pressure of ever measuring population and growing demand of land for non-agricultural uses, it is necessary to grow more food, feed and fibre from less and limited area of land. Food production is mainly governed by soil fertility, weather, climate, insect peat and disease and management. Early maturing and highly productive variety of crops, vegetables, fruits and other are grown under intensive agriculture to get more production in lesser time and lesser land area. This directly and remarkably uptake more nutrients from soil. Also, multiple cropping in a single year mine more nutrients from soil as compared to single cropping in a year. This will deplete the soil from plant nutrients which will decrease yield in future. Therefore, management of soil fertility is essential for long sustainable agriculture. Soil fertility is mainly related with soil nutrition. It is an inherit capacity of soil to supply nutrients to plants in adequate amount and in suitable proportion. A good fertility soil provide essential plant nutrients in optimum for entire life span of plants.

Essentiality of Plant Nutrients

More than ninety elements are present in plants. Advancement in analytical techniques may indicate reuse or even more elements in plants. But it is not necessary that all the elements present in plants are essential for their nutrition. Arnon established three criteria to know the essentiality of plant nutrients. These are.

1) Plant cannot complete their life cycle without that element.
2) That element is directly essential for growth of plants.
3) No other nutrient can substitute that nutrient for plant nutrition. However, there are some exceptions of these criteria.

Recent investigations indicate that second point cannot be accepted because Mo may be substituted by V, CI by Br, K by Rb and Ca by Sr. Also, Nicholas (1961) proposed Functional Nutrient for any mineral element that functions in plant metabolism whether its action is specific or not. Thus, Na, Co, V and Si are also included in functional nutrients in addition to 16 essential elements.

Plant nutrient is an element absorbed by plant which is essential for the completion of their life cycle. Carbon, hydrogen, oxygen, nitrogen, phosphorus, potassium, calcium, magnesium, sulphur, iron, copper, zinc, molybdenum, boron and chlorine are sixteen essential and useful elements for all plants. Besides these some nutrient elements are required by some specific plant species. These are beneficial plant nutrients. Sodium, cobalt, vanadium, selenium, silicon are beneficial plant nutrients.

Water, air, soil, manure and fertilizers provide nutrient elements to plants. Water and air provide carbon, hydrogen, oxygen and soil, manure and fertilizers provide remaining nutrient elements to plants. Nutrients absorbed in more quantities or nutrient content of matured plants is more than 500 mg per kg, such nutrient is called as macronutrient. Nitrogen, phosphorus and potassium are macronutrient. Calcium, magnesium and sulphur come under secondary nutrients. Remaining elements iron, manganese, zinc, copper, molybdenum, chlorine and boron are absorbed in very low quantities and contents of these is less than 100 mg per kg in matured plants. Except boron all these nutrients are essential for animals. Beside these animals also require sodium, iodine, selenium, and cobalt. Different facts about plant nutrients are presented in table 3.16 (modified from numerous sources)

Table 3.16: Some facts about plant nutrients elements

Element	Molecular weight	Discovered by	Year of discovery	Essentiality for plants discovered by	Year of discovery of essentiality	Forms in which taken by plants
Carbon (C)	12.011	Since ancient	Since ancient	Sash	1882	CO_2 (by leaves)
Hydrogen (H)	1.0079	Cavendish	1766	Sash	1882	HOH, H^+
Oxygen(O)	15.999	Priestley	1774	Desassure	1804	CO_2 (by leaves), O_2, OH^-, CO_3^{2-}
Nitrogen (N)	14.00067	Levoiseur	1792	Rutherford	1872	NO_3^-, NH_4^+
Phosphorus (P)	30.9738	Brond	1669	Posturnuk	1903	$H_2PO_4^{2-}$ HPO_4^{2-}
Potassium (K)	39.098	Davy	1807	Skinner	1890	K^+
Calcium (Ca)	40.08	Davy	1807	Samhorsmar	1855	Ca^{2+}
Magnesium (Mg)	24.305	Davy	1808	Wilstater	1906	Mg^{2+}
Sulphur (S)	32.06	Since ancient	Since ancient	Patterson	1911	$S0_4^{2-}$, SO_2 by leaves
Iron (Fe)	55.84	Since ancient	Since ancient	Griss	1844	Fe^{2+}, Fe^{3+}
Manganese (Mn)	54.938	Sheeley	1774	Wortrund	1897	Mn^{2+}
Zinc (Zn)	65.38	Since ancient	Since ancient	Maz	1915	Zn^{2+}
Copper (Cu)	63.54	Since ancient	Since ancient	Branchley	1914	Cu^{2+}
Molybdenum (Mo)	95.94	Hem	1782	Arnon & Stout	1939	MoO_4^{2-}
Boron (B)	10.81	Lusac & Kenai	1808	Augulhome	1910	BO_3^{2-}
Chlorine (Cl)	35.453	Sheeley	1774	Brover, et.al.	1959	Cl^-

Functions of Plant Nutrients in Plants

Carbon, hydrogen and oxygen are 94% of dry weight of plants. These are main components of carbohydrates, proteins and fats and provide energy for growth and development of plants.

Nitrogen is main constituent of protein, amino acids, enzymes, nucleic acid and lipid of plants. It promotes multiplication of cells and chloroplasts and synthesis of carbohydrates. Protein synthesis stops due to the lack of nitrogen. Therefore, plants remain stunted.

Phosphorus is the part of adenosine diphosphate and adenosine triphosphate of plant and phosphoprotein (lecithin). It takes part in energy transfer. Phosphorus is main constituent of DNA, RNA, phospholipid, phosphatide and nucleic acid. It is present in structural composition of cell membrane, chloroplasts and mitochondria. It makes roots stronger and moves to storage organ of plant at the end of growing season.

Potassium regulates photosynthesis. It promotes carbohydrate synthesis and translocation, proteins synthesis and permeability of cell membrane. Hydration condition of protoplasm remain correct due to potassium. It is regulator of osmotic pressure and activator of enzymes.

Calcium is an essential component of cell wall and provides and stability to cell membrane. It promotes fruit maturation. Many enzymes of plant remain active due to calcium. It neutralizes organic acids formed by metabolism. It accumulates in wood and bark.

Magnesium is an important component of chlorophyll and a part of chromosomes and some protein of plants. It is a catalyst for many enzymes related with carbohydrate metabolism.

Sulphur is a component of some plant proteins and amino acids (methionine, cysteine). It increase quantity of oil in oil seeds crops. It is constituent of some enzyme cofactor.

Iron is the component of cytochrome and heme and non heme enzymes. It takes part electron transformation of enzymes. It is conductor of oxidation reduction processes. It is regulator of nitrogen fixation and nitrogen reduction.

Manganese takes part many oxidation reduction systems of plants and forms oxygen during photosynthesis. It is constituent and activator of some enzymes.

Zinc regulates the activities of cysteine and enzymes. It is also related with auxins like growth regulators. It helps in synthesis of chlorophyll.

Copper is an essential component of oxidising enzymes. It acts as a catalyst for respiration. It is constituent of enzymes ensuring lignin synthesis.

Molybdenum is main constituent of nitrate reductase enzymes. Therefore, helps in nitrogen fixation and utilisation by plants.

Boron helps in the formation of polysaccharides Thus, helps in development of the plant cells. It also regulates translocation of sugar and helps in chloroplasts synthesis.

Chlorine activates oxygen production system during photosynthesis and affects carbohydrate metabolism.

Deficiency Symptoms of Nutrients on Plants

Deficiency symptoms of plant nutrients depend upon the mobility of nutrient in the plant system. Nutrients having higher mobility show deficiency symptoms on older leaves. Due to higher mobility such nutrients go towards younger leaves or apical buds. Therefore, older leaves become deficient and show deficiency symptoms. Nitrogen, phosphorus, magnesium and manganese have higher mobility. Calcium, iron, copper and boron have lower mobility in plant system. They remain in lower parts and older leaves of plants.They do not move towards young leaves or upper parts of plants. Therefore, deficiency symptoms of these nutrients arise on younger leaves of plants.

Deficiency symptoms arise when amount of nutrient is below than the critical limit. These deficiency symptoms are very specific for any nutrient element and can help to assess the deficiency of that particular nutrient. However, sometimes due to weather and climate conditions and insects, pests and disease infections, the deficiency symptoms for particular nutrient element are unrecognizable and confusive. Therefore, laboratory analysis of deficient plant should be done to confirm the deficiency of nutrient element. When the deficiency is confirmed, the fertilizers should be applied to remove the deficiency. Deficiency symptoms of nutrients on plants are given here.

Nitrogen

Deficiency of nitrogen stops the growth if plants and reduces tillering. Branched plants develop lesser branches. First of all, deficiency symptoms of nitrogen arise on older leaves of plants and they became lighter green. In the excess deficiency leaves look dry and became erect. Lesser root development and small fruits are other deficiency symptoms of nitrogen.

Phosphorus

Deficiency of phosphorus stops plant growth. Plants become stunted and bushy in appearance. Low phosphorus deficiency plants show more darker colour than normal plants. Anthocyanin producing plants become purple due to phosphorus deficiency. Older leaves start drying from upper portion and later dried portion become brownish red.

Potassium

Potassium deficiency symptoms appear on older leaves. First, leaf top dries and then leaf edges. If potassium is not much deficient leaves may wilt without showing deficiency symptoms. There are less tillering and branching. Highly deficient soil may give dark fire colour in whole crop.

Calcium

Interveinal chlorosis of young leaves is one symptom of calcium deficiency. Except two or three new leaves all older leaves fall. Plants attain spindle shape. Plant may lodge due to poor root system. Tip of leaves dry and later on dried parts become reddish brown in colour. Lesser flowers and fruits dropping also symptoms of calcium deficiency. New leaves of terminal become hooked.

Magnesium

Magnesium deficiency stops the growth of plant. There is interveinal chlorosis among older leaves but veins remain green. Sometimes early fall of leaves.

Sulphur

Deficiency of sulphur appear on young leaves and they become light green. Later on, leaf edges dry from upper to lower side.

Iron

Plant become stunted due to iron deficiency. Greenness of leaves reduces and then they become white. In the excess deficiency tips and edges of leaves dry up and died. Tomatoes and other fruits do not attain normal colour after ripening

Manganese

Plants become stunted due to manganese deficiency. But there is no effect on tillers and branches of plants. Newer leaves are small, thin with palish green colouration. There is interveinal chlorosis which starts from tip and increase towards base of the leaves. Brown spots in the separated cotyledons of green pea is one major symptom of Mn deficiency.

Zinc

Plants become stunted and attain bushy appearance due to zinc deficiency. Internode distance become less. Therefore, leaves emerge nearer on branches. Except mid rib colour of leaves become greenish yellow. Monocots develop yellow stripes in between veins. Late tillering and late maturity of crop is also associated with zinc deficiency. Lower leaves of plants of citrus family remain normal but upper leaves become smaller and smaller. Growth of apical bud stops. Therefore, upper view of whole branch looks like flower of rose. Disease rosette or mottle leaf of citrus family, little leaf of apple and peas, white bud of maize and khaira of paddy are due to zinc deficiency.

Copper

Copper deficiency develop bluish colour on leaves and chlorosis on the leaf tip. Young leaves do not open and rolled just like needle. Sometimes upper part of leaf become just like needle but inner part grow normally. Copper deficiency causes diseases exanthema in lemons and summer die back in apple.

Molybdenum

Symptoms of molybdenum deficiency are somewhat like to deficiency symptoms of nitrogen. Plants develop chlorosis. Leaves of cauliflower etc. look like whiptail. Leaves of citrus etc. show yellow spots which are more at lower part of leaves.

Boron

Deficiency of boron reduces the growth of upper portion of the plant. Plants do not develop and show lesser number of branches. Leaves are thick and brittle. Heart rot of sugar beet, cracking of tomatoes, internal cork and bark measles of apple and top sickness of tobacco are some disease of plants caused by boron deficiency.

Chlorine

Top of young leaves develop white spots. Leaves are twisted due to these spots. Sometimes white spots cover larger area of leaves. Such area dries up and fall due to necrosis.

Amount of Nutrients in Plants

Water, organic matter, mineral matter are three main components of living plants. Relative amount of these are not similar in plants. But water is maximum and mineral matter is minimum. Generally, green live plants contain 70% water, 27% organic matter and 3% mineral matter. Also, after drying of green live plants, the dry matter has lesser mineral matter, as compared to organic matter. But mineral matter is too much important as organic matter synthesizing capacity of plants depends on this. Physiologically and practically part of mineral element is most important in plant. This is true with the use of fertilizers and manure in crop production.

Factors Affecting Amount of Nutrients in Plants

Amount of different nutrients in plants depends upon four factors. Genetic characters of plants, availability of nutrients in soil, plant part and plant age.

Genetic Characteristics of Plants

Absorption of different nutrients by root is a specific genetic character of plant, hence amount of nitrogen and potassium in plants is approximately ten times more than phosphorus and magnesium. Similarly amount of phosphorus and magnesium is 100 to 1000 times more than micronutrients. Generally, this is time for all plants. But different plant of each family have a remarkable difference in their nutrient content which is governed by the genetics of plant. Dicotyledons plants have more amount of divalent cations (calcium and magnesium) as compared to monovalent cations (potassium). Cation exchange capacity of roots of plants does not have any distinct effect on the absorption of cations

Availability of Nutrients in Soil

Availability of nutrients in soil directly affects the amount of nutrient in plants. But there is a relation between concentration of nutrient in soil solution and its absorption by plants. At lower concentration this relation simple linear and curved at higher concentration. Tissues of plants should have a minimum concentration of nutrients below which plants die. The minimum level or concentration of nutrient is called critical level. Critical level is different for each nutrient. Critical level for micronutrients many times lower than macronutrients. Deficiency of different nutrients can be known by chemical analysis of whole plant and accordingly fertilizers application is recommended.

Plant Part

There is significant difference in the amount of nutrients among different plant parts. Leaves, roots and stem have more difference in nutrient content as compared to fruits, seeds and tubers which have lesser differences. This is because plants have a tendency to store sufficient nutrients and organic matter in reproductive and storage parts, even vegetative parts have lesser amount of these.

Plant Age

Younger plants and younger organs have more amount of nitrogen, phosphorus, and potassium. Older plants and their organs have more amount of calcium, magnesium, iron and boron. In the initial growth period nutrient concentration increases because during this period growth rate is lesser than nutrient absorption rate. But in the next stage due to higher growth rate, there is dilution of nutrients in plant parts. In the annual crops nutrient absorption decrease after flowering and translocation of nutrients from vegetative organs to reproductive organs increases remarkably. Nitrogen and phosphorus are more translocated as compared to other nutrients.

Flow of Nutrients from Soil to Plant Roots

Before absorption of plant nutrients by roots, it is necessary that other nutrient should reach to root in dissolved form or root should go up to the nutrients and take them. Activities of reaching of nutrients up to roots or moving of roots up to nutrients run simultaneously. During germination seed provide nutrient to seedlings and radical enters in the soil and take nutrients from there. Plants become completely dependent on soil for nutrients, till nutrients of seeds lasts. Mass flow and diffusion are two processes for flowing nutrient from soil to roots and through roots interception process root go up to nutrients.

Mass Flow

Some nutrients are present in irrigation and rain water. When this water come in to soil, nutrients dissolved in it also come in soil and dissolve more amount of nutrients present in soil. Evaporation pumps plant water into the atmosphere. This creates a potential gradient by which soil water moves towards the roots. By this way nutrients dissolved in water also come to the roots. Nutrient carrying capacity by mass flow depends upon amount of moving water and amount of nutrients dissolved in it. If both these are more mass flow of nutrients will also more.

Diffusion

Continuous uptake of nutrients by plants deplete the nutrient status near roots. Also, mass flow process can not completely provide nutrients up to roots. At this stage diffusion process carries nutrients up to plant roots. There is an equilibrium between concentration of nutrient ions present in soil solution and ionic concentration of these nutrients present on soil particles. When more amount of nutrients are taken by roots, concentration of these nutrients near roots is too much depleted. This creates a concentration gradient. Due to this gradient, nutrients diffuse from places of higher concentration i.e., soil particles to lower concentration i.e., plant roots. Diffusion of ions goes on till a new equilibrium is established. Diffusion on of nutrients ions from soil particles to roots obey Fick's law of diffusion ions.

Root Interception

In this process roots physically move towards nutrients. Root tips are highly active for taking up nutrients. When roots move forward in the search of water and nutrients, they meet newer soil particles. They absorb nutrients present on soil particles through contact exchange process. By this way when roots further develop, they cover a large volume of soil. Roots exchange with the nutrients present on soil colloids by the ions (mostly hydrogen) on their surface. This process is given here,

Absorption of Nutrients by Plants

Leaves, stem and roots of plants can absorb plant nutrients. But mainly roots absorb the nutrients and send them towards internal organs of plants. Entry of nutrients from root surface into the root is a metabolic process which is selective for nutrient absorption.

Plant leaves can absorb different nutrients. Absorption of micronutrients is more by leaves as compared to roots. Stomata absorb carbon dioxide, rain water and oxygen during respiration. Nutrients absorbed by leaves react quickly. Therefore, nutrients sprayed on leaves show their effect very earlier as compared to nutrient applied in soil. Nutrients required in lesser quantities by plants should supply by foliar spray. Foliar spray of lesser quantities of nutrients are more beneficial as compared to soil application.

Entry of Nutrient Ions in Roots

Passive absorption in which metabolic energy is not required and active absorption in which metabolic activity is required are two main processes of entry of nutrient ions in roots.

Passive absorption

Main component of epidermis of roots are cellulose fibres and peptic material. Cellulose fibres acts as sponge in which solution is diffused continuously. Volume of tissues in which dissolved ions can diffuse according to concentration gradient is called as apparent free space. In the passive absorption ions are diffused from soil solution to apparent free space. This process is performed due to concentration gradient and interrelation among ions without any expenditure of energy. Therefore, it is called as passive absorption.

The driving force of passive absorption is electrochemical potential. This potential depends upon the ratio between concentration of solution and charge present in it. Cytoplasm in the cell membrane is negatively charged and out of this soil solution is neutral. Even concentration of any ion is same/similar in the cell membrane and solution present out of this, an electrochemical potential difference is produced.

This potential difference acts as a driving force for ions present in soil solution to cross cell membrane and to move inside. Sufficient uptake of calcium, magnesium and potassium may take place only due to electrochemical potential difference of both sides of cell membrane. This is called as passive uptake because this is due to physical forces and there is no consumption of energy.

Active Absorption

Active absorption is the uptake of ions in opposite direction of electro chemical potential difference produces between solutions present inside and outside of cell membrane. Uptake of ions like nitrate, potassium, sulphate etc. take place in opposite direction of electrochemical potential difference. Active uptake requires energy which is produced by roots through respiration.

It is supposed that outer layer of epidermis of roots have biological membrane. This membrane has some molecules which carry ions inside the membrane. These molecules are called as carriers and this process of uptake is called as carrier concept. These carriers (molecules) have specific binding site for different ions for selective absorption of ions.

Energy is required for formation of carriers which is available in the form of ATP. Carrier attaches the ion present on roots surface and diffuses into cell membrane. Inside the cell an enzyme separates ion from the carrier. Now carrier again goes outside of cell membrane to start second cycle of ion uptake.

Active uptake may be in opposite direction of concentration gradient. This is selective for ions. There may be different carriers on different binding sites for different ions. Sometimes few ions have competition for a particular binding site, for example, calcium and strontium. This competition results mutual antagonism in the absorption of nutrient ions. Potassium has another binding site. Therefore, potassium has no competition.

Factors Affecting Nutrients Absorption by Plants from Soil

Nutrients absorption by plants is a multi-step process. This is affected by numerous soil and plant factors. The soil factors are availability of plant nutrients, physical condition of soil, availability of water, temperature, soil pH and concentration of other ions. The plant factors are, nature and expansion of root surface, age of plant and growth rate and transpiration and respiration rate etc. These factors are described here in short.

Availability of plants nutrients in soil: Absorption of nutrients by plants mainly depends upon their availability in soil. Availability means amount of nutrient already present in soil solution and amount attached with soil particles which can easily come in soil solution.

Physical condition of soil: If availability of nutrients is more but physical condition of soil is not good for roots elongation and expansion, then nutrient absorption is needed. Hard layers or pan in soil, more clods and pebbles in soil and bad soil structure reduces root growth, elongation and expansion which ultimately reduce nutrient absorption.

Availability of water: Water is medium for movement of nutrients. If water is in short supply movement of nutrients from soil to root system reduces. This decreases absorption of nutrients. Excess of water in soil reduces oxygen status and decreases carbon dioxide removal from soil. This condition, except rice, reduce growth and activity of roots. Presence of oxygen is essential for respiration of roots as respiration provides metabolic energy for nutrient absorption.

Temperature: Absorption of nutrients is a biological process which is influenced by temperature. If temperature is less than 15°C then rate of nutrient absorption reduces remarkably.

Soil pH: Nutrient absorption rate depends upon pH of soil solution. Generally, absorption rate of anions like nitrate, phosphate etc. is higher at weakly acidic conditions. Ionic forms of phosphate are also governed by pH. The ratio of HPO^{2-}_4 / H_2PO_4 increases with increase in pH and absorption rate decreases. Highest absorption of phosphate by rice plants takes place at pH 5-6. Acidic conditions have more concentration of hydrogen ions. This increases permeability of cell membrane which decrease its cations absorption capacity. Therefore, absorption of cations is higher at neutral and alkaline pH.

Concentration of other ions: Ions of similar properties have competition during their absorption by roots, for example, absorption of zinc reduces in the excess of iron and absorption of potassium decrease due to excess of calcium.

Nature and expansion of root system: Root morphology govern intake and nutrients absorption after making contacts with soil particles. Depth of root system, its expansion and number of root hair etc. come under root morphology. Root morphology is a genetic character but it is affected by environment also. Dicot plants have deep tap roots which can absorb water and nutrients from deeper layer of soil as compared to shallow roots of monocot plants. Roots of perennial plants are deeper as compared to annual plants. Root system of agricultural crops remain up to 50-100 cm depth. Thus, they absorb nutrients from deeper layers of soil. Root system expanded as per the age of plants.

Number of ions exchange sites of root surface remarkably influence the nutrient absorption. Ion exchange capacity and cellular exchange site of dicot plants are higher than monocot plants. Therefore, dicot absorb more calcium. Instead of whole root system only tip portion of root is active to absorb nutrients. Nutrients absorption capacity of roots gradually decrease with the increase in age of plant.

Rate of transpiration: rate of transpiration acts as a driving force for absorption of water and nutrients dissolved in it. This driving force causes the movement of water along with dissolved ions towards roots. Absorption of any nutrient is not necessary which has reached to root surface, but there is decrease in nutrient absorption with the decrease in transpiration.

Rate of respiration: metabolic energy is required for active absorption of nutrients. This is produced from oxidation of sugar during respiration. Any obstacle in the respiration process has adverse effect on nutrient absorption. Long continuous water stagnation in the fields develop pale yellow colour in the maize, wheat etc crop. This is due to lesser absorption of nutrients.

Translocation of Nutrients from Roots to Stem and Leaves

Nutrients move to xylem vessels after absorption from where they translocated to stem and leaves along with water. Mass flow along with water is main process of translocation of nutrients. Mass flow translocation of nutrients is mainly governed by transpiration rate and nutrient absorption rate of roots. Decrease in transpiration rate also decreases nutrient translocation rate.

Translocation of nitrogen in the xylem is in the form of nitrate, ammonium and amino acids. Nitrate reduction process partly takes place in the root which produce amino acids. Woody plants have more nitrate reduction capacity as compared to herbaceous plants. Glutamine and asparagine amino acids and their precursor glutamate and aspartate are present in xylem sap. Concentration of these organic nitrogenous compounds in xylem sap depends upon the nitrate reduction capacity of plant roots. Amino acids are rapidly absorbed by tissues nearby xylem. Glutamate and aspartate are not absorbed by stem but they directly reach to the leaves and utilized for protein synthesis there.

Like amino acids, phosphate and potassium are absorbed by cells nearby xylem. But other ions are absorbed slowly. Therefore, they translocate to leaves and accumulate there. Excess amount of these nutrients than the requirement is accumulated at the edge and tip of the leaves. Such accumulation up to toxic level can burn them.

Soil Fertility Evaluation

There are greater difference among total and available and exchangeable quantities of nutrients present in soil. Soil fertility evaluation is essential to accurately assess the amount of nutrient that plant can uptake directly or which is available for the plants out of total quantity of nutrients. Also, this is necessary for long term sustainable agriculture that nutrients taken up by previous crops should be added to soil as per crop requirement.

There are so many methods for soil fertility evaluation. Deficiency symptoms of nutrients on plants, plant analysis, soil analysis and biological methods are commonly employed for soil fertility evaluation.

Deficiency Symptoms of Nutrients on Plant

The deficiency symptoms are very specific for anyany nutrient element and can help to assess the deficiency of that nutrient. However, sometimes due to weather, climatic conditions and insect, pest and disease infestations, the deficiency symptoms for nutrient element are unrecognizable and confusive. Therefore, laboratory analysis of deficient plants should be done to confirm the deficiency of any nutrient element. When the deficiency is confirmed, the fertilizers should apply to remove the deficiency.

Deficiency symptoms of plant nutrients depends upon the mobility of nutrient in the plant system. Nutrients having higher mobility show deficiency symptoms on older leaves. Due to higher mobility such nutrients go forward to younger leaves or apical buds. Therefore, older leaves become deficient and show the deficiency symptoms. Nitrogen, phosphorus and manganese have higher mobility whereas calcium, iron, copper and boron have lower mobility in plant system. Low mobility nutrients remain in lower parts and older leaves of plant and do not move toward younger leaves or upper parts of plants. Therefore, deficiency symptoms of these nutrients arise on younger leaves of plant. Deficiency symptoms of the nutrient on plants are given earlier.

Plant analysis: Two system are employed for plant analysis method namely determination of total quantity of nutrients and plant tissue analysis.

Determination of Total Quantity of Nutrients

Whole plant or specific part of plant mainly leaves are analysed for this method. Dry combustion or wet digestion are used to bring plant sample in liquid form. This liquid is analysed for nutrient by the standard method. The analytical results indicate critical nutrient concentration. The critical nutrient concentration is the amount of nutrient in plant or specific part below which that nutrient is deficient or less for the optimum growth of plant.

Plant Tissue Analysis

This is an easy and quick method. A qualitative amount of nutrients present in plants can be done in field itself. This method is mostly used for nitrogen, phosphorus and potassium estimation. In this method cell sap of stem, leaf, midrib or leaf blade is used. Plant part is dissected with knife or pressed to obtain cell sap. This cell sap is treated with specific chemicals indicated for that nutrient. The change in colour of cell sap is matched with standard chart and deficiency of nutrient is known. In maize etc. stem is cut obliquely by knife and chemical is sprayed directly on cut portion for reaction. The colour change is matched and amount of nutrient is predicted.

Soil Analysis

Plant analysis gives qualitative as well as quantitative amount of nutrients present in plant. Similarly, soil analysis also gives actual and accurate amount of nutrient present in soil. Soil analysis can be done to know total as well as plant available quantities of nutrients. However, for supplying nutrients to crop only plant available quantity is determined. This test is done before sowing of crop so the nutrient can be applied according to soil test. There are three step of soil analysis or testing namely, collection of soil samples, analysis of soil samples and fertilizer recommendation according to soil analysis or soil test. These are described here in short.

Collection of Soil Samples

The three-dimensional nature of the soil was emphasized in the first practical, Sampling of such a soil which varies to a great extent from place to place is quite problematic. Since analysis is done on such a sampled soil. It is necessary that the soil sample should be representative of that area. The importance of a true representative sample can be realized from the fact that only a minute quality (say 1-10g) of the large soil mass of the field is used for the analysis in the laboratory. Unless one is sure of representative and proper sampling, the results obtained in the laboratory analysis will be of no use under the field conditions. Therefore, the soil sample must properly represent the area.

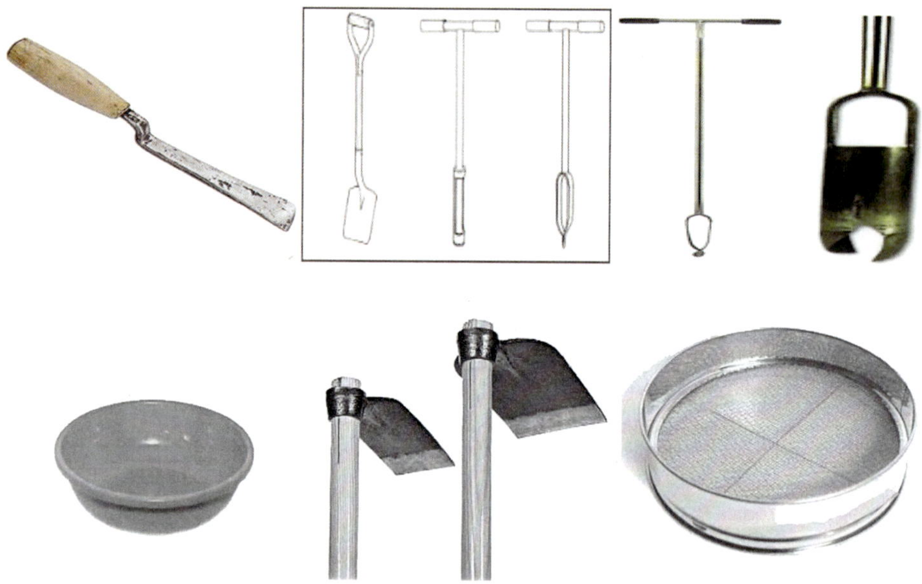

Fig. 3.53: Soil Sampling tools

A field can be considered as a single sample, if it is uniform in all the respects. Variations in slope, colour, texture, crop and management should be observed and separate sets of composite samples must be collected from each of such area. Sampling of recently fertilized plots, bunds, channels, marsh and places near trees, wells, compost pits and fertilizer storage should be carefully avoided.

Soil sampling tools: Depending on field conditions and the objective of sampling, select proper sampling tools. Figures of some soil sampling tools are given here like khurpi, belcha, different augers, spade, tasla and sieve

Soil sampling procedure: Based on difference in soil type, colour, crop growth or slope, divide the area in different homogenous parts.

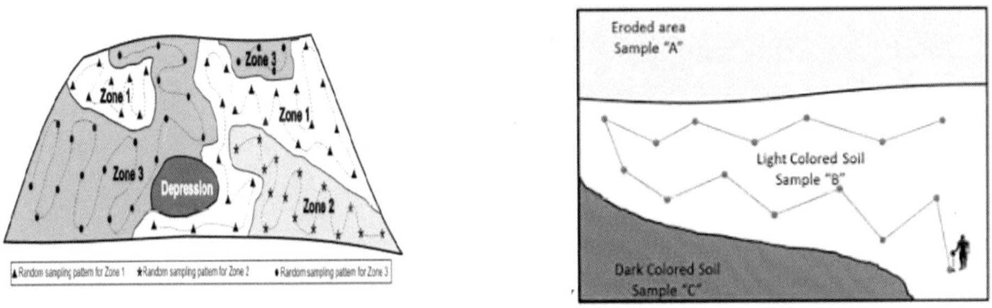

Fig. 3.54: Four different somewhat firm parts (not zones) one small pond and small farm building and show random soil sampling points at any one part

At the sampling site of the part, remove the surface litter with *khurpi* or spade and with the help of the sampling tool collect one 15 cm deep sample in a bucket or tasla. If a *khurpi* is used, make a 'V' shape cut up to 15 cm. depth and collect the soil. This sample is known as 'primary' sample.

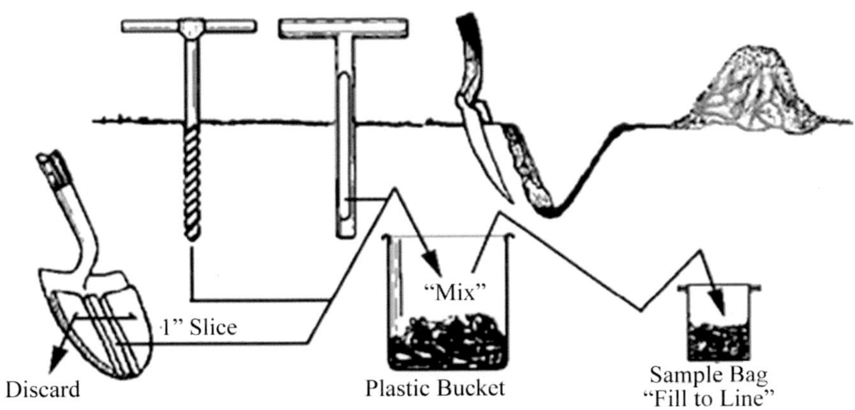

Fig. 3.55: Soil Sampling and processing

Collect such primary samples from different points of the selected part of area in a Zigzag fashion. Such primary samples should have approximately the same weight.

Soil sample processing and storage: After bringing the sample from the field, spread it on a dry tray. Be sure that the tray has the proper label. Do not keep tray where fertilizer or manure is stored. Allow the samples to dry in shade (on a dry rack) for 24-48 hours. So that the samples become air dry. Using a mortar and pestle or a wooden roller grind the samples and pass it through a 2 mm sieve.

Transfer the sieved material in a polythene bag, put one label indicating details of soil samples inside the bag and finally place it in a cardboard carton. Label the carton properly with the details of soil sample.

2. Analysis of Soil Samples

Use appropriate and recommended analytical methods for particular property of soil from book and Laboratory manuals of soil science (Bharadwaj, Venkatesh,1999 and 2002).

3. Fertilizer Recommendation

On the basis soil testing and farmers choice for targeted yield, fertilizer recommendation are given for different crops. Generally, farmers choices for soil testing are for maximum yield, maximum profit, maximum income or return for rupees invested or minimum beneficial amount of fertilizer to be applied in soil for any crop. All these are known by soil test- crop response correlation.

Biological Methods of Soil Fertility Evaluation

Just like chemical reagents are used in chemical analysis methods, biological methods of soil fertility evaluation use higher plants and other microorganisms to determine amounts of plant nutrients present in soil. Although biological methods give accurate result but due to much laborious and time taking, these are not used for routine analysis of soil. Brief description some biological methods of soil fertility evaluation are given here.

1. Higher Plant Methods

Field experiment, use of indicator plants, Mitscherlich pot methods and Neuber technique are come under higher plant methods.

Field experiment: Statistically designed simple and complex field experiments are conducted at farmer's field and research stations. Graded doses of nutrients are applied and tested as treatments to different crops. When similar results are obtained for atlast two years, then required doses of nutrients are recommended for different crops based on these experiments.

Use of indicator plants: Some plants show deficiency symptoms of nutrients very earlier than other plants. These indicator plants are grown at probable deficient areas and according to severity of deficiency symptoms, the amount of nutrients predicated. Indicator plants for different nutrients are as, Cauliflower and Cabbage for nitrogen and calcium, Rape for phosphorus, Potato for potassium and magnesium, Alfalfa and Lucerne for sulphur, Jowar for iron, Oats for magnesium, Paddy and Maize for zinc, Barley and Wheat for copper, Lucerne and Berseem for Molybdenum, Cauliflower and Sunflower for boron and Potato and Sugar beet for chlorine.

Mitscherlich pot methods: This is like to field experiment method. The only difference is that experiment is conducted in green house in small pots instead of larger fields. This technique was developed by Mitscherlich for determining N, P and K requirement of oats crop. Results of pot experiment cannot be completely utilized for fields due to use of lesser volume of soil, change in natural properties of soil during filling of pots and hindrance in normal growth of roots in pots. These directly affect the results of this method.

Neubaur technique: Excess number of plants are grown in small quantity of soil for its complete extraction of nutrients. These plants are analysed to know the amount of nutrient present in soil. Neubaur developed this technique for determination of phosphorus and potassium.

In one 11 cm diameter and 7 cm deep glass plate, 100 gm soil, mixed with 50 gram nutrient free sand is filled. In it 100 plants of rye grass are grown for seventeen days. After that whole plants with roots are analysed for phosphorus and potassium and amount of these nutrients in soil is determined.

Microbiological Methods

In these methods fungi and bacteria are grown in very small amount of soil to determine soil fertility. These methods give results in very short period. *Azotobacter* method. *Aspergillus niger* method and carbon dioxide evolution methods are three main microbiological methods.

***Azotobacter* method:** S.N. Winogradski indicated that microorganisms are also adversely affected by nutrient deficiencies. Azotobacter bacteria is highly sensitive for the deficiency of acidity, phosphorus and potassium. Therefore, Azotobacter can be utilized to determine the deficiency of calcium, phosphorus and potassium. Based on information of Wingradsky, Sacret and Steward added starch in the soil to be tested and divided the same in three parts. In one part phosphorus, in second part potassium and in third part phosphorus and potassium both were added. These soils are mixed with Azotobacter culture and incubated for 72 hours. After this period growth of Azotobacter colonies were measured and status of phosphorus and potassium were assessed accordingly.

***Aspergillus niger* method:** *Aspergillum niger* fungus is a good indicator of amount of potassium, phosphorus and magnesium in soil. In this method 2.5 gm soil is added to culture solution of spores of *Aspergillus niger* fungus. This mixture is incubated for four days. After that the growth of mycelium is the indication of potassium status of soil.

Carbon dioxide evolution method: Respiration of all the organism produces CO_2. Higher evolution of CO_2 gas indicates higher number of organisms present in soil. But in nutrient deficient soil, activity of organism and CO_2 evolution by them are reduced.

Mineral Nutrition of Crop Plants

In the nutrition of plants, quality of produce has utmost importance. The quality of produce depends upon the plant species, intended use of produce, prevailing physical conditions, environmental and climatic situation and chemical parameters of growing media. These are considered the most important components to decide of overall quality of produce. The mineral nutrition of plants or crops influences their processing, sensory, nutritional quality and aimed to improve sustainability, food safety and security.

Globally the actual and hidden hunger of nutrients and the increasing awareness of the importance of nutrition for good health of human beings and animals, the judicious use of mineral nutrients to crops is present day need of agriculture for increasing productivity, profitability and quality. However, the mineral nutrient requirements to achieve maximum yield may not always coincide with those to attain the best quality, higher quality and higher return per rupee invested. Here we are describing various aspects of individual plant nutrients.

Nitrogen Nutrition of Plant

Major sources of nitrogen in soil is organic matter. Only 2-3 percent of total nitrogen is available to plants and remaining 97-98 percent is in the forms of very complex organic compounds. These compounds are not easily decomposable, but during ploughing complex organic compounds are changed in to NH_4^+ and NO_3^-. $NH_4^+$and NO_3^- are absorbed by plants for their nutrition. Rain water also provides 2-20 kg. nitrogen per hectare in soil. Symbiotic nitrogen fixation, green manure, farm yard manure, compost, deoiled cakes and nitrogenous fertilizers provide appreciable quantities of nitrogen to soil. Nitrogen is lost from soil by plant harvest, denitrification, volatilization, leaching and erosion. The erosion loss of nitrogen may be up to 20 kg/ha/ per year.

Ammonium fixation: Ammonium fixation is different than nitrogen fixation by microorganisms. In ammonium fixation soluble or exchangeable ammonium ions are fixed by clay minerals due to contraction of the inter layer space. These fixed or adsorbed ammonium ions are again exchangeable only after the expansion of interlayer space. Ammonium fixation reduces the availability of nitrogen to plants and microbes.

Factors affecting ammonium fixation: Numerous factors affect the ammonium fixation in soil these factors are, Type of clay mineral, Organic matter, Soil texture, Soil temperature and moisture, Presence of potassium ions in soil, and Presence of nitrogen and carbon in soil. These are briefly described here,

Type of clay mineral: Vermiculite is dioctahedral. Wedge zone present in it is too much selective for ammonium and potassium fixation. Montmorillonite and illite clays also fix ammonium. But maximum fixation is in vermiculite followed by illite. Octahedral spheres of expanding clays have a size of 1.40 nm. The ionic radius of ammonium ions is somewhat like to this size. Therefore, ammonium ions enter in to the interlayer spaces and fixed there. The wedge zones close the clays. This closure strongly and tightly fixes the ammonium ions.

These fixed ammonium ions may be replaced by lattice expanding ions like calcium, magnesium, sodium and hydrogen. But lattice contracting ion like potassium, rubidium and caesium cannot replace the fixed ammonium.

Organic matter: Organic matter also fixes ammonium. It is supposed that Hydroxyl (OH^-) group of organic matter takes part in it. More amount of oxygen and higher pH of soil enhance NH_4^+ fixation by soils.

Soil texture: Coarse clay fixes more ammonium as compared to fine silt and sand. Subsurface soil fixes more ammonium as compared to surface soil because of more clay in subsurface soil.

Soil temperature and moisture: Soil temperature and moisture also govern ammonium fixation. Moist soils have lesser ammonium fixation. Freezing of soil increases ammonium fixation and oven drying of soil increases ammonium fixation tremendously.

Presence of potassium ions in soil: Presence of potassium ions in soil reduces ammonium fixation. This is because of lesser ionic radius of potassium ions (1.33 n.m.) than the size of octahedral spheres of clays. Therefore, potassium easily enters in the spheres and occupies them. Hence lesser spheres remain for ammonium to be fixed. Application of potassium is recommended in the ammonium fixing soils prior to the application of ammonium. This creates more availability of ammonium in soil. Because uptake of potassium by plants expands the interlayer spaces of clays. Therefore, ammonium ions present in these spaces come into the soil solution by diffusion.

Presence of nitrogen and carbon in soil: Increased concentration of nitrogen in soil and carbon in soil organic matter also increases ammonium fixation. Surface soil has very low amount of fixed ammonium but in sub surface soil of it may about half of total nitrogen. Ammonium remains in an equilibrium with exchangeable ammonium. Therefore, recently fixed ammonium brings native fixed ammonium into soil solution by displacing it.

Phosphorus Nutrition of Plants

Sometimes it has been seen that phosphorus deficiency symptoms are seen in plants even phosphorus in soil is more than thousand kilogram per hectare. This may be clarified after understanding the reactions of phosphorus in different soils. In soil major sources of phosphorus are farm yard manure, plant residues and green manures, natural inorganic and organic compounds present in soil and fertilizers. Plants absorb phosphorus in the forms of $H_2PO_4^-$ and $HPO^{2-}_{4.}$

Natural mineral of phosphorus present is apatite. Fluorapatite $[Ca_3 (PO_4)_2]_3$ CaF_2, Chlorapatite $[Ca_3 (PO_4)_2]_3$ $CaCl_2$, Hydroxyapatite $[Ca_3 (PO_4)_2]_3$ Ca $(OH)_2$, Carbanatoapatite $[Ca_3 (PO_4)_2]_3$ $CaCO_3$ are four main types of apatite. Besides these iron phosphate $Fe_3(PO^-_4)_2$ and aluminium phosphate, $Al_3(PO^-_4)_2$ are also present in soils. Organic compounds of phosphorus present in soils are phytin, Inositol, nucleic acid, phospholipids etc.

Reaction of phosphorus in acidic soil: The solubility of added monocalcium phosphate in to the acidic soil is reduced. This is called as phosphate retention. Phosphate retention may be due to precipitation of insoluble compounds in soil solution, reaction with the hydrated sesquioxide and reaction with clays. These are described here in short.

Precipitation of insoluble compounds in soil: Excess amounts of iron and aluminium are found in soil solution or in exchangeable form. These retain more amount of phosphorus by forming complex iron and aluminium orthophosphate. This complex is insoluble in water. $H_2PO^{2-}_4$, HPO^{2-}_4 and PO^{3-}_4 are three ionic forms of orthophosphoric acid (H_3PO_4). More $H_2PO^{2-}_4$ ion is present below pH6.7, more $H_2PO^{2-}_4$ in between pH 6.7 to 9.0 and more PO_4^{3-} ion is found above pH 9.0. Solubility of calcium salt of orthophosphoric acid is like $Ca(H_2PO_4) > CaHPO_4 > Ca_3 (PO_4)_2$.

Precipitates of iron and aluminium compounds are formed with $H_2PO_4^-$ like

$Fe^{3+} + H_2PO_4^- + 2H_2O \longrightarrow Al (OH)_2 .H_2PO_4 + 2H^+$

Soluble insoluble

$Al^{3+} H_2PO_4^- + 2H_2O \longrightarrow Al (OH)_2. H_2PO_4 + 2H^+$

Soluble insoluble

In acidic soil concentration of iron and aluminium is more than calcium. Therefore, above reactions form insoluble complex compounds of iron and aluminium with phosphorus. Therefore, amount of soluble $H_2PO_4^-$ is very less in soil and phosphorus deficiency symptoms arise on plants.

Reaction with hydrated sesquioxide: Most of the soils contain sufficient amount of hydrated oxides of iron, aluminium, titanium and manganese. These are in colloidal forms and phosphate ions colloidal forms and phosphate ions attract on these and adsorbed on its surface as Iron and Aluminium phosphates. Although initially this is only surface reaction but later on other compounds are formed due to direct precipitation.

Reaction with clays: Phosphate directly attached with clays, first by substituting OH^- group of aluminium octahedron and secondly by making clay – calcium phosphate bond. Clays of low silica sesquioxide ratio (SiO_2:R_2O_3) retain more phosphate ions. There is anion exchange between phosphate and hydroxyl ion. Phosphate adsorbed in soils by anion exchange is replaced by applying arsinate, hydroxide and oxalate ions. This will increase the availability of phosphorus. Anion exchange reaction is more in ordinary clays and kaolinitic (1:1 type) clays OH^- ion of Al-OH group is substituted by phosphate ion as-

$$Al\,(OH)_3 + H_2PO_4 \longrightarrow Al\,(OH)_2\,.H_2PO_4 + OH$$

Reaction of phosphorus in alkaline soils: Phosphorus reactions in acidic and neutral soils are quite different than reaction in alkaline soils. Reactions of calcium and pH are more in alkaline soils. Therefore, concentrations of di and triphosphates are more than monophosphate and solubility of di and triphosphate is less than monophosphate. Generally, phosphate makes hydroxyapatite type compound. The phosphate ions precipitate on the surface of calcium carbonate in the soils containing free calcium carbonate. Such precipitation depends upon area of surface and concentration of phosphate in solution. Initially this is a surface action but later precipitation is like mass action and rate of precipitation depends only on the concentration of phosphate in solution.

Clay saturated by calcium ion retains more phosphate as compared to clays saturated by sodium and other monovalent ions. The concept of clay -Ca- H_2PO_4 linkage is considered at pH less than 6.5 in soil but in alkaline soils clay – Ca- H_2PO_4 can directly make the precipitate.

Retention of phosphorus in soil: Retention of phosphorus in soil directly influences its availability. Factor like type of the clay, Soil pH, time of reaction, temperature and organic matter directly affect the retention of phosphorus.

Type of clay: 1:1 type clay (kaolinite) retain / fix more phosphorus as compared to 2:1 type clay (montmorillonite). Therefore, soils having more kaolinite also fix or retain more phosphorus. This is because hydrated oxides of iron and aluminium are more in soils rich is kaolinite clay. Also, soils having more clay minerals fix more phosphorus as compared low clay mineral soils.

Soil pH: Soil pH values directly influence the phosphorus availability. Soils with pH below 5.5, there is reaction of phosphate with hydrated oxides of iron and aluminium or precipitation of iron and aluminium phosphate. These reduce the phosphorus availability or fix more phosphorus. Between soil pH 5.5 to 7.0, there is more availability of phosphorus. Above pH 7.0, the presence of ions of calcium and magnesium and their carbonates form insoluble precipitates of phosphorus. Therefore, availability of phosphorus is decreased. But if pH is more than 7.0 due to sodium ion the phosphorus availability does not decrease.

Time of reaction: The availability of phosphorus gradually reduces with the increase in time of reaction. This is due to the change of form of the added phosphorus in soil. Some soils retain more phosphorus. In such soils repeated application of phosphorus is necessary to maintain the phosphorus supply to each crop. But in lesser phosphorus retaining soil single application of phosphorus in the crop rotation is sufficient.

Temperature: In the tropical soils phosphorus retention is more than temperate soils because high temperatures increase the rate of P transfer to strongly bond forms and thus decrease P concentration in solution.

Organic matter: Phosphorus retention is less in organic matter rich soils. Addition of green manure or FYM increase availability of phosphorus for next crop. Decomposition of organic matter produces carbon di oxide gas, which form organic acids. These acids increase the availability of phosphorus. Organic matter also forms Phosphohumic complexes with phosphorus. Phosphohumic complexes are more absorbed by plants. Humate ions replace phosphate ions which increases phosphorus availability. Humus makes its layer on particles of sesquioxide. This inhibits sesquioxide to make insoluble phosphate compound of phosphorus.

Potassium Nutrition of Plants

Major sources of potassium in soils are potassic minerals. Potash feldspars, muscovite and biotite minerals provide potassium to soils. Biotite provides maximum potassium to plants followed by muscovite and then potash feldspar. Besides these minerals some secondary or clay minerals like montmorillonite group, illite or hydrous mica, chlorite, vermiculite and other interstratified minerals also provide potassium.

Plants absorb potassium from soils solution as exchangeable potassium ion. Normal soils contain abort 2.40 precent potassium. This is quite higher than phosphorus content (0.11%) in soil. Therefore, potassium deficiency is not generally observed in soil as compared to the deficiencies of nitrogen and phosphorus. Fine textured soils contain more potassium as compared to coarse textured soils. Amount of readily available potassium for plants remain very less even total potassium is more in soil. Therefore, higher total potassium soils also very well respond for applied potassium. Unavailable (90-98% of total potassium), slowly available (1-10% of total potassium), and readily available (1-2% of total potassium) are three forms of potassium present in soils. These forms are under an equilibrium with each other. Loss of potassium from soils take place by plants uptake and leaching.

Potassium of primary minerals is called as unavailable potassium which gradually converts in to exchangeable and readily available potassium. Potassium present in mica and feldspars is unavailable potassium. In the crystalline form its solubility is very less. However, it becomes available due to decomposition caused by weathering as.

$2KAlSi_3O_8 + 2H_2O + CO_2 \rightarrow HAlSi_3O_8 + K_2CO_3 + 4SiO_4$

$K_2CO_3 \rightarrow 2K^+ + CO_3^{2-}$

This K is used by the plants or microbes, drain out through water or change into slowly available form.

In soil unavailable potassium is changed to available potassium and simultaneously reverse of it also takes place. Therefore, excess amount of available potassium in soil is changed into unavailable forms.

In soil, water soluble and exchangeable potassium is readily available to plants. Slowly available potassium is not generally extracted by chemicals in the laboratory but available to plants after some time. However, it is more available than the potassium present in primary minerals. Slowly available potassium is also called non – exchangeable potassium.

There is an equilibrium between soil solution potassium and exchangeable potassium. Therefore, potassium added to fields increases its concentration in soil solution. After some time, potassium from soil solution comes on colloid complex. Similarly, absorption of soil solution potassium by plant roots decreases its concentration in soil solution. Therefore, exchangeable potassium comes in soil solution to maintain the equilibrium.

Potassium fixation in soil: Potassium ion is also fixed in soil as phosphorus ion and unavailable to plants for some time. Potassium is fixed or retained in between the silica tetrahedron layers of 2:1 type of clay. The oxygen spherule of silica tetrahedron layer has size similar to potassium and ammonium ions. Therefore, these are fixed in 2:1 type of clays.

Factors affecting potassium fixation in soil: Many factors affect the fixation of potassium in soil. These are described here in short:

1. **Type of clay:** Potassium is fixed only by 2:1 type clay like montmorillonite, vermiculite etc. 1:1 type of clays do not fix the potassium. Although organic matter holds the exchangeable potassium but does not fix it.

2. **Alternate wetting and drying:** Alternate wetting and drying changes the exchangeable potassium in to slowly available potassium form and reduces its exchange capacity also.

3. **Temperature:** Higher temperature rapidly increases the potassium fixation process.

4. **Freezing and thawing:** Freezing and thawing process in wet soil remarkably influence the release of fixed potassium and the fixation of exchangeable potassium. The fixed potassium changes to exchangeable potassium. But

soil rich in exchangeable potassium and illite clay mineral show some fixation of exchangeable potassium due to freezing and thawing.

As given here added calcium ions substitute the fixed potassium from the potassium rich soil colloid just like cation exchange,

(Soil colloid) K K NH_4 K + $CaSO_4 \rightarrow$ (Soil colloid)NH_4 Ca K + K_2SO_4

Calcium rich soil colloid adsorbed more potassium from soil solution. If soil colloid has more adsorbed calcium and hydrogen, then potassium of added potassium chloride substitutes calcium as given here,

(Soil colloid) Ca H + 2KCl $\rightarrow$ (Soil colloid) H K K + $CaCl_2$

Therefore, calcium rich soils often show potassium deficiency after addition of potassic fertilizer. Addition of lime increases the base saturation of soil. This decreases the losses of exchangeable potassium.

Calcium Nutrition of Plants

Calcium in the form of Ca^{2+} ion present in soil solution or as exchangeable calcium is available for plants. Soil calcium has equilibrium among mineral calcium, exchangeable calcium and water-soluble calcium. Dolomite, calcite, apatite and calcium, feldspars are calcium minerals.

Weathering of calcium minerals provides calcium ions. Calcium ions are either absorbed by plants and microorganisms, adsorbed on colloidal complex, lost in drainage water, or form secondary compounds of calcium, which again exists as precipitate.

Generally, coarse textured soils of humid regions contain less calcium as they are derived from low calcium rocks and fine textured soils derived from high calcium rocks contain more calcium. In dry areas irrespective of soil texture, soils contain more calcium because lesser rain fall accumulate secondary precipitates of calcium carbonate and calcium sulphate.

In high rainfall areas, carbon dioxide forms carbonic acid with rain water. Carbonic acid dissolves calcium and leach out it as calcium bicarbonate to form acidic soil as,

(Soil colloid)Ca Ca H + $2H_2CO_3 \rightarrow$ (Soil colloid)H H H Ca + Ca $(HCO_3)_2$

In acid areas, irrespective of soil texture, calcium is found in more amount. This is because, low rain fall accumulates secondary deposits of calcium carbonate and calcium sulphate.

Deficiency of calcium can be correlated by adding calcite or dolomitic lime or gypsum in to soil according to soil testing.

Plant availability of calcium in soil: Many factors govern calcium availability such as quantity of exchangeable calcium, calcium saturation, Type of clay and nature of absorbed cations.

1. **Exchangeable calcium:** Amount of calcium in relation to other cations is more important than the amount of total calcium present in soil.
2. **Calcium saturation:** Calcium saturation also governs the calcium availability.
3. **Type of clay:** In 2:1 type clay higher degree of calcium saturation can provide calcium to plants but in 1:1type clay lower degree of calcium saturation can also provide sufficient calcium to plants. In montmorillonite 70 % calcium saturation is essential for normal growth of plants. But in Kaolinite only 40 – 50 % calcium saturation is sufficient for normal growth of plants.
4. **Nature of adsorbed cations:** Different cations adsorbed on soil colloidal complex does not remain adsorbed with same energy. Cations adsorbed with more energy do not move easily towards plants, whereas which are adsorbed with less energy can move easily. The energy by which cations are adsorbed on colloidal complex is called as lyotropic series. Hydrogen is adsorbed with more energy and sodium is adsorbed with less energy.

Magnesium Nutrition of Plants

Magnesium is available for plants in Mg^{2+} ion form from soil solution and exchangeable ions. Minerals of magnesium, exchangeable magnesium and water-soluble magnesium forms present in soil are under an equilibrium.

Weathering of minerals of magnesium (Biotitic, Chlorite, dolomite, olivine and serpentine) produces magnesium ions. These magnesium ions either absorbed by plants and microorganisms, leached by drainage water or make precipitate of secondary minerals.

Behaviour of magnesium in soil is just like behaviour of potassium and calcium. Absorption of magnesium by plants depends upon amount of magnesium in soil. Amount of magnesium saturation, amount of another cations and nature of clay minerals.

In humid region mostly coarse textured soils have deficiency of magnesium because magnesium chloride and magnesium sulphate drain out with water. Magnesium saturation in soil is also lower down at places of high rainfall and use of low magnesium fertilizers. Use of calcium compounds in high rainfall areas to neutralize the acidity of soil creates magnesium deficiency in crops. It is due to spoil of calcium magnesium ratio and imbalance between these two nutrients. Excess amount of soluble magnesium is fixed in acidic conditions of 2:1 type of clay.

Sulphur Nutrition of Plants

Plants absorbs sulphur as sulphate anion and some amount as sulphur dioxide gas. In soils sulphur is mostly present in organic matter which is decomposed by *Thiobacillus* bacteria in to sulphate ions. Sulphate ions are more retained in soil as compared to nitrate and chloride.

Sulphur is present in many important compounds like cysteine, methionine, amino acid, glutathione, thiamine, biotin etc. It is mobile from older leaves to new leaves. Organic form of sulphur is found in many plants like onion, garlic, cabbage and turnip etc.

Deficiency of sulphur can be removed by the additional of gypsum or sulphur powder. Sulphur bearing fertilizers like ammonium sulphate, potassium sulphate and super phosphate also provide sulphur to the soil along with their respective nutrient element.

Zinc Nutrition of Plants

Zinc is an important micronutrient. It reduces the growth of plants in many areas. Zinc deficiency has been noticed in lemon, orange, apple, pear, apricot trees etc. Yield of maize, rice, wheat and soybean also reduces due to zine deficiency.

Zinc is present in soil as oxides and relatively insoluble salts, exchangeable forms and water-soluble forms. These forms are under an equilibrium. This equilibrium is pH dependent. When pH increases, zinc comes under insoluble forms and when pH decreases, it changes into soluble forms. Zinc deficiency does not arise in the soils having pH less than 6.0. Soluble zinc applied in soil is adsorbed on colloidal complex. After some time, it is entrapped in clay lattice and does not release with salt solution. However, dilute hydrochloric acid releases it.

Conditions of Zinc deficient soil: Plants contain very low amount of zinc which is generally 10 – 100 mg kg^{-1}. This varies according to crop, growth condition and variety. Normally less than 20 ppm zinc is an indicator of deficiency during initial period of annual crops. Critical limit of zinc is variable according to soil and crop but it can be ascertained by soil testing. Normally 1-2 kg ha^{-1} DTPA extractable zinc is considered as critical level.

Many soils conditions which show strong possibilities of zinc deficiency are 1. soil formed from low initial zinc rocks liked granite and gneiss. 2. Alkaline and calcareous soils 3. Reclaimed saline and alkaline soils, eroded and levelled lands 4. Low organic matter soils 5. Shallow soils with hard layer 6. Phosphorus enriched soils 7. Soils applied with more nitrogenous fertilizers 8. Submerged soils like paddy soils having more activity of iron due reduced conditions.

Copper Nutrition of Plants

Earlier copper was considered as poison for plants but spray of copper containing mixture on plants shows beneficial effects on their growth and yield. Copper is essentials element for fungi, yeast, green plants and animals. In soil copper is present mineral Cu, hydrated oxides and complex Cu, exchangeable Cu and water-soluble Cu forms. All these are under an equilibrium.

In soil copper is present in Cu^{+} and Cu^{2+} ionic forms. The redox potential of soil governs the presence of two ionic forms of copper in the soil. Redox potential is pH dependent and affected by the formation of copper complexes and chelates. Clay minerals also adsorb the copper in non-exchangeable forms. Divalent copper is utilized by plants. Therefore, exchangeable and water-soluble copper are available to plants.

Condition of deficient soil: The soil condition related with copper deficiency are 1. Sandy soils mainly of high rainfall areas 2. Alkaline and calcareous soils, 3. Soils very rich in organic matter like peat 4. Soils with much more accumulation of phosphorus 5. Soils with more application of nitrogenous fertilizers 6. Soils rich in zinc 7. Soils irrigated with saline water.

Iron Nutrition of Plants

Iron is an essentials plant micro nutrient. It is available in sufficient quantity in soils, but its availability depends upon soil manganese, copper, zinc, organic matter, pH, aeration etc.

Different forms of iron present in soil under an equilibrium are Mineral iron, hydrated oxides iron, exchangeable iron and water-soluble iron. Iron present in soil Fe^{2+} and Fe^{3+} ionic forms and adsorbed on soil colloids. Ferrous iron is available to plants from soil colloids and soil solution.

Condition of deficient soils: Deficiency of iron for plants may be due to soil condition like, 1. Less iron soils having plenty of iron free minerals 2. Alkaline calcareous soils 3. Carbonate and bi carbonate rich soils 4. Acidic soils with low organic matter status 5. Highly acidic soils 6. Soils having more quantity of heavy elements 7. Poorly aerated soils with the accumulation of carbon di oxide gas 8. Very low or very high soil water content 8. Too much build-up of phosphorus.

Manganese Nutrition of Plants

Manganese is a micronutrient required by plants during growth and maturity. Different forms of manganese in soil are under an equilibrium. These forms are

Higher manganic oxides, hydrated manganese oxides, exchangeable manganese and water-soluble manganese.

Higher soil pH forms higher manganic oxides and decrease in soil pH increases availability of manganese. Available manganese is classified into water soluble, exchangeable, reduced, active etc. Active manganese is the total of water soluble, exchangeable and reduced manganese. Reduced Manganese contains all higher manganic oxides which are reduced by oxidizing compounds like hydro quinol and quinol. Water soluble and exchangeable manganese are in ionic form (Mn^{2+}) and easily or readily available to plants. Mn^{2+} ions mostly present in soil solution, some adsorbed on soil colloids and very few present as non- exchangeable forms of Mn.

Conditions of deficient soil: The deficiency of manganese in soil is related with 1. Sandy, slight acidic and alkaline soil and alkaline calcareous soils 2. Naturally poorly drained soils 3. Soils rich in iron, copper or zinc 4. Soils poor or rich in organic matter content 5. Soils irrigated with saline water 6. Soils with low soil temperature and low sun light intensity.

Molybdenum Nutrition of Plants

Micro nutrient molybdenum is required in very lesser quantity by plants. At the time of deficiency addition of very low amount of molybdenum is sufficient for many years. In soil mineral and organic matter Mo, Soil complex Mo, Exchangeable Mo and Water-soluble Mo are various forms of Molybdenum. These forms of molybdenum remains in an equilibrium.

In soil molybdenum is in anionic from. Therefore, it is adsorbed by the soil colloids in very specific condition. Molybdenum is immovable in soil. The equilibrium shifted towards Water-soluble Mo side with increase in soil pH.

The plants absorb molybdenum in anionic form and water soluble and exchangeable Mo is available for plants. Molybdenum acts as a catalyst for *Azotobacter chroococcum* bacteria which fixes atmospheric nitrogen.

Condition of deficient soil: Actually, molybdenum availability in soil is pH dependent. Increase in soil pH increases the molybdenum availability. Therefore, alkaline soils are not deficient in molybdenum. Various soil condition enlisted here have strong possibility of molybdenum deficiency. 1. Soil developed from acidic sand stone 2. Acidic soils of higher rainfall area having more organic matter 3. Soils having more amount of active.

Boron Nutrition of Plants

Micro nutrient boron affects the growth of sugar beet, alfalfa, clover, cauliflower, apple, grapes, apricot, sunflower, aster etc. plants. Various forms of boron in soil remain under an equilibrium. These forms are mineral tourmaline, organic matter B, inorganic salts B and water-soluble B. Orthoborate (BO_3), metaborate (BO_2) and tetraborate (B_4O_2) are the forms of boron in soil. Boron is movable (mobile) in soils as borates are very negligibly adsorbed by the soul colloids.

Conditions of deficient soil: Boron may be deficient with the soil condition like 1. Soils formed from acidic igneous rocks where Boron has almost leached 2. Alluvial and podzol soils of high rainfall areas 3. Acidic soils with higher organic matter 4. Neutral and alkaline soils specially soil with free lime 5. Soils with very low organic matter and soils irrigated with boron free irrigation water.

Chlorine Nutrition of Plants

Plants take up chlorine (Cl) as the chloride (Cl^-) anion. It is active in energy reactions in the plant. Most Cl- in soils comes from salt trapped in parent materials, marine aerosols and volcanic emissions. Chlorine (Cl) occurs predominantly as Cl^- in soil and plant. It is an essential micronutrient of higher plants and participates in several physiological metabolism processes. Its functions in plant growth and development include osmotic and stomatal regulation, evolution of oxygen in photosynthesis and disease resistance and tolerance. At adequate levels of supply, Cl improves the yields and quality of many crops such as onions and cotton if the soils are deficient in this nutrient. When excessive, Chlorine can be as a major component of salinity stress and toxic to plants.

Time of Fertilizers and Manure Application

Fertilizer is a costliest input of agriculture. It is necessary that crops and trees should utilize maximum amount of applied nutrient so the nutrient loss should be minimum. This will help to get maximum benefit for unit expenditure on fertilizers and minimum adverse effect on environment

Therefore, it is necessary that nutrients as fertilizers or manures should be applied at proper time according to soil characteristics, climate, crop requirement, nutrient absorption capacity etc.

Time of fertilizer and manure application: The basis of appropriate time for fertilizer and manure application is to provide optimum nutritional conditions for the growth and developments of plants. Many views may be in this regard, but basic aim of all these is to provide optimum nutrition to the crops. Three major view in this regard are:

1. Crop fertilization view in which fertilizer efficiency may be increased by applying the fertilizer at the time when crop can utilize maximum amount of applied nutrient

2. Soil fertilization view in which expenditure on fertilizer is minimized by using lesser quantity of fertilizer and fertilizer application is done at the time of more and easy availability of labours.
3. Crop rotation fertilization is also a view in which fertilizer requirement of different crop of the rotation and effect of fertilizer on yield is considered. For example, in paddy- wheat rotation, if enough phosphorus is given to wheat crop, then phosphorus application to next paddy crop is not much needed. It indicates that fertilizer should be applied to that crop of rotation which is more responsive.

Right Time of Fertilizer Application

Right time of fertilizer application depends upon many factors like plant nutrient, soil, climate, crop etc. Among plant nutrients nitrogen is needed by crops in their whole growth period. But its requirement is changed along with the growth stage. In the initial stage, nitrogen requirement is low but it is maximum at extreme growth of crop and later it reduces. The nitrogenous fertilizers are highly water soluble. They are lossed through leaching. Therefore, application of whole amount of nitrogen as a single dose is less effective. It should be given in two or three split doses of as per need of the crop. However, less soluble or slow-release fertilizer can be used as single dose.

Major amount of phosphorus is required by plants at their initial growth stages. Phosphatic fertilizers are less water soluble and slowly available. Therefore, whole amount of phosphatic fertilizer is applied as a single dose at the time of sowing of the crop. Just like nitrogen, potassium is also required by crops for their whole life period. But potassic fertilizers release potassium slowly and slowly as per crop need. Therefore, full dose of potassic fertilizer is applied at the time of sowing. All the organic manures and amendments slowly release their nutrients. According to these considerations, time of manure and fertilizer application may be 1. Before the sowing of crops 2. At the of sowing crops 3. After the sowing of crops.

Before the sowing of crops: Farm yard manure (FYM) compost, cakes, green manure and other organic manures are applied before the sowing of crops. Because plants nutrients are not in available forms in these bulky organic manures. After application in the soils, microorganisms decompose these and change complex forms of plant nutrients into the plant available forms. This process takes place at least form 10 -15 days to two months under different condition. If these manures are applied at the time of sowing or during the crop growth, their effect is negligible or even negative. This is because microorganisms decomposing these manures utilize plant available nutrients, mostly nitrogen, for their growth and development.

Normally FYM and compost are applied about 3-4 weeks and cakes about two weeks before sowing. These are evenly broadcasted on the soil surface and mixed into the soil by ploughing. Proper moisture is maintained even irrigation is done and some amount of nitrogenous fertilizer is applied to increase the rate of decomposition.

Chemical fertilizers are not applied before the sowing. Because their available plant nutrients are fixed by soil colloids and changed into unavailable forms. Also, their nutrients may be drained out through infiltration and run off.

At the time of sowing of crops: Fertilizers at the time of sowing of crops is generally applied in sandy soils, soils situated at slopy lands and higher rain fall areas. Also, under poorly fertilized soils with more weeds, fertilizer application at the time of sowing can provide sufficient nutrients for better crop growth at initial stages. Fertilizer application at the time of sowing also saves time, money and labour.

Half dose or two third dose of nitrogenous fertilizers and full doses of phosphatic and potassic fertilizers are given at the time of sowing of crops. Much earlier application of phosphatic and potassic fertilizer causes fixation of these in soil.

Fertilizers like urea and muriate of potash adversely affect the germination if they come in contact with the seeds. Therefore, these should be used in such a way that seeds should not come in direct contact of fertilizers.

After the sowing: After sowing of the crop, fertilizers are broadcasted or foliar sprayed in the standing crop according to the need of vegetative growth of the crop or to remove the deficiency of nutrient.

In the sandy soils of the heavy rain fall areas whole amount of nitrogenous fertilizers are applied two- three time after sowing of the crop instead of applying full dose at the time of sowing. This is done to save, the nitrogenous fertilizers from leaching losses. In the unirrigated rainfed areas fertilizers are applied during crop growth just after the rainfall.

Methods of fertilizer and manure application: Soil Testing report also indicates the nutrient (fertility) status of the soil. Any deficiency of the nutrient is fulfilled by its source, generally by fertilizer. Appropriate methods of fertilizer application depends upon many factors like soil characteristics, availability of fertilizers and its cost, methods of sowing and line to line and seed to seeds or plant to plant distance. Keeping in mind all these factors different methods of fertilizers application are described here:

1. Basal dressing of fertilizers: This is the simplest method for the application of solid fertilizers. In this method fertilizers are homogenously broadcasted on soil surface and later on properly mixed in soil by light ploughing or harrowing. This method saves time, money and labour in manure and fertilizer application.

Basal dressing of fertilizer is generally done prior to ploughing of field or sowing of the seeds. Most of the farmers of the world use basal dressing method of fertilizer application. Generally broadcasting is done by hands. But equipments can also be used. At the time of sowing fertilizers are applied by this method in the following conditions:

1. For the distribution of fertilizers in more quantity than the capacity of seed drill or fertiseed drill.
2. For the equally distribution of necessary quantity of fertilizers in the whole field and for mixing it equally to the shallow depth in the soil.
3. For the application of citrate soluble phosphatic fertilizers (dicalcium phosphate, basic slag) and low soluble phosphatic fertilizers (bone meal, rock phosphate) in the neutral or more acidic soils.
4. For the application in more quantity on account of very much shortage of macro or micronutrients in the soil.
5. In crops being sown by broadcasting method.
6. In the pasture.
7. For nitrogenous fertilizers.

Application of fertilizer by broadcasting method increases the contact of fertilizers with the soil. In this condition possibility of fixation of phosphatic and potassic fertilizers is increases so that crop response may be decrease. Therefore, specially in the acidic, alkaline and clay soils phosphatic and potassic fertilizers should not be applied by broadcasting method. In the low fertility soils this method also is not beneficial for application of small quantity of fertilizers.

In sandy soils or sloping fields of more rainfall areas broadcasting of fertilizers may drain out nutrients with water or by leaching is more. Here small quantity of fertilizers along with micronutrients should be used. Because this method is not suitable to use these separately. In the alkaline soils broadcasting of nitrogenous fertilizers enhances the loss of nitrogen as ammonia gas. It is obvious that broadcasting of fertilizers is most simple, less expensive and in some conditions most effective but it has some limitations indicated above. So, this method does not remain beneficial.

Top dressing in standing crops: According to need of crops at sowing time the use of full quantity of fertilizers to supply mineral nutrients always is not beneficial. So, after sowing at various stage of vegetative growth of standing crops, split doses or amount of total required quantity of fertilizers are given. In standing crop only those fertilizers are applied, in which nutrients are soluble in water and they easily move in soil. This thing is applicable fully with nitrogenous fertilizers. Top dressing of these fertilizes in the standing crop is found beneficial. In ordinary conditions application of phosphatic and potassic fertilizers in the standing crop has not given good results. Broadcasting is done by hands in the standing crop by farmers.

Top dressing can also be done by implements and in specific situations by aeroplanes. In the following situations use of aeroplanes may be more beneficial:

1. When a small quantity of fertilizers is to be top dressed on too much big terrain is necessary, e.g., application of micronutrients.
2. When application of fertilizers with insecticides and pesticides is to be done by aeroplanes.
3. When saving of time is essential and labour is unavailable.
4. When application of fertilizers is to be done in inaccessible hilly terrain, pastures, reserved forests etc.

In our country top dressing of nitrogenous fertilizers in the standing crops is more popular. Top dressing may be more profitable and the probable nutrient losses can be avoided if following points are considered,

1. At the time of top dressing of fertilizers on the standing crops, moisture should not be present on leaves of the plants. They should not wet. Particles of fertilizers cling on the moist leaves, so that leaves are scorched. Therefore, to avoid this, top dressing of fertilizers is suitable in the clear weather and in the afternoon.
2. At the time of top dressing of fertilizers, sufficient moisture in the soil is essential. In the areas of dryland farming where surface soils are dry, application of solid fertilizers by this method should be avoided.
3. After topdressing of nitrogenous fertilizers leaving open it on the surface of soils, results in nitrogen loss as ammonia gas. In the alkaline soils this activity has special importance. Therefore, after top dressing of fertilizers mixing it in the soil by inter- cultural operations is beneficial.
4. At the time of application of fertilizers by this method, wherever is possible, crops should be weed free. Otherwise, weeds will go profusely.
5. At the time of vegetative growth of crops application of fertilizers in the standing crops should be done. Fertilizers should not be applied after flowering and fruiting of crops.

Methods of side dressing: Some crops are sown separately in rows, e.g., potato, corn, sugarcane, sorghum, tobacco, cotton and vegetable. The fulfilments of requirement of nutrients, at the time of vegetative growth of these crops, instead of broadcasting uniformly in the whole field, side dressing of fertilizers along the rows is more suitable. Thus, sprinkling or dredging of fertilizers in the side of rows is called method of side dressing.

Methods of dibbling and drilling: When plants to plants distance of the crop is more, the fertilizers are applied around each plant and mixed in soil. Similarly, fertilizers are poured or filled in holes made by the help of pegs and covered with soil. These two methods save the quantity of fertilizers beyond expectations without adversely affecting growth and yield. If the pegs and hole method of applications of fertilizers is done by hands then it is called as dibbing. But application of fertilizers by these methods is time and labours consuming. When this work is done by automatic implements of special type, then it is called drilling (placing fertilizers with the help of drill). Fertilizer use efficiency of this method can be increased by taking care of all those factors which have been described in the method of top dressing in the standing crop.

2. Placement method: In this method fertilizers are placed at any specific depth, but it is not essential that fertilizers are placed absolutely near the seeds and plants. Placement method is equally beneficial for applications of solid, liquid and gaseous fertilizers. Various implements have been developed for placements of fertilizers and with the help of them fertilizers can be placed at required depth before sowing. Shallow placements of solid fertilizers can be done by attaching a funnel with drill, plough or with the help of manual hoe. This method of application of liquid fertilizers or dry ammonia is done by special implements. Five placement methods of fertilizer are, Plough sole placement, Deep placement, Subsoil placement, Proximal (adjacent) placement and Drill placement.

(a) Plough sole placement- In this method fertilizers are placed at depth of plough sole with the help of plough. This method of fertilizer application is adopted in such areas where few centimetres thick top layer of soil becomes dry due to shortage of moisture, but at the depth of plough sole, soil remains moist. By applying phosphatic and potassic fertilizers by this method, process of fixation can be lessened. For field having hard layer of clay at the depth of plough sole (furrow) this method is also found beneficial.

(b) Deep placement- In this method placement of fertilizers is done at more depth than plough sole. Placement of nitrogenous fertilizers reduces the loss of ammonia and nitrogen as gas. Also, quantity of fertilizers to be lost with water is lessened. Because of regional tests, National Rice Research Institute, Cuttack has confirmed beneficial results of deep placements of nitrogenous fertilizers in paddy. Good results are obtained with phosphatic fertilizers also.

(c) Subsoil placement- In this method fertilizers are placed in subsoil by the help of very deep ploughing or sub soiling. In more rain fed areas acidic subsoil reduces availability of nutrients. In such areas growth of roots of plants stops. Then subsoil placement increases availability of nutrients for good growth and developments of plants. Subsoil placement is best suited and most successful for application of phosphatic and potassic fertilizers in above maintained conditions.

(d) Proximal (adjacent) placement- In this method of placements, fertilizers are placed in the soil near seed or plants. When availability of fertilizers is short or application of fertilizers in less quantity is desirable for more crop response, this method is found more beneficial. Several methods of proximal placement are available which are given here:

(i) Drill placement: In this method, fertilizers and seeds both are sown simultaneously. Its main purpose is to place fertilizer too close of seeds.

When quantity of fertilizers is limited, soil fertility is low and moisture in soil is available in abundance, placement of fertilizers by this method increases the extent of profit. In adverse conditions fertilizers should not be applied by this method. Because low moisture and excess fertilizer thickens soil solution. Direct contact of seeds with thick soil solution adversely influences germination of seeds, especially when potassium salts are also present in fertilizers mixture. Therefore, application of fertilizers in small quantity by this methods is found safe. Application of phosphatic and potassic fertilizers can be done in crops of wheat, sorghum, millets, barely etc. by this method and for pulses this method is not more beneficial.

(ii) Side band or row placement: Fertilizers are placed as a band on one or both sides of seeds or planted rows in side or row placement method. The placement is done by special implements. Fertilizer band is placed 5-centimeter away from rows of seeds or plants and 2-5 centimetre deep in soil. Such placement is also possible by traditional local plough. Band placement of fertilizers in wheat, maize and other cereal crops is suitable where development of lateral roots begins immediately. Also, in crops where tap roots develop immediately placement of some fertilizers below seeds is beneficial. Root growth in ground nut, sugar beet and potato is limited in comparison with maize. Thus, placement of fertilizer below seed is suitable for appropriate growth of plants.

When distance between plants in rows is very less, as sugarcane, potato, maize, cotton, tobacco, etc. application of fertilizers is done by this method. Occasionally some plants in row die or dry by many reasons. In such condition, fertilizers are not applied in the vacant space. Such fertilizer bands are called broken bands and this altered method is called broken band placement. Placement of fertilizers by this method is applied in such crops which are planted individually at distance, as tomato, chillies, eggplants, and other vegetables.

(iii) Basin placement: When distance between plants is one or more meters, fertilizers are placed at certain depth near or around plants in circle. This method is called basin placement. In our country in orchards of apple, pear, papaya, banana, orange, lemon etc. fertilizers are applied by this method. This method of placement of fertilizers is proper for more wide spreading crops where plant to plant distance is more than one meter, as watermelon, muskmelon, bitter gourd, bottle gourd, ribbed gourd, pumpkin etc. In this method depth of placement of fertilizers is decided by distance between plants to plants and basic characteristics of crops and properties of fertilizers. Placement of phosphatic fertilizers is done at more depth than nitrogenous fertilizers. Placements of fertilizers in wide spreading crops is done at more space than erect growing crops.

3. Carrier fertilizer solution: After transplanting of vegetables plants, if dilute solution of fertilizers instead of water is applied, plants establish quickly and initial growth of plants becomes faster. Fertilizer solution of this type is called carrier solution or carrier fertilizer solutions. Generally, ratio of nitrogen, phosphorus and potassium is kept 1:2 :1 and 1:2:2. Only water-soluble fertilizers e.g., ammonium phosphate, urea, potassium, chloride and potassium sulphate are selected for making carrier solution.

Concentration of solution depends on quantity of nitrogen in fertilizer mixture. If 25 present nitrogen is present in fertilizer mixture, its one kg quantity in 400 litre water is safe and nitrogen in fertilizer mixture is only 3 precent then up to 10 kg mixture in 400 kg litre water may be dissolved. At the time of making solution, it is necessary to consider that it should not be more acidic (pH less than 5), especially when its application is to be done in acidic and sandy soils.

Factors Influencing Fertilizer Placement

The efficient use of fertilizers is governed by proper placement of these in proximity of plants Improper placements of fertilizers has adverse effects on plants number and growth of crops and ultimately on yields and fertilizer efficiency of crop. Success of fertilizer placement method depends on three factors,1. Effect of root system of the crop. 2. Soil conditions and 3. Reactions of soil and fertilizers.

1. Effect of root system of crops: Morphology and physiology and extent of development of roots differ in various crops. If tendency of rooting in initial stages of crops is known, determination of satisfactory fertilizer placement method for crops becomes easy and convenient. When tap roots develop quickly, placement of some fertilizer below seeds is beneficial. Lucerne, tomato, ground nut, Sugar beet etc. are some examples of such crops.

Contrary of this if, lateral roots in crops as maize, wheat and cereal crops develop quickly, then side band placement of fertilizers remains appropriate. In the period of two weeks, the limited root system develop in tobacco and cotton. Therefore, for proper growth placement of fertilizers below seeds is appropriate. Potato also has limited root system which is distributed in small portion. Therefore, the absorption of nutrients from fertilizers by potato crop is more in comparison of other crops like maize etc.

Major part of roots is distributed up to upper 15-25 cm depth in the soil. Roots of some crops reach up to depth of one or two meters. However, active nutritive area is also surface soil and use fertilizers in active area of roots is appropriate and beneficial. In fruit trees it is most important.

2. Soil conditions: Generally, soil conditions do not influence placement method and efficiency of nitrogenous fertilizers because in general conditions ammonia and nitrate present in soil are available when water becomes available. But in more rainfed or irrigated areas appreciable loss of nitrate by leaching from sandy soil is possible. If root system of the crop is limited, this leaching increases tremendously. In such conditions foliar spray of urea is more beneficial. In submerged area, urea or ammoniated fertilizers are nitrified in thin oxidised layer, where by being originated nitrate goes below in reduce layer, where its denitrification occurs. If in these soil conditions nitrogenous fertilizers set up directly in the reduced layer, then loss of nitrogen can be lowered by denitrification.

3. Reactions between soil and fertilizers: There is deep relation between water solubility and placement of phosphatic fertilizers. It is presumed that localized placement of water soluble phosphatic fertilizers is nearly double effective in comparison with placement by broadcasting method of it. In soils having high phosphorus fixing capacity, as acidic and calcareous soils, this difference is found more than neutral and sandy soils.

Potassium is less movable than nitrate in soils, but it is more movable than phosphorus. Application of potassic fertilizers in small quantity can help its fixation by clay aqueous micaceous. Problem of potassium fixation in soils of heavy texture having alkaline nature is more in comparison with sandy soils. Therefore, band placement of potassium in heavy soils is more beneficial. Methods of broadcasting of fertilizers in sandy soil is as much effective as method of band placement.

Fertilizers with sulphur and magnesium may be broadcasted easily, because fixation of these does not exist. Micronutrients are required in very small quantity, so mixing of these with elemental fertilizers or foliar sprays of these are most convenient.

Strong fixation of zinc, iron and manganese occurs in soil. So, band placement of these is more effective than broadcasting. In view of variety of the crop and expected quantity of nutrients, application of boron is done by broadcasting, side band or foliar spray method. Requirement of molybdenum is in very small quantity. So, seeds treatment @ 0.02 Kg / ha is best method of its application. Furthermore, its application as a foliar spray may also be done.

Methods of Fertilizers Application in Fruit Trees

Just like vegetables, flowers, other plants etc. fruit trees require good nutrition to grow, develop and to produce plenty of fruits. Fruit trees are perennial and remain in orchards for several years. They continuously remove adequate amount of plant nutrients without timely replenishment of them. Thus, there is partial or complete depletion of available nutrients in the soil of the orchard. This depletion of nutrients ultimately causes reduction of growth and productivity of fruit trees.

Obviously, gardeners require to replenish these nutrients externally through the application of manures and fertilizers. But each type of fruit species has its own nutrient requirement and soils of each orchard have inherent capacity to supply nutrients to plants. Therefore, it is very necessary to assess fertility status of soil, by soil testing and then calculate the quantity of fertilizers or manures to be applied in soil. When amount of fertilizer is known then it is necessary to decide the appropriate method of its application. Fertilizers are applied by broadcasting method, straight band or strip placement method and basin placement method. All these are briefly described here,

Application of fertilizers by broadcasting method- Just like other crops in fruit trees also fertilizers are broadcasted on soil surface and then mixed in soil by ploughing etc. The fertilizers are broadcasted up to the canopy of each tree. Limitation and utility of this methods are like to fertilizer broadcasting in cereal crops. Generally, application of nitrogenous fertilizers is done by this method. Weeds are checked easily by no broadcasting of fertilizers in vacant space between two fruit trees. This method is not beneficial for phosphatic and potassic fertilizers.

Straight band application: In this method phosphatic and potassic fertilizers are placed in bands between two rows of trees. Generally, placement of these fertilizers is done in 40 -50 cm wide and 10-20 cm deep bands. In this method first year bands are made in length and second year in width. Use of tractor driven fertidrill is suitable for placement of fertilizers. This method is beneficial for big orchards and planted forests and at scarcity of labourers. However, this method is not much useful when fruit trees and their expansion of canopy are small.

Basin (Ring) placement: It is more useful method for application of phosphatic and potassic fertilizers. In this method fertilizers are placed making 30-45 cm wide and 10-5 cm deep drain around plants in a circle.

Circles for fertilizer placement from stem is made nearer to maximum activity area of sucking roots for more effective fertilizer use. Experiments conducted with the help of radioactive elements indicated that maximum activity of sucking roots is up to the half distance of expansion of tree from stem. Then activity decreases with increasing distance towards expansion and about 40-45 cm distance towards expansion from stem activity of sucker roots is minimum. Also, maximum activity of suckers (sucking root) is found at 10 -15 cm depth from soil surface. Therefore, fertilizers placement should not be done very close to stem or at outer circle of expansion of tree and not more than 10 -15 cm depth.

In majority of fruit trees on time placement of phosphatic and potassic fertilizers is done in the month of December with full quantity of fertilizers. But application of nitrogen by this method has no special advantage.

Inoculation of fertilizers by syringe: Inoculation of fertilizers by syringe method can be used in both soil and plants. In foreign countries special system have been developed for this purpose. In India also this method has successfully tested and found suitable for nutrients application in fruit trees. The technique of this method of fertilizer application has been developed in the Horticulture department of G.B. Pant University of Agriculture and Technology.

In this method dilute solution of fertilizers in water is made. In any condition concentration of solution should not be more than that of concentration of nutrient solution used for cell culture. Holes are made in stem at least two places in opposite direction. Depth of holes is kept a little less than from radius of stem. A plastic tube with outer circle equal to circle of holes is fully entered in this hole and each hole is made airtight. Another end of the plastic tube is connected to the bottle of fertilizer solution. Before connecting it, tube is filled with solution putting it slowly in the bottle and tied to of tree. After fastening bottle with branches, it is hanged at tree. Solution enters slowly and slowly into stem and spread in whole tree by xylem vessels. This method is not a new one, but at present this method is used only for scientific experiments and research centres because it is more complicated and requires sophisticated technical know-how. However, in acute deficiency of micronutrients, fruit growers can use this method under the supervision of agricultural scientists to overcome micronutrient deficiency.

Fertilizer application with irrigation water: Application of liquid fertilizers and water-soluble solid fertilizers is possible with irrigation water. This method is also called as Fertigation. In special circumstances it is useful method for reduction in labour and production expenditure. Its utility depends on soil quality, method of irrigation, quality of irrigation water and quality of available fertilizers.

Efficiency of fertilizers applied with irrigation water is mainly influenced by soil texture. Application of fertilizers by this method is found more beneficial in sandy soil of coarse texture in comparison to clay soil of fine texture. This fact is especially true for nitrogenous fertilizers. However, probable leaching losses in sandy soils may be reduced by using small quantity of nitrogenous fertilizers several times and applying the same in active area of roots of plants. This method cannot be beneficial in soils of fine texture in comparison of other methods.

Utility of fertilizer application by this method also depends on methods of irrigation. Surface irrigation or drain method does not uniformly distribute water in whole field. So, application of fertilizers by irrigation water will also not uniform. This is more important in uneven and sloping fields. Comparatively there is more accumulation or deposition of nutrients at entrance area of water in row sown crops and irrigation by drains.

Contrary to this nutrient in irrigation water can be applied by sprinkler system or drip method. This will increase fertilizer use efficiency, saving of time and labour and distribute uniformly fertilizer in whole field. At the time of irrigation by drip method, mixing of fertilizers in the water can save about 50 % fertilizers. Dissolving the fertilizer in irrigation water and homogenously sprinkling them is like to spraying of nutrient mixed water in the field. This method saves extra expenditure to be incurred in labours used for fertilizer application. But anhydrous ammonia and ammoniated nitrogen solution cannot be applied by this method. Its reason is loss of ammonia as gas and adverse effect of free ammonia on crops. This method is not found suitable for micronutrients.

Selection of fertilizers for use with irrigation water is to be done very carefully. The quality of irrigation water has special important. Inappropriate selection of fertilizer may create problem of choking of irrigation system (drip or sprinkler system).

Taps of irrigation drains become choked or clogged by using anhydrous ammonia in nitrogenous fertilizers or ammoniated nitrogen solution. As a result of more pH of water, dissolved substances in irrigation water precipitate and deposit in drip or sprinkler system.

Application of ammonia up to 30,000 mg per litre is found safe when quantity of magnesium is up to 10 mg per litre and electrical conductivity 0.2 millimhos per centimetre in irrigation water. Quantity of calcium and magnesium salts up to 30 mg/ litre and electrical conductivity 0.8 millimhos in irrigation water, ammonia only up to 1000 mg/ litre as nitrogen concentration can be applied. If application of free ammoniated solution is not possible due to quality of irrigation water, other sources of nitrogen should be selected and urea is a best option.

Application of superphosphate in phosphatic fertilizers with irrigation water is not found beneficial. Its main reason is less solubility in water and precipitation of soluble phosphorus. Solution of orthophosphoric acid mixed with irrigation water can be applied. It should be mixed with irrigation water in such quantity that pH value of water may be less than neutral. Consequently, the possibility of obstruction or clogging of irrigation system will be minimum.

Application of some soluble salts of orthophosphoric acid, as ammonium orthophosphate fertilizer with irrigation water is possible. Quantity of calcium in irrigation water influences directly on applicable quantity of these fertilizers. Mixing of more than 7.5 percent ammonium orthophosphate (8-24-0) in irrigation water having 200 mg calcium / per litre produces precipitation. Appropriate amount of Ammonium polyphosphate can also be mixed with irrigation water after considering quality of water to avoid problem of precipitation.

Major potassic fertilizers easily dissolve in water. So, application of potassic fertilizers is comparatively easily. As far as problem of micro nutrients is concerned, proper selection of fertilizer is essential according to quality of irrigation water. Chelates of iron, zinc, manganese and copper can be applied successfully with irrigation water. Specially in plants, rose etc. grown in gardens, use of balanced solutions of chelates with major nutrients in irrigation water is more beneficial.

Foliar spray method: In this method dilute solution of fertilizers are applied as a spray on leaves of crop. First in the year 1916 application of water-soluble iron salts was sprayed to treat chlorosis disease. Now foliar nutrition is considered as an important record of achievement and progress in the field of agricultural production, Generally, foliar spray is more effective for absorption of nutrients in comparison to application in soil. In various crops the difference in effectiveness was observed up to 2:1 to 100:1. So, more crop response and production can be obtained by per unit nutrient applied by foliar nutrition as crop response occurs immediately. Therefore, plants swiftly recover from nutrient deficiency and attain their average growth. However, this method is not beneficial for total supply of main nutrients requirement by plants. Because multiple times foliar sprays are needed during entire growth period of crop, results more expenses for fertilizer application. Furthermore, possibility of scorching of leaves increases. But need of micro nutrients is in less quantity for which supply one or two sprays are sufficient.

In all plants absorption of nutrients by leaves is possible, but rate of absorption depends on nature of fertilizers and variety of crop. As for example absorption of urea becomes quickly, but absorption of phosphatic fertilizers is very slow speed. Nearly half quantity of spray of urea on leaves of maize, tomato, apple and pineapple is absorbed within 1-6 hours whereas absorption takes place within 12 to 36 hours in sugarcane, potato, tobacco etc. crops.

Safe concentration of foliar spray can be obtained by increasing the fineness of spray. However, 15 to 20 kg/ ha quantity of urea can be applied conveniently at a time. Also keep in mind that biuret in urea should be less than 1 percent because it is a toxic substance for plants.

Foliar spray is effective when leaf area is more like in fruit-trees, ornamental gardening crops, vegetables and pre flowering stage of annual crops. In condition of to be low number of leaves and to be short their size, because of less surface area of leaves absorption of nutrients lessens. So foliar spray method does not take place fully of application of fertilizers in soil. It will be appropriate to recognise as a supplement method. Foliar spray method of crop nutrition is especially beneficial in conditions like,

1. In unirrigated rainfed areas where moisture is in scarcity in soil, contracted root growth due to any obstacle, diseased or damaged roots make roots of plants unable to absorb nutrients from soil in sufficient quantity.
2. After application of fertilizer if nutrients are changed in unavailable form like salts of zinc, iron and manganese.
3. There is a possibility of unavailability or leaching of fertilizers in slopy lands and sandy soil.
4. Nutrients are to be applied in very small quantity, as small quantity of micro nutrients or beneficial elements.
5. Nutrients are to be mixed in agricultural chemicals as spray.
6. If response of fertilizer application appears very late and is required immediate result of fertilization as in fruit- trees.
7. If maturity stage of crop is nearer and by application of nutrient will improve the quality of product, as before cutting of sugarcane if foliar spray of potassium dihydrogen phosphate is done, percentage of sugar increases and speedy growth rate of trees occurs.

Preparation of Multinutrient Fertilizer Mixture

The application of individual nutrient increases cost of labour, transport, storage etc. and to avoid this, fertilizer mixtures are used. Fertilizers mixture is defined as mixture of two or more straight fertilizers. Straight fertilizers are those which supply only one primary plant nutrient, namely N, P or K. The complete fertilizer means the fertilizer that contains all three primary nutrients, (N, P and K). Complex fertilizers contain two or three primary plant nutrients of which two primary nutrients are chemical combined. A mixed fertilizer is a mixture of two or more straight fertilizers. A mixture of Nitrogen (N), Phosphorus (P) and Potassium (K) is generally known as NPK mixture. Three elements that are critical and essential to plant survival and growth are added in the mixed fertilizer. Mixed fertilizer or fertilizer mixture is also a physical mixture and thorough mixing of straight fertilizers is done manually or mechanically.

Some rules should be followed for successful preparation of mixed fertilizers. It is necessary to fully know about physical and chemical properties of various fertilizers. Also, extensive information about method of maintenance of mixture and changes that due to effect of temperature and moisture of atmosphere.

However, for making multi-elemental fertilizer mixture by directly mixing various fertilizers following should be considered.

1. Ammonical nitrogen fertilizers (ammonium sulphate, ammonium nitrate) should not mixed with alkaline substance, as lime, basic slag, rock phosphate, calcium cyanide, because mutual reaction among them will release nitrogen as ammonia gas.
2. Water soluble phosphatic fertilizers as triple superphosphate and ammonium phosphate should not mix with those fertilizers in which free lime is found, because this will convert some part of soluble phosphorus into water insoluble form.
3. After preparation of mixture by mixing readily soluble and hygroscopic fertilizers e.g calcium ammonium nitrate, urea and potassium salts should be applied immediately in the crop because after sometime clods are formed in the mixture.
4. Fertilizers having low acidity e.g., superphosphate, reacts with nitrate or chloride with fertilizers produce (generate) acid. Thus, sacks of jute are decay and fertidrill is also badly affected.

Calculation of Quantities of Fertilizers for Fertilizer Mixture

According to need mixed fertilizer may be prepared by farmers at their farm by mixing required fertilizers with one another. Mixture of fertilizers should be prepared on brick build floor or polyethene sheet. The required tools are spade, weeding hook, sieve, wooden hammer and weighing scale. Mixture should be made when it will be used and not more than one day before its use. When it is necessary to keep for more duration than to avoid clods formation and reduction in quality mix it with dry loam or sandy soil.

Now, according to need, fertilizers should be weighed separately. If there are clods, break them by pounding by hammer or wooden hammer and make it fine. Then first one fertilizer of maximum quantity should be spreaded. Then fertilizers to be mixed are placed one above the other in layers. Then turn all layers of fertilizers with the help of spade or weeding hook.

Calculation of quantity of various fertilizers: Calculation of quantity of various fertilizers is done according to need and according to quantity of nutrient present in various fertilizers. Suppose that 1000 kg fertilizers mixture of 10-8-10 grade is to be prepared. The available fertilizers are, urea (46% N), Superphosphate (16% P_2O_5) and muriate of potash (60%K_2O). How much quantity of each fertilizer will be needed separately?

First, for 100 kg. mixture calculation of quantities of above-mentioned fertilizers is done in such a way that 10 kg. nitrogen, 8 kg. phosphorus and 10 kg. potash could be available. Again, after making 10 times quantity of each fertilizer are mixed with one another. Fulfilment of remaining weight done by mixing proper supplements (dry friable) soil, sand, ash etc.

To know quantity of fertilizers separately after multiplying desired percent (grade) of that nutrient in the mixture by 100 it is divided by percentage quantity of nutrient in fertilizers.

In 100 kg. fertilizer mixture,

Quantity of nitrogenous fertilizer(urea) $= \frac{10 \times 100}{46} = 22$ kg.

Quantity of phosphatic fertilizer (super phosphate) $= \frac{8 \times 100}{16} = 50$ kg.

Quantity of potassic fertilizer (muriate potash) $= \frac{10 \times 100}{60} = 17$ kg.

Total = 89 kg.

Quantity of supplement = 11 kg.

Total =100 kg.

Thus, for 1000 kg. fertilizer mixture of 10- 8- 10 grade 220 kg. urea, 500 kg. super phosphate, 170 kg. muriate of potash and 110 kg. supplement (dry soil, sand etc.) are needed.

Preparation of Fertilizer mixture with two nutrients in one fertilizer: If any fertilizer contains two nutrients and we must use that for making fertilizer mixture. In such condition quantity of fertilizers should be calculated by method given here,

Firstly, quantity of such fertilizer in which two nutrients are present should be calculated supposing that more quantity required as a main nutrient and containing another nutrient also. Then by deducting quantity of second nutrient from its full requirement should be calculated that how much nutrient is to be taken from second fertilizer of that nutrient? Afterwards quantity of fertilizers of other nutrients should be calculated. According to need mixture should be prepared after mixing all fertilizer and filler in it.

Suppose, in any crop 120 kg. nitrogen and 60 kg. P_2O_5/ ha is to be applied from urea and diammonium phosphate (DAP). It is supposed that DAP has 18 percent nitrogen and 46 percent P_2O_5, then first for phosphorus quantity of DAP will be calculated by method given here,

46 kg P_2O_5 is obtained from 100 kg DAP

1 kg P_2O_5 will be obtained from 100 / 46 kg DAP

60 kg P_2O_5 will be obtained from 60x100/46 kg DAP = 130.5 kg DAP

Therefore, to apply 60 kg P_2O_5, 130.5 kg DAP will be needed.

Now how much nitrogen will be obtained from DAP?

It will be calculated as,

100 kg DAP gives 18 kg nitrogen

1 kg DAP will give 18/100 kg nitrogen and

130.5 kg DAP will give 18x130.5/100 kg nitrogen = 23.49 =23.50 kg nitrogen

Because 23.5 kg nitrogen will be obtained from DAP so after deducting 23.5 kg from 120 kg nitrogen, remaining 96.5 kg nitrogen will be applied by urea as,

46 kg nitrogen is obtained from 100 kg urea

1 kg nitrogen will be obtained from 100/46 kg urea

96.5 kg nitrogen will be obtained from 100x96.5/46 kg urea = 209.8 = 210 kg urea.

Therefore, 210 kg urea and 130.5 kg DAP will be required to supply 120 kg nitrogen and 60 kg P_2O_5.

Integrated Plant Nutrient Management System (IPNMS)

Integrated plant nutrient management system or scheme deals with the combined application of mineral fertilizers and organic manures to meet the nutritional requirement of crops. The adverse effects of the long-term application of mineral fertilizers on crop productivity and sustainability of crop production can be avoided by following integrated plant nutrient management. It also leads to improve soil health and mitigation of global warming. This is also deals with safest disposal of crop residues to produce high-quality organic manures for maintaining soil fertility.

Why IPNMS is Essential

Essentiality of IPNMS is due to decline in productivity of the crop due to decrease in nutrient supply and cost of fertilizers. Ineffective utilization of the nutrients by the crop. Increase in deficiencies of P, S and Zn due to more use of Nitrogenous fertilizers. Decline in Soil Organic Matter status is associated with continuous application of nitrogen fertilizer. Change in soil pH and increase salt concentrations. Continuous Change in land use pattern from the forest ecosystem to agro ecosystem and then urban and industrial ecosystem and degradation of soil fertility. IPNMS is also essential for reducing climate change, global warming and environmental pollution.

Objectives of IPNMS

Among numerous objectives from farmer to farm level, the salient objectives of integrated plant nutrient management are,

1. To increase the availability of nutrients from all available sources in the soil during crop growing season.
2. To minimize the inorganic fertilizer requirement.
3. To synchronize the demand of nutrients by the crop and supply of the nutrients from all sources.
4. To optimize the beneficial functions of the soil biosphere in the soil.
5. To minimize the losses of nutrients through volatilization, denitrification, surface runoff, leaching etc.

Components of the IPNMS

Fertilizers, organic manures, green manures, crop residues and biofertilizers are the main components of the IPNMS.

1. **Fertilizers:** Fertilizers is the most important component of IPNMS. The dependence on fertilizers has been increasing constantly because of the need to supply large amounts of nutrients in intensive cropping with high productivity. Various constraints like global price hike, more demand and short supply of fertilizers and raw materials are leading to a big gap between supply and demand. While organics and biofertilizers are expected to bridge a part of this gap, efficient use of fertilizers in narrowing the nutrient supply gap also needs greater emphasis.
2. **Organic Manures:** Organic manures like urban compost, FYM, crop residues, human excreta, city refuse, rural compost, sewage-sludge, press mud and other agro-industrial wastes have large nutrient potential. Compost and FYM have traditionally been the most important manures for maintaining soil fertility and ensuring yield stability. Other potential organic sources of nutrients such as non-edible oil cakes and wastes from food processing industry are also there. Moreover, there are several industrial by-products and municipal wastes with fair nutrient potential. The vermicompost, biogas slurry, Kunap water can also successfully be incorporated in IPNMS. There is need to integrate these sources depending on their availability in different crops and cropping systems after properly evaluating their nutrient contents.
3. **Green Manures:** Green manures are well placed to increase and restore soil fertility since ancient periods. Leguminous green manure has ability to fix atmospheric Nitrogen. Leguminous crops should be essentially included

in the crop rotation. The legume may be for green fodder in rice-wheat-cowpea green fodder or for pulses in paddy- lentil and wheat-urd, wheat-moong etc. These legumes may be grown as third crop of annual crop rotations. This will increase cropping intensity of crop rotation, fertility of soil and income of farmers.

4. **Crop Residues:** Crop residues have several competitive uses and may not be always available as an ingredient of IPNMS, yet in the regions like North-West India where mechanical harvesting is practiced, a sizeable quantity of residues is left in the field, which can form a part of nutrient supply. There are large amounts of residues of other crops like potato, sugarcane, vegetables, etc., which are practically wasted in most cases.

   Although cereal crop residues are valuable cattle feed, these could be used to supplement the fertilizers wherever available in excess of local needs. Stubbles left in the field in traditional harvesting methods range from 0.45 to 1.5t/ha in case of different crops. When mechanical harvesting is done, this amount is much greater. Stubbles of coarse cereals like sorghum, maize, pearl millet etc. which are difficult are normally collected and burnt during land preparation causing significant loss of plant nutrients.

5. **Biofertilizers:** Biofertilizers are the materials containing living or latent cells of agriculturally beneficial microorganisms that play an important role in improving soil fertility and crop productivity due to their capacity to fix atmospheric N, solubilize/mobilize P and decompose farm waste resulting in the release of plant nutrients. The extent of benefits from these microorganisms depends on their number and efficiency which, however, is governed by a large of soil and environmental factors.

Procedures of IPNMS

Procedures or methodology of IPNMS depend on various factors and situations. However, a generalized theme is that first select, at least one month prior, the field in which IPNMS is to be applied because some of the components are to be applied quite before sowing of seed.

Collect all the information of this field like type of soil, crops grown, previous cropping history including green manure and leguminous crops grown and their yield potential, quantity of manure and fertilizers applied, weather and climate, source of irrigation water and water quality, socioeconomic status of beneficiaries etc.

Collect soil samples from the selected field and test them for various soil properties including available plant nutrient status. This will help to decide the amount of nutrients to be applied through IPNMS.

Collect the information about locally grown crops and their varieties particularly that crop and variety to be used for IPNMS.

Finalize the crop, variety of crop and nutrient requirements of crop/variety. Here we select, for example, wheat crop and its variety DBW 222. The N, P and K requirement of DBW 222 is 150, 90 and 60 kg/ha, respectively.

After the soil testing of field, the calculated N, P and K dose for this variety was 120, 60 and 40 kg/ha, respectively.

The beneficiary farmer want to use FYM and Azotobacter in wheat and he has about 10 tons of FYM.

Now considering 0.5% nitrogen in FYM and fixation of nitrogen by Azotobacter is 20 kg/ha/year. The IPNMS planning for nitrogen only will be as,

Total amount of nitrogen available from FVM $= 0.5 \times \frac{10000}{100} = 50\,\text{kg}$

Total amount of nitrogen fixed by Azotobacter = 10 kg during six months

Total amount of nitrogen available from FYM and Azotobacter = 50 + 10 = 60 kg

Amount of nitrogen to supplied by inorganic fertilizers = 140 - 60 = 80 kg

Now see this actual benefit in nitrogen only, inform this result to nearby farmers and start demonstrations of IPNMS on other farmers' fields with various combinations of locally available organic sources of nutrients. The other benefits of IPNMS are worthwhile and beyond expectations of anyone engaged in agriculture etc.

Processes and Cycles of Molecules and Elements in Soil

The optimization of nutrient cycling is essential for the sustainable productivity of agricultural systems. The matter is never destroyed in nature. It changes its form and properties from one to other under a natural cycle. Cycles of molecules and elements are defined as the biogeochemical pathways by which they are transformed and moved through various states by geological, chemical and biological processes.

The six most common elements associated with organic molecules are carbon, nitrogen, hydrogen, oxygen, phosphorus and sulphur. These are changed in to the other chemical forms or remain in the atmosphere, on land, in soil, in water or under the Earth's surface for longer periods. The biogeochemical cycles of molecules and elements on Earth are mainly governed by weathering of rocks and minerals, soil erosion, water movement, the subduction of the continental plates etc. Here Matter cannot be created or destroyed in a chemical reaction - or a biological reaction, for that matter. This means that elements are simply incorporated into different compounds, which are then transformed into other compounds, and, only if energy is available to drive the reactions, they may be changed to different compounds, probably repeating the process. In other words, elements are recycled in Nature and re-used by living organisms. In many cases, living organisms use these elements so much that they cannot get enough of them elements - they are thus a limiting factor for growth - and so the rate at which they are recycled is critical. This is especially true for plants, but also for the whole food chain of organisms which depend on them. An ecosystem consists of an environment together with all the naturally occurring organisms in it. The elements and compounds which are the most essential for the growth of living organisms are said to have their own (interlinked) cycles, but similar cycles exist for elements of "lesser" importance. These are transformation processes, in that chemical elements are moved from one compound to another. Cycles of Carbon, Nitrogen, Phosphorus, Potassium, Sulphur, Micronutrients etc. in environment are described:

1. Carbon cycle

Presence of carbon in the soils, oceans, atmosphere, animals, plants and human beings enforces to say that carbon cycle is the life cycle. The carbon cycle is the biogeochemical cycle by which carbon is exchanged among the biosphere, pedosphere, geosphere, hydrosphere and atmosphere of the Earth. The carbon cycle may be defined as a natural cycle in which carbon di oxide, is fixed in flora by photosynthesis or by chemosynthesis, consumed by animals and human beings in the form of carbohydrate, proteins and fat and finally returned to the atmosphere through the respiration or through death and decomposition of bodies of plants and animals. Carbon cycle is the continuous transformations of organic and inorganic carbon compounds by plants and animals among the soil, plants and the atmosphere.

Chemical compositions of all the organisms show a significant amount of carbon. Organisms acquire the carbon from carbon di oxide gas. However, the mode of acquisition may be different. The chemoautotroph micro-organisms directly absorb CO_2 from the atmosphere. The chlorophyll bearing vegetations in the presence of CO_2 change it into carbohydrates through photosynthesis. The organic compounds (carbohydrates, proteins and fat) serve as the food for human and herbivores. Human beings and carnivores also utilized some herbivores as their food. The remains or leftovers of food materials, residues of plants and animals and other organic waste come into the soil in the form of organic matter. Plants roots and organisms remain in the soil and lock up the carbon for longer period. By this way CO_2 of the atmosphere is converted in to different simple and complex organic compounds.

Human beings, animals and plants liberate CO_2 gas during respiration. Organic matter present in soil is decomposed by various micro-organisms and produce

CO_2, CH_4 and CS_2 gases. All these gases are released in to the atmosphere after increasing some CO_2 status of soil air. Carbon di oxide present in soil or dissolved in rain water forms carbonic acid (H_2CO_3). Carbonic acid reacts with the elements present in soil and forms their carbonates and bicarbonates. Bicarbonates are more soluble Thus, utilized by plants for their nutrition or drain out from soil with excess water. Bicarbonates also produce oceanic proteins after reacting with nitrogenous compounds present in seas and oceans.

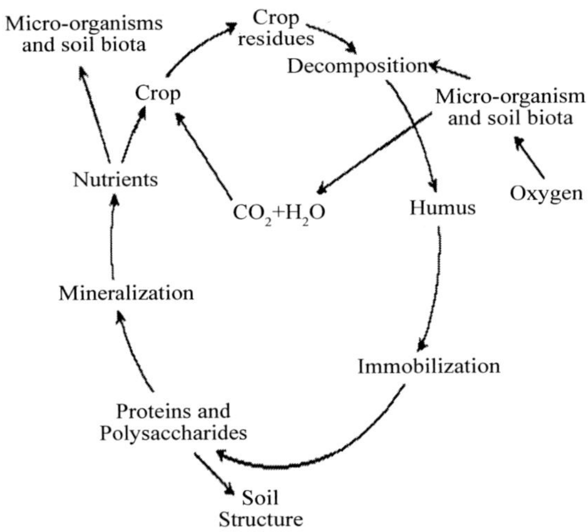

Fig. 3.56 Carbon cycle www.fao.org/3/a0100e/a0100e04.htm

Buried carbon in the form of coal, natural gas, petroleum and lignite and on the surface source of carbon e.g., peat and wood are used by humans and industries for heating purposes produce tremendous amount of CO_2 gas. This increases the CO_2 status of atmosphere remarkably. Lime, limestone and coral-reef also release the CO_2 into the atmosphere through various transformations and processes.

2. Nitrogen Cycle

Nitrogen Cycle is a repeating biogeochemical process through which nitrogen is converted into many forms, consecutively passing from the atmosphere to the soil, water, plants, animals and microorganism and back into the atmosphere. Air is an unlimited store of nitrogen gas. This nitrogen reaches in the soil by natural photochemical and electrochemical actions.Various asymbiotic and symbiotic microbes also enrich soil with nitrogen by fixing atmospheric nitrogen. Industrial fixation of nitrogen and their applications in soil as fertilizers are other means to provide it in soil. In soil nitrogen is used in nutrition of plants after transforming in to nitrate.

Proteinaceous residues of dead animals and vegetations firstly transform in to ammonia and then in nitrate. Process of transformation from protein to ammonia is called as ammonification. Process of transformation from ammonia to nitrate is called as nitrification. Plants use nitrate for their nutrition. If nitrates are more than plant's need, they drain out with drainage water or can reach in lower layers of soil. In anaerobic conditions, nitrates are again changed into nitrogen gas and can reach in atmosphere. Process of changing nitrate to nitrogen gas is called denitrification.

After ammonification, ammonia is adsorbed and fixed on organic and inorganic colloids present in soil. This fixation is of such type so that microbes also cannot use this ammonia. Ammonia is also volatized and reached to atmosphere from sandy, calcareous and alkaline soils due to higher temperature. Soil erosion by water and air also lost the nitrogen along with soil. Processes by which nitrogen reaches to plants and soils. Similarly various processes escapes atmosphere. Thus, this cycle of nitrogen is completed in environment.

Soils also have their own cycle of nitrogen. This cycle is little different but not separate or apart from main cycle of nitrogen. In this cycle, nitrogen changes into organic forms from inorganic forms by biological processes and vice-versa. This change controls availability or unavailability of nitrogen to plants. Change of nitrogen from organic to inorganic form is called mineralization. Mineralization increases availability of nitrogen to plants. Nitrogen used by plants is called Assimilated Nitrogen. Process of change inorganic nitrogen to organic nitrogen is called immobilization. Immobilization decreases availability of nitrogen to plants. Figure of nitrogen cycle is given here:

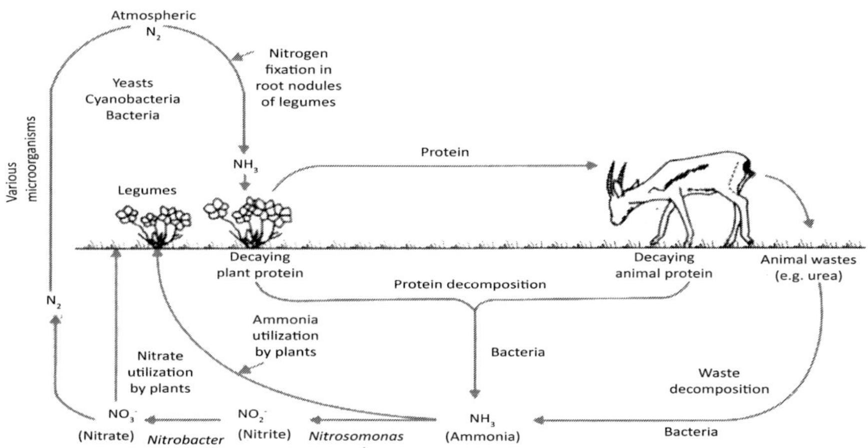

Fig. 3.57: Nitrogen Cycle

Processes of ammonification, nitrification, denitrification and nitrogen fixation by colloids are parts of internal cycle of nitrogen in soil. Internal nitrogen cycle of soil is a large and main component of environmental nitrogen cycle. Brief description of this cycle is given here:

1. Ammonification

In ammonification ammonia is produced from nitrogenous organic matters by biochemical processes. In soils heterotrophic bacteria and protozoa perform this process. These microbes produce proteinases, peptidases, amino acid dehydrogenase, amino acid oxydase and other such enzymes. Ammonia is produced from proteins, peptides and amino acids by actions of these enzymes. First of all, process of ammonification initiated from proteins and peptides where proteinase and peptidase enzymes, respectively, break peptides bonds of these and produce amino acids. Now amino acid dehydrogenase and oxydase enzymes act on amino acids and produce ammonia gas.

Besides proteins and peptides biochemical decomposition of nucleic acid also takes place and ammonia gas is produced. Nuclease enzyme produces mononucleotides from nucleic acid. Nucleosides are produced from mononucleotides by the action of enzyme nucleosidases and nucleotidases. Nucleosides are the structural sub unit of nucleic acids e.g RNA and DNA and nucleotides are building blocks of RNA and DNA. Formation of purine and pyrimidine from nucleosides is due to the action of nucleosidase enzyme. Amidohydrolase and Amidinohydrolase enzymes, respectively act on purine and pyrimidine and produce ammonia.

Ammonia also is produced from amino sugars and urea. Amino sugars are converted in to glucasamine-6-phosphate by various enzymes. The action of Glucasamine-6-phosphate isomerase enzyme upon glucasamine-6-phosphate forms ammonia. Urea is applied in soil as fertilizers. Urea and uric acid are also found in urine of various animals and excretions of insects. Urease enzyme acting upon the urea makes ammonia and carbon dioxide gases as given here:

$$C = O\ (NH_2)_2 + H_2O_2\ NH_3{}^+ CO_2$$

Ammonification is a most important biochemical process. This is not only the first half of nitrogen mineralization, but also provides necessary ammonium nitrogen to many plants. Requirement of nitrogen of any crop in the whole life time depends on source and nature of organic nitrogen present in soil and ammonification capacity of prevailing microbes. Often it is seen that root exudates of crop plants provide energy to microbes to increase the rate of ammonification according to their need. In this context contribution of protozoa and nematodes is also important. Both these organisms eat microbes and plant residues and after digesting proteins they themselves release ammonia.

Ammonia produced by ammonification returns in the atmosphere only in some special conditions. In normal conditions ammonia is mixed with various acids and water found in soil and makes ammonium ion. Ammonium ions are used for nutrition of plants or microbes. It can be adsorbed on soil colloids. Also, in next half part of the nitrogen cycle it can be converted in nitrate by nitrification process.

2. Nitrification: Biological oxidation of ammonia/ammonium to nitrite and subsequently to nitrate is called as nitrification. Change of nitrogen from highly reduced state (NH_4) to highly oxidized state (NO_3) is known as nitrification. Basically, nitrification is a biological process performed in two steps. But in soil it may be a chemical process also because of oxidation of nitrogen by different chemical reactions of soils.

In 1890 Winogradski discovered biological nitrification. He observed that two groups of bacteria perform nitrification process. First group of bacteria is performing oxidation of ammonium in nitrate and second group oxidizes nitrite in to nitrate. In the processes of conversion of nitrogen into nitrate hydrogen ions are produced which naturally decrease soil pH.

Microorganism

Process of nitrification is performed by autotrophic bacteria. These bacteria are called nitrifiers. In addition to autotrophic bacteria, heterotrophic fungi, actinomycetes and bacteria also nitrify soils. Ammonium is converted into nitrate by *Nitrosomonas* and nitrite into nitrate by *Nitrobacter*. Primarily *Nitrosomonas europee* and *Nitrobacter winogradski* take part in this process. These bacteria are chemoautotroph, gram negative and belong to Nitrobacteriaceae family.

Nitrosomonas are rod shaped. These have mobility in straight line. *Nitrobacter* are irregular lobular shaped and mobility is of different types. These are somewhat big in shape. Because of chemoautotroph, these get carbon for their cells from carbon dioxide gas. Also, for the reduction of carbon dioxide gas they obtain energy from the reduction of inorganic compounds of nitrogen. Besides both bacteria other chemoautroupotroph bacteria are also found in nature which perform nitrification. These genuses are *Nitrosococous, Nitrosopira, Nitrospina, Nitrococcus, Nitrosolobus* and *Nitrosovibrio.*

In addition to autotrophic bacteria other heterotrophic microbes, which can perform nitrification are *Arthrobacter, Azotobacter, Aerobacter, Bacillus, Pseudomonas* and *Proteus* bacteria, *Streptomyces* and *Nocardia* actinomycetes and *Aspergillus* and *Penicillium* fungi. Although nitrification rate of these heterotrophs is very less than autotrophs, but these are also important because they continuously work in harmful conditions for the benefit of autotrophs.

Factors Affecting Nitrification

Primarily *Nitrosomonas* and *Nitrobacter* perform nitrification in soil, so here factors are described in their context. The efficiency of nitrification depends upon the various environmental factors that include 1. Quantity of ammonium in soil, 2. Amount of oxygen in soil 3. Soil aeration, 4. Soil moisture 5. Soil temperature 6. Soil reaction, 7. Use of crop protecting chemicals.

1. **Quantity of ammonium in soil:** Presence of ammonium in soil is very much essential for nitrification. Presence of organic residues having wide carbon: nitrogen ratios reduce the amount of ammonium in soil. So, nitrification also decreases. But excess of ammonium reduces activity and growth of *Nitrobacter*. NO_2 in soil is essential for *Nitrobacter* to convert it into NO_3.
2. **Amount of oxygen in soil:** Nitrification is an oxidation process. It is performed by aerobes. Presence of excess of organic residues which decompose quickly reduces oxygen in soil, because heterotrophs use more oxygen. In such conditions process of nitrification becomes slow. In the absence of oxygen nitrification is impossible. Therefore, quick diffusion of oxygen into the soil from atmosphere is most essential.
3. **Soil aeration:** Supply of oxygen to microbes depends on soil aeration. In the good soil aeration CO_2 gas present in soil is exchanged with the atmospheric O_2 gas. Therefore, good aeration increases the rate of nitrification. Nitrification occurs at the optimum aeration porosity between **50-60%** of the total soil porosity. It has been observed that nitrification would not take place if Eh is less than +200 milli volt.
4. **Soil moisture:** Aeration is related directly with the quantity of moisture in soil and increase in moisture decreases air in soil. In saturated soil all the pores are filled with water. In such conditions lack of oxygen has unfavourable effect on nitrification. It does not mean that water is not essential for nitrification. Nitrifiers also require water for their lives. Therefore, nitrification occurs more in irrigated areas than dry areas. Generally, nitrification occurs satisfactorily at one half to two third of water holding capacity. Fungal nitrifiers perform well under water stress.
5. **Soil temperature:** The nitrification rate is a function of temperature range between 5° and 35°. The optimum range of temperature for the growth of nitrifying bacteria is between 28°C and 36°C which may raise up to 42° C for Nitrobacter. Maximum activity is between 15° C to 35° C. The nitrification rate drops sharply below 15° C and is reduced by 50 per cent at 12° C temperature. Nitrification completely stops at temperature more than 50˚C and less than 5˚C. Spring season is best for nitrification as temperature and moisture are in optimum condition. Reference: The Influence of

Temperature on the Nitrification Rate - Nitrogen ...*https://www.climate-policy-watcher.org/nitrogen-removal/the-influence...*

6. Soil reaction: Soil pH directly influences nitrification. The nitrifiers are pH sensitive and thrive well between 6 to 8 pH and pH value 5 to 8 of soil is suitable for nitrification. Nitrification is reduced very much at pH less than 6 and stops completely at pH less than 5. In such soils nitrification can be increased by adding lime in soil.

   Activity of *Nitrosomonas* is sufficient at 7 to 9 pH. But activity of *Nitrobacter* reduces at 8 pH. Thus, *Nitrosomonas* can work in more alkaline environment. Therefore, in more alkaline conditions first part of nitrification is continuously going on and second part goes on very slowly or stops completely. This results the accumulation of nitrite (NO_2) in alkaline soils.

7. Use of plant protecting chemicals: Plant protection chemicals applied in recommended doses have no any unfavourable effect. Nitrifiers are more sensitive than other bacteria, which are also found in soil. Therefore, little more quantity of any chemical also can reduce their population for short period. The unfavourable effect of methyl bromide fumigant has been observed on nitrifiers.

Biochemical Process of Nitrification

Nitrifying bacteria possess ammonia monooxygenase, hydroxylaminc oxidoreductase and nitrite oxidoreductase enzymes to complete the nitrification process. In the nitrification oxidation of ammonium in nitrate occurs. In this process ammonium of negative (-3) valency is oxidized and converts in to nitrate positive (+5) valency. In this whole process transformation of 8 electrons occur in two steps.

Nitrification completes in two steps. In first step, valency changes from (-3) to (+3) i.e. ammonium to nitrite. In second step valency (+3) is changed to valency (+5), when oxidation of nitrite takes place to form nitrate. First step is performed by *Nitrosomonas* which is the most important bacteria associated with this step. In this ammonium oxidation process enzyme ammonia monooxygenase takes part to form hydroxylamine (NH_2OH) and enzyme hydroxylamine oxidoreductase converts hydroxylamine to nitrite. The equations are as:

$$2NH_4^+ + 3O_2 \rightarrow 2NO^-_2 + 2H_2O + 4H^+$$

$$NH_3 + O_2 \rightarrow NO_2^- + 3H^+ + 2e^-$$

$$NH_3 + O_2 + 2H^+ + 2e^- \rightarrow NH_2OH + H_2O$$

$$NH_2OH + H_2O \rightarrow NO^-_2 + 5H^+ + 4e^-$$

In the second step the process is hydrolytic oxidation linked with cytochrome in which nitrite oxidizing bacteria *Nitrobacter* transform nitrite to nitrate. Enzyme nitrite oxidoreductase takes part in this part. In this transformation of two electrons are transferred and energy is produced. The equation is as:

$$NO^-_2 + H_2O \rightarrow NO^-_3 + 2H^+ + 2e^-$$

In this process there is transfer of 6 electrons and hydrogen ions are produced. These hydrogen ions lower the soil pH. It is also supposed that nitrous oxide (N_2O) gas is also produced. This gas goes off into the atmosphere. Plants and microbes use nitrate formed by nitrification for their nutrition. These nitrates can be washed away by drainage water or can be leached in lower layers of soil. Their mixing in ground water can pollute ground water. Also nitrates can again converted in to nitrogen gas by various chemical and biochemical processes can go into the atmosphere.

Nitrification Inhibition

In agriculture the management of applied fertilizer nitrogen (ammonium, nitrate and urea) is most essential because of its loss into the environment in different forms through ammonia volatilization, denitrification or leaching. Thus, whole amount of applied nitrogen is not available to the crop. The nitrate (NO_3^-) leaching and the nitrous oxide (N_2O) emissions are of great concern. Generally, it is seen that greater part of nitrates formed by nitrification goes out from soil with drainage water or ground water. Also, a sizable part of it emits in the air as nitrogen gas. In the fields of paddy aforesaid losses occurs very much because of more moisture and anaerobic condition. So, nitrification is considered as a harmful activity instead of a beneficial process.

Some natural and synthetic, artificial or commercial substances are used to prohibit the nitrification in soil. So the nitrogen as ammonium ions can be preseved on soil colloids which will be utilized by plants for their nutrition as per need. Substances that delay nitrate production by depressing the activity of *Nitrosomonas* bacteria to convert ammonium into nitrate are called as nitrification inhibitors. Among natural substances neem cakes, pongame or fever nut (karanj) cake, tea waste (tea poly phenol) and other similar substances inhibit nitrification very easily.

In Karanj a crystalline substance namely karanjin is found. Furan ring present in karanjin primarily acts as a nitrogen inhibitor. There are many synthetic substances used as nitrification inhibitors. But most used are 2-chloro-6-(trichloromethyl)-pyridine (Nitrapyrin as N-serve), dicyandiamide (DCD) and 3,4-dimethylpyrazole phosphate (DMPP). These compounds suppress microbial activity from few days to weeks. Names of other synthesized nitrification inhibitors are thiourea, guanilthiourea, sulphathiazole, 2-amino-4-methyl-trichloromethyl-1, 3, 5 triazene and 6-amino-1, 2, 4 triazole hydrochloride.

These are used in very small quantity (nearly half kilogram per hectare). Money spent on these is very negligible in context of conservation of nitrogen as ammonium. Efficiency and effective period of nitrification inhibitors mainly depend on type of soil, soil moisture, climate and cultural practices. In general, nitrification inhibitors are more effective in sandy soils or soil low in organic matter, fertilized with Cu and Zn and exposed to low temperatures.

3. Denitrification

Denitrification is a part of the biological nitrogen cycle and it is reverse of nitrogen fixation. The already fixed nitrogen present in soil goes to the atmosphere by denitrification process in the form of various gases. Thus, in fixation, soils get nitrogen, the denitrification loses it. This depletes soil fertility and reduces crop production. Reduction of nitrite and nitrate in oxide of nitrogen or nitrogen gas by microbiological actions or by chemical reducing agents is called as denitrification. In addition to nitrite and nitrate, ammonium and amines may also convert into nitrogen gas.

Denitrification in soil may be of two types e.g., biological or chemical. This process is mainly performed by heterotrophs particularly facultative anaerobic bacteria. Denitrification provides respiration and acquires energy for these bacteria. Submergence of soil, poor water drainage and insufficient aeration or anaerobic conditions are favourable for denitrification.

Microorganisms

Denitrification is mainly done by facultative, anaerobic chemoheterotrophic bacteria. These bacteria are *Achromobacter*, *Bacillus, Thiobacillus, Enterobacter, Micrococcus, Pseudomonas, Rhodopseudomonas, Spirillum, Proteus, Aerobacter* and *Flavobacterium*.

Factors Affecting Denitrification

Nitrogen losses from soil by denitrification depend on several factors. In these factors, 1. Presence of nitrate ions in soil, 2. Quantity of organic matter, 3. Soil aeration, 4. Quantity of moisture, 5. Soil pH and 6. Temperature are most important

1. **Presence of nitrate ions in soil:** Presence of inorganic nitrogen in the form of nitrate in soil is necessary for denitrification. Organic matter where nitrate is in abundance enhances denitrification.
2. **Quantity of organic matter:** Organic matter content of the soil is a very important factor. The more the organic content in the soil, more is the growth of denitrifiers and more and rapid denitrification. The quantity of

organic matter present in soil is the only source of nutrition for the bacteria. Readily available organic matter like plants roots or root exudates increases drnitrification. Nitrates present in the organic matter with wide carbon: nitrogen ratio are assimilated by microorganisms before denitrification starts. So the denitrification is very less or negligible.

3. **Soil aeration:** Anaerobic conditions are very much essential for denitrification. Denitrification begins only in oxygen free environment, but nitrate required for denitrification is formed in the presence of oxygen. So, aeration influences denitrification in both ways. Ploughing in soil with poor aeration enhances denitrification, because by ploughing in such soil increase partial aeration, which produces nitrate. Afterwards in anaerobic condition these nitrates are converted in nitrogen gas.

4. **Quantity of moisture:** Excess of water in soil inhibits supply of oxygen to denitrifiers. This increases the denitrification. As soon as quantity of water increases in soil, simultaneously denitrification of nitrate also increases. When water comes in soil by rain or irrigation then reduction of already present nitrates starts. Denitrification stops completely when water in soil becomes less than 60% of water holding capacity. Denitrification is also possible near water filled microspores of soil having good drainage conditions and decomposing organic matter. At such places shortage of oxygen and presence of water-soluble organic matter help in denitrification.

5. **Soil pH:** Soil pH has positive relationship with denitrification. Denitrifying microorganisms are more sensitive towards hydrogen ions. At low pH denitrification decreases and it is very low below pH 4.8. pH more than 6 is essential for the formation of nitrogen. Denitrification is vey much rapid and at pH range 8.0 to 8.8. Denitrification is maximum in neutral to slightly alkaline soils. In acidic soils nitrous oxide gas is half of the total denitrification.

6. **Temperature:** Hot environment encourages denitrification. The best temperature for denitrification is 25˚C or more. At temperature below 10˚C rate of denitrification rate is very much low. Denitrification stops completely at 2˚C temperature.

Biochemical Process of Denitrification

The existing knowledge about denitrification indicates that denitrifying bacteria perform this process with the help of various reducing enzymes. These bacteria use oxygen of nitrate, nitrite and nitrogen oxides, in the absence of free oxygen, as hydrogen receptors.

The general equation of denitrification process converting nitrate to gaseous nitrogen is as:

$2\ NO_3^- + 10\ e^- + 12\ H^+ \rightarrow N_2 + 6\ H_2O$

Bacteria perform denitrification for two tasks. In first task they do reduction of nitrate in ammonia as $NO_3 \rightarrow NO_2 \rightarrow NH_3$ and use ammonia for the formation of their cells. This is called assimilatory nitrate reduction.

In second task reduction of nitrate happens gradually. In this process nitrate accepts hydrogen. It is called dissimilatory nitrate reduction. It is actual denitrification. Dissimilatory nitrate reductase enzymes perform this task. The stepwise equations are,

1. The conversion from nitrate to nitrite performed by nitrate reductase enzyme.

   $NO_3^- + 2\ H^+ + 2\ e^- \rightarrow NO_2^- + H_2O$

2. The conversion from nitrite to nitric oxide performed by nitrite reductase enzyme.

   $NO_2^- + 4\ H^+ + 2\ e^- \rightarrow 2\ NO + 2\ H_2O$

3. The conversion of nitric oxide to nitrous oxide performed by Nitric oxide reductase enzyme.

   $2NO + 2\ H^+ + 2\ e^- \rightarrow N_2O + H_2O$

4. Finally nitrous oxide is converted into nitrogen by nitrous oxide reductase enzyme.

   $N_2O + 2\ H^+ + 2\ e^- \rightarrow N_2 + H_2O$

In this process of nitrate reduction can stop at any step. So these gases can go in the atmosphere at any time. Generally, losses of N_2O and N_2 in soil are more. In very much acidic condition when soil pH is less than 5, ratio of N_2O and N_2 is equal. In such condition losses of N_2O are more. Because N_2O gas is more soluble in water, so it can also drain out with drainage water.

Chemical Denitrification

Generally, in soils the action of change from NO_2 into NO_3 is more rapid than change from NH_4 into NO_2. Therefore, the possibility of accumulation of NO_2 in soil is very less. But sometime due the use of ammonia or ammonical fertilizers considerable quantity of ammonia accumulates in soil. Excess quantity of ammonia is harmful for *Nitrobacter*. Therefore, nitrites are not converted in nitrates. Consequently, more nitrites accumulate in soil.

Excess of NO_2 in soil is harmful for plants. Also, NO_2 is very unstable and reacts with various compounds to form nitrogen gas. In slight acidic condition when nitrite meets ammonium salts or amino compounds, then nitrogen gas is produced. Also, reactions of nitrite with lignin, phenols and carbohydrates components of organic matter in soil form nitrogen gas.

Phenols in them are very much active and react with nitrite to make nitrophenols. These react with HNO_2 and converted in N_2O and N_2. Since this action is performed by chemical agencies and no living organism take part in it, so this is chemical denitrification. In chemical denitrification process, chemicals like acids are used as reducing agents to form ammonia or elemental nitrogen. Often this process also requires catalysts, high heat and pressure. Formation of gaseous nitrogen or oxide of nitrogen by nonbiological process is called as chemical denitrification.

Process of chemical denitrification with various compounds is shown by equations as:

1. with ammonia, $NH_3 + HNO_2 \rightarrow NH_4NO_2 \rightarrow 2H_2O + N_2$
2. with urea, $CO\,(NH_2)_2 + HNO_2 \rightarrow 3H_2O + CO_2 + 2N_2$ and
3. with amino acid, $R.NH_2 + HNO_2 \rightarrow R.OH + H_2O + N_2$

4. Volatilization of Ammonia

Nitrogen may be lost from the soil in different ways and volatilization of ammonia is one of them. Ammonia volatilization reduces the yield of crops and increases the expenditure on the use of ammonical nitrogenous fertilizers. Volatilization losses of ammonia occur from dung and urine of animals in pastures and sheds, excreta of poultry in poultry farms and urea, ammonium sulphate, ammonium nitrate and urea ammonium nitrate like fertilizers. In these, nitrogen is found in the form of ammonia. In favourable situations this ammonia goes in the atmosphere by volatilization.

Factors Affecting Ammonia Volatilization

Volatilization of ammonia is affected by many factors. These factors are 1. Soil texture, 2. Soil cation exchange capacity, 3. Soil pH and calcareousness 4. Temperature, 5. Drought and moisture content in soil, 6. Wind velocity 7. Application of ammoniated fertilizers or organic manure. These factors are described here:

1. **Soil Texture:** Volatilization of ammonia takes place more from sandy soil as compared to silty, clayey or peaty soils. On account of less quantity of colloids in sandy soils ammonium does not adsorb by them. So, it goes to the atmosphere.
2. **Soil cation exchange capacity:** Low cation exchange capacity increases ammonia volatilization. In soil having more cation exchange capacity

ammonium ions are adsorbed by soil colloids and in lieu of ammonium other cations are released in soil solution.

3. **Soil pH and calcareousness:** Alkaline pH and more calcareous soils increase ammonia volatilization. As pH of soil increases, volatilization of ammonium from soil also increases. Possibility of volatilization of ammonia increases at higher pH due to hydrolysis of urea. Ammonia volatilization is enhanced by calcareous soils.

4. **Temperature:** Higher temperature increases increases ammonia volitilization. Solar radiation increases the temperature and increase in temperature increases ammonia volatilization. It has been observed that 50 % of applied nitrogen is volatilized as ammonia at a temperature of 30°C as compared to 35 % at 25°C temperature. Therefore, ammonical ferizers or manures should not apply during period of higher temperature.

5. **Drought and moisture content in soil:** Drought increases ammonia volitilization. Higher moisture content in soil through rainfall or irrigation directly decreases the volatilization of ammonia. Excess water improves infiltration of the ammonia into the soil. Generally, ammonia volatilization decreases with an increase in relative humidity due to lesser space is available for ammonia in the atmosphere.

6. **Wind velocity:** Higher wind velocity increases ammonia volatilization. The ammonia volatilization rate increases when wind velocity increases up to 9 km/h. There is no consistent increase in ammonia volatilization with the wind velocity increased from 9 to 15 km/h.

7. **Application of ammonical fertilizers or organic manure:** Application of ammonical fertilizers or organic manure on soils surface enhances volatilization of ammonia. Significant volatilization of ammonia can occur within the first 24 hours after the soil application of manure. Among the method of application, it was found that incorporation of manure into soil is one of the most efficient methods for decreasing ammonia volatilization.

Equations of Ammonia Volatilization

Ammonical fertilizers or ammonia producing materials applied are converted in to their carbonates and hydroxides either due to enzymes, anions or prevailing conditions present in soil and finally to ammonia gas as:

$$CO(NH_2)_2 + 2H_2O \rightarrow (NH_4)_2CO_3$$

$$(NH_4)_2CO_3 + 2H^+ \rightarrow 2NH_4^+ + CO_2 + H_2O$$

$$NH_4^+ + OH^- \rightarrow NH_4OH$$

$$NH_4OH \rightarrow NH_3 + H_2O$$

3. Phosphorus Cycle in Soil

Importance of phosphorus can be understood because of its absence any process of energy transfer cannot be taken place in living organisms and by virtue of phosphorus difference between living matter and dead protoplasm can be established. Main mineral of phosphorus is apatites. Phosphatic fertilizers made from apatites are applied in the soil. Microorganisms and plants take this inorganic phosphorus from the soil. Animals and humans take phosphorus from plants with their food. In the plant, animal or human the phosphorus is incorporated into DNA like organic molecules. The residues, wastes and remains of dead plant and animals contain phosphorus as organic compounds. This is immobilization of phosphorus.

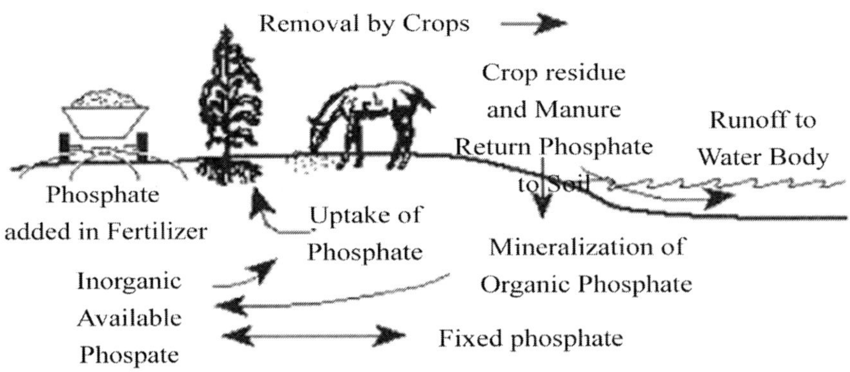

Fig. 3.58: Phosphorus Cycle in Soils (*Source*: Busman et al., 1997)

The organic compounds of phosphorus can be made available to plants by bacteria. Bacteria by releasing phosphorous from organic compounds, enrich soil with inorganic forms of phosphorus and make phosphorous available to plants. This is phposphorus mineralization. By this way immobilization and mineralization are two main processes related with phosphorous in soil. Phosphorus in soil can go in water bodies through surface runoff and leaching. A figure of Phosphorus cycle in soil is given here:

Phosphorus Immobilization

Phosphorus cycle is a biogeochemical process showing the movement of phosphorus and its different forms through the lithosphere, hydrosphere and biosphere. The immobilization of inorganic phosphorus is the reverse reaction of mineralization. Microorganisms convert inorganic phosphates to organic phosphates during immobilization and incorporate them into their living cells. Microorganisms require useful and available compounds of phosphorus for their cells and uptake some quantity of mineralized phosphorus. The assimilation of phosphorus by microorganisms reduces quantity of plant available phosphorus in the soil. Therefore, the yields of the crops may be reduced due to the lesser availability of phosphorus. If decomposing organic matter is low in phosphorus or carbon: phosphorus ratio is wide then some part of mineralized phosphorus will be immobilized.

So, mineralization and immobilization both processes run simultaneously and which side the process will go depends on carbon: phosphorous ratio of decomposing substance. If carbon: organic phosphorous ratio is 300 or more than this, then there will be complete immobilization and inorganic or mineral phosphorus will be converted into organic phosphorous. If this ratio is 200 or less, then there will be complete mineralization. If this ratio is in between 200 and 300, then phosphorus will remain as such and there will be neither mineralization nor immobilization. If quantity of phosphorous in decomposing materials is more than 0.3%, then phosphorous will be mineralized and released.

Phosphorus Mineralization

Phosphorus mineralization is a process by which organic phosphorus in soil is converted into inorganic phosphorus with the help of soil microbes. The major source of organic phosphorus in soil is the organic matter. Organic P is mostly in the form of inositol phosphate (soil phytate) and other organic P compounds are phospholipids, phosphoprotein and nucleic acids. The amount of organic phosphorus may be in the range 30–60% of the total P of soil. However, this may be much more (about 85%) in organic soils. Inositol phosphates, nucleic acids and phospholipids are described here:

Among all these compounds inositol phosphate is major source of organic phosphate in soil. Inositol phosphates are a group of mono to hexaphosphorylated inositols. These play crucial and important roles in different functions of cells. Hexaphosphate ester (inositol hexaphosphate) is called as phytic acid and phytate is its salt. Calcium and magnesium salts of phytic acid are called as phytin. Major source of these in soils are higher plants and microorganisms.

Nucleic acids are naturally occurring chemical compounds. These are found in all living cells. These make up the genetic material and carry basic information in the cells. Nucleic acids are biopolymers or large biomolecules which are essential for all lives. These are made up of nucleotide monomers. Nucleotides have three components namely a nitrogenous base, a 5-carbon (Pentose) sugar and a phosphate group. The two main classes of nucleic acids are deoxyribonucleic acid (DNA) and ribonucleic acid (RNA) which have deoxyribose sugar and ribose sugar, respectively.

Phosphoglyceride is major fraction of phospholipids found in soils. Plant residues, animal waste and microbiological masses are main source of these in the soil.

Bacteria, actinomycetes and fungi use phospholipids for their phosphorous source. During decomposition these heterotrophic microorganism separates phosphate out of phospholipids. Here it will be appropriate to mention that phospholipids are insoluble in water but they are soluble in organic solvents and benzene, ether or chloroform.

Microorganisms

In soil numerous bacteria, fungi and actinomycetes mineralize organic phosphorus. Among them *Bacillus, Streptomyces, Pseudomonas, Arthrobacter, Proteus, Serratia, Micrococcus Aspergillus, Rhizopus, Penicillium* are more important as they produce phosphatase enzyme.

Factors Affecting Phosphorus Mineralization

Major factor influencing the mineralization of phosphorus are 1. Temperature, 2.Moisture, 3.Aeration, 4.Reaction, 5.Quantity of phosphorus and other nutrients. 6. Cultivation and 7. Microorganisms.

1. **Temperature:** Generally, it is noticed that mineralization of phosphorus increases when temperature is more than 30°C and optimum temperature is 35°C.
2. **Moisture:** Soil moisture in proper quantity is also essential for phosphorus mineralization. Too much moist or too dry condition is not suitable and up to 50-70% water holding capacity of moisture in soil is favourable for phosphorus mineralization. Alternate drying and wetting in soil increase phosphorus mineralization. It is supposed that alternate dryness and wetting breaks water stable aggregates. Breaking of aggregates provides more surfaces for the action of microorganisms.
3. **Aeration:** Effect of soil aeration on phosphorus mineralization is not clear yet. Usually anaerobic condition of submerged soil enhances mineralization of phosphorus.
4. **Reaction:** The pH which is favourable for the growth of microorganism is also favorable for phosphorus mineralization. Phosphorus mineralization enhances by the liming of acidic soils.
5. **Quantity of phosphorus:** Application of inorganic phosphorus in soil increases the mineralization of organic phosphorus. It is supposed that solubility of organic phosphorous increases in the presence of inorganic phosphorous.

   Therefore, it becomes more effectual or sensitive towards mineralization. If there is shortage of nitrogen in soil, then plants receive nitrogen from mineralization of organic nitrogenous compounds. During this process mineralization of phosphorus also occurs, because mineralization of nitrogen and phosphorus is nearly similar processes. Effect of sulphur on phosphorus mineralization is same as nitrogen.

6. **Cultivation:** Mineralization of organic phosphorus is more in virgin soil than cultivated soil. So there is more quantity of organic phosphorus in virgin soil. Also shortage of organic phosphorus near roots of plants is the indication of the more mineralization near roots. Its main reason may be root's exudates which probably also contain easily assimilable glucose. Glucose enhances the activity of phosphorus mineralizing microorganisms.

   Also roots of some plants excrete or emit phosphatase enzyme. Phosphatase dephosphorylates organic phosphorus. Dephosphorylation process removes phosphate groups from a molecule by the phosphatase enzyme. A well-known example of the dephosphorylation is the conversion of ATP to ADP and inorganic phosphate.

7. **Microorganisms:** Mineralization of the phosphorus also depends on the number and activity of phosphorus mineralizing microorganisms in the soil. These microorganisms produce phosphatase enzyme which separates phosphorus out of organic phosphorus. Several microorganisms produce many times more phosphatase enzyme in phosphorus deficient soils.

Some microorganisms found in the soil decompose organic and insoluble phosphorus compounds to enhance the availability of phosphorus to plants. These are *Bacillus, Pseudomonas* and *Rhizobium* bacteria, *Penicillium* and *Aspergillus* fungi and arbuscular mycorrhiza. Phosphatase activity of plants having mycorrhiza is two times more than the plants devoid of mycorrhiza. Naturally, in soils having mycorrhiza organic phosphorus mineralization will be more.

Biochemistry of Phosphorus Mineralization

Generally, all enzymes which mineralize phosphorous are termed as phosphatases. The phosphomonoesterase, phosphodiesterase and inorganic pyrophosphatase are also in it. Phosphomonoesterase separates phosphate from inositol phosphates and nucleotides. Phosphodiesterase decomposes phospholipids and nucleic acids. Inorganic pyrophosphatase enzyme is ubiquitous in nature, associated with energy related reactions and availability of from super phosphates.

Mineralization of RNA and DNA nucleic acids takes place by ribonuclease and deoxyribonuclease enzymes, respectively. These enzymes perform depolymerization of nucleic acids. This action separates phosphorous from nucleic acids. Generally, action of depolymerization is done by heterotrophs. The hydrolysis of the phytic acid or phytase by phosphatase enzyme and separation of phosphorus from it as given here in an equation:

$$C_6H_{18}O_{24}P_6 + 6H_2O \rightarrow C_6H_{12}O_6 + 6\ H_3PO_4^-$$

Phytic acid Water — Inositol Phosphorus

Reference: Girmay Kalayu 2019 Phosphate Solubilizing Microorganisms: Promising Approach as biofertilizers...*https://www.hindawi.com/journals/ija/2019/4917256*

4. Potassium Cycle in Soil

The potassium cycle is the biogeochemical cycle showing its movement throughout the Earth's lithosphere, biosphere and hydrosphere. Potassium comes in soils from primary minerals like micas and potassium feldspar**.** These minerals slowly weather and make the potassium available for plant's uptake or to be present in soil solution.

Bacillus, Pseudomonas, Burkholderia, Paenibacillus bacteria, *Aspergillus, Mucor* and *Penicillium* fungi and Actinomycetes decompose aluminosilicate minerals present in soil. The mechanisms of decomposition are acidolysis, chelation, exchange reactions, complexolysis, production of organic acids etc. This decomposition releases potassium present in these minerals. The released potassium is used by microorganisms and plants. The microorganisms also use potassium released after decomposition of organic matter for their cells. Potassium released after the death and decomposition of microbial cells is also available to plants. A generalized figure of Potassium cycle is given here

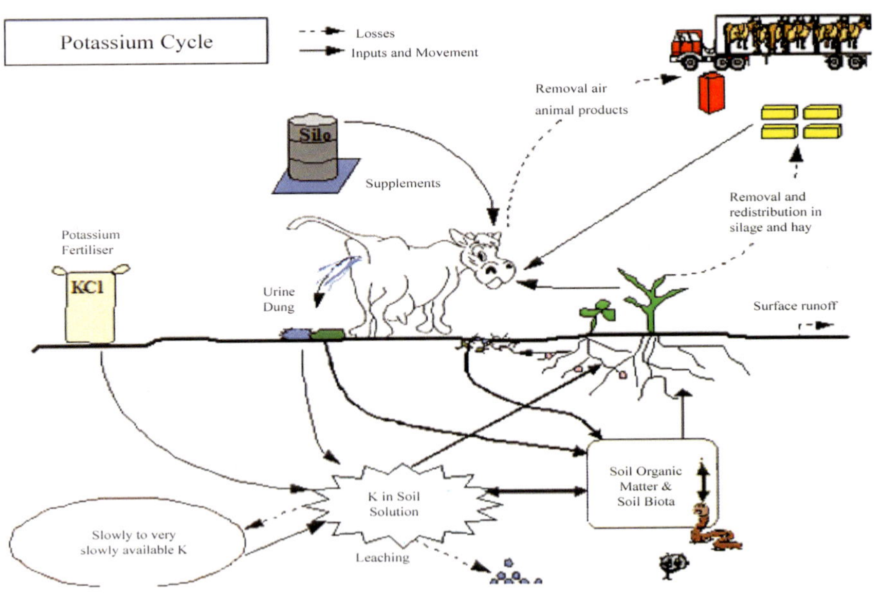

Fig. 3.59: Potassium Cycle

Sulphur Cycle in Soil

The sulphur cycle is a biogeochemical cycle showing the movement of sulphur through different atmosphere, biosphere and lithosphere. Sulphur present in primary minerals as sulphide is converted into sulphate by weathering. Plants use sulphate in their nutrition and convert it into organic sulphur. Plant residues, animal residues, wastes, chemical fertilizers and irrigation and rain water are major source of sulphur.

The action of microorganism on organic sulphur can again convert it into sulphate which can be used in nutrition of microorganism and can be a part of humus. Conversion of organic sulphur into sulphate is called as mineralization. Sulphur assimilated by microorganisms comes in the process of immobilization of sulphur.

Mineralization and immobilization of sulphur is just like mineralization and immobilization of nitrogen and phosphorus. In addition of mineralization and immobilization processes two other processes related with inorganic sulphur are simultaneously run in the soil. These processes are inorganic sulphur oxidation and organic sulphur reduction. Brief description of these four processes is given here:

1 Sulphur mineralization

Organic compounds of sulphur found in the soil are quite diversified. Sulphur is found in various forms in the bodily tissues of plants, animals and microorganisms and their wastes and residues. Amino acids namely cystine, cysteine and methionine appreciably contain sulphur. Moreover, sulphur also occur in vitamin B and Taurine.

After mineralization microorganisms use some part of inorganic sulphur (sulphate) released from aforesaid substances. Remaining part of sulphur is left in the soil. Plants use these remaining or leftover parts of sulphur in soil. Finally, the remaining inorganic sulphur goes out from the soil by erosion and leaching.

Biochemistry of Sulphur Mineralization

In well aerated soils, sulphates are produced by microbiological actions on cystine. This process in soil is performed by various ways. First, cysteine is formed from cystine. Then cysteine is converted in to cysteine disulphoxide. After this cysteine sulphinic acid is formed. This acid is converted in β-Sulphinylpyruvic acid. There by sulphite and finally sulphates are formed. Aforesaid process for cystine is indicated in chemical equation:

$C_6H_{12}N_2O_4S_2 \rightarrow C_3H_7NO_2S \rightarrow C_6H_{10}N_2O_6S_2 \rightarrow C_3H_7NO_4S \rightarrow$

Cystine Cysteine Cysteine disulphoxide Systeine sulphinic acid

$C_3H_4O_5S \rightarrow SO_3^{2-} \rightarrow SO_4^{2-}$

β- Sulphinylpyruvic acid Sulphite Sulphate

In anaerobic decomposition of cystine, hydrogen sulphide gas may also be formed along with pyruvic acid or serine. This can be shown by following two equations:

$$C_6H_{12}N_2O_4S_2 \rightarrow C_3H_4O_3 + NH_3 + H_2S$$

Cystine Pyruvic acid Ammonia Hydrogen sulphide

$$C_6H_{12}N_2O_4S_2 \rightarrow C_3H_7NO_3 + H_2S$$

Cystine Serine Hydrogen sulphide

Decomposition of Methionine ($C_5H_{11}NO_2S$) also takes place in many ways. In addition to methyl mercaptan (CH_3SH) which is also called methanethiol, ammonia and dimethyl disulphide (CH_3SSCH_3) are final products of methionine.

Microorganisms

Microbial organic sulphur mineralization to sulphate is very poorly understood yet. However, it is supposed that aerobic and anaerobic microorganisms which are able to produce sulphatase enzyme conduct this process. Among these thermophilic firmicutes bacteria are most important.

Factors Affecting Sulphur Mineralization

Mineralization of organic sulphur is significantly influenced by1. Temperature, 2. Moisture, 3. Soil reaction, 4. Quantity of $CaCO_3$ in soil and 5. Quantity of organic matter in soil.

1. **Temperature:** Rate of mineralization of sulphur depends directly on temperature. Maximum rate of mineralization occurs between 25°C to 35°C. Temperature less or more than this range shows adverse effect on mineralization of sulphur.

2. **Moisture:** Sixty percent moisture of water holding capacity is appropriate for mineralization of sulphur. Mineralization is reduced if moisture is more or less than 60% in soil. More moisture develops anaerobic conditions in soil which are not good for mineralization of sulphur. At low moisture (10-20 %) cells of microorganisms do not grow and develop significantly. So, mineralization of sulphur is reduced appreciably. In the conditions of alternate wetting and drying formation of sulphate increases which is the indication of more mineralization of sulphur.

3. **Soil reaction:** Generally moderate alkaline state of soil is appropriate for mineralization of sulphur. Mineralization is highest if soil reaction is in between 7.4 to 8.2. Generally, pH less than 6.1 and more than 9.1 low down the mineralization.

4. **Quantity of $CaCO_3$ in soil:** In acidic soils application of calcium carbonate enhances mineralization of sulphur. Generally, the effect of calcium carbonate this is visible when quantity of calcium carbonate in soil is 6% or more. Main reason for more formation of sulphates from application of calcium carbonate may be due to more growth of microorganisms. More growth to microorganisms may increase mineralization of organic matter. Also with the enhancement of soil pH more quantity of adsorbed sulphates may be released.
5. **Quantity of organic matter in soil:** The increasing the quantity of organic matter in soil also increases sulphur mineralization. The experiments indicated that by adding forage of Egyptian clover in soil mineralization of sulphur had been increased and by adding straw of wheat immobilization of sulphur took place. Therefore, besides the quantity of organic matter, types of organic matter also remarkably influence the formation of sulphates. Carbon: sulphur ratio of plant residues to be added in the soil totally controls gain or less of sulphates in the soil.

Sulphur Immobilization

Mineralization and immobilization processes of sulphur run simultaneously. Therefore, their quantification is very difficult. Wide carbon: sulphur ratio of organic residues immobilizes the sulphur in soil. More carbonaceous organic matter encourages the growth of microorganisms. These microorganisms use inorganic sulphur for their cells. Immobilization of sulphur also increases with the increase in the quantity of carbohydrate in the soil. It is due to assimilation of inorganic sulphur by the microorganisms developing on carbohydrates. Later on when activities of microorganisms reduce, then quantity of inorganic sulphur begins to increase again. So, the mineralization and immobilization of sulphur is like to the mineralization and immobilization of nitrogen up to some extent.

Carbon: nitrogen: sulphur ratio of decomposing organic matter has great importance in the immobilization of sulphur. If this ratio is too much wide then immobilization will be more. Generally, Carbon: nitrogen: sulphur ratio of soils also remains stable like carbon: nitrogen ratio. Usually this ratio is considered 130:10:1.3. This ratio determines behaviour of these elements in the soil.

Oxidation of Inorganic Sulphur

Sulphide, elemental sulphur and thiosulphate present in soil are very readily oxidized by microorganisms. This microbiological oxidation occurs very swiftly than chemical oxidation. Microorganisms use energy produced by oxidation. Numerous factors which govern inorganic sulphur oxidation are variability in oxidation rates among soils, climatic conditions and agronomic practices.

Oxidation of thiosulphate is stimulated using KH_2PO_4, urea, superphosphate and potassium sulphate and inhibited by $CaCO_3$. However, in well aerated soils moisture at field capacity and 25°C to 40°C temperature is good for oxidation of inorganic sulphur. Oxidation rates are lower in clayey soils.

Oxidation of inorganic sulphur reduces pH of soil. If sulphur element is applied directly or through fertilizers in the soil, then it is converted in to sulphate by microbiological activities. Hydrogen ions are also produced. Thus, in saline and alkaline soils its pH value can be reduced using sulphur element. Acid sulphate soils are also formed due to oxidation of sulphur. Oxidation of pyrite (FeS_2) forms sulphuric acid which too much reduces soil pH (sometimes 3.5 to 4.0). Mottles of yellow or variegated colours are found in these soils.

Biochemistry of Oxidation of Inorganic Sulphur

Oxidation of inorganic sulphur is mostly done by bacteria of *Acidithiobacillus* and *Thiobacillus* spp. on the compounds of inorganic sulphur and elemental sulphur by the enzyme known as sulphite oxidoreductase. In this oxidation transformation of sulphur can be up to -2 to -6 (sulphate). These reactions can be expressed by the equations,

Acidithiobacillus thiooxidans

$2S + 3O_2 + H_2O \rightarrow H_2SO_4$

Sulphur Sulphuric acid

$Na_2S_2O_3 + 2O_2 + H_2O \rightarrow 2NaHSO_4$

$Na_2S_2O_3 + 2O_2 + H_2O \rightarrow 2Na_2SO_4 + H_2SO_4$

Sodium thiosulphate Sodium sulphate

$2Na_2S_4O_6 + 7O_2 + 6H_2O \rightarrow 2Na_2SO_4 + 6H_2O$

Sodium tetrathionate Sodium sulphate

Acidithiobacillus ferrooxidans

$12FeSO_4 + 3O_2 + 6H_2O \rightarrow 4Fe_2\ (SO4)_3 + 4Fe(OH)_3$

$2S + 3O_2 + H_2O \rightarrow H_2SO_4$

$Na_2S_2O_3 + 2O_2 + H_2O \rightarrow 2Na_2SO_4 + H_2SO_4$

Thiobacillus thioparus:

$5Na_2S_2O_3 + 4O_2 + H_2O \rightarrow 4S + 5Na_2SO_4 + H_2SO_4$

$Na_2S_2O_3 + Na_2CO_3 + ½O_2 \rightarrow 2S + 2Na_2SO_4 + CO2$

Elemental sulphur (S) produced by these processes deposit as globules on the outer surface of the cells of *Thiobacillus thioparus*.

Thiobacillus novellus:

$Na_2S_2O_3 + 2O_2 + H_2O \rightarrow 2NaHSO_4$

Thiobacillus denitrificans:

$5S + 6KNO_3 + 2H_2O \rightarrow K_2SO_4 + 4KHSO_4 + 3N_2$

Microorganisms

Oxidation of inorganic sulphur compounds in soils is done by Heterotrophs and autotrophs. Among heterotrophs primarily *Arthrobacter, Bacillus, Flavobacterium, Micrococcus* and *Pseudomonas* bacteria, *Streptomyces* actinomycetes and *Aspergillus, Penicillium* and *Microsporum* fungi are included. In addition to *Thiobacillus Beggiatoa, Thiothrix, Thiospirillum, Thiocystis* and *Sulpholovs* also oxidize inorganic sulphur up to some extent.

In autotrophs *Acidithiobacillus* and *Thiobacillus* bacteria primarily oxidize inorganic sulphur compounds in soils. Genus *Acidithiobacillus* has seven species out of which two namely *Acidithiobacillus thiooxidans* and *Acidithiobacillus ferrooxidans* are main species. These bacteria are included in family Acidithiobacillaceae. Genus *Thiobacillus* has three species. Among these *Thiobacillus thioparus* and *Thiobacillus denitrificans* are placed in family Thiobacillacaeae. These bacteria do not form spores. These are 1-3μ long and 0.3μ thick. These are rod shaped and gram negative. Brief description of some main species is given here:

Acidithiobacillus thiooxidans

This is very much acid tolerance bacteria. The optimum pH for its good functioning is 2 to 3. This is obligate autotroph. Sulphuric acid is produced by the action of this bacterium which too much lowers the pH. In addition to elemental sulphur, thiosulphate and tetrathionate, this bacterium is also functional on iron pipes, concrete and pyrite.

Acidithiobacillus ferrooxidans

It is also very much acid tolerance and obligatory autotroph as *Acidithiobacillus thiooxidans*. This is also obligatory aerobic. This primarily oxidizes iron in addition to elemental sulphur and thiosulphate.

Thiobacillus thioparus

Thiobacillus thioparus cannot tolerate acidity at all. So, it functions excellently in neutral soils. Its growth stops completely at pH less than 4. Globules of elemental sulphur are deposited outside of its cells. This aerobic and obligate autotrophic bacterium mainly oxidizes thiosulphate.

Thiobacillus novellus

Thiobacillus novellus is now reclassified as *Starkeya novella* and placed in family Xanthobacteraceae. *Starkeya novella* is a chemolithoautotrophic and methylotrophic bacterium isolated from soil. *Starkeya novella* can oxidise thiosulphate. Neutral to slight alkaline pH is optimum and may develop on soil organic matter also.

Thiobacillus denitrificans

It likes neutral to slight alkaline pH. It functions by several ways in both aerobic and anaerobic conditions. In the presence of oxygen it oxidizes sulphur and in the absence of oxygen it uses nitrates as hydrogen (electron) acceptors for the oxidation of sulphur. As a result of this process nitrates are converted in gaseous nitrogen.

Reduction of Inorganic Sulphur

Strong odour like rotten eggs emit at the places which are continuously under muddy or swampy conditions. This odour is due to emission of hydrogen sulphide gas. This gas is produced by microbiological reduction of sulphur under anaerobic conditions. Mainly reduction of sulphate occurs in sulphide during reduction of inorganic sulphur.

Anaerobic conditions due to lack of oxygen and excess of water and anaerobic micro niches present within a larger aerobic system, excess of organic matter, higher temperature and slight acidic to neutral pH enhance reduction of sulphur. At the same time use of nitrate, ferric and manganic salts lower the reduction. Sulphur reducing microorganisms also transform pipes and wires of iron buried in moist soils into black coloured ferrous sulphide. During this action corrosion of these pipes and wires takes place and they are destroyed.

Biochemistry of Inorganic Sulphur Reduction

Similarly, to the use of nitrates during oxidation of carbon by denitrifying bacteria, sulphur reducing bacteria use oxygen of sulphate. The sources of energy to perform this process are various carbohydrates, organic acids and alcohols. Mainly two enzymes namely ATP sulphurylase and dissimilatory sulphite reductase perform the process of inorganic sulphur reduction. The sulphate is activated before it can be used as an electron acceptor. The activation is done by the enzyme ATP sulphurylase.

ATP and sulphate form APS (Adenosine pentaphosphosulphate). This APS is reduced to sulphite which is further reduced to sulphide. The whole process requires two molecules of the ATP. The enzyme dissimilatory sulphite reductase catalyses the last step sulphate reduction. Also, few sulphate reducing microorganisms perform the anaerobic oxidation of Methane. However, the simpie equations of this process are given here:

$$2CH_3CHOHCOOH + SO_4^{2-} \rightarrow 2CH_3COOH + H_2O + S^{2-}$$

Lactic acid Sulphate Acetic acid water Sulphide

$$2CH_3CHOHCOONa + MgSO_4 \rightarrow 2CH3COONa + MgCO_3 + H_2O + H_2S$$

Sodium lactate Magnesium sulphate Sodium acetate Magnesium carbonate water Hydrogen sulphide

$$CH_4 + SO_4^{2-} \rightarrow HCO_3^- + HS^- + H_2O$$

Methane Hydrosulphide

Inorganic sulphur reducing microorganisms are considered to be among the oldest forms of microbes, Group of sulphate reducing microorganisms or sulphate reducing prokaryotes consists of sulphate reducing Bacteria and sulphate reducing Archaea. These can perform anaerobic respiration utilizing sulphate as terminal electron acceptor and reduce sulphate to hydrogen sulphide. Microorganisms which reduce sulphate into sulphide are completely anaerobic. Microorganisms are in two genera named *Desulphovibrio* and *Desulphotomaculum*. These both bacteria are found in abundance in lands with standing water and mud.

Desulphovibrio desulphuricans grow in profusely in slight acidic soil (pH 5.5). So in such soils there is more accumulation of sulphide. It is also supposed that *Desulphovibrio* deposits elemental sulphur after formation of sulphate from sulphide.

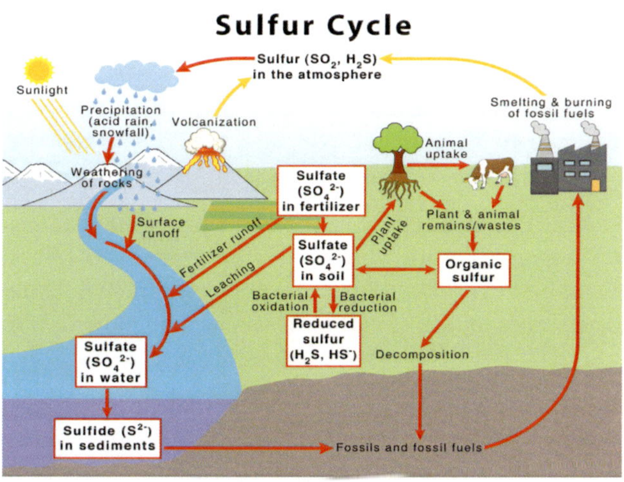

Fig. 3.60: Sulphur Cycle

Cycling of Micronutrients

Micronutrients present in soils are transformed by microorganisms and control their availability or unavailability for plants. In this context release of micronutrients by decomposing organic matter and mineralization of them, assimilation and immobilization of micronutrients for cell growth by microorganisms, oxidation and reduction of micronutrients and chelate formation of micronutrients with biochemical matters are main activities to be done by microorganisms. Organic acids and polyphenols are produced during various microbiological decompositions in soil. These substances may form several types of complexes with micronutrients. These complexes slowly and slowly supply micronutrients to plants as per their need.

Plants also require micronutrients for normal growth and development as well as their involvement in various enzymes and other physiologically active molecules. Micronutrients are important for gene expression, biosynthesis of proteins, nucleic acids, formation of growth substances, chlorophyll and secondary metabolites alongwith metabolism of carbohydrates and lipids.

Like higher plants and other organisms, microorganisms also need micronutrients for their nutrition. If enough quantities of micronutrients are not present in soil, then they will utilize micronutrients released by mineralization process. Later, after the death of microbial cells the micronutrients will be mixed again in soil. Thus, like other elements processes of mineralization and immobilization of micronutrients also go on simultaneously and continuously. Here cycling of some micronutrients is briefly described.

Cycling of Iron: Oxidation and reduction of iron are two major activities in soils. The biological cycling of Fe^{2+} is done by iron oxidizing and reducing microbes. On account of this activity availability of elements is reduced. Reduction of iron also occurs in anaerobic conditions. Generally, oxidation of iron is performed by *Thiobicallius ferrooxidans* that converts ferrous in ferric.

The cycling of iron are also abiotic processes. These include rusting of iron in which $Fe^{2}+$ is abiotically oxidized to $Fe^{3}+$ in the presence of oxygen and the reduction of $Fe^{3}+$ to $Fe^{2}+$ by iron sulphide minerals.

Cycling of Manganese: Manganese occurs as two types of ions namely manganous (Mn^{2+}) and manganic (Mn^{4+}). Manganous is only found in the soil at acidic pH (less than 5.5). After oxidation at alkaline pH (about 8.0) it is converted in Manganic oxide. Plants can use only divalent manganese and tetra valent manganic is not useful for plants.

Cycling of Zinc: Zinc is essential to life and perform an important role in biological processes of all living organisms. Several soil bacteria such as *Pseudomonas, Bacillus, Gluconacetobacter, Burkholderia, Acinetobacter, Serratia, Flavobacter* and *Enterobacter* are already reported as zinc solubilizing bacteria.

https://www.researchgate.net/publication/215600713_Microbial_Zinc...

Cycling of Copper: Copper (Cu) is a micronutrient and somewhat similar to other nutrients it is also immobilized by microorganisms, it is mineralized before available for plant uptake, then taken up by plants or it may exchange on soil particle surfaces. The immobilized Cu in plants can be remobilized during Cu deprivation. Plant Cu acquisition, transportation, and remobilization for growth are managed by several Cu-uptake proteins.

Soil Organic Matter

The mineral matter, organic matter, water and air are four components of the soil. Soil organic matter is any material produced originally by living organisms (plant or animal) that is returned to the soil and goes through the decomposition process. At any given time, it consists of a range of materials from the intact original tissues of plants and animals to the substantially decomposed mixture of materials known as humus. Organic matter is also the residues of plants and residues and wastes of the animals. All these deposit on the surface of the soil or mixed in the soil by various activities and processes. Soil organisms turn these residues and wastes from organic forms to the inorganic forms. The inorganic forms are again utilized by the plants or by some micro-organisms. Thus, soil organic matter consists of fresh to decomposed residues and wastes of plants and animals, cells and tissues of soil organisms and substances synthesized by them.

Sources of soil organic matter: Plant, animals and organic manures are three sources of organic matter in soils. Plants are major and primary source of soil organic matter. Plants add organic matter in to soils by their roots, shoots, twigs, branches, leaves etc. Grass clippings, straw and stubble of crops and vegetable peelings provide organic matter to the soils. Animals are supposed as secondary source of organic matter in soil. Numerous soil organisms like insects, millipeds, nematodes, microorganisms etc. provide a considerable amount of organic matter to the soil. Farm yard manure, compost, vermicompost, green manure, fish meal, oil cakes, cocopeat, paper clippings etc. are the good source of organic matter in the soil.

Influence of organic matter on soil properties: Organic matter beneficially influences the physical, chemical and biological properties of soil.

1. Influence on physical properties: Organic matter increase soil aggregation by binding sand, silt and clay particles. It reduces the destruction of soil aggregates by water. Organic matter improves soil structure to granular which increases porosity, aeration and permeability. It reduces soil erosion by increasing water infiltration, percolation and water holding capacity of soil. Organic matter can absorb water up to 80-90% of its weight whereas clays absorb water up to 15-20% of their weight. Therefore, soil rich in organic matter have higher water holding capacities.

Organic matter lying on the soil surface in form of plant residues acts as a mulch which forms a physical barrier to wind and water erosion. This mulch also serves as an insulator and helps in the maintenance of soil temperature. Organic matter reduces the stickiness and surface crusting of soils to make them easier for seedbed preparation.

2. Influence on chemical properties: Soil organic matter increase the soil's CEC. It retains and supply essential nutrients such as calcium, magnesium, potassium etc. as and when required by crops. Soil organic matter holds 10 to 1,000 times more nutrients as compared to same amount of soil minerals. It improves the buffering capacity of a soil to resist pH change. In long run it accelerates decomposition of soil minerals to convert their nutrients in the plant's available forms. In acid soils organic matter turns soil P readily available to plants and organic acids produced during organic matter decomposition reduces soil alkalinity.

Humic acid, fulvic acid and other intermediary compounds produced during organic matter decomposition in the soil also hold on and protect certain nutrients particularly metallic elements by making their chelates. Chelate word is derived from a Greek word Chela meaning Claw. Chelate particles are smaller than the humus particles. In general, chelates hold elements more strongly as compared to the binding by positive and negative charges. Phosphorus and other micronutrients like iron, zinc, copper and manganese form chelates with organic ions. Nutrients present in chelated forms are not fixed in soil and they are slowly available to plants.

Chelates made with amino acids, simple aliphatic acids and fulvic acids are more soluble as compared to chelates made with phenolic, alcoholic, enolic groups and humic acids are insoluble or less soluble. In the chelate compounds central metallic ion is called as particle and attached compound with this ion is called as ligand. The figure of simple chelates of copper and iron with amino acid are given here:

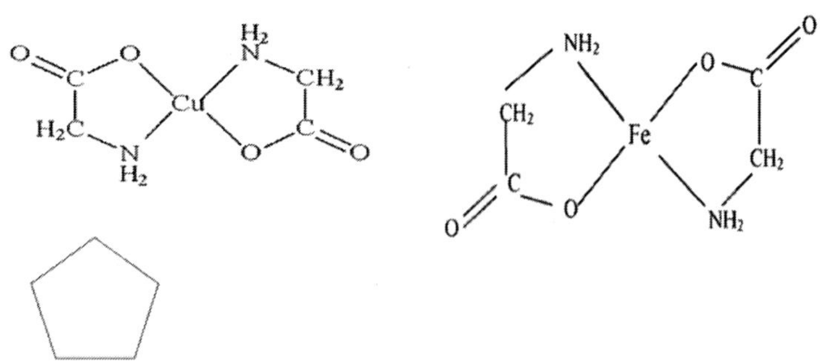

Fig. 3.61: Copper and iron chelates

3. Influence on biological properties: Soil organic matter is the food for the living organisms in the soil. Organic matter serves as a source of energy for the microbes. It is a reservoir of nutrients, hormones, antibiotics etc., which are essential for plant and different soil processes. Fresh Organic matter is the food for earthworms, ants, rodents etc. It enhances soil microbial biodiversity and activity which reduces diseases and pests. It increases pore space through the actions of soil microorganisms. This enhances infiltration and reduces runoff.

Composition of soil organic matter: A very wide literature is available on soil organic matter emphasing various aspects of its composition etc. Soil organic matter consist much diversified components in different proportions and stages of decomposition. These components are, 1. Raw plant residues and microorganisms (1 to 10 %) 2. Active organic matter (10 to 40 %) and 3. Resistant or stable organic matter i.e., humus (40 to 60 %). However, soil organic matter is mainly composed of 1. Dissolved organic matter 2. Particulate organic matter 3. Humus and 4. Resistant organic matter. These are described here:

1. Dissolved organic matter is Generally, less than 5% of total soil organic matter and its size is less than 45µm in solution. Its reverse time is many minutes to few days. It consists of soluble root exudates, simple sugars and decomposition by-products.

2. Particulate organic matter is 2–25% of total soil organic matter. Its size is 53µm–2mm.Its reverse time is 2–50 years. It consists of fresh or decomposing plant and animal residues with identifiable living organisms and cells.

3. Humus is 50-65% of total soil organic matter. Its size is less than 53µm. Its reverse time is 10s to 100s years. It is made up of older, decayed organic compounds that resist to decomposition.

4. Resistant organic matter is up to 10% of soil organic matter. Its size varies from less than 53µm to 2mm. Its reverse time varies from 100s to 1000s years. Relatively inert material, chemically resistant materials and organic remnants like charcoal come in this fraction.

Compositions of plant residues directly influence the composition of soil organic matter. The water content of most plant residues is in the range of 60% to 90% which is around 75% moisture and 25% dry matter. The dry matter is made up of Carbon (10-12%), Oxygen (9-10%), Hydrogen (1.5-2.5%), N (1-2%) and mineral matter (1-3%). The mineral matter comprises of phosphorous, sulphur, potassium, calcium, magnesium, micronutrients etc.

Plant tissues, the main source of soil organic matter, contain Carbohydrates (Celluloses 20-50%, Hemicellulose 10-30%, Starch, and Sugar 1-5%), Proteins

1-15%, Fats, waxes, tannins 1-10%, Lignins 10-30%, Inorganic residues (mineral matter) etc.

Decomposition of soil organic matter: Various organic compounds of plants and animal residues are continuously under decomposition. At one end fresh organic residues are added in soil and simultaneously soil microorganisms decompose and convert them in new forms. Organic natter decomposition is enzymatic biochemical process that includes the physical breakdown and biochemical transformation of complex organic molecules into simpler organic compounds and inorganic molecules, elements and ions.

Processes of soil organic matter decomposition: This is a gradual and complex oxidative process which starts with the flourishing of different heterotrophic and saprophytic microorganisms on the dead and decayed organic residues. In this process both chemical and biological processes take part to change organic residues into humus. The first process is enzymatic oxidation which produces carbon dioxide, water and energy. Simultaneously a series of specific reactions works to release, mineralize and immobilize the nutrient elements. Finally. the compounds formed through the transformation of the original compounds or newly synthesized by microbes produce humus.

First of all, microbes flourish or grow well on the materials which can quickly provide ample amounts of carbon and nitrogen for their rapid growth and development. Normally simple sugars, amino acids and simple proteins can quickly provide ample amounts of carbon and nitrogen. Hence, at the start these are decomposed first of all. In the initial stage of decomposition bacteria and fungi are more effective and later actinomycetes start work. Maximum decomposition is attained within 24-48 hours if optimum conditions of moisture, nitrogen, pH, aeration and temperature are available in soil.

Production of carbon di oxide gas during the decomposition is the indication of the rate of decomposition. Sometimes organic acids are also produced which are harmful for plants under anaerobic conditions. Meanwhile rate of decomposition increases due to increase in the amount of some enzymes. Now autooxidation and polymerization of phenolic compounds starts. Microorganisms start to increase their population tremendously by absorbing nutrients particularly carbon and nitrogen. Thus, release, mineralization and immobilization of nutrients of organic residues go on. Ultimately several normal and specific biochemical processes produce a dark brown to black coloured amorphous substance called as humus.

The decomposition takes place from fast decomposable rate organic compounds to slow decomposable rate organic compounds as given here

1. Very fast decomposable: Sugars, starches, simple proteins etc.
2. Fast decomposable: Crude Proteins, hemicelluloses, celluloses etc.

3. Slow decomposable: Lipids (Fats, oils etc.).
4. Very slow decomposable: Waxes, resins, lignins etc.

Decomposition of individual organic compounds: Numerous organisms present in soil consume plants, living organisms, dead and decaying plant or animal residues and wastes including excreta. Each organism plays a different role in organic matter decomposition. Larger soil organisms shred organic material creating greater surface area and reduce the size of organic material. This helps soil microorganisms to start decomposition of soil organic matter. Here description of decomposition of individual organic compounds is given:

Sugar: Sugars are the most abundant organic compounds and they are monomers of all polysaccharides. Sugars are decomposed by microorganisms. The process starts with the breakdown of sugars into glucose monomers. The glucose is decomposed in aerobic as well as in partially oxidized conditions. In the aerobic condition the reaction is as:

$$C_6H_{12}O_6 + 2O_2 = 2CO_2 + 2H_2O$$

Sugar + Oxygen = Carbon dioxide + Water

In partially oxidized conditions glucose is decomposed in to various acids like acetic, formic, citric, lactic or ethyl alcohol etc. The reactions may be represented as:

$$C_6H_{12}O_6 + 2O_2 = 2CH_3COOH + 2CO_2 + 2H_2O$$

$$C_6H_{12}O_6 + 3O_2 = C_6H_8O_7 + 4H_2O$$

$$C_6H_{12}O_6 + 2O_2 = 2C_2H_5OH + 2CO_2$$

Starch: Starch is the reserve food of plants and is present in tubers, bulbs, rhizome, seed etc. In starch hundreds of glucose molecules are bound together by α-1,4-glucoside bonds. Amylose and Amylopectin are two components of starch. There is little difference in these two. In amylopectin, besides the α-1,4- glucoside bonds of glucose molecules, side chains are attached through α-1,6- glycoside bonds.

Aerobically, the microbes in the presence of amylase enzyme convert starch in to maltose. Maltose is converted in to different low molecular weight organic acids by the enzyme α-glucosidase and in the form of pyruvic acid it goes to TCA cycle. Starch is converted in to methane, acetic acid, lactic acid, butyric acid etc. under anaerobic conditions. Starch decomposing microorganisms are *Micrococcus, Flavobacterium, Pseudomonas* and *Clostridium* (bacteria) *Fusarium, Aspergillus, Molds* and *Trichoderma* (fungi) and *Nocardia, Streptomyces* and *Micromonspora* (actinomyces) etc.

Proteins: Proteins are very large and complex molecules composed of one or more twisted and folded strands of basic units called as amino acids. Amino acids contain nitrogen, carbon, hydrogen and oxygen. Proteins are the structural components of plants and other living organisms. The human body can biosynthesise some amino acids. The microorganism's breakdown and decompose proteins into polypeptides and later on into simpler amino acids by Proteolysis process. Proteases enzymes are responsible for the breakdown of proteins. This process does not occur at a uniform rate. Therefore, some proteins are degraded during early decomposition, while others are degraded during later stages of decomposition. Proteolysis is influenced by moisture, temperature and microorganisms.

These amino acids are directly utilized by the microorganisms for their growth and development and also are a substrate for other organisms. Aerobically proteins are decomposed to ammonia, carbon dioxide, various organic acids, and alcohol. Anaerobic decomposition produces various amino acids and other compounds with foul smell. The crude and residual proteins of soil are hydrolysed by numerous microorganisms.

Proteolytic organisms are mostly bacteria which decompose protein. The examples are *Pseudomonas, Bacillus, Clostridium, Propionibacterium, Streptococcus, Staphylococcus* etc.

Hemicelluloses: Hemicelluloses are constituent of the plant cell wall and make 20 – 30% of the dry weight of wood. These are water soluble polysaccharides consist of β-(1,4)-linked structures. The sugars found in hemicellulose are glucose, xylose, galactose, mannose, rhamnose and arabinose and their acids like glucuronic acid and galacturonic acid. Hemicellulose cross-links with either cellulose or lignin to strengthen the plant's cell wall. The composition of hemicellulose in softwood and hardwood is different. The softwood mainly contains mannose, galactose and lignin while the hardwood mainly consists of xylan and acetyl.

A group of hemicellulases enzymes (L-arabinanases, D-galactanases, D-Mannanases and β-D-xylanases) take part in the microbial decomposition of hemicellulose. Most of the hemicellulases are extracellular. These enzymes convert hemicellulose in to different pentoses and hexoses.

Hemicellulose decomposing microorganisms are *Clostridium, Bacillus, Acetobacter, Sporocytophaga*, *Streptococcus spp.* etc. (bacteria) and *Fusarium, Aspergillus, Alternaria, Botrytis* and *Trichoderma* (fungi) and *Streptomyces* (actinomyces) etc.

Microbial Degradation of Hemicellulose (Enzymes, Steps)

https://microbenotes.com/microbial-degradation-of-hemicellulose

Cellulose: Cellulose is the most abundant carbohydrate in the biosphere in which many D-glucose molecules linked together in a linear chain of β-1,4- glucoside linkage. The mechanical strength of the plant cell is attributed to the structural properties of cellulose as it can retain a semi-crystalline state of aggregation. Cellulose occurs in the cell walls of plants to provide their structure. Cellulose is a source of food to a wide variety of organisms except human. Human cannot digest cellulose but it is the source of fibre for them.

Enzyme Cellulases decompose it in to molecules of glucose. Cellulases consist of three enzymes, exo- β -1, 4-glucanases, endo- β -1, 4-glucanases and β -1, 4-glucosidases. These enzymes act in a sequence and synergistic system to decompose cellulose in to glucose. Under aerobic conditions there is complete oxidation of cellulose. It decomposes in to glucose. Glucose is converted in Carbon dioxide and water as: $C_6H_{12}O_6 + 6O_2 = 6CO_2 + 6H_2O$. Glucose is the main source of carbon for many microorganisms. In the anaerobic conditions there is incomplete oxidation and cellulose is decomposed in to various acids like acetic, formic, butyric, lactic and ethyl alcohol along with carbon dioxide gas.

Cellulose decomposing or cellulolytic microorganisms are mainly and bacteria and fungi. *Bacillus, Pseudomonas, Vibrio, Acinetobacter, Cellulomonas* and *Clostridium* are bacteria, *Aspergillus, Fusarium, Trichoderma* and *Neurospora are* fungi and *Nocardia* and *Streptomyces are* actinomyces.

Lipids: The term lipid is derived from the Greek word “lipos” which is referred to animal fat or vegetable oil. But now term lipid has broad meaning which includes fats, oils, waxes, cholesterol, triglycerides etc. Lipids can be defined as fat, oil or wax that is soluble in alcohol but not in water. Lipids are an important part of living cells. Lipids are the main constituents of plant and animal cells along with carbohydrates and proteins.

Lipoxygenase and lipase are two main types of enzymes which decompose lipid. Lipoxygenase is common in the legumes. Lipase also works on the soil organic contaminants. It helps in the reduction of pollutants from the polluted soil by the processes like esterification, aminolysis, hydrolysis etc. Lipids are hydrolysed to glycerol and fatty acids by lipase. Glycerol is converted in to carbon di oxide and water.

The lipids decomposing microorganisms are: *Clostridium. Acinetobacter, Bacillus* etc.

Lignin: Lignin is one of the main components of plant cell wall. The term lignin is derived from the Latin word ‘lignum’ which indicate the relationship between lignin and coniferyl alcohol units. It is a natural phenolic polymer with high molecular weight, complex composition and structure. The primary building blocks of lignin may be hydroxycinnamoyl alcohols (or monolignols), coniferyl alcohol and sinapyl alcohol with very low amounts of *p*-coumaryl alcohol.

Lignin biosynthesis extensively contributes to plant growth and development, resistance to lodging and various biotic and abiotic stresses. The precursors of lignin synthesis in nature include p-coumaryl alcohol, coniferyl alcohol and sinapyl alcohol. Lignin is composed of phenylpropane units heterogeneously crossed linked together by the chemical linkages of alkyl-alkyl, alkyl-aryl and aryl-aryl groups.

Numerous transformations occur during microbial decomposition of lignin. These are very much complex and incompletely understood. Ligninases enzymes namely Manganese peroxidases (MnP), lignin peroxidase (LiP), versatile peroxidise and laccases take part in lignin decomposition. These enzymes attack the lignin and convert the phenolic group to free-radicals and these radicals lead to lignin depolymerization. First of all the depolymerization of aryl and biaryl compounds takes place. This decreases the length of the lignin polymer and later on decomposed into smaller molecules. The depolymerized smaller molecules of lignin are used by different microorganisms. Finally, the CO_2 and other essential molecules are formed due to mineralization processes and simultaneously these are utilized by the microorganisms.

Lignin is most resistant for microbial decomposition and microorganisms also require hydrogen peroxide (H_2O_2). Certain bacteria, actinomycetes and fungi of the white-rot group and other completely decompose lignin in to carbon dioxide and water. However, there is incomplete decomposition also. Different microbial population dominates at various stages and has distinct roles in the decomposition of lignin. Bacteria usually act on lignin along with fungi.

Among the lignin decomposing microorganisms in soils, bacteria are *Mycobacterium, Pseudomonas, Caulobacter, Rhodobacter, Yersinia, Campylobacter* etc. The white rot fungi of Ascomycetes, Deuteromycetes and Basidiomycetes groups particularly *Ganoderma* and *Heterobasidion, Rhizoctonia, Aspergillus, Neurospora* and *Trichoderma, Fomes, Agaricus, Armillaria, Polyporous* etc. also decompose lignin. Actinomycetes, *Streptomyces* etc. also decompose lignin.

End products of soil organic matter decomposition: Organic matter produces CO_2, NO_3^-, $H_3PO_4^-$, SO_4^{2-} and water etc. under aerobic conditions whereas under anaerobic conditions due to incomplete oxidations non-beneficial organic acids, CH_4, NH_4^+, H_2S etc. are produced. End products of some elements present in soil organic matter decomposition are given here:

Element	End product
Carbon	CO_2, CO_3^{2-}, HCO_3^-, CH_4, C
Nitrogen	NH_4^+, NO_2^-, NO_3^-, N_2 (gas), N_2O (gas)

Phosphorus	$H_3PO_4^-$, $H_2PO_4^-$, HPO_4^{2-}
Potassium	K^+
Calcium	Ca^{2+}
Magnessium	Mg^{2+}
Sulphur	SO_4^{2-}, SO_3^{2-}, CS_2, H_2S, S
Others	H_2O, O_2, H_2, H^+, OH^- etc.

Factors Affecting Organic Matter Decomposition

A team of macro to microorganisms are always present in soil and factors affecting all these also affect soil organic matter decomposition. Some important factors are described here

1. **Presence of organisms in soil and their activities:** This is most important factor of soil organic matter decomposition. Many bacteria, fungi and other organisms feed on the organic matter of the soil. Soil organisms with the high level of activity decompose fast and more organic matter. Any factor affecting the microbes and other living organisms in soil will also affect the rate of organic matter decomposition.

2. **Temperature:** The rate of decomposition is more rapid in the temperature range of 30° to 40°C. At temperatures below or above this range, the rate of decomposition is markedly retarded because most of the organisms are mesophiles. Appreciable organic matter decomposition also occurs at 25° C. Soil organic matter decomposes faster in the summer. Areas with whole year warm climate will decompose more organic matter annually than areas with prolonged cold winters. Similarly organic matter decomposition is much higher in tropical areas and takes only one fourth times as compared to temperate areas. In the cold zones there is less organic matter production and decomposition. Hence there is more organic matter accumulation.

3. **Soil moisture:** Soil moisture affects the decomposition of organic matter directly and indirectly. A too dry soil directly decreases organic matter decomposition as all the organisms present in soil cannot survive without water. Indirectly, a wet soil causes slower decomposition because water is filled in all the pores of the soil. This reduces air and oxygen supply to the microbes and only anaerobes can decompose. Therefore, adequate soil moisture should be present in soil for the proper decomposition of organic matter. This may be slightly more than the field water-holding capacity of the soil. Alternate wetting and drying of soils like irrigation of orchards and gardens enhances organic matter decomposition as aerobes and anaerobes work simultaneously.

4. **Aeration and oxygen supply:** Organic matter decomposition is basically aerobic process. Therefore, good aeration is necessary for the proper activity of the organic matter decomposing microorganisms. Under anaerobic conditions decomposition is very slow and incomplete. About 70% less decomposition takes place in anaerobic condition as compared to aerobic condition.

   Oxygen is required by the microbes to decompose organic matter efficiently. Atmospheric oxygen enters in the soil. So, the easy entry of air into the soil is utmost essential. Soil with a loose friable structure allows more and easy entry of air in to the soil for quick decomposition of organic matter. Oxygen diffusion rate (ODR) is about ten thousand times less in water as compared to air. So little bit more wet condition significantly reduces the organic matter decomposition. It is supposed that about 5% oxygen level is good for decomposition.

5. **Size of organic matter particles:** Smaller, pieces of organic matter increase its decomposition. The smaller sized organic particles are easily and quickly utilized by the microbes. A shredder is useful for chipping or shredding most landscape refuse and is essential if brush or sticks are to be composted. A low-cost method of reducing the size of fallen tree leaves is to mow the lawn before raking. Wind-rowing the leaves into long narrow piles one foot high will make the shredding process more efficient. If the mower has an appropriate bag attachment, the shredded leaves can be collected directly. However, grinding is entirely optional.

6. **Soil pH:** The rate of organic matter decomposition is highest in neutral soils. Slightly low pH in acidic range greatly reduces organic matter decomposition as compared to slightly high pH in alkaline range. Mostly 6 to 8 pH is optimum for organic matter decomposition which is severely reduces below 4.5 and above 8.5 pH.

7. **Quantity of clays in soil:** Soils with high clay content do not provide easy access for air accumulation in soil. This causes lesser activity of microbes. This results in slower and less organic matter decomposition. There is less organic matter decomposition in fine textured (clayey) soils as compared to coarse (sandy) textured soils. Lesser organic matter decomposition increases organic matter accumulation. Black colour of black cotton soils of India is also due to more adsorption of organic matter by them.

   Higher adsorption capacities of clays reduce availability of organic matter for microbial action. Clays also adsorb intermediary compounds formed during decomposition to restrict or slowdown the further decomposition processes. Decrease in organic matter decomposition is more in soils with montmorillonite as compared to soils with kaolinite or illite. This is due to higher CEC of montmorillonite which adsorbs more materials. Organic matter present in between two layers of montmorillonite remains ineffective for microbial activities for longer periods. Hence late and lesser decomposition takes place.

8. **Amount of nutrients in soil:** Organic matter decomposing microorganisms also require nutrients for their growth and development. More amounts of carbon, phosphorus and nitrogen are required by them. Microbes require nitrogen for the synthesis of amino acids and proteins required to form new cells. However, too much high N supply can also reduce the decomposition rate.

9. **C: N ratio of organic matter:** C: N ratio is defined as the ratio of the mass of organic carbon to the mass of total nitrogen in a soil or organic matter. It is the relationship between organic carbon and nitrogen content of soils, plants, microbes, sediments, humus, compost etc. There are approximately fixed C: N ratios of different organic materials. C: N ratios of different organic materials are given here:

Table 3.16: C: N ratio of different organic materials

Description	C: N ratio	Description	C: N ratio
Bacteria	5:1	Forest litter	30:1
Actinomycetes	6:1	Green rye	36:1
Humus	10:1	Maize stalk	40:1
Fungi	11:1	Sugar cane bagasse	50:1
Soil	11:1	Wheat straw	80:1
Alfalfa	12:1	Rice straw	100:1
Cow dung	19:1	Rye straw	200:1
Clovers	20:1	Saw dust	350:1

The C: N ratio, 10:1, is generally, used for the calculation purposes of most agricultural soils. Organic matter from various plant materials has very wide C: N ratio, say, 20-200:1 whereas C: N ratio of microbes is quite narrow (5-11:1). The wide C: N ratio means that organic materials have more carbon and less nitrogen. Similarly narrow C: N ratio indicates less carbon and more nitrogen.

Suppose C: N ratio of microbes present in soil is 8:1 i.e., nine kg microbes have one kg nitrogen and C: N ratio of wheat straw is 80:1 i.e., eighty-one kg wheat straw also has one kg nitrogen. It means nitrogen content of the wheat straw is about eleven times lesser than the microbes. If the wheat straw is added in to the soil, then the population of straw decomposing microbes will start to increase and they will consume the nitrogen present in straw. But due to narrow C: N ratio, microbes require much more nitrogen during decomposition which will not be provided by wheat straw.

Therefore, to overcome the nitrogen deficiency microbes will utilize previously present nitrogen in soil to synthesize their N containing cellular amino acids, proteins, enzymes, DNA etc. Thus, the nitrogen which should be available to plants is now unavailable to them and they will suffer from nitrogen deficiency. If the soil is already deficient in nitrogen, then the supply of nitrogen to microbes will be reduced. This will reduce the rate of organic matter decomposition and this will only depend on the nitrogen present in wheat straw. Now it is clear that C: N ratio of added organic wastes or residues in the soil has an important place in their decomposition.

C: N ratio also regulates two important and opposite processes, mineralization and immobilization, of nitrogen availability in soils. Mineralisation refers to the microbial conversion of organic forms any element to inorganic or mineral or nutrient forms which are readily taken up by plants. NH_4^+, NO_2^-, NO_3^- are the mineralized forms of nitrogen. Immobilisation is the conversion of inorganic form of any element or compounds to organic forms by microbes or plants. This is opposite of mineralization. Nitrogen present in wheat straw and microorganisms is immobilized nitrogen.

There will be more mineralization of nitrogen if C: N ratio of organic materials added into the soil is less than 20:1. This will increase the amount of mineral nitrogen in soil. Amount of mineral nitrogen will decrease in soil, if C: N ratio of organic materials added into the soil is more than 30. Because there will be immobilization of nitrogen. There will be neither mineralization nor immobilization and no any change in nitrogen status of soil if C: N ratio is in between 20 to 30. The C: N ratio in the range of 20-24:1 is supposed to be best for maximum decomposition of organic matter because it maintains equilibrium between mineralization and immobilization processes.

Maintenance of Organic Matter in Soil

Traditionally since ancient times farmers understood the benefits of soil organic matter in agriculture. They efficiently know that how their soils are well supplied with the organic material. This is evidenced that ancient farmers mostly relied on crop residues, manures, legumes and neem for enriching soil fertility and soil organic matter. Many ancient literatures like Krishi - Parashara, recommended compost preparation from cow dung. The dried, powdered cow dung is placed in pit for decomposition. Kautilya mentioned the use of cow dung, animal bones, fishes, milk etc. as manure. Surapala described the ancient practice of preparing liquid manure (kunapa) prepared by boiling a mixture of animal excreta, bone marrow, flesh, dead fish in an iron pot and then add it to sesame oil cake, honey and ghee.

It is supposed that photosynthesis annually produces about 55 billion (b/bn) tons organic carbon in the whole world. If factor of 1.724 is multiplied with it then amount of organic matter produced will be about 95 b tons. Such a huge amount is added annually then all the soils were too much enriched with organic matter. But this is not possible because microbes always decompose it. Simultaneously increase in mean annual temperature remarkably decreases its quantity. Therefore, maintenance of a steady quantity of organic matter in soil is essential for productive farming. This may be possible when organic materials are added in to soil time to time to recover the losses caused by microbes and temperature.

Maintenance of organic matter in soil can be achieved by changing in agricultural practices from tillage up to crop residue management. Because in the whole world crop production has Generally, depleted soil organic matter status and adversely affected the soil fertility. The agricultural practices which can help in the maintenance of organic matter in soil are:

1. **Tillage:** Mould-board and disc ploughs cause rapid decomposition of soil organic matter and create plough pans below the ploughing depth. Minimum, reduced, zero or no tillage systems leave more residues on soil surface and found useful for maintaining and increasing soil organic matter. Direct seed drill sowing is also a good option allowing for leaving stubble residues on top of the soil. Organic matter decomposes less rapidly under reduced tillage systems. No tillage systems are also useful to increase and maintain soil organic matter and can change agricultural soils into carbon sinks.

2. **Crop cultivation:** Crop cultivation has an important role in the maintenance of soil organic matter. Cereal crops after harvest left remarkable amounts of organic matter in the soil as their roots and stubbles. Cereal crops due to their wide C: N ratio increase more organic carbon and less nitrogen in soil. Buried roots are self-mixed in soil and easily approachable for microbial activities as compared to stubbles lying on soil surface. This is true for all the crops also.

   Generally, leguminous crops like sun hemp, sesbania and cowpea are grown for green manuring. Green manure crops after ploughing into the soil increase soil organic matter. But due to their narrow C: N ratio they are very rapidly decomposed to give rise slight increase in organic carbon as most of the carbon is escaped in to atmosphere as CO_2 gas. But they considerably increase nitrogen status of soil. Regular application of green manure will increase soil organic matter up to some extent but build-up of soil organic matter will depend upon climate, tillage etc.

Cover crops, perennial crops, inter cropping and crop rotations also help to maintain soil organic matter. Perennial grassy pasture as well as annual grasses can effectively increase soil organic matter as perennial grasses provide root and leaf litter and annual grasses are the source of roots. The wide C: N ratio and low nitrogen of grasses slow down their decomposition. This helps to maintain organic matter for longer periods in soils. Inter cropping and crop rotations also maintain organic matter in soil up to their extent.

3. **Broadcasting of manures:** Frequent broadcasting of bulky organic manures also increase and maintain soil organic matter. Long term applications of different organic manures increased soil organic carbon from nine to 388 percent over control in different soils.

4. **Application of organic fertilisers and biochar:** Application of organic fertilisers in large amounts can maintain soil organic matter status. But small quantities of organic fertilizer do not have any significant effect on soil organic matter status. Biochar is charcoal that is produced by pyrolysis of biomass in the absence of oxygen. Biochar can very stably store carbon in the land for centuries. Thus, it can maintain soil organic matter also.

5. **Use of concentrated organic manures:** Some most common concentrated organic manures are mustard cake, groundnut cake, sesame cake, cotton seed cake, neem cake, mahua cake, karanj cake etc. and blood-meal, fish manure, bone meal etc. These manures decompose very slowly. Therefore, these maintain organic matter in soil for longer periods.

6. **Crop residue management:** A judicious residue management system should be adopted for maintaining organic matter status of soil. Crop residue management practices should be utilized for annually produced substantial quantities of crop residues of rice, wheat, sorghum, pearl millet, maize etc. to maintain organic matter of soil, Generally, crop residues are managed by adopting their removal from field, burning in the field and incorporation in the soil. Among these incorporation in the soil is the best practice. Mulching of crop residues is also a good practice for maintaining soil organic matter.

4

Vulnerability of Soils

Vulnerability of Soils

Soil vulnerability, Soil degradation, soil vulnerability index

Water and wind erosion, depletion of soil organic matter, loss of plant nutrients, salinity, alkalinity, acidity, pollution and contamination, compaction, sealing, crusting, water-logging, aridity, lowering of water table, over grazing, deforestation, landslide, loss and change in soil biodiversity, Abandoned mines etc.

Soil vulnerability, Soil Degradation, Soil Vulnerability Index

The terms vulnerable and vulnerability are now bites of every one and used for everything. Vulnerable means to the possibility of attack or harm by any mean. Human beings are vulnerable to Covid-19, computers are vulnerable to virus, softwares are vulnerable to hackers and attackers to control system, environment is vulnerable to pollution. Vulnerabilities refer to the lack of ability to resist the effects of the aggressive environment. Soils are vulnerable to numerous threats like water and wind erosion, depletion of soil organic matter, loss of plant nutrients, salinity, alkalinity, acidity, pollution and contamination, compaction, sealing, crusting, water-logging, aridity, lowering of water table, over grazing, deforestation, landslide, loss and change in soil biodiversity etc. However, this book will describe some of the most pertinent vulnerabilities for soils.

Soil vulnerability: Soil vulnerability is defined as the degree and capacity of one or more ecological and environmental factors or stress to which soils are likely to be harmed. It is a complex concept which requires the identification of multiple factors and land management systems at different temporal and spatial scales.

Soil degradation: Soil degradation is defined as a change in the soil health status resulting in a diminished capacity of the ecosystem to provide goods and services for its beneficiaries.

Soil vulnerability index: The Soil Vulnerability Index (SVI) is an index proposed by the USDA NRCS to identify areas that are inherently vulnerable to runoff-induced sediment and nutrient loss and to leaching-induced loss of nutrients in percolation and subsurface flows. So, this is mainly related with soil erosion by water.

Human is living in two worlds. First world is of soil, water, air, plants and animals. Nature freely provided this world to him. Second world is of social, economic and political institutions made by themselves. Needs, greed, rules and regulations of this second world are consumption oriented. We are consuming and utilizing natural resources of first world in such a way that our standard of living is uplifted tremendously but qualities of environment are reduced remarkably. Agriculture also does not remain untouched. On the account of needs of development and pressure of ever-increasing population, utilization of land into non- agricultural works is continuously increasing. Consequently, land available for agriculture is shrinking.

Intensive agriculture is one solution to meet the requirements of increasing population and decreasing land. In intensive agriculture it is unavoidable to use more irrigation, more fertilizers, more plant protection chemicals, and improved seeds etc. Irrigation is essential for crops, but it may also produce salinity and alkalinity in soils. Application of fertilizers increases yields, but without soil testing unscientific use of these pollute soil and underground water. Same condition is with insecticides etc. Bhopal gas tragedy and presence of D.D.T. residues in mother's milk are also ill effects of insecticides. When modern and intensive agriculture has significantly contributed to solve problem of food grain scarcity, it has also degraded soil.

Soil degradation is the decline in soil condition caused by its improper use or poor management, usually for agricultural, industrial or urban purposes. It is a serious environmental problem. Soil degradation is defined as a change in the soil health status resulting in a diminished capacity of the ecosystem to provide goods and services for its beneficiaries. Generally, soil degradation is linked with land degradation. Land degradation is the temporary or permanent lowering of the productive capacity of land due to physical, chemical or biological factors.

It covers the various forms of soil degradation, adverse human impacts on water resources, deforestation, lowering of the productive capacity of rangelands etc. Globally, about 25 percent of the total land area has been degraded. Scientists recently warned that 24 billion tons of fertile soil was being lost per year, largely due to unsustainable agriculture practices. Globally, 3.2 billion people are affected by land degradation, especially rural communities, small holder farmers and the very poor.

In India Land degradation mostly hurts farmers and forest dwellers. More than half of the degraded land in the country is either rainfed farmland, responsible for the food security of the country, or forest land that offers the best defence. About 30 per cent of our geographical area is already affected by land degradation. It is definitely among India's most pressing environmental problems. Almost all Indian states have recorded an increase in degraded land in the past 15 years and the most rapid increase being noted in the biodiversity-rich north-eastern states. About

82% of India's degraded land lies in just nine states and UT namely Rajasthan, Maharashtra, Gujarat, Jammu & Kashmir, Karnataka, Jharkhand, Odisha, Madhya Pradesh and Telangana.

Recently National level land degradation mapping is taken up by ISRO along with partner institutions, under Natural Resources Census (NRC) mission of DOS/ ISRO, towards generating information on land degradation. Two cycles of land degradation mapping at 1:50,000 scale, for the time frame 2005-06 and 2015-16, using 23m resolution (multi-temporal & multi-spectral) IRS data. Regular updates and information on degraded land is essential for the Government to plan necessary measures to tackle degradation processes and to plan for reclamation programmes. The task involves adaptation of uniform classification standards along with common geospatial database framework for generating reliable land degradation information. Land degradation map of India (generated using LISS-III data of 2015-16) depicting major land degrading factors is given in Fig. 4.1.

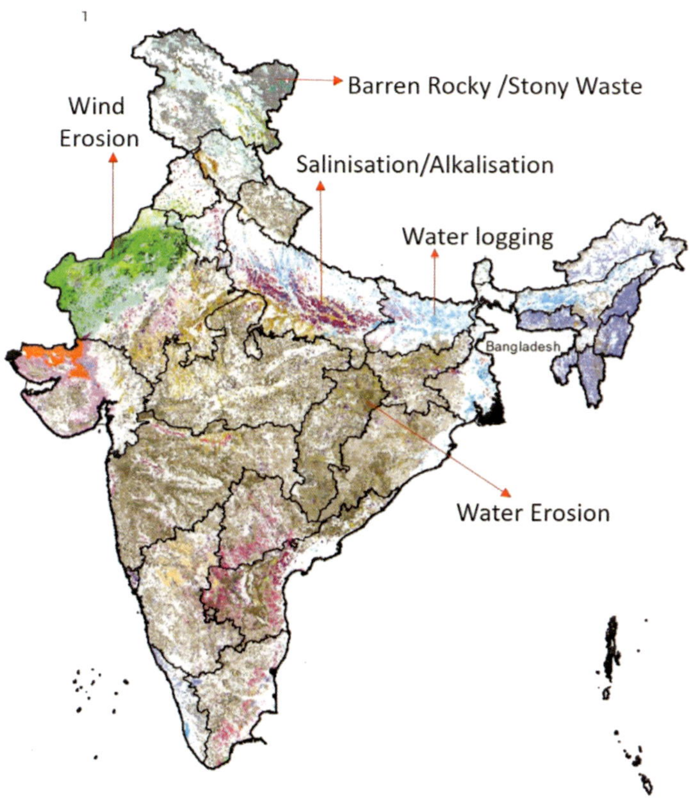

Fig. 4.1: Land degradation map of India (generated using LISS-III data of 2015-16) by ISROLand Degradation - ISRO
www.isro.gov.in/earth-observation/land-degradation

Based on severity or seriousness of factors causing soil vulnerability can be written in this order: water and wind erosion, depletion of soil organic matter, loss of plant nutrients, salinity, alkalinity, acidity, pollution and contamination, compaction, sealing, crusting, water-logging, aridity, lowering of water table, over grazing, deforestation, landslide, abandoned mines etc. Some important factors from these are being discussed here in context of Indian agriculture and soils.

Soil Erosion by Water and Wind

Soils are a fundamental natural resource, part of ecosystem and the basis for all terrestrial life. It is situated between atmosphere and hydrosphere. It is physical receptor of various substances and helps in the flow, transformation and up to some extent storage of energy on the Earth. Any type of alteration cannot be rectified easily. In the extreme undesired alteration, the soil can move to any other place from its original place. Formation of one centimetre soil layer takes hundred and thousands of years. But a violent blast of air or slightly heavy rain remove it very easily from its original place. Thus, efforts of years of factors and processes of soil formation wastes in the twinkling of any eye. Therefore, to conserve this invaluable legacy of the nature is very much necessary for pleasant future of coming generations.

Fig. 4.2: Soil loss showing open roots due to erosion and fire (Photo Dr. Venkatesh Bharadwaj)

Under an estimate of the Indian Institute of Remote Sensing (2015), that in India 147 million hectares land is degraded out of 329 million hectares of total geographical area. Under this broad figure, 94 million hectares are affected by water erosion, 16 million hectares by acidification, 14 million hectares by flooding and 9 million hectares from wind erosion. Twenty nine percent of the eroded soil is lost in the sea while 61 percent is just relocated. Also, another estimate was that nearly half area (1500 lakh hectares) of total 3280 lakh hectares of total geographical area is badly affected by water and wind erosion and out of which 690 lakh hectares area have been critically deteriorated by erosion including 320 lakh hectares by wind erosion. Gullies and ravines occupy 40 lakh hectares. Thirty lakh hectares land is deteriorating under shifting cultivation.

It is obvious from aforesaid data that problem of soil erosion in our country is very horrendous. Eventually, what is this soil erosion? Soil erosion is the removal of the upper surface of soil from its previous place by water, rainfall, wind or gravity. Soil erosion detach, transport and deposit soil particles at new place. Soil erosion is a work and for every work energy is required. Energy for the soil erosion is provided by agents of soil erosion. Agents of soil erosion are rain drops, flowing water, thawing snow and blowing air. These agents provide motion to soil. Denudation is another word similar to erosion. Denudation means formation of sediments in unit area of any watershed. Withdrawal of suspended solids and dissolved solutes in unit time from the outlet of watershed is called denudation rate.

Effect of Soil Erosion

Soil erosion is a serious problem. Its most common effect is physical movement of soil particles by water and wind and deposition of these in new places and reservoirs. This deposition may develop new soil surface. Soil formation process continues along with the natural erosion. Natural erosion has the coordination with soil formation and regeneration. But due to very much pressure on land, natural balance existed between soil formation and erosion is deteriorated and this imbalance produces serious soil erosion. Soil qualities are changed by erosion and productivity is adversely affected. In urban areas erosion not only damages natural environment of there and also spoils urban infrastructure of transport system (roads and rails). In this way beneficial effect of erosion is only deposition of soil material at new sites, otherwise its harmful effects are more.

Major effects of the soil erosion are the loss of soil of top soil, thereby making the soil unsuitable for the crop cultivation. Loss of water and nutrients are directly related with soil loss. The increasing soil loss also increase loss of water and nutrients. This ultimately deplete the fertility of the soil. Soil erosion change soil texture, structure and many other soil properties including water-holding capacity. Soil erosion can cause significant alteration to the natural relief of the land by making deep valleys of different shapes. The most famous examples of erosion include the Grand Canyon, which was worn away over the course of tens of millions of years by the Colorado River, Antelope Canyon, Arizona, USA, this rock structure is formed due to the erosion of the sandstone called Navajo. Rainwater, during floods, gushes through the basin area, causing the hard rock gaps underneath to smoothen. The immense speed of this abrasion gives it a glossy and wavy look.

Sedimentation of water, siltation of water bodies like rivers, reservoirs etc. due to soil erosion cause flooding. Excessive sediments in water bodies decrease their effective volume and carrying capacity. These make them overflowing to generate floods.

Decrease in yields is another ill effect of soil erosion. This may be due to uprooting of shallow rooted plants and crops, scarcity of water or washout of plant nutrients along with soil. Soil erosion particularly wind erosion can generate health problems. Blowing dust and finer soil particles can cause irritation in eyes and asthmatic problems when they enter into eyes and lungs. Brief description of these is given here:

Soil and water loss: When soil does not absorb rain water completely, it begins to flow on the soil surface as runoff. The runoff after detaching soil particles them and increase erosion loss of both soil and water. Transfer of soil by erosion may be from few centimetres to hundred kilometres according to runoff and wind velocity. Generally, after erosion coarse fraction is deposited quite earlier and fine fractions reach far away by flowing or flying.

Every year about 53330 lakh tons soil is destroyed by erosion in our country. You know that formation of a centimetre soil takes hundred to thousand years. If a centimetre soil is destroyed by runoff or violent wind, means that hard work of hundreds to thousand years of nature is destroyed. It is observed that a small stream having only 4 km drainage area can destroy about two-hectare land by flowing away about 15 thousand tons sand, gravels and stones. A study of Sukhana lake of Chandigarh having catchment area of forty kilometres indicated that 50 percent of rainfall is destroyed as runoff and about 50 tons sediments per hectare per year are deposited in catchment area. This type of loss of soil and water is a serious problem and efforts to stop this are too much necessary.

In hilly areas erosion causes the great loss to soil and water. The area which is devoid of vegetation and barren has much more runoff capacity and 30 percent of rain water is waste as runoff. Naturally more runoff causes more soil loss which may be up to 80 tons/hectare/year. Clayey soils of this area develop cracks on drying after rainfall. These cracks allow rapid flow of runoff water through them. The rapidly flowing runoff makes cracks wider and deeper by eroding their sides, surfaces and bottoms.

Loss of nutrients: Organic matter and nutrients are present in more quantities in the upper layer of soil. First, soil erosion erodes upper layer of soils means greatest loss of organic matter and nutrients. The loss of organic matter and nutrients by erosion has remarkable effect on soil texture, capacity and intensity of soil nutrient pool, soil fertility and soil productivity. Sometimes eroded material contains appreciably more plant nutrients than its original soil. In our country the annual loss of plant nutrients by erosion is about 54 to 84 lakh tons. Soil Conservation Secretariat of Utter Pradesh has reported (Table 4.1) that loss of nutrients also increases with increasing slope percentage.

Table 4.1: Effect of slope percentage on loss of nutrients

Percent of Slope	Loss of nutrients (Kg / ha)			
	Organic matter	N	P_2O_5	K_2O
0.5	8.6	0.7	2.3	11.7
1.5	13.9	1.4	3.8	17.1
3.0	46.0	4.6	9.0	43.1

Change in soil texture: Rain water very rapidly breaks and scatter unstable soil aggregates. This breaking and scattering separate sand, silt, clay, organic matter and other binding material present in aggregates. These separated matters go ahead floating with runoff water. Later on, these separated matters begin to settle down due to reduction in quantity and speed of runoff water. First of all, coarse fractions settle and fine fraction and organic matters move far away floating in the water. Clay fraction remains in suspended condition up to sufficient distance. Thus, after erosion, separation of soil particles of any one texture generates many soil textures like sandy at nearer place, then silty at middle place and finally clayey at faraway place from the place of erosion. Therefore, stable texture of surface soil without erosion does not remain cultivable after erosion due to its change in various fractions. After erosion the opened subsoil does not has soil properties as good as surface soil. Therefore, the yield potential such subsoil is very much low. However, these soils can be made productive after sufficient expense of time, labour and money.

Wind also carry and sift eroded soil particles to different distances i.e., coarse particles very near and fine particles at distant places. Naturally in new settings of such sifted particles difference in soil texture is observed. A soil which recorded 42% silt and 15.4% sand before wind erosion and after wind erosion when these soil particles settled again, silt and sand in drift soil was 15% and 58.2%, respectively. Then the mechanical analysis of this dust of soil indicated 62.5% silt and 14.3% sand. So, after sorting of primary fractions wind converted them in sand abundant and silt abundant fractions.

Sedimentation or silt deposition in reservoirs: Soil erosion is the major cause of reservoir sedimentation. Sedimentation of reservoirs is a complex process depending upon numerous natural and anthropogenic factors. Soil particles with runoff water reach in various reservoirs through rivers, rivulets, streams etc. The incoming runoff water disturbs the quietness of water of reservoirs and increase settling speed of particles. Thus, particles settle in the bottom of reservoirs rapidly which reduces their effective depth and water storage capacity. Sediments are transported and deposited particles which were obtained from rocks, soil and organic matter. Silting is the accelerated process of deposition of aqueous sediments, garbage, organic matter etc. in the beds of rivers, lakes, flood plain, seas and other water bodies which reduces their depth.

There are many losses as a result of reduction in water storage capacity of rivulets, rivers, reservoirs etc. The chances of floods are increased and water supply and electricity production from reservoirs decrease. If sedimentation rate is quite high then efficient age of reservoirs is remarkably reduced than its designed age. So, it is not possible to get complete return out of money invested.

Floods: The rising of problem of floods in various areas is also an ill effect of soil erosion. When sedimentation of rivers becomes very great, then effective volume for the purpose of water flow and water storage of them reduces. In such condition at the time of more rainfall embankment of rivers break because of heavy flow of water and water as a flood spread on all sides. Rivers in which frequent floods take place should be taken care at their catchment area itself for controlling the sediment generation. About 15 most flood prone countries of the world, India is highly vulnerable and top on the list and more than 40 mha is flood affected. Floods are a recurrent phenomenon, which cause huge loss of lives and damage to livelihood systems, property, infrastructure and public utilities.

Reduction in yield of crops: Violent runoff and wind cause harm to the standing crops which ultimately reduce yield. Because of intense runoff soil below the roots is eroded and flow away. Therefore, plants don't have any support and they fell down easily by light blow of wind. Increasing runoff remarkably increases soil loss and declines crop productivity as indicated in Table 4.2.

Table 4.2: Effect of increasing runoff on loss of soil and reduction in crops yield

Run off (mm)	Soil loss (ton/ ha)	Yield of grains (kg/ha)	
		Maize	Wheat
163.3	11.7	2350	4600
221.1	24.1	2281	3650
304.1	34.3	1868	3390

It is clear from table 4-2 that about two times increase in runoff, loss of soil increased to about three times. Also, reduction in yield of maize was 482 kg / ha and reduction of yield of wheat was 1270 kg/ ha. Soil erosion has more harmful effect on wheat than maize.

Types of Soil Erosion

Soil erosion may also be defined as transporting away the weathered materials like soil, pebbles, boulders, sand, glacial till, etc. by wind, water, ice or other means to different places. Soil erosion may be natural or manmade. Natural erosion is also called geological erosion. It is a normal and unavoidable activity that does its work from geological periods by geological processes. Its major contribution is construction and distribution of various types of soils on Earth. As a result of long-term activities of geological processes there is continuous wear and tear of all the mountains and they are being converted in various plains, valleys and river plains. Thus, today we see numerous types of landforms. These all are the result of activities of geological erosion. Genesis of Sindhu valley and alluvial plains of Ganga –Yamuna are, origin of fertile plains is a result of natural erosion.

General action of natural erosion is accelerated due to various activities of human beings such as farming, ranching, forestry and construction. The unscientific use of soil, accelerated removal of soil organic and mineral matters or removal of natural vegetation cover from soil surface erodes soil much rapidly. This type of erosion is called as manmade or accelerated erosion. This erosion causes breakdown of soil aggregates. In the consequence of this breakage of soil aggregates, soil particles are displaced very easily by water and air. It indicates that the unnecessary interference of man in the well-established equilibrium among soil, climate and vegetation, generate accelcrated erosion.

Classification of Soil Erosion

Generally, soil erosion is classified on the basis of its agents whose description is given here:

Soil Erosion

1. Water
    a. Rain i) Rain drop or Splash
    b. Flowing water i) Sheet ii) Rill iii) Gully iv) Stream
    c. Sea i) Coastal
    d. Ice i) Glacier
2. Gravity i) Land slide
3. Wind

1 Water Erosion

Removal and transportation of soil by water is called as water erosion. This primarily takes place by water present in rain, rivers, rivulets, stream or sea. On the basis of these it is divided into six types. These types are: rain drop or splash erosion, sheet erosion, rill erosion, gully erosion, stream erosion and coastal erosion.

Rain drop erosion: Drops of rain falling from the sky behave just like small bombs colliding on uncovered or bare soil. This is the first stage in the erosion process in which soil particles are displaced and soil structure is destroyed. Thus, it is called rain drop erosion. The power of falling rain drops also scatter soil particles. Collision of rain drops to soil produces spatter or jump. This spatter or jump displaces affected soil paricles from their previous place.

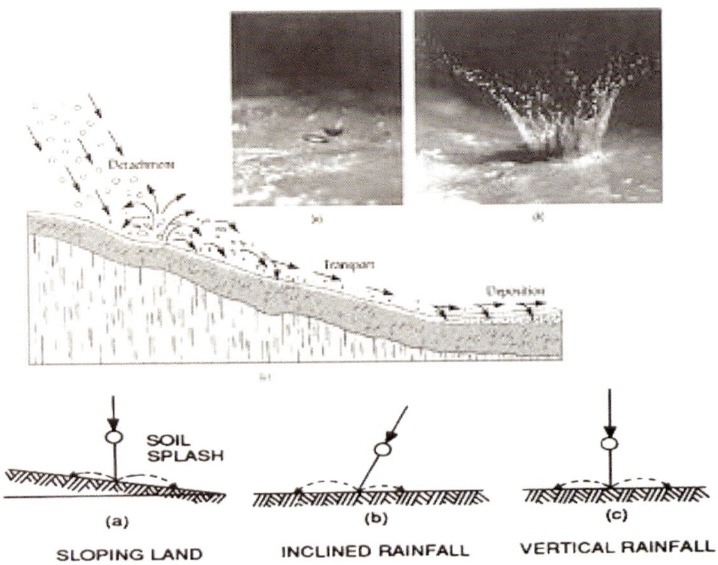

Fig. 4.3. Types of Erosion (Source: http://www.cep.unep.org/pubs/Teschreports/tr41en/image11.gif)

Separation of minute particles of soil by the effect of rain drops on the soil and their airborne movement is called splash erosion. This definition clearly indicates that drops of rain only separate soil particles from their original place but movement in particles is provided by the wind blowing at that time. If wind velocity is more then particles will go far away in the direction of air. The separated soil particles will equally deposit around the crator but they will deposit more towards slope on slopy land. The whole effect of splash erosion is to break large clods or soil aggregate in small particles or aggregates, to remove soil particles from its previous place and to produce turbulence in thin layer of water present on soil surface. The whole process is depicted in depicted in Figures 4.1

Splash erosion depends primarily on soil properties, topography, vegetational covering, thickness of water layer on the soil and rain. The rain includes its quantity, size of rain drop, shape of rain drops, speed and direction and rainfall chiefly influence erosion. A violent rain can displace up to 225 metric tons of soil from one hectare area.

Raindrop characteristics: The raindrop size, shape, weight, velocity, size distribution and direction of fall are the main characteristics of raindrop regarding soil erosion.

Size of raindrops: Size of drops of rain may vary between 0.1 mm to 8 mm. Drops of more than 5 mm size are unstable and these are divided in small drops in the air due to colliding themselves or other particles. Drops larger than 2.5 mm can produce erosion. Therefore, 2.5 mm drop diameter is representative of erosion producing rains. As the size of drops increases rate of entry of water in soil decreases.

Shape of raindrops: Generally, it was observed that the raindrops up to 1 mm diameter are in spherical shape. Shape of rain drops is not always spherical or simply stream lined. But beyond 1mm size the drops become flattened from the bottom and oval in cross-section. Stream lined means a form or surface that produces very little resistance to a flow of air or water. Drops of rain are deformed due to dissimilar resistance produced by wind on them.

Weight of raindrops: The density of water is 1 g/cc. Accordingly, the smallest mass of a raindrop produced during a drizzle is 0.004 mg and the largest raindrops produced during a heavy storm is 300 mg.

Velocity of raindrops: The velocity of raindrops may be 9-13m/s. The velocity of raindrop is governed by many factors like size of raindrop, shape of raindrop, weight of raindrop, height of rainfall, gravity, frictional drag and density of air. Drops larger than 2.5 mm size can erode soil. The velocity of these dropds increases with the increase in the height of rainfall. This is clear from an experiment that the velocity of raindrops was 3m/s at 0.5 m height of rainfall which gradually increased to 4.0, 5.2 and 7.4m/s at the heights of 1.0, 2.0 and 20.0 meters, respectively.

Direction of fall: Generally, splash erosion on levelled soils does not occur more seriously. On such soils when drops fall upright, scattering of soil particles takes place equally in all directions. In such case jump of minute particles can reach up to 0.6 meter high and 0.5 to 1.5 meter away from the centre of hit. In sloping soils most of the part of the jumped particles goes towards slope. So, at slopy areas splash erosion is more harmful.

Detachment of soil particles continuously increases with the increase in the size of soil particles and transport of particles continues to decrease. It indicates that large particles of sandy soils can be detached easily but they will not go to farther distances. Also, small particles of clay soil will not be detached easily by drops of rain, but after detachment their transport is possible longer distances.

Cognition of rain erosivity: Erosivity is potentiality of soil erosion by water, wind, gravity etc. It can be cognized by following formula:

$E = F\ (I \times t \times mv^2 / A)$ where

IE = Erosivity

F = Function of

t = Time

m = Mass of drop

A = Transverse area of drop

V = Speed of falling drop

Erodibility or Erodability is another term use in this context which means sensitivity or non-resistance towards erosion of any soil. Park etal. (1982) has expressed displacement of soil particles by rain on the basis of Newton's Second Law of Motion. By the help of this formula splash towards slope can be calculated. The formula is:

$Wd \cos\alpha = Ws\ Vs \cos + Wd\ V_R \cos$ Where

Wd = Mass of drops

α = Effective angle

Ws = Mass of displaced soil particles

Vs Incipient velocity of sediment

θ = Angle of slope

$V_{R=}$ Effective velocity of drop

β = Repulse angle

As a result of splash erosion on account of wash out of minute particles of soil, erosion pavement is formed. Erosion pavement means the layer of coarse fractions like sands, gravels, pebbles or rocks on the surface after removal of minute particles by erosion. It further protects underlying soil layer from erosion

Sheet erosion: Among all types of soil erosion sheet erosion is most harmful, because its effect does not visible immediately. This is also called as sheet wash. In the sheet erosion a uniform thin layer of top soil from a wide area is removed by runoff water. Such type of removal of the top soil in the form of a layer or a sheet is very difficult to visualize. Flow velocity, flow volume, physicochemical properties of soil etc. are very remarkably affect sheet erosion.

Actually, very small channels called as inter-rills carry soil particles detached by raindrops as a very thin layer of the sheet flow. Generally, the sheet erosion occurs in common agricultural soils having moderate slope. Sheet erosion is more effective and harmful in places having loose and shallow surface soils overlying on hard subsoils.

A thin layer of fertile soil is eroded in each Monsoon. Repeated removal of surface of soil at each Monsoon or year arises a situation that infertile subsoil of below comes above. Subsoil is not better for plant nutrition and development. A large part of soil particles separated by rain drop erosion are displaced by sheet erosion. By this away rain drop erosion and sheet erosion both go on simultaneously.

Regions of sheet erosion: UP, Uttarakhand, MP, CG, Bihar, Jharkhand, Rajasthan, Gujarat, Haryana, Maharashtra and Tamil Nādu. Black soils recorded annual soil loss up to 64.5 tons per hectare by sheet erosion. In our country about 40 lakh hectares land is affected by sheet erosion. This whole area is divided in four main regions:

1. Western coastal region is extended from upper valley area of Godavari in north to Tinnevali in south. Pune, Satara, Kolhapur, Dharwad, Tumkur, Mysuru, Nilgiris, Coimbatore, Palghat, Raichur, Ernakulum, Kottayam and Quilon districts are included in this region.
2. Eastern coastal region which is extended from Palani hills of Tamil Nādu to valley areas of Mahanadi of Chhattisgarh.
3. Region of shifting cultivation in which Assam, Tripura, Manipur, Nagaland etc. are included.
4. Area of Aravali Hills other than area affected by Gully erosion.

3. Rill Erosion: Rill erosion takes place when water of rainfall is not completely soaked by soil and it is accumulated on bare or recently ploughed soil surface in much more quantity. This water begins to flow towards slope. Due to the flow of water on soil surface narrow, smaller and shallow (not more than few centimetres deep) channels or gullies just like fingers are formed, which are called as rills. Generally, rills are aligned somewhat perpendicular to the slope and develop in a series of parallel rill lines. A network of these rills joined together and extended in whole field. These rills are destroyed easily by moderate ploughing. Erosion caused by these rills is rill erosion.

Flow of water and soil in rills: The rills flow overland as well as rain water directly falling in them. Therefore, sufficiently more water is present in rills than sheet erosion. Depth of water in rills may be 50 times and velocity of water may be 10 times more than sheet flow of water. The enhanced depth and velocity of water significantly increase its power of sheer and scour. Therefore, with the flow of water soil already chopped from bottom and sides also reach in the flow. By virtue of the turbulence produced by raindrops, displaced soil particles remain in suspension and discharge with rill flow. As soon as water flows out of rills, its cutting capacity of soil also continue to increase tremendously. Consequently, the rills become deeper and wider. Soil separated by sheet erosion and chopped by rills also pour into rills and flow out from field. So, gravity and seriousness of rill erosion is much more than raindrop erosion and sheet erosion.

Subsurface or Internal rills etc.: Occasionally rills also are formed inside the soil i.e., internal rills. In addition to rills, excess of water on the soil surface enters in the soil by seeping. This water intensively passes through internal rills, cracks, burrows, cuts etc. present in the soil. Thus, tubules and thin tunnels are formed in the soil. Less resistance soil materials present in these is eroded swiftly. Therefore, this swiftness increases the diameter of tubules and tunnels. Occasionally several tubules and tunnels mingle together and their roofs are sink. Thus, erosion resulting by tubules and tunnels is called tunnel erosion as classified by some scholars. But it is similar to rill erosion because rills are on the soil surface and underground. Therefore, it is right to include tunnel erosion in rill erosion itself. Figure of rill erosion is given here:

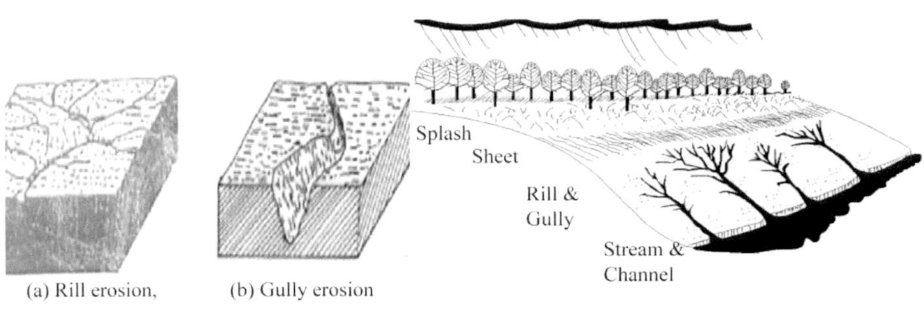

Fig. 4.4: Types of Erosion *Source*: http:/www.cep.unep.org/pubs/Techreports/tr41en/Image11gif

4. Gully erosion: On the excessively slopy lands heavy rainfall and higher runoff cut and scour more quantity of soil and flow that speedily into small shallow channels. Also, rills and tunnels existing in the subsurface soil are sunk to make thin and long pits and craters on the surface of soil. If the passage and ways of agricultural implements, animals, bullock carts etc. are not properly maintained in time, then these are also converted in wider channels. These channels, so formed, are such wide and deep that these cannot be removed or destroyed by normal ploughing. These wide and deep channels are called gullies.

Also, several rills gradually join in the fields and make gullies. Sometimes, depth of these gullies may be up to 16 to 33 meters. Gully is such a valley or ravine which is formed due to erosion and strong but intermittent flow of water. In heavy rain gully acts as drainage channel as water after rain flows very speedily through it. Gullies are so deep that cannot be put right by common cultural operations. But they provide hinderance and interruption in these operations. Rate of gullies erosion depends on character of runoff producing rain, area of water supply, drainage, soil properties and size and shape of gully.

Stages of gully formation: Formation stage, developmental stage, vegetative growth stage and final stage are four main stages of formation of gullies.

1. **Formation stage:** In this initial stage run off moves towards slope, surface soil is eroded by its rub (scrap). If surface soil is heavy and compact, it resists erosion and action of erosion takes place gradually.

2. **Developmental stage:** In the initial stage of development the channels formed so far because of water flow become wide and deep. Due to a pace of flowing water weak parent material together with 'C' bed of soil also run off. Occasionally spring (fall) is produced in the gully.

3. **Vegetative growth stage:** In this stage natural vegetation and grasses begin to grow. Therefore, gullies seem to be strong and compact and erosion begins to decrease.

4. **Final stage:** In this stage gullies retain a stable gradient level. Slopes of walls of gullies become permanent. Good growth of natural vegetation and grasses appears. Therefore, formation and development of new top soil begins.

Process of formation of gullies: In the formation of gullies several processes operate together or alternately. In these waterfall erosion and channel erosion are most important. First of all, when sheet erosion producing runoff forcefully falls at a particular place i.e., gully head of the sloping land, then a pit or hole is formed there. When such pit enlarges then water begins to fall there with more force. Thus, a small spring or waterfall is formed. This is waterfall erosion. In this fall speedily passing water scour, erode and wash away the bottom and sides, top soil as well as subsoil to form a deep and wide channel.

After the end of rainy season exposed banks or edges of soil begin to dry. The soil becomes loose and loose and intermixed due to the nature and properties of different minerals present in the soil and repeated drying and wetting. All these form cracks in the walls of channels. In the next rainfall more amount of water is filled in these cracks. This helps in mass movement of soil. Ultimately gully becomes wider. The soil particles in gully mixed with water wash away rubbing bottom or bed and sides of gully. The continuous rubbing, cutting and washing away make gullies deeper.

Classification of gully: Gullies are classified on the basis of their size or depth, shape, continuation, drainage area etc. Although these classifications are not universally accepted because various states have their separate classification systems.

Size or depth of Gully: Gullies having size or depth less than 1 meter are called small gullies, having size 1 to 5 meters are called medium gullies and more than 5-meter size or depth come under large gullies.

Shape of gullies: Gullies show three types of shape. Two resemble to English letters 'U' and 'V' and third shape is trapezoidal. Type of erosion, age of gully, climate and soil are main factors influencing shape of gully.

'U'shape gully: U-shaped gullies are formed when both the surface soil and subsurface soil have the same soil material and resistance against erosion. When subsurface soil is eroded as easily as the surface soil and runoff flow undermines both then gully banks are collapsed and nearly vertical walls are developed on each side of the gully. The cross section of the gully is like U.

Airborne (aeolian) and water borne (alluvial) parent materials are basically loose and and deeply uniform. These can be eroded very easily. In such conditions soil from top to bottom is uniformly eroded by water and washed out. Flowing water in the bottom of gullies erodes walls from downwards and make lower sides of walls hollow. On account of not withstanding the weight of soil present on the hollow side walls, these sides collapse down in the rainy season. So due to uniform erosion at upper and lower parts, U-shaped gullies are formed.

'V' shape gully: Contrary to 'U'shaped gullies, the V-shaped gullies are developed when the surface soils are soft and loose and subsoils are hard and compact enough to resist the rapid scouring and cutting of soil by the runoff. As resistance to erosion increases with depth, the width of open cut decreases accordingly and it is narrowest at the bottom of the gully. This results in formation of V-shaped gully. Also surface soil is easily eroded in more amount by runoff where as due to resistance towards erosion subsurface soil erodes in little amount. So uppermost part of exposed top soil becomes continuously wider while lower part of subsoil remains very much narrower. Thus, cross section of the gully looks like V with very much slope gradient of walls.

Trapezoidal gully: Trapezoidal gullies are formed when the gully bottom contains too much erosion resistant material in comparison to the surface soil. Therefore, below the bottom of gully, the subsoil layer does not erode and becomes flat. Thus, the further depth of gully is completely stop or cease. The figures of shape of gullies are given here:

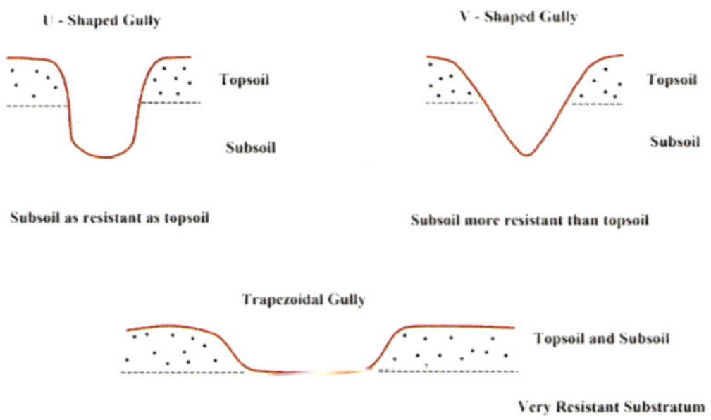

Fig. 4.5: Shape of Gullies

Continuation of gullies: Continuous gullies and discontinuous gullies are two types on the basis of continuation.

Continuous gullies: Continuous gullies consist of many branches. A continuous gully has a main gully channel and many well developed or under developed branches. Continuous gullies are development on gentle foot slopes in areas of saturation, drainage paths with a large contributing area, erodible duplex soils i.e., texture contrast soils derived from mudstones and poor vegetation cover due to overgrazing.

Discontinuous gullies: Discontinuous gullies are developed on hillsides after landslides. They are also called as independent gullies. Discontinuous gullies occur more on rolling slopes areas where the surface becomes less frequently saturated, drainage path with a smaller contributing area, more stable and shallower soils developed from mudstone and overgrazing. Discontinuous gullies do not have a distinct junction with the main gully or stream channel at the initiation of their development.

Drainage area: Gullies are also classified because of their drainage area. Gullies are small if drainage area is less than 2 hectares, medium 2 to 20 hectares and large more than 20 hectares.

Depth of gullies: Very small and shallow gullies are up to 3 meter deep, bed width less than 18 meter and slope of walls are nearly 3 percent, small gullies are also

3 meter deep, bed width more than 18 meter and slope of walls vary from 8 to 15 percent, medium gullies are 3 to 9 meters deep and slope of walls vary 8 to 15 percent. Deep gullies are 10 meters deep, bed width less than 18 meter and more than 15 percent slope of walls. The narrow gullies are more than 9 meter deep with varying bed width and slope of walls.

Besides these, there are active and inactive gullies also. Active gullies are found in plain areas and show enlargement in dimensions with time. The enlargement depends upon soil properties, land use and volume of runoff passing through the gully. Inactive gullies are found in rocky areas and their dimensions remain same with time.

Areas Affected by Gully Erosion

Gully erosion is one of the most important and devasting types of water erosion that causes the destruction of lands, property, natural habitats etc. The area most affected by gully erosion in our country are classified as extremely severe gully erosion area, severe gully erosion area, medium severe gully erosion area and normal gully erosion area. These are briefly described here:

Extremely severe gully erosion area: Alluvial soils situated in arid and semiarid region are most adversely affected by gully erosion. Deep ravines and rugged grounds are present in this area. Area having very much intense gully erosion and deep rugged grounds is triangular in shape. This area is mainly situated along with banks of Chambal and Yamuna rivers. Besides these rivers, areas of Betwa, Ghassan, Kain and Banas like tributaries are also affected by gully erosion. Very much degraded area of Chambal and Yamuna rivers is egg shaped. A glimpse of this area can be seen between Dholpur and Morena on Agra – Gwalior Road and between Etawah and Bhind on Lucknow – Gwalior Road. However, this whole area is extended from Mathura to Prayag Raj (Allahabad) and Bhilwara.

Geologists indicated that severity of gully erosion in this area is by virtue of continuous bulging of land surface. Main reason of this bulge is changes occurring in lower surface of land. As a result of these changes more heat is accumulated at any one place which produces expansion at various layers of land. The land becomes anywhere high and anywhere low due to this heterogenous expansion and form ravines. Ravines of Yamuna Rivers are cultivable due to their more width as alluvial matter has U shaped wider gullies. But in case of Chambal River area due to hard subsoil V shaped gullies are formed. These gullies are narrower at bottom or gully bed. Hence, unsuitable for cultivation. Only permanent and stable vegetations can be grown.

Severe gully erosion area: The first severe gully erosion area consists western sub-Himalayan region and some part of Gujarat. Thousands of water streams from Himalayan area enter in Doon and Bhabar area through steep slopes. On account of passing through steep slope, soils of upper layer and sides are eroded, washed out and ultimately form gullies. Here main reason of erosion is bulging caused by land formation. Here about 50 km wide and 650 km long Doon and Bhawar strip is attached with Shivalik hills. This strip is extended from Haldwani to Jhelum River. Here soil matter is too much porous due to the presence of coarse sand, gravel and stony pebbles stone.

The second severe gully erosion area is situated in Gujarat from southern bank of Tapti River to the Sabarmati valley area at the boundary of Rajasthan and surrounding valley area along with surrounding area of Mahi River flowing in the region of Narmada and Sabarmati rivers. Generally,, this whole area is present in the plain part of the alluvial soils.

Medium severe gully erosion: In our country area of medium severe gully erosion is situated in Chhota Nagpur, upper parts of valleys of Mahanadi and Son Rivers. Medium severe gully erosion is mainly visible in Ranchi, Hazaribagh, Nrisinghpura, Jalgaon, Budhlana, Amaravati and Malwa region.

Normal gully erosion: Normal gully erosion area exists in Uttar Pradesh in interfluves of Ganga -Yamuna in the vicinity of Kali, Hindan and Sengar rivers, in the washlands of these rivers and in interfluves of Ganga - Ghaghara on banks of Sai and Gomati rivers. In Maharashtra at Beed and Osmanabad districts on right bank of Godavari and at Nasik, Aurangabad, Parbhani districts on left bank and districts of Nanded, Chandpur and Nagpur in the vicinity of Wardha and Pradhiyan rivers are affected by normal gully erosion. In Telangana districts of Hyderabad, Nalgonda and Medak are also affected by normal gully erosion.

5. Stream erosion: Cutting and scouring of soil of banks of streams and rivers and washing up of eroded soil in their water is called stream erosion. In this erosion trees grown on banks are uprooted or soils below their roots is eroded and wash out. The excess of water flow also breaks the embankments. So, danger of flood arises. In cultivated fields more loss is occurred by this erosion. Velocity and direction of water flow, depth and breadth of stream, soil texture and vegetations grown on banks affect stream erosion.

Water flowing above the sides of streams and rivers continuously cut the soil. Also, soil continues to cut by scouring between surface water and layers beneath it. Thus, stream erosion continuously run together in both surface and subsurface soil and stream bottom. Severity of this erosion increases by cutting of trees, grazing of animals and ploughing on the sides of streams. By virtue of this erosion streams and rivers change their routes and deposit silts and sediments at new sites.

Transport of eroded soil in water current: The soil particles displaced by water from their original place come in water current or flood water and move from one place to another. This movement or transport of soil particles in the thin layer of runoff made by rain drops, rills, gullies or streams takes place by three actions. These actions are suspension, saltation and bed load movement.

Suspension: Displaced soil particles remain in suspension flowing or in stagnating water. These particles settle in bottom of stream following Stock's law. Soil particles settle more speedily in stagnated calm water than flowing water. Soil particles remain in suspended condition in flowing water for sufficient time before settling in bottom. These particles are called as suspended soil particles.

These particles come in suspended condition by two ways, 1. Rain drops falling on flowing thin current of water produce splash. 2. When water passes with violent current on soil surface, then soil particles rise above from below and begin to flow with current of water. Generally, sides and bottom of natural currents are rough. This roughness produces hinderance in flow of water. On account of this rate of flow is more in the middle than sides and bottom. This difference in flow rate produces several layers of flow velocities in flowing water. In these layers of flow velocities suspended soil particles move with various speeds.

Flow velocity, sediment concentration and sediment discharge are related with the depth of current. Sediment discharge depends upon velocity and concentration. This relationship is depicted in figure 4.6. On account of being more concentration in the bottom of current its velocity is lowest. Similarly discharge of fine sediments is highest just above the bottom as compared to coarse sediments. Although in the other layers of velocity fine particles are more uniformly suspended in comparison to coarse particles.

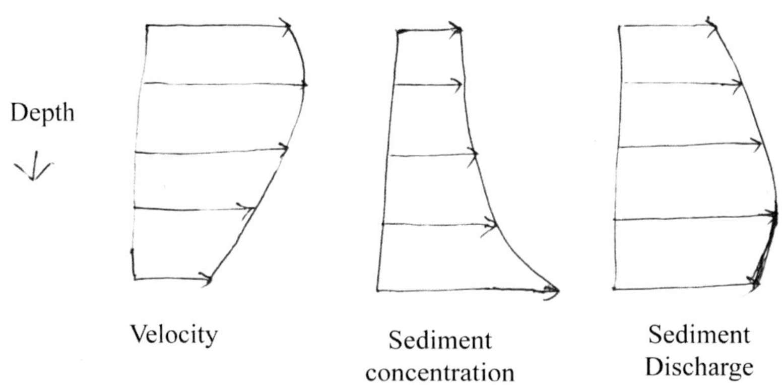

Fig. 4.6: Relationship among flow velocity, sediment concentration and sediment discharge

Saltation: Saltation can also be called as frisk of soil particles. Jump and bounce of soil particles by energy provided by flowing water is called saltation. In saltation movement of sediments occurs due to skip or jump with gush of soil particles and push or roll of particles by flowing water. Activity of saltation Generally, occurs in such areas, where soil particles jump on the sides of river, stream, brook or flowing water. Height of jump of soil particles is directly proportional to the ratio between particle density and density of fluid medium. Saltation transport is considered less important in comparison to transport of total sediments produced by soil particles.

Bedload movement: Bedload is that sediment which moves nearly in continuous contact with bottom of flow or at the sediment-water interface, such as a stream bed or tidal flat. In bedload movement sediments are pushed or rolled by the force of flowing water in the bottom. Bedload movement of unigrannular particles having 0.35 to 5.7 mm diameter and 1.87 to 2.6 specific gravity can be calculated by given formula:

$$Vt = 0.152\, d^{4/9} (G-1)^{1/2}$$

where

Vt = Threshold velocity, m/s

d = particle diameter, mm

G = specific gravity of the particle.

6. Coastal erosion: Coastal erosion is defined as the break and loss of the soil or sediments etc. from the littoral zones due to the action of waves, currents, tides, wind-driven water etc. Coastal erosion begins due to the continuous thrust of sea waves on coastal line. Thundering waves coming from sea side swiftly collide with seashore and continuously erode coastal lands. Therefore, vertical cliffs are formed in coastal regions.

Severity of coastal erosion increases tremendously by tide. Tide enhances soil erosion by tearing or breaking of littoral areas. Rivers deposit silt and mud in nearby terrains. But oceanic water waves break and erode these and embankment of rivers of coastal area. Ultimately whole area is destroyed or converted in wasteland. The reasons behind coastal soil erosion are, energy of oceanic waves, climate change, heavy littoral drift, construction of dams in oceanic area, sand and coral mining and dredging etc.

Coastline of India: Indian Coastline is facing soil erosion problem and about 45.5% of our coastline is affected by it in varying magnitude. At present revised Indian coastline length is 11,084.50 km among related states and territories as 1. Gujarat 2,125.82 2. Maharashtra 896.98 3. Karnataka 309.59 4. Kerala 597.43 5. Tamil Nadu 1,064.98 6. Andhra Pradesh 1,272.55 7. Odisha 667.12 8. West Bengal 662.90 9. Goa 181.48 10. Daman & Diu 63.10 11. Puducherry 42.11 12. Andaman & Nicobar Islands 3,078.52 13. Lakshadweep & Minicoy Islands 121.92 km.

The West Bengal coastline has the maximum erosion (60 percent) followed by Puducherry (56 per cent); Kerala (41 per cent); Tamil Nadu (41 per cent); Andhra

Pradesh (28 per cent); Gujarat including Daman and Diu (26 per cent); Odisha (26 per cent); Karnataka (24 per cent); Maharashtra (22 per cent) and Goa (19 per cent).

Coasts of India: West coast and East coast are two coasts of India. Western coast (ghat) is extended from Gulf of Khambhat to Kanyakumari and from north to south it is divided into Konkan coast, Karnataka coast and Kerala coast. Narmada and Tapi are major rivers of west coast. East coast is in between the Eastern Ghats and the Bay of Bengal and extends from the Ganga delta to Kanyakumari. Deltas Mahanadi, Godavari, Krishna and Cauvery is one major feature of East coast.

Monsoon and coastal soil erosion: Western and Eastern both coasts are affected by coastal soil erosion and monsoon is one major factor. In rainy season seas and oceans become more turbulent or disturbed and cause excess soil erosion. In western coastal area mainly south- western monsoon is responsible for soil erosion. Towards south of eastern coastal area, north eastern monsoon shows its erosive effect. Erosion is also is more effective on the outfall of Mahanandi. Coastal area in between Hooghly and Krishna rivers is very much sensitive towards erosion as it is affected by both south- western and north - eastern monsoons. Opposite to this plateau of Kathiawar protects area of Khambhat. Therefore, in this area coastal erosion by monsoon is very less.

7. Glacier erosion: When voluminous glaciers roll over from heights towards slope, they smash up vegetations, fields, rocks, houses etc. come in their way. Also, they displace huge amounts of soil during their movement, which subsequently flow away by sheet erosion. The Indian Himalayas has an area about 5,33,000 km^2 (about 16.2% of country's total geographical area) and forms the northern boundary of the country. It extends over Jammu & Kashmir, Himachal Pradesh, Uttarakhand, Sikkim, Arunachal Pradesh, Meghalaya, Nagaland, Manipur, Mizoram, Tripura, Assam and West Bengal and from foot-hills in the south (Siwaliks) the region to Tibetan plateau in the north (trans-Himalaya). Most of the area is covered by snow-clad peaks, glaciers of higher Himalaya, dense forest cover of mid-Himalaya. India has 16,627 small and large glaciers which covers a permafrost area of 40,563 sq. km and 2 to 9 percent area is affected by erosion of these glaciers.

2. Gravity: Gravity is responsible for landslide or creeping erosion and water helps in this process. Landslide is defined as the very rapid collapse of huge mass of soil or rocks or both from a mountain slope or cliff. Activities like deforestation, forest fire, animal grazing, construction work, mining, cultivation etc. spoil the stabilized balance among land, climate and vegetation of the slopy area. This imbalance increases the landslide. Strength of soil and water table both influence the landslide and vegetative cover influences these both.

There is more possibility of landslide in such mountainous areas where slope material is very new, transient and fragile. In these areas heavy rain to (64.5 to115.5 mm per day) and more seepage of water makes whole slope super saturated with water. If rock below the slope is impermeable, the whole soil material of above is converted in a semifluid mud. In such condition whole material present above the impermeable rock comes down due to its own weight. Occasionally pits situated above the slope are filled with water. This water seeps slowly and slowly and macerate the land below the slope. Ultimately the macerated bases of the slope sank and whole material present above the slope slides down altogether.

In Himalayan and southern peninsular areas of our country incidents landslides and creeping erosion frequently take place. This results the great loss of public and property and control of these is a crying need. Two photographs of landslides at northern and southern hills, respectively are given.

Northern Hill Southern Hill

Fig. 4.7: Landslides on Northern & Southern Hills (Photo by Dr. Venkatesh Bharadwaj)

Components of hydrological cycle related to soil erosion

Water present as clouds in the atmosphere comes on land as rain and distributed in various parts by many processes and means. Some part of rain water is detained on leaves, some part enters in the soil, some part of it flows on soil surface as runoff water and some goes again in the atmosphere by evaporation and transpiration.

Water entering in the soil is mingled or mixed in underground water by percolation and lateral seepage. This underground and surface runoff water reaches in the sea through rivers and streams. Water evaporated from water bodies and evapotranspired from vegetation reaches in the atmosphere again. There again clouds are formed and it rains. So, this cycle goes on continuously. The hydrological, hydrologic or water cycle is defined as the continuous circulation of water throughout the earth and its atmosphere by various components and processes like evaporation, condensation, precipitation and transpiration. The hydrological cycle has many components. Here few components related to soil erosion are described and depicted in figure 4.8.

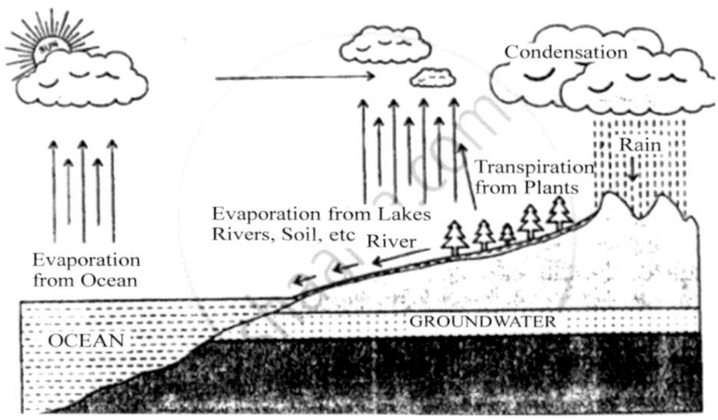

Fig. 4.8: Hydrological cycle

Leaf interception: Leaf interception prevents rainfall from immediately reaching to the soil. Leaves catch rainfall on their surface when it is rain on the soil having vegetative cover. When rain falls, the first drops of water are intercepted by the leaves and stems of the vegetation. This is called as interception storage. The interception storage or drops of water remain stuck there due to forces of adhesion and cohesion, until a water layer of sufficient thickness is formed. In the state of the continuity of rainfall, water present on any leaf will remain stable till water falls from that leaf.

Leaf interception reduces the velocity of raindrops because the leaves are coming in between soil and raindrops. The reduced velocity of raindrops has lesser erosive power and increased time to reach rain water on soil surface. Quantity of rain water accumulated on leaf surface may vary from 0.25 mm to 9.14 mm. Its average value is supposed to be 2.5 mm for agricultural crops and 1.33 mm for grasses. Leaf interception reduces run off up to some extent which decreases soil erosion. In the areas of dense forests and bushes leaf interception can withheld up to 25 percent of annual rainfall.

Water accumulation in small pits of the field: Before to flow as a runoff on soil surface water remainder from leaf interception is filled in small pits present in the field. But this state comes after the saturation of soil surface with water. Water filled in pits evaporates in the atmosphere or seeps in the soil, but its surface runoff does not occur. It is necessary to increase accumulation of water in the soil to reduce surface runoff as water accumulation on soil surface reduces runoff. Runoff also increases with the increase in the slope of soil whereas water accumulation capacity decreases. At seven percent slope this capacity is reduced to half. Water accumulation can be increased by tillage operations and agronomical methods. For example, in ideal conditions water can be accumulated 6.25 cm in furrows formed on contour line and 5 cm on level bench terraces. In annual crops accumulation capacity is found maximum after preparation of field which begins to reduce later on and at the time of harvesting it is minimum.

Infiltration: Entry of water in the soil surface is called infiltration and downward movement of water in the soil profile is called percolation. Raindrops directly fell on soil surface or excess water of leaf interception both reach on the soil surface and enter in the soil. Infiltration of soil water depends primarily on soil porosity. But size and number of pores also influence the infiltration. In sandy soils infiltration is more than silty and clayey soils. Runoff decreases with the increase in infiltration rate.

Soil surface becomes very compact due to wandering of animals and working of heavy farm machines, which reduces the rate of infiltration. Improvement in the soil physical conditions increases infiltration and reduces runoff. Soil aggregates and clods broken by strokes of raindrops are dispersed in small and fine particles and mixed in soil water. This water having fine particles is entered in soil pores and seal them. These sealed surface soil pores reduce infiltration of water. So, sealing of surface soil pores can be reclaimed by mulching or tillage to increase infiltration.

Fig. 4.9: Sealing of Soil Surface by Raindrops

It is seen in the experiments that infiltration rate of encrusted soil is very much less whereas it is remarkably higher in tilled fallow land and in old permanent pastures. In soils where higher vegetational root system of plants is extended vastly in the soil and more plant residues provide sufficient organic matter to the soil. Organic matter increases soil aggregate stability. Stability of aggregates is helpful to enhance infiltration.

Soil water and underground water: Accumulated water in soil is divided in two parts. First part is saturated state or underground water and second part is unsaturated state between soil surface and underground water. Water of unsaturated state Generally, moves upward due to capillary potential but during rain it changes to saturated state and moves downwards. If permeable surface soil is present upon the impermeable subsoil, then infiltrated water does not cross subsoil. In such condition surface soil is immediately saturated with water. Also flow of water instead of going downward begins to move laterally. Water begins to gush out after some distance through lower surface of sloping soils due to this lateral movement.

Evaporation and transpiration: Evaporation of water occurs from the soil surface due to the heat of sun. Likewise, water from surface of leaves of vegetation evaporates as vapour in the atmosphere. Evaporation of all water from leaf surface is not transpiration because leaf intercepted water also evaporates. Single term for both these activities is evapotranspiration. Combined loss of water as evaporation from the soil or land surface and transpiration from plants during a definite period from any area is called as evapotranspiration. As a result of evapotranspiration water moves towards the atmosphere.

Evaporation dissipates more than 70 percent of rainfall in the atmosphere. Evaporation becomes less in the moist and mountainous area than arid areas. Precipitation up to cent percent in arid areas and 50 to 70 percent in mountainous and humid areas is moved into the air by evapotranspiration. Evapotranspiration is very much important for crops because this activity influences crop growth along with quantity of run off and depth of infiltration. Sufficient warmth, less vapour pressure in the air than evaporative surface and continuous availability of water on evaporative surface are essential for evapotranspiration at any area.

Run off: Water flows over land as surface water is called as runoff. Drainage of water in runoff occurs from the upper surface of soil and this extruded water is merged in any stream, rivulet or reservoir. Runoff has great contribution in the displacement, transport and deposition of soils. Runoff depends on intensity of precipitation and infiltration rate. If rainfall is stormy and infiltration rate is less, then runoff will be quick and more.

In the formation of runoff first of all a thin layer of water is formed on the soil surface and begins to move downwards of slope. During this movement water starts to fill in pits coming in the path. When all these pits are filled completely, water begins to flow outside and enters in small channels. Now many small channels join togather and attain a shape of rill.

The flow of rills falls in small gullies. Gullies release their water in any main stream. This continuity of runoff fluctuates according to intensity of precipitation. Also, runoff is specially affected by factors related to watershed, as type of catchment area, physical nature of soil, length and gradient of slope and density and type of vegetation covering.

Measurement of Water Erosion

The exact measurement of soil erosion by water from any area is very difficult, but it is very necessary to know actual loss. Measurement of erosion can be done from valleys of large rivers and watersheds to the level of very small fields. But in each situation with the purpose of measurement, methods applied and equipments used will be different. Generally, erosion measurement can be divided in three scales according to size of area:1. large scale, 2. medium scale and 3. minute or small scale.

1. **Large scale:** Size of area of large scale or level may be from hundreds to thousands of kilometres. In this level measurement is done in view of geographical, ecological and regional problems. Measurement of soil erosion in various rivers, mountains or ecological areas in large level measurement. Study of sedimentation in various rivers is a major example of large level. Prediction equations for rivers are prepared by studying effects of origins and transport rate of sediment and climate, rocks, minerals and land utilization and such other factors. Here ton/km/ year and other such large units are used as unit of measurement.
2. **Medium scale:** Area of measurement on medium level is from some hectares to thousand hectares. Measurement of sheet and rill erosion, study of pollution of natural water resources by agricultural chemicals and crop erosion correlations are included in this level. Equations are prepared to study effects of various farming systems, topography and precipitation. Its units are kg/ha/year/or ton/ha/year.
3. **Minute scale:** As it is clear by name, size of minute level is about a hectare. In this level study of soil type in context of rainfall and runoff, raindrops, sheet and rill erosion is done in very small area. Study of erodibility of soil and factors affecting it, effect of nature of slope and sediment origin, effect of interculture methods and cropping systems on runoff formation and erosion control etc. are included in this class. Unit of this level is also very small: gram/m^2/year/, kg/m^2/year or kg/ ha/year etc.

Approaches of soil loss measurement: Approaches of soil loss measurement include 1. Reconnaissance methods, which are cheap, simple and depend on the measurement of changes in soil surface levels. 2. Volumetric methods which involve three-dimensional observations of the lengths and cross sections of rills, gulleys etc. and to estimate the amounts deposited in outwash fans, catchpits or reservoirs. 3. Aerial photography is also a useful approach. 4. Runoff plots are expensive, usually less effective and difficult to assess the results from these plots due to yearly variation in natural rainfall and from storm to storm.

The cost of construction and operation increases greatly with the size of runoff plots, because more sophisticated sampling and recording equipment is required. Catchment studies are desirable for studying the effects of major changes in land use. but while the measurement of large rates of runoff is required, it is more difficult to obtain reliable and accurate estimates of sediment movement. Artificial rainfall simulators can eliminate these variations and speedy data collection.

Calculation of runoff: In the calculation of runoff, velocity of water, water level, discharge and variations in these in context of time are measured and for continuously observing these many sophisticated instruments and equipments are now available. Few important instruments are current meter, floats and pendulum, various types of weirs, flumes etc. Estimation of total quantity of sediments with runoff is essential to know the actual amount of erosion. Various types of instruments and equipments are available for taking samples of sediments. Today to estimate the quantity of sediments turbidity meter and neutron or gamma probe are used.

Prediction of soil loss by water: Measurement of soil erosion is very much expansive and time consuming under various soils, crops and management conditions. Observations are fluctuated with time, place and conditions. All these does not provide the actual or correct data. Long term experiments are conducted for several years to estimate accurate soil loss and then any result is obtained. But from last several years many attempts to predict soil loss are being done with the help of some basic factors and stages affecting soil erosion.

First, in 1933 Baver gave an empirical formula to predict soil erosion. Afterwards in 1940 Zing in context of length and steepness of slope gave formula of soil erosion. In 1945 Alison gave formula to study effect of rain on surface erosion. Later on, Musgrave (1947) modified the formula of Alison after including characters of land.

After this Wishmear and Smith (1958) proposed Universal Soil Loss Equation (USLE) on the basis of soils and circumstances of Mid - Western America, but its factors are standardized up to the sufficient extent for various soils and circumstances of India. The equation is as:

$A = R\,K\,LS\,C\,P$ where

A = computed soil loss,

R = the rainfall-runoff erosivity factor,

K = the soil erodibility factor,

L = the slope length factor,

S = the slope steepness factor,

C = the cover management factor, and

P = the supporting practices factor

The factors LS, C and P are dimensionless hence, unit of A is the multiplication of the units of factor R and K only. Factors A, R and K have dimensions, their units are usually given in U.S. customary units. These should be converted to SI units with the help of appropriate conversion factors, for example A is tons per acre per year in U.S. customary units. It should be Multiplied by the conversion factor of 2.242 to get the values in SI units. Likewise, R and K should also be converted in SI units by using conversion factors. However, brief description of these factors is given here:

Rainfall factor (R): It is also called as rainfall erosion index. This refers capacity of rain which erodes the unconserved soil. This index indicates erosive effect of each storm. Erosive effect depends upon intensity and volume of rain. Many researchers calculated rainfall erosion indexes on the basis of rainfall of periods 5, 10, 15 and 30 minutes on various places. They found rainfall of 30 minutes is most correct. The kinetic energy of each stormy rainfall and maximum intensity of rain in 30 minutes expressed in cm per hour are used for R factor. This index corresponds to the potential erosion risk in a given region where sheet erosion appears on a bare plot with a 9% slope. Total amount of all such storms during one year provides annual index and average indexes of 20-22 years is used for the estimation of soil loss.

In reference to soil erosion intensity of rain is more important than amount of rain. More annual rain, which falls on the soil as many light rains, is less harmful, because due to light rains water gets time to enter in the soil. But less annual rain, which falls as torrent in lesser numbers shows more erosive effect on the soil, because in such rain at once more water fills on soil surface. This water has less time to enter in soil and before entering in the soil it takes form of runoff. So, after harvesting crops or at the time of preparation of field violent rainfall on uncovered surface is more harmful than standing crops. Therefore, seasonal distribution of rainfall is more important in erosion.

Soil erodibility factor (K): This factor means inherent susceptibility of soil towards runoff and rainfall i.e., how much tolerant or sensitive is soil towards K factor. More tolerance less erosion. Inherent erodibility in soil depends on its physical, mineralogical and biological properties. Infiltration capacity and aggregates stability influence soil erosion. Soil erosion reduces with increase in infiltration and size and class of clods. Soil texture, quantity of organic matter, type and quantity of swelling clays and presence of impermeable layer influence 'K' factor. Erodibility reduces with the increase in coarse and rough texture and organic matter. Resistance of particles present on soil surface towards beating action of rain also conserves soil during runoff. Soils having clays with excess of hydrous oxides of iron and aluminium are more tolerant towards torrent.

Soil erodibility (K) factor often deviates between 002.-0.69. It is less (0-0.2) in soils having more infiltration rate, good water drainage, more sand, friable clay with excess hydrous oxides of iron and aluminium. In soils having medium infiltration and medium structural stability, this factor is medium (0.2-0.3). Soils with low infiltration capacity are easily eroded. So, K factor of these soils is high

(0.3). The preferred method (Goldman, et.al.1986) for determining K factor is the nomograph method. This is mathematically written as:

$$K_{fact} = (1.292)\ [2.1x10^{-6}\ f_p^{1.14}\ (12\text{-}P_{cm}) + 0.0325\ (S_{struc} - 2) + 0.025\ (F_{perm} - 3)]$$

where

f_p = particle size parameter

P_{om} = percent organic matter

$S_{struc=}$ soil structure index

f_{perm} = profile-permeability class factor

$P_{clay\,=}$ percent clay.

In the formula the factor (1.292) is needed to; convert K_{fact} from the English units to the metric units.

The soil structure index, S_{struc}, is equal to: 1 for very fine granular soil; 2 for fine granular soil; 3 for medium or coarse granular soil; 4 for blocky, platy, or massive soil.

The profile-permeability class factor, f_{perm}, is equal to; 1 for very slow infiltration; 2 for slow infiltration; 3 for slow to moderate infiltration; 4 for moderate infiltration; 5 for moderate to rapid infiltration; 6 for rapid infiltration.

Reference: https://mepas.pnnl.gov/mepas/formulations/source_term/5_0/5_ 32/5 _32.html

K factor of universal soil loss equation is cognized by experiments on 22.2 meters long and 9 percent sloping continuous fallow and denudated plots (reference plots). For the purpose of various soils of India K factor is like that: silty clay loam of Dehradun 0.30, laterite soil of south India 0.34 and alluvial soil of Vadodara 0.059.

Topography (LS): Combined factor of length (L) and steepness (S) of slope is called topography (LS) factor. When other stages and conditions are equal and same then topography factor refers ratio of soil loss between 22.2 meters long and having 9 percent slope plots (reference plots) and actual field. By this reason this is a dimension less variable. L and S factors can be cognized by following formulas:

$$L = (1/22)^x$$

Where

L= length of slope (meters)

X = a constant which is 0.5 at slope more than 4 percent. On 4 percent slope 0.4 and on less than 3 percent slope 0.3 is considered

$$S = (0.43 + 0.30_s + 0.043_s^{\,2}) / 6.574$$

S = steepness percent of slope

S = slope percent of area

Velocity of runoff increases with increase in steepness of slope. So that erosion increases. Theoretically it is considered that if speed increases two-fold, then erosive force of water increases four- fold. Also, in suspension 32 times more matter and 64 times large soil particles can be transported. In an area of 1056.1 mm rainfall at Dehradun, it was observed that with the increase in slope percent, runoff and soil loss also increase accordingly Table 4.4.

Table 4.4: Effect of various slopes on soil loss

Percent of slopes	Runoff		Soil loss
	Mm	Percent of rainfall	(ton / ha)
0.5	139.4	13.2	6.6
2.5	163.4	15.5	11.7
4.5	221.1	20.9	24.1
9.5	304.1	28.8	34.3

Length of slope also has important role in erosion. Quantity of transported water increases as inclined sloping area as is extended. It has been seen that if length of area having 9 percent slope is doubled, then runoff and soil loss increase 1.6 times and 2.6 times, respectively.

Crop management factor (C): This factor is called cover and management factor also. This factor is indicator of incorporating effect of soil cover, crop rotation, mixed farming, crop period, productivity level, cultural practices, plant residue management, mulching etc. It is used to determine the relative effectiveness of soil and crop management systems in terms of soil loss prevention. The C factor is a ratio between the soil loss from land under a specific crop and management system to the soil loss from weed free, tilled and continuously fallow land.

Vegetative covers minimize soil erosion in several ways. They reduce the force of falling raindrops by leaf interception, minimize runoff, reduce velocity of flow on soil surface by producing blockage in water current, increase soil aggregation and porosity by plant residues and roots, increase biological activities in soil, reduce quantity of water available for runoff, increase water storage capacity of soil etc.

In cultivable lands C factor is between 1.00 to 0.01. This is high (nearly 1.00) in soil having scanty cover and bare soils and low (nearly 0.01) in dense forests and soils having more plant residues. Forests and grasses are best natural soil conservation agents. Cover of forage crops and pastures is very less dense. Medium dense cover crops as wheat and oat stops surface losses of soils, while row crops like maize, soybean, potato and other are helpful in soil erosion because of their less covering of soil.

C factor is quite variable and changes according to soil situations. In runoff plots, C factor for wheat and maize is 0.73. C factor for mung bean, ground nut and cowpea are 0.465, 0.374 and 0.137, respectively. C factor for various crop rotations like potato-potato 0.82 to 1.00, for potato – barley 0.93 and potato - fallows 0.83 were recorded.

Protection practice (P): It is also called as support practice factor. This factor is the ratio between any specific protection or support practice and soil losses up and down of slope i.e., cultivation practice in the direction of slope. This factor is the indication of impressibility of various protection practices like contour line, strip cropping, ridge furrow system, terracing and other such engineering practices. Protection given by cover and crop management in sloping land should be helped by other such practices which can reduce erosion by minimizing speed of runoff. P Factor were 0.321 and 0.76, respectively for contour cultivation and bench-terraces in runoff plots.

Limitations of USLE: Although universal soil loss equation is very much useful, but it has some limits, because this is not capable to estimate erosion on gullies, banks of rivers and more than 20 percent sloping lands. This only can estimate soil losses of sheet erosion and rill erosion.

This equation is quite successful for those places, where exact values of R, K, C and P factors are available. Since according to geographical and ecological system soil, vegetation and rainfall fluctuate, so soil loss can be estimated only when exact values of these factors are available for each place. Also, this equation cannot be used for watersheds having uneven slope.

Revised Universal Soil Loss Equation (RUSLE): The Revised Universal Soil Loss Equation (RUSLE) is an upgrade of USLE to overcome the limitations of USLE up to some extent. RUSLE can be used on cropland, disturbed forestland, rangeland, construction sites, mined land, reclaimed land, military training grounds, landfills, waste disposal sites, and other lands where rainfall and its associated overland flow cause soil erosion.

RUSLE was first introduced in the USDA, Soil and Water Conservation Service in 1993. RUSLE maintains the same empirical equation given earlier for USLE to compute sheet and rill erosion as: A=R K LS C P.

The major changes in RUSLE as compared to USLE are in the values given for erosion as modified by vegetative cover, better calculations of the slope (LS) factors and more advanced computerization. RUSLE also gives more credit to the ability of surface residues to reduce erosion and residues incorporated in the soil near or on the soil surface.

USLE assumed that runoff was uniform over the catchment, but RUSLE takes better into account that some runoff is channelled into rills and gullies. RUSLE also captures better than USLE that long rains can saturate the soil, leading to reduced intake and greater erosional runoff. In contrast with USLE, RUSLE can handle converging and diverging terrain and considers areas with net sedimentation.

An additional change incorporated in the RUSLE is to consider for rock fragments present on and in the soil. Rock fragments on the soil surface are treated like mulch in the C-factor. K factor is adjusted for rock present in the soil profile to account for effects of rocks on permeability and ultimately in runoff.

3. Wind Erosion

Just like water erosion, wind erosion also displaces, transport and deposit soil. The wind erosion occurs due to kinetic energy of flowing wind. Wind erosion is also a serious environmental problem not less severe than water erosion. High velocity winds strike the uncovered soil with increasing force. Fine, loose and light soil particles blown from the soil surface are taken many kilometres away. Thus, causing a great damage to the crops and other. Uncovered or with very scanty vegetative cover, shortage of rain, excess of temperature, presence of fine single grained soil fraction, extensive terrain and violent wind velocity encourage the activity of erosion.

Flowing wind shows drag effect on soil surface. This drag is proportional of square velocity of floating wind in proximity with soil surface.

$D = KV^2$

where

D= drag

V= wind velocity

K= coefficient

Coefficient (K) depends on roughness of soil surface and height of measuring wind velocity.

Wind velocity profile: Wind velocity has a very specific profile. Wind velocity is very much less or zero on soil surface due to friction between soil and wind. The height at which the wind velocity is zero or theoretically zero, is called roughness height. Lowest velocity occurs close to the ground and increases in proportion to the logarithm of the height above the soil surface. Change in wind velocity with height is called as velocity gradient. This gradient in wind velocity determines the amount of wind force. A profile of wind velocity is formed on soil surface due to various wind velocity gradients. With the help of data of wind velocity at any height, wind velocities at other heights can be calculated by formula given here.

$$\frac{V_1}{V_2} = \frac{\log Z_1 - \log Z_0}{\log Z_2 - \log Z_0}$$

where V_1 and V_2 are winds velocity on various heights. Z_1 and Z_2 are heights from surface and Z_0 is roughness height.

Wind velocity can be zero at any point on soil surface, but at a relative height, this becomes speedy and turbulent. This speedy and turbulent velocity produces such force which displaces soil particles. Large clods, very heavy soil particles or small soil particles in aggregation present on soil surface can tolerate turbulent wind velocity up to any extent. But small and light particles cannot tolerate this. So, their transport begins in the wind.

Wind erosion in India: Wind erosion is commonly observed in arid and semi-arid areas where the precipitation is inadequate, e.g., Rajasthan, Gujarat, Punjab and Haryana and some part of Uttar Pradesh. Sand dunes are found in this area. This area is more affected by wind erosion. Wind erosion is also found in some parts of Tamil Nadu, Karnataka and Maharashtra. Wind erosion Generally, occurs in summer season before monsoon. In summer season violent wind blows up soil in more quantity far away.

In desert area of Rajasthan many scientists have been estimated loss of soil by wind erosion. In Jodhpur area soil loss was 50 to 420 kg/ha/day, while in Jaisalmer it was 500 kg/ha /day. Soil loss by wind erosion in Bikaner and Chandan areas of Rajasthan was 615 and 325 ton / ha, respectively during 75 days and very high exponential correlation between wind velocity and soil loss was also observed at both places. Total annual production of dust world-wide from soils and sediments estimated to be 61 to 366 million tons. It was also noticed that wind erosion is related with monsoonal rainfall. Less dust storms come in next year of good monsoon because good vegetative cover develops on the soil. This saves soil by the action of violent winds.

Mechanism of Wind Erosion

In the mechanism of wind erosion two forces, Lift and Drag and two steps namely movement or Transportation and Deposition are associated. These are briefly described here:

Lift: Lift of soil particles from previous place is under Bernoulli effect. The pressure of wind is very low when blowing very speedily. So, soil particles begin to move in direction of pressure differential. Lighter and loose soil particles are lifted by wind from the surface in the movement initiation process.

Drag: Energy is needed for the movement or carry forward of soil particles lifted by Bernoulli effect. This energy is provided by drag. Drag is directly related with wind speed i.e., more speed, more drags. Consequently, more drag, more soil erosion.

Movement or Transportation: The movement of soil particles takes place after their detachment from previous place by the impact and cutting action of wind. The erosive wind is turbulent at all heights except very near to the soil surface. Most part of force exerted by wind is retarded by soil particles or other present on the soil surface. But wind may lift lighter and loose soil particles from the surface in the initiation process. Therefore, minimum threshold velocity of wind is needed to initiate the movement of soil particles.

The wind velocity on which most of the soil particles are transported is known as minimum liquid threshold velocity. When wind velocity increases from this minimum, then relatively larger particles of soil also begin to move or transport and a situation arises when all the soil particles are moved. The critical wind velocity of this stage is called as maximum liquid threshold velocity.

When once movement of soil particles begin, then particles are collided or impact to each other and remain in movement. It means that impact threshold wind velocity required to maintain the movement is more than liquid threshold velocity essential to start the movement. When essential liquid threshold wind velocity is obtained for any particular soil and its specific surface conditions, then rate of movement of erodible soil particles is power function of wind velocity. Generally, as indicated by Cook and Warren (1973) that when air velocity reaches at critical stage of 4.08 meter / second then rate of soil transport is in the ratio of cubic (power of three) of wind velocity and it can be shown as:

$$Q_1 = C_1 Vt^3$$

where

Q_1= transport of soil from unit area

Vt= critical shear velocity of air above threshold limit

C_1= proportionality constant depending upon soil condition and surface covering.

Processes of soil particles transport: Transport of displaced soil particles by wind is done by three processes namely saltation, suspension and surface creep.

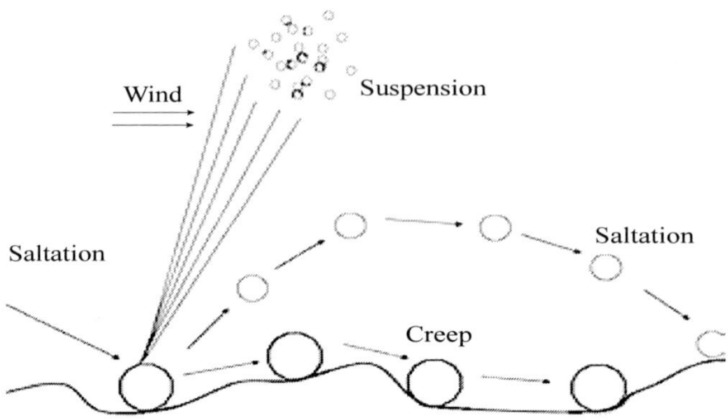

Fig. 4.10: Process of Soil Particles Transport.
Sources http://commons.wikimedia.org/wiki/File:Saltation-mechanics.gif)

Saltation: Saltation is a main process of soil particle movement. Saltation moves soil particles of 0.1 to 0.5 mm diameter and 50 to 70 percent weight of total separated soil particles are transported by saltation. Moved soil particles in the saltation can cause severe damage to soil surface and vegetation. Pressure of wind speed on soil particles causes their movement from soil surface and when adequate speed is obtained, they begin to hop and leap. Activity of hopping and leaping of particles is also done by mutual impact of soil particles.

In saltation in the beginning soil particles jump vertically, some particles jump only up to some distance, but some particles according to leap velocity raise above up to 30 cm or more and travel about four times longer in distance than height. The height of the jump varies with the size and density of soil particles, roughness of soil surface and wind velocity. During saltation transported soil particles can revolve @ 200 to 1000 rpm. These raised or leaped particles on account of upward acceleration are balanced by gravity and they flow or fly parallel for some moments. In final stages due to more gravity than drag soil particles come down on the soil surface with the angle of 6 to 12^0.

Suspension: Fine particles of soil suspended in the wind also move just like movement of soil particles suspended in water. Generally, transport of particles smaller than 0.1 mm occurs by suspension, but at higher wind velocity some large particles also come in suspension. Around 3 to 35 % of total soil weight is lost by the suspension process of soil transport by wind erosion. In summer season moving dust storms are a good example of transport of soil particles by suspension. When required wind velocity for the displacement is obtained, then displaced soil particles are raised upwards by wind and impact with other particles. Those particles which reach at height more than 90 cm are transported to very long distance and their transport is done by suspension.

When sedimentation force becomes higher than forces keeping soil particles in suspension, then soil particles begin to come downwards. In such situations these fine particles deposit on leaves of trees and plants, buildings and other and soil surface and make their surface slippery and sliding.

Surface creep: Roll and slide of soil particles with soil surface after coming in contact with the particles in saltation process is called surface creep. Surface creep adversely affects soil surface and soil aggregates. Transport of larger erosive soil particles of 0.5- 2.00 mm are transported by surface creep. Generally, 7 to 25 percent of total soil weight is carried away by surface creep which depends upon soil particle size, grades of soil particles, wind velocity and the distance along the eroding area.

Amount of soil transport: The amount of soil moved by wind erosion is primarily influenced by winds of variable velocity and direction. These produce eddies (An air or water current circularly moving reverse to the direction of the main current, just like whirlpool) and cross-wind currents which lift and transport soil. The amount of soil moved or transported depends on the median of diameter of soil particles and difference in threshold and actual wind velocity. The amount of soil moved by some influencing parameters can be calculated by given equation:

Amount of soil moved = $(V - V^t)^3 / D^{0.5}$

where

V = actual wind velocity,

V^t = threshold velocity

D = diameter of soil particle

Deposition: Deposition of soil particles occurs when wind slows down or gravitational force becomes greater than the forces holding particles in the air. This often happens when the wind has to move over or around obstacles like rock, trees, vegetation, ditches, benches etc. Raindrops may also take dust out of air. When wind slows, then soil particles of different types and size are deposited. The largest particles are deposited first. When the wind deposits sand, then small hills of sand i.e., sand dunes are formed and when the wind deposits fine particles of silt and clay, the deposits called as loess.

Measurement of wind erosion: Methods of measurement of wind erosion are still developing as compared to measurement of water erosion. No precise method of measuring wind erosion has been developed so far. However, various dust collectors, remote and in-place sensors, wind tunnels, sediment samplers and microrelief surveys before and after each storm are used. Soil loss can be calculated with the help exposed roots of plants and trees by observing their exposure before and after each erosion. These observations are helpful in the estimation of soil loss during any specific period.

The erosion pin or marker is a simple method for estimating the soil loss by wind erosion in specific areas. However, this method has many problems like the influence of human activities, low estimation accuracy and inapplicability to large soil loss by wind erosion. In soil surface graduated markers are fixed or inserted. As soil will be eroded, marker will be come above from the soil surface.

In another method soil particles flowing in the wind are collected and soil loss is determined. In this method soil loss by saltation, suspension and surface creep can be separately calculated.

Estimation of soil loss by wind: Time to time various wind erosion equation (WEQ) is formulated to estimate the soil loss by wind erosion. An equation similar to universal soil loss equation is given here. It can be used for estimating soil loss by wind. However, the evaluation of the constants in the equation for wind erosion is comparatively difficult than the universal soil loss equation.

The equation is E = IRKFCWDB

where

E = soil loss by wind erosion,

I = soil cloddiness factor,

R = surface cover factor,

K = surface roughness factor,

F = soil textural class factor,

C = local wind condition factor,

W = field width factor

D = wind direction factor, and

B = wind barrier factor.

The another WEQ formulated with primary wind erosion factors by N. P. Woodruff and F. H. Sidooway (1965) and used in USA and other countries is given here:

E = f (I C K LV) where

E = average soil loss from an area (t/ha/yr.),

f = function of

I = soil erodibility index (t/ha/yr.),

C = climate factor

K = roughness of ridges on soil surface

L = length or distance of related area (m)

V = vegetative cover factor.

Brief description of components of this equation is given here.

Soil erodibility (I): Soil erodibility means the soil loss from such area, which is unconserved and /or unpreserved whose soil is loose and smooth and whose top surface is not hard. The "I" factor is indication of stability of the soil aggregates against breakdown by tillage and abrasion from wind erosion. Erodibility of soil depends on size of particles and stability of soil aggregates. As size of particles and stability of aggregates reduces, soil erodibility increases rapidly. If the percentage of dry nonerodible surface soil aggregates larger than 0.84 mm in diameter are more then wind erosion will be less. When there is shortage of large particles than 0.84 mm diameter in the soil then value of soil erodibility by wind erosion increases quickly.

Climate factor (C): In the climate factor wind velocity and moisture directly influence soil erosion. Near surface winds can remarkably increase soil erosion by removing top soil of an area. The rate of soil movement varies directly as the cube of the wind velocity and inversely as the square of the effective moisture or the P-E index. The climatic factor for any location can be computed from the equation:

$$C = 34.483 \frac{V3}{(P - E)2}$$

C = climate factor

V = mean annual wind velocity on standard height of 30 feet (9.14 meter) for any place

P-E = Thornthwaite's P-E ratio = 10(P/E) = 115 (P/T -10)

P = rainfall

E =Evaporation

Roughness of soil surface: Random roughness (clods) and oriented roughness (ridges) reduce soil loss by wind erosion. Formation of clods also influences erodibility. If more clods are formed, then erosion by wind will be less because resistance produced by clods present on soil surface reduces wind velocity. Along with this when wind velocity is reduced, soil material brought by blow off from other places is also deposited in the bottom of the clods after colliding with these. Loam, silty loam, clay loam, clayey and silty clay loam are such soils in which more clods are formed, so there is less soil loss by wind. Across the wind ridges reduce soil loss by wind erosion.

Length across related area: Any area will be more affected by wind erosion if the length of that area is more. Therein if unconserved and unpreserved area is more, then more will be evil effect of wind erosion. Area length factor "L" is called unconserved or unpreserved distance in length of related area towards direction of wind. It is expressed by L=Df-Db, where in Df is total length of area and Db is preserved distance.

Vegetative cover: Vegetative cover reduces very much the velocity of wind blowing in close contact of soil surface. So, vegetative cover has an important place in stopping wind erosion. Quantity, type and layout of vegetative cover directly influence wind erosion.

Revised Wind Erosion Equation (RWEQ)

The revised wind erosion equation (RWEQ) makes annual or period estimates of wind erosion based on a single event wind erosion model. It includes factors for wind and rainfall, soil roughness, the erodible fraction of soil, crusting and surface residues. The WEQ (Wind Erosion Equation) was reported by Woodruff and Siddoway in 1965. Some changes in WEQ were made after 1965, but no field data were available to support or confirm that the changes were scientifically correct or valid. The RWEQ (Revised Wind Erosion Equation) was reported in 1998 (Fryrear et. al.). RWEQ incorporates advances in wind erosion science after 1965. The various components of RWEQ were developed from controlled laboratory and field studies and the actual transport model was calibrated using field data of transport mass. The average soil erosion can be determined from transport mass and field length.

Once the transport mass equation was developed, transport mass could be estimated for any site where wind erosion was a significant problem. Actual field transport mass was used to test RWEQ in a wide variety of field conditions in 20 states. RWEQ field data was collected in regions with 191 to 1255 mm annual rainfall, and the elevation at the validation sites varied from 141 to 1341 m. Sand content in the surface soil varied from 10 to 87%. Erosion measurement periods were from three to thirty-six months. While most of the sites were 2.6 ha. circles other sites were a 52 ha. circle and 36 to 128 ha. rectangular fields.

The RWEQ model is useful to assess the effects of various cropping systems, wind barriers, soil roughness, hills and residue management systems to determine the daily, weekly or annual erosion. In addition, the daily transport mass can be used to identify periods when crop injury is most likely to occur and the protection required. The RWEQ model will operate with a minimum of input data. However, it needs good weather data to obtain good estimates of soil erosion.

Factors Influencing Soil Erosion

Factors influencing erosion are climate, topography, soil, vegetation and management. Soil erosion cannot be taken place even in the presence of agents and factors. If naturally a little erosion took place is considered a part of soil development. Various activities of human, as indiscriminate cutting and logging of forests, mining and digging, construction of roads and buildings, unlimited grazing by animals and to the extent of preparation of fields for crop cultivation also encourage soil erosion. The ecological parameters influencing the agents of soil erosion also influence erosion. In this context climate, topography, vegetation and soil properties are major factors. These factors may be described separately for water erosion and wind erosion, but for the sake of convenience these are briefly described under soil erosion:

Climate: Among various parameters of climate, rainfall and wind velocity are two such parameters directly influence soil erosion. Other parameters of climate such as evapotranspiration, temperature and relative humidity indirectly influence soil erosion.

Soil erosion increases or decreases according to quantity of rain, intensity and distribution of rain and size of raindrops. Quantity of rain influences whole water balance of soil as well as also determines quantity of runoff. Erosive runoff depends upon quantity, duration and intensity of rainfall and weather condition. It is seen that if rainfall is 1.25 cm/ day or less then runoff is very less or negligible. When rainfall is more than 5 cm/ day then runoff occurs continuously.

If size of raindrops falling on soil surface is large, then their erosive effect will be more. Soil particles are separated from soil surface by the hit of large raindrops and soil erosion begins. At the time of rainfall if wind is blowing, then erosive effect of rain increases tremendously. Violent wind increases kinetic energy of rain, so that drops of rain hit on soil surface with more force. Blowing wind changes the angle of falling raindrops and instead of straight falling angular or oblique falling is more harmful. Oblique drops can displace more soil that of straight drops.

Violent wind velocity displaces and transports more soil. Sand dunes deposited by wind instead of being stable move in the direction of wind. So, every time sand deposits at new place. Sand dunes are formed from sand grains due to the interaction between wind and soil. In summer season in desert area of our country wind speed reaches up to 20-27 km/hour. Threshold velocity should be 14 km/ hour for initiation of soil movement. Generally, this velocity remains continuously available from April to June months. Also, in summer season on account of less vegetative cover, erosion of unconserved soil is more in these months than other months of year. Rainfall also decreases wind erosion, because vegetative cover begins to increase.

Topography: Degree, gradient or percent, shape and length of slope and size and shape of the watershed are few topographic features that influence soil erosion. On steep slopes, runoff water is more erosive, and can more easily transport detached sediment down the slope. On longer slopes excess accumulation of overland flow tends to increase rill erosion. Concave slopes are less erosive than convex slopes. On normal levelled lands runoff water is very less, because greater quantity of water is absorbed by soil and some is filled in pits etc. present on soil surface. The deviations due to microtopographic conditions develop different microclimates which generate differences in floral and faunal abundance and diversity, soil moisture, temperature and organic matter to ultimately affect soil properties.

Slope percent: It has been reported that with the increase of slope of land, velocity of flowing water and its sediment carrying capacity percent increases many times. Effect of slope percent on soil loss and runoff has shown in table 4.5a

Table 4.5: Effect of slope on soil and runoff losses

Slope percent	Soil loss (tons/ha)	Run off loss percent
0.5	3.0	34
1.5	9.4	42
3.0	17.6	51

These data of annual report, Utter Pradesh Soil Conservation Secretariat (1976) clarify that with the increase in slope percent (0.5 to 1.5) soil loss is increased about three times and with 1.5 to 3 percent slop soil loss is doubled. If slope percent is six times (0.5 to 3 percent) then runoff percent increases one and half times. It indicated that adverse effect of slope is more on soil loss as compared to runoff. Central Soil and Water Conservation Research and Training Institute (1986) of our country have given similar data in table 4.5b.1with 1056.1 mm rainfall.

Table 4.5b: Effect of slope on soil loss and runoff loss

Slope percent	Runoff		Soil loss(ton/ha)
	Mm	Rain percent	
0.5	139.4	13.2	6.6
2.5	163.3	15.5	11.7
4.5	221.1	20.9	24.1
9.5	304.1	28.8	34.3

These data proved that with the increase in percentage of slope, runoff and soil loss both increases. Also, adverse effect of slope is more on soil than runoff.

Generally, it is seen that when velocity of flowing water increases twice, then it receives 64 times large particles, 32 times more material in suspension and achieve 4 times more erosive force. Slope percent or gradient of land has negligible effect on splash erosion but increases enormously rill erosion.

Length of slope: In addition to percentage of slope, length of slope also has specific effect on soil loss. Soil and runoff losses increase with the increase of slope length. Although length of slope is less effective than slope gradient or percentage. Velocity of runoff increases with the increase of slope. This increased velocity cooperates in separation and transport of soil. Rill erosion is more on long slopes than short slopes, because increase in length of slope, total discharge also increases. But for the purpose of soil loss estimation, it is necessary to consider soil type, steepness of slope, soil and crop management, rainfall characteristics etc. also.

However, length of slope always does not increase soil erosion. Increase in length of gentle slope helps in the deposition of large particles with the reduction of soil loss. This is true as indicated data given in table 4.6. Effect of length of slope on losses of soil and runoff can be changed by help of land utilization, crop rotation and cultural practices.

Table 4.6: Effect of length of slope on soil losses

Length of slope (meter)	Soil loss (ton/ ha)	Runoff loss (Percent of rainfall)
18.3	1.8	26.4
36.6	1.5	24.4
54.9	1.1	21.0

Shape of slope: Shape of slope also directly influences soil erosion. Shape of slope may be uniform, convex, concave or complex. Often it is seen that convex shape of slope raises the velocity of runoff and ultimately increases soil separation and transport. Contrary to this velocity of runoff is reduced on concave shape, so that deposition of soil particles increases. Thus, soil erosion caused by slope gradient and length of slope is significantly influenced by shape of slope. Surface runoff and soil erosion is also affected by aspect of slope due to microclimatic and microtopographic effects.

Soil properties: Sensitivity towards erosion of any soil depends on its properties. In these properties physical, chemical and mineralogical, all three types are included. If other conditions are similar only soil properties can increase or decrease soil erosion up to thirty times.

Size of soil particles: Size of soil particles play an important role in the formation and transport of sediments. Necessary threshold force also depends on particle size for separation and movement of particles. If particles are large in size, more force is required for their transport. Similarly, for fluid drag of large particles by wind or water, more force is required than small particles. Generally, soil particles of nearly 0.1- 0.5 mm size are eroded easily. Loamy sand with more particles between 10- 100 microns size, is the most vulnerable soil erosion by wind. More clayey soil is much stickier, better-structured, and hence more resistant. Coarse sand and gravelly or rocky soils are also more resistant, since the particles are too heavy to be removed by wind erosion.

Size and stability of soil aggregates: Size and stability of soil aggregates also influence soil erosion. Mutual bonding of soil particles, dispersion of these in water, change in soil surface by hitting raindrops and water entry in soil depend on soil structure. If soil aggregates are not mutually and firmly tied than they will very easily disperse in water, invading raindrops will easily quickly break soil aggregates and if water will not enter in soil easily, then by increase in runoff, erosion will also increase.

Generally, large size soil aggregates resist more soil erosion. If any soil has more quantity of water stable aggregates as well more aggregates of high mean weight diameter, then there will be less erosion of that soil. There is less erosion of soils where soils have more quantity of aggregates of 0.25 to 0.5 mm. Unevenness of soil and tendency of formation of clods also influence erosion. Uneven or rugged soils are more resistant towards wind erosion than soils having smooth surface. Generally, soils having fine tilth are more sensitive to both water and wind erosions.

Crust formation on soil surface restricts easy entry of water in the soil which increases runoff and soil erosion. Soil compaction by increasing runoff also increases soil erosion by water, but it helps in control of wind erosion. There is less wind erosion in compact soil than loose soil. Soil erosion begins to reduce with the increase in the quantity of organic matter in the soil. This is due to improvement and stability in soil structure, increase in soil porosity, water infiltration rate and water holding capacity.

Silica-sesquioxide ratio of soil: Silica-sesquioxide ratio of soil influences erosion. Laskar and Govindarajan (1980) indicated that more Silica-sesquioxide ratio reduces aggregation in soil. So, soils rich in silica-sesquioxide ratio are more sensitive towards sheet and gully erosion.

Presence of clay mineral in soil: Various clay minerals present in soils have different effect on soil erosion. Red and yellow soils of our country are less eroded than black soils. Its main reason is presence of kaolinite, halloysite and aqueous oxides of iron and aluminium in red and yellow soils. These clays and minerals are very much resistant towards dispersion. In black soils smectitic clay minerals are found. These minerals have quality of swelling in the presence of water and shrinkage in absence of water. Due to this repeated swelling and shrinkage wide and deep cracks are formed. Runoff water flows quickly through these cracks and allows movement of soil particles to cause soil erosion.

Presence of cations on exchange complex of soil: Presence of cations on exchange complex of soil also influences soil erosion. If there is excess of sodium than calcium in soil, then erosion of soils will also be more. Its main reason is that sodium disperses soil immediately. More dispersion increases saltation and rill erosion. Permeability of soils having high quantities of sodium is very much less than soils having calcium. Because of less permeability water enters in soil in less quantity. So, more amount of water is available for runoff. There will be more soil erosion because of more runoff. More amount of calcium + magnesium in alluvial soils of Utter Pradesh has negative correlation with erosion i.e., with the increase in quantity of Ca+Mg ions erosion of soil is reducing.

Vegetation: Natural vegetation, trees and plants save the soil by hit of raindrops. The vegetations intercept enough rain water above themselves and left less available water for runoff. Vegetations reduce velocity of water flow and force of wind is reduced by vegetation. So, soil erosion is reduced. Terribly of soil erosion is reduced according to density of vegetation, type, height etc. In soils devoid of vegetation water and wind both erosions are visible in most terrible form.

Land use and management: Improper use of land remarkably increases soil erosion. Ploughing of field etc. also encourages soil erosion. More ploughing opens soil and oxidizes organic matter present in it. Low available organic matter decreases soil porosity and low porosity increases runoff. Therefore, erosion is increases. Ploughing also forms a hard layer at the depth of plough in the soil. This hard layer reduces infiltration of water in the soil. Consequently, runoff increases and so more erosion of the soil.

In view of reduction in erosion the very best option of land utilization is forestry. Lowest erosion is found in forest land than pastures, perennial crops, cultivable lands and uncovered lands. But irregular logging or cutting of forests, shifting cultivation, irregular grazing by animals etc. increase soil erosion.

In land management besides the influence of soil and climate, here and there partnership of various social, economic and financial agencies horribly enhances problem of soil erosion. No one care thing which belongs to all. So, land which is used by all nearby residents as common forests, pastures or other public utilities is not properly cared and managed. Such land gradually becomes prey of erosion and other bad conditions.

In the economy of a country, agriculture, forestry and animal husbandry have specific importance. In context of soil conservation appropriate management of these is very essential. On account of to stop soil erosion by agriculture, lesser use of machines to cultivate lands, lesser interculture operations, fallow with vegetation, mixed farming and agroforestry should be adopted. In case of forestry, special attention should be paid on cleaning of selected land, high level forestry, planting of perennial, woody and timber trees with dense canopy and controlled grazing. Similarly in the animal husbandry selected cleaning, establishment of pastures with the help of less interculture operation and growing of leguminous forages in pastures is most proper to stop soil erosion.

Depletion of Organic Matter in Soil

Depletion of organic matter or organic carbon in soil is second most important vulnerability of soils. In addition to individual nutrients like N, P, K etc. soil organic matter (SOM) or soil organic carbon (SOC) plays an important role in depletion and restoration of agricultural production. Soil organic carbon is more concerned as it is a major portion of large molecules of soil organic matter. Both soil organic carbon and soil organic matter feed microbes in soil, allow them to perform nutrient cycling functions and positively influence soil properties. Soil organic carbon is not considered as plant nutrient but considered as a major input for crop production. Originally it comes from the atmosphere in the form of carbon containing plant remains and residues that provide carbon fixed by them in soil. In addition, by far the largest pool of nitrogen in soils is held in N atoms within many types and sizes of soil organic molecules, and also within the bodies of soil microbes.

Dalwai (2023) indicated to reporters at Nagpur that SOC is the main component of the soil organic matter and gives soil its water-retention capacity, structure, and fertility Soil organic carbon content fell from 1% to 0.3% in 70 years in India and such a drastic fall in the SOC content affects the productivity of soil as micro-organisms do not survive, a key factor to provide nutrients for plants. Intensive cultivation of crops without providing proper compost to the soil is the reason behind the fall in SOC content. Farmers should reduce their over-dependence on pesticides and fertilisers. Biofertilizers and compost can increase the SOC level of the soil. Presently about 63 Pg soil organic carbon has been reduced to 22.72 ± 0.93 Pg,

The knowledge of OC in terms of its amount and quality in soils is essential to sustain the quality and productivity of soils. The first ever estimate of organic carbon (OC) stock in Indian soils was 24.3 Pg (1 Pg = 1015 g) based on 48 soil series considering of a few major soils. The present OC stock has been estimated as 22.72 ± 0.93 Pg,in the first 150 cm depth of soils. Soils with their geographical distribution through the country were used for computing soil organic carbon (SOC) stock in different depths in various physiographic regions. This has more relevance in soils of the tropical and subtropical parts of the globe, including the Indian sub-continent. SOC stock appears to be a single parameter that can help effectively in prioritizing the area for restoring soil health.

Reliable mapping of soil organic carbon (SOC) and Soil Inorganic Carbon (SIC) densities and estimates of their pool size are important from global warming perspective to understand the sequestration potential and losses. This study by using various remote sensing parameters indicated that the SOC and SIC densities up to 100 cm depth or paralithic contact (whichever is shallower) were estimated for 1198 soil samples located across India using a stratified random sampling that integrated land use, soil, topography and agro-ecological regions.

Using Random forests (RF) based spatial prediction procedure with climatic, land cover, rock type, soil type, multi-year NDVI, irrigation status as independent input variables, models for predicting carbon density at 250 m spatial resolution were developed. For modelling with RF algorithm, about 898 soil profile observations (75% observations) were used, while the rest of 300 (25% of total observations) were used for validation. It was observed that the data distribution of sample points does not have significant influence on RF model predictions. The relationship between observed and predicted values was characterized by Mean Squared Deviation (MSD) and Root Mean Squared Error (RMSE) parameters. The SOC, SIC and total soil carbon pool size of India has been estimated at 22.72 ± 0.93 Pg,12.83 ± 1.35 Pg and 35.55 ± 1.87 Pg, respectively, which are comparable to previous studies while providing first spatially explicit 250 m map of their distribution. The spatial distribution indicates that majority of the carbon stock resides in the northern part of India. The soil carbon stock of eastern India has contribution from organic carbon, while the western portion has contribution mainly from inorganic carbon.

Loss of Plant Nutrients

Soils are a healthy, nutrient-rich growth medium for plants that provide plant parts containing most of the elements required for human life when they are consumed. Sometimes I think that food I am taking now a day is not nutritive as earlier which I was taking in my childhood. This is the observation of many people. This is because of soil depletion of plants nutrients. Therefore, crops grown decades ago were much richer in vitamins and minerals than the varieties most of us get today. Crop removal of nutrients, soil erosion and modern intensive agricultural methods have stripped increasing amounts of nutrients from the soil in which the food we eat grows. It is very true that plant nutrient status of soil directly influences the nutritional value of produce.

Nutrient depletion of soils is a widespread soil degradation phenomenon that occurs because of soil erosion (it is the topsoil in which generally most soil nutrients are present that erodes fastest) but also because of poor management practices, such as slash and burn and other subsistence agricultural practices that do not replenish the nutrients taken out the soil by the crops. Note that nutrient mining is only considered in agricultural areas. No nutrient depletion is expected under other land uses like forestry and pastures.

Negative nutrient balances in most Indian soils not only mirror poor soil health, they also represent severe on-going depletion of the soil's nutrient capital, degradation of the environment, and vulnerability of the crop production system in terms of its ability to sustain high yields. In the prevailing regime of widespread negative nutrient balances, it is difficult to foresee positive nutrient balances in most parts of India, even when all available sources of plant nutrients are deployed, unless their quantity and efficiency is raised substantially. Depleted soils cannot be expected to support bumper crops or high growth rates.

Nutrient balance sheets in most soils of India have been deficient and continue to be so. This is primarily because nutrient removals by crops far exceed the nutrient additions through manures and fertilisers. For the past 50 years the gap between removals and additions has been estimated at 8 to 10 M t N+P_2 O_5 +K_2O per year. This has been the case in the past, at present, and this will likely continue. The soils are becoming depleted and mining soils of their nutrient capital, leading to a steady reduction in soil nutrient supplying capacity. On top of this deficit are the nutrient losses through various other means. For example, nutrient losses through soil erosion are alarmingly large, but are rarely considered.

Soil Salinity, Sodicity, Acidity

Wealth of any nation are his soil, water, air, living beings and other substances which are produced there or other useful things are manufactured by them. Civilization and culture relate to his soil. Destruction of the soil destroyed many civilizations of the world during various periods.

Soil salinity: Soil salinity is an index of the concentration of salts in soil and is usually expressed as electrical conductivity (EC). Soil salinization is a process by which there is build-up of salt concentration in soil to such a level that impacts on the agricultural production, environmental health, and economics and quality of life. Soil salinization involves a combination of processes like evaporation, salt precipitation and dissolution, salt transport, ion exchange etc. Soil salinization has become one of the major environmental and socioeconomic issues globally and this is expected to be exacerbated further with climatic change.

Article Open Access Published: 18 November 2021

Global predictions of primary soil salinization under changing climate in the 21st century Amirhossein Hassani, Adisa Azapagic & Nima Shokri *Nature Communications* volume 12, Article number: 6663 (2021) Cite this article

General properties of saline soils are pH Less than 8.5, Electrical conductivity (dS/m) More than 4 and Exchangeable sodium percent Less than 15

Development of soil salinity: Soil salinity generally develops under low rainfall and high temperature conditions. Therefore, due to higher evapouration soluble salts deposit on the soil surface. Poor drainage conditions and banks of rivers and canals also help in the development of soil salinity. Factors developing salinity are, parent material, ground water, irrigation and ocean.

Parent material: Soluble salts are produced during weathering of minerals. In humid regions due to excess rainfall salts move towards lower depth of soil or mixed in ground water. But due to low rainfall and more evapotranspiration, lesser water is available for the leaching of salts. In arid areas rainfall is 150 to 500 mm whereas annual evapotranspiration ranges from 1500 to 5000 mm. Thus, lack of rainfall remarkably influences the formation of saline soils.

Ground water: Ground water also helps in the development of soil salinity. As and when ground water moves upward, the salts dissolved in it also come at the soil surface. When ground water moves downwards, the salts accumulate in the soil horizons, because they are adsorbed by the soil particles. Capillary rise also helps in the accumulation of salts. There is deficit of water at the soil surface due to the excess evaporation. The ground water moves upwards through capillaries to overcome this deficit. Simultaneously, soluble salts also come upwards and deposited at the soil surface. Salinity of ground water also increases with the increase in the dryness of the area.

Irrigation water: More quantity of soluble salts in irrigation water also increases salinity of soil. In addition to this irrigation greatly increases water table which is helpful to produce salinity. Similarly in canal irrigated areas, farmers do not take care of quantity of irrigation after opening the channel of canal in their fields and generally field is over flooded with water. If irrigation water is rich in salts, then process of salinity development will be very much intense and abrupt. The effluent water is also released by various industrial factories, in which excess salts are dissolved. If this water is used for irrigation, then this makes soil more saline. If this water mixes with underground water, then this will enhance the concentration of salts there. These salts will come above and deposit on the soil surface.

Salinity produced by irrigation with saline water can be calculated by following equation given here:

$$\Delta EC_e = \frac{1-e\left(-\frac{1}{nd}\right)}{I/nd} EC\ iw \quad \text{Where}$$

ΔEC_e = electrical conductivity of saturated extract of soil (milli mhos /cm)

I = depth of irrigation water (cm)

n = Soil porosity

d = depth of root area (cm)

EC iw = Electrical conductivity of irrigation water (milli mhos / cm)

Ocean: In coastal areas on account of tides, salts continue to accumulate on soil surface. In coastal soils quantity of soluble salts may be 3 to 6 percent. Saline soils situated at western coastal area of Maharashtra has formed by tides. Rivers joining in the sea bring salts with them. These salts produce salinity in soils situated at confluence of rivers and sea in coastal areas of West Bengal, Orissa, Andhra Pradesh and Tamil Nādu. Generally, heavy cyclones come in months of April and November. At the time of cyclone sea water extends on nearby lands to cross embankments of sea water. Because salinity of sea water is very much, so these soils also become saline.

Growth of plants in saline soils: Excess of soluble salts in the soil adversely affect plants. In saline soils plants germinate late. Germination percent also decreases. Rate of growth and development is very much less than normal conditions. Generally, distribution of salts in a field is very dissimilar. Salts are accumulated very much on some depressed places. Consequently, at these places seed do not germinate at all or after germination plants die. This results very small or lesser crop stand here and there.

Leaves of plants growing in saline soil are somewhat thick. These are brittle to some extent. Leaves show dark green colouration. Thin layer of wax is deposited on leaves of plants of mustard family and some grasses. So, their colour seems green with some blue tint.

Osmotic pressure of soil liquid of saline soils is 1.37 to 24.37 atmospheres. It is sufficiently more than common soils. Less than 4.3 atmosphere osmotic pressure has not adverse effect on plants. But as osmotic pressure increases roots of plants face difficulty in water absorption for normal development. Between 4.3 to 5.2 atmosphere growth of plants is retarded and on more than 8.2 osmotic pressure germination of seeds also becomes impossible. At equal osmotic pressure different soils, excess chlorides more adversely affect plants than soils having excess sulphate. So, chlorides are more harmful than sulphates.

In soil salinity or in saline soils if salts are present in more quantities, then they will also remain under imbalanced quantities also, where as in normal soils salts occur in less and balanced quantities. The unbalanced quantity any one ion is more in quantity and another is less in quantity. This unbalanced quantity of salts stops growth of plants by lowering the availability of calcium and micronutrients. Occasionally in soils toxic elements are also found. Excess of boron, lithium and fluorine retard growth of plants. There is no bad effect of excess of selenium and molybdenum on yield of forage, but health of animals is badly affected by feeding of such forage.

Soil alkalinity and soil sodicity: There are abundance of alkalis in the soil is called as soil alkalinity and excess of exchangeable sodium in soils refers to soil sodicity. Both these termed as alkaline soils and sodic soils respectively also as salt affected soils. These soils also show excess of sodium carbonate and bicarbonate. Major properties of salt affected soils are shown in table

Table 4.7: General properties of salt affected soils

Soil Type	pH	Electrical conductivity (dS/m)	Exchangeable sodium percent
Sodic soils	8.2 to 11.0	Less than 4	More than 15
Saline- sodic soils	Less than 8.5 sometimes more	More than 4	Less than 15

Formation of sodic soils: In sodic soils, there are excess of carbonates and bicarbonates of sodium. In these soils exchangeable sodium percent also is more than 15 and pH values of these also is more than 8.5.

In these soils, formation of carbonates and bicarbonates of sodium, their translocation and deposition is a natural geochemical process. Later, unscientific use of soils has encouraged this problem. In the presence of carbonic acid water, minerals having lesser stability is weathered immediately. Thus, weathering of aluminosilicates of sodium and potassium of form soluble carbonates and bicarbonates of sodium and potassium. Formation of sodium carbonates and bicarbonates specially take place in those rocks and minerals, which arc of neo volcanic period. Separation of sodium carbonates and bicarbonates produced by aluminosilicates and feldspars takes place by alkaline hydrolysis. When alkaline hydrolysis repeatedly occurs in the nature, then at lesser depth deposition of carbonates and bicarbonates of sodium forms sodic soils. In the depths of soils as alkalinity reduces accumulation of aforesaid salts is also reduced, because action of alkaline hydrolysis also continues to decrease. Finally below the substrata availability of sweet water is a specific property of alkaline soils.

In India sodic soils are found or sodicity develops in areas having low rainfall (550 to 990 mm annual rainfall). In rainy season in low lying areas of that region because of incomplete drainage weathering products accumulated. After rain soil solution becomes concentrated due to excess of evaporation. So, sodium absorption ratio of soils also increases. In addition to this pH also increases. Increase in sodium absorption ratio of soils cooperates in formation of sodic soils. In sodium enriched soils, sodium hydroxide is formed by alkaline hydrolysis. This sodium hydroxide reacts with atmospheric carbon dioxide or carbon dioxide produced by biological activities and makes sodium carbonate. If aforesaid activities continuously goes on without break then sodic soils are formed.

In the formation of sodic soils, cation exchange process also takes part. Sodium present in soil solution is adsorbed there in the exchange of other cations present on soil colloids (clays). In this activity mainly calcium is replaced. At excess pH, replaced calcium is precipitated. By virtue of precipitation of calcium, pebbles are formed in the soil. In pebbles chiefly calcium carbonate exists. In soils as soon as concentration of exchangeable sodium becomes more than soluble salts as sulphates and chlorides of calcium, magnesium or potassium, then solubility of these salts decrease. Now as a result of carbonation these salts are converted in carbonates and are precipitated. In the soil when these salts are precipitated then formation of pebbles begins.

Growth of plants in sodic soils or soil sodicity: Like saline soils, also in sodic soils plants germination is delayed and reduced. A hard layer exists on surface of sodic soils. So, sufficient germination of crops does not take place. Growth and development of plants is also less than normal soils. Its reason is less permeability of these soils, so movement of water also decreases. As a result of this plants do not receive required water to complete their life cycle. By virtue of stagnation of water aeration is badly reduced. Naturally its evil effect is plants' lodging.

Excess of sodium carbonate and bicarbonate ions present in soils produces unwanted effects on plants. Colour of plants becomes light green and yellowish. Roots become thick and black. Excess sodium starts scorching of leaves. Excess of exchangeable sodium reduces availability of calcium to plants. More pH of soil decreases availability of phosphorus, zinc and iron present soil and increases availability of boron. So, because of ionic imbalance growth and development of plants are adversely affected.

Soil acidity: Soil acidity is a condition in which the soil pH is lower than a neutral pH (less than 7). Soil acidity is a useful concept relating to pH, aluminium toxicity, base saturation, cation exchange capacity, cationic ratios, lime requirement, nutrient availability etc. Generally, those soils whose pH is less than 7.00 are called acidic soils. These soils are more responsive towards application of lime. If 1:1 soil water suspension has pH 5.5, these soils are included in the class of acidic soils.

Effect of soils acidity on growth of plants: Generally, soil acidity influences growth of plants by two ways. Firstly, due to acidity, activity of plants roots is affected. Secondly, soil acidity affects soil qualities, by which plants are affected ultimately. Main effect of acidity is due to hydrogen ions. Necessary pH limits for plants have been described earlier. But aluminum ions with hydrogen ions also influence the growth of plants.

Generally, aluminum is not an essential element for plants. But very small quantity of it influences health of soils. Beneficial effect of aluminum has been seen on lucerne, red clover, grasses, maize and tea. But on account of excess aluminum roots become short and thick. Many other effects of toxicity of aluminum also have been seen. It adversely affects cell division, respiration, DNA synthesis, phosphorylation, sugar, calcium and water uptake.

It is supposed that if pH of soil is less than 5 or aluminum has more than 10 percent of cation exchange capacity than toxicity of aluminum can be found in the soil. Symptoms produced by excess of aluminum in acidic soils somewhat like to symptoms of deficiency of phosphorus or calcium. So, it is very difficult to know correct facts without soil or plant testing.

In addition to toxicity of aluminum excess or toxicity of manganese and iron in acidic soils is also found. At soil pH less than 5, accumulation of manganese begins in plant tissues which adversely influences normal metabolic activities. Because in acidic soils more quantity of exchangeable manganese is available than normal soil. With decrease in pH of soils quantity of Fe^{2+} begins to increase. In acidic state a Fe^{3+} begins to convert in Fe^{2+}. However, plants uptake iron as Fe^{2+} ions, so in plants growing in acidic soils symptoms of excess or toxicity of iron are also seen.

Deficiency of exchangeable bases is found in acidic soils. This deficiency becomes more pronounced with decrease in soil pH. This result deficiency of calcium and magnesium in the plants. In acidic soils deficiency of phosphorus and molybdenum also occurs. Excess of iron and aluminum increases phosphorus fixation of phosphorus. It is seen that acidic soils (pH 4.2 to 5.6) of Sikkim had been fixed 90 percent quantity of applied phosphorus. So, in such situation availability of phosphorus to plant is reduced. Plants also face the deficiency of sulphur because of increasing fixation of sulphate ions and decreasing oxidation of sulphur.

Fomation of acidity in soils: In genesis and development of soil acidity or acidic soils. leaching of parent material, uptake of calcium and magnesium by plants, use of acidic fertilizers, acid rain, use of organic manures, alumina, silicate clay mineral and some microbial activities take part.

Leaching of parent material: Generally, soils developed from granite, liparite and quartz rocks are of acidic nature. pH of water of streams flowing through quartzite rocks is often acidic. So, this water can make acidic to normal soils. In acidic soils kaolinite clay mineral is found in abundance. This mineral is found generally in the presence of neutral to acidic pH and contain small quantity of exchangeable and soluble bases.,

In areas where mean annual rainfall is more than 100 cm generally acidic soils are developed. In light textured alluvial soils leaching is found more because of very excess rainfall. Due to more leaching deficiency of basic ions occurs in the soil because basic ions move out of soil and in the lower layers of soil.

Uptake of Calcium and Magnesium by plants: There is more possibility of increasing soil acidity in the soils where such crops are grown continuously which uptake more quantity of calcium and magnesium. So, application of these elements as fertilizers is essential. Generally, wheat takes 13 kg calcium and 9 kg magnesium per hectare from soils. Potato and tobacco uptake calcium (130-165 kg/ ha) and sugar beet uptake 56 kg magnesium oxide / hectare. So, crops like potato, tobacco and sugar beet will immediately exhaust soil bases. When saturation of base ions on soil complex will be less, then to saturate the charge of complex, hydrogen and aluminum ions will be adsorbed. So, soil will become acidic.

Use of acidic fertilizers: Continuous application of ammonium sulphate and ammonium nitrate fertilizers produce acidity by decreasing pH of the soil. Hydrogen ions are produced during nitrification of ammonium sulphate and ammonium nitrate in the soil. It is expressed by following equations,

$$(NH_4)_2\ SO_4 + 4H_2O \rightarrow 2NO^-_3 + 4H^+ + 2H_2O + SO^{2-}_4$$

Ammonium sulphate

$$NH_4NO_3 + 2H_2O \rightarrow 2NO^-_3 + 2H^+_2 + H_2O$$

Ammonium nitrate

It is clear from these equations ammonium sulphate produces hydrogen ions which displaces calcium, magnesium and potassium. Such displaced elements move out of soil by leaching and to neutralize hydrogen produced by ammonium sulphate and ammonium nitrate 7.2 kg and 3.6 kg calcium carbonate, respectively, per kilogram nitrogen used will be needed. In comparison of both fertilizers effect of urea is less to increase acidity. Generally, effect of potassic fertilizers is partly acidic.

Plants uptake of potassium from potassium chloride and potassium sulphate or fixed on clays and to neutralize acidity produced by chloride and sulphate ions, it will be needed 1.1 kg. calcium carbonate per kilogram potassium used.

Acid rain: Burning of fossil fuels (petrol, diesel, kerosene etc.) sulphur dioxide, carbon dioxide, nitrogen monoxide etc. gases are produced. These gases after mixing with water vapours present in atmosphere gradually produce sulphuric acid, carbonic acid and nitric acid. In rainy season these acids come on the earth with rain water and help in forming soil acidity or acidic soils. If calcium carbonate is present in the soil, then it with the help of sulphuric acid will be converted in calcium sulphate. Because calcium sulphate is a soluble salt. So, with flowing water it will go out of soils.

$$CaCO_3 + H_2\ SO_4 \rightarrow CaSO_4 + 2H_2O + CO_2$$

Use of organic manures: In organic matter or humus carboxylic, phenolic, amino etc. organic group are present. These groups are helpful in binding hydrogen ions. Saturated with hydrogen ions these groups act as a weak acid. After dissociation these weak acids produce hydrogen ions, which reduce pH of the soil. In Kerala state black-coloured acidic soils are found which have pH up to 3.9. These soils are very heavy with 10- 40 and 20 to 30 percent iron and aluminium resperatively. Acidity of these soils is due to anaerobic decomposition of organic matter. Similarly in mountainous areas of Utter Pradesh, northern area of Bihar, Assam and some coastal areas genesis of acidic soils or soil acidity is due to accumulation of organic matter.

In acidic soils a specific equilibrium between hydrogen and aluminum ions is found. When by any reason hydrogen ions increase in the soil, then some dissolved aluminium come out or released from aluminosilicate clay minerals. By gradual hydrolysis of this aluminium, hydrogen ions are released which makes more acidic pH of soil. This process can be explained by following equations.

$$Al^{3+} + H_2O \rightarrow Al\,(OH)^{2+} + H^{+}$$

$$Al\,(OH)^{2+} + H_2O \rightarrow Al\,(OH)_2^{+} + H^{+}_{2}$$

$$Al\,(OH)_2 + H_2O \rightarrow Al\,(OH)_3 + H^{+}$$

Microbial activities: During microbial respiration carbon dioxide gas is produced. This gas mixing with water makes carbonic acid. By dissociation of this acid hydrogen and bicarbonate ions are produced. Bicarbonate mixing with calcium make calcium bicarbonate, which is being partially soluble can come out of soil. Remaining hydrogen ions in soil increases acidity of soil. Anaerobic decomposition of organic matter in soil produces various kinds of organic acids. Nitrification of ammonium ions increases hydrogen ions in soil. By nitrification of an ammonium ion produces two hydrogen ions which develops soil acidity.

Soil Pollution and Soil Contamination

"Global soils are under great pressure," said Qu Dongyu, head of the UN food and agriculture organisation. "This thin crust of the Earth's surface, the soil, supports all terrestrial life and is involved in many key ecosystem services that are essential to the environment and to human health and wellbeing." Inger Andersen, head of the UN environment programme (Unep), said: "Soil pollution may be invisible to human eyes but it compromises the food we eat, the water we drink and the air we breathe. Pollution knows no borders – contaminants move through soil, air and water. "It is time to reconnect with our soils, as it is where our food begins," she said. "Soil pollution should no longer be a hidden reality. Let us all be part of the solution to soil pollution."

The future for soils looks "bleak" and their state is as at least as important as the climate emergency and the destruction of the natural world above ground, according to the scientists behind another UN report on soil biodiversity, published in December. Since the Industrial Revolution, about 135 bn tonnes of soil has been lost from farmland and, given that it takes thousands of years for soils to form, urgent protection and restoration of the soils that remain is needed, the scientists said.

The new UN report concludes: "Soil contaminants can have irreparable consequences on human and ecosystem health." The biggest source of soil pollution varies by region, it found. The biggest problem is industrial pollution in western Europe and North America, farming in Asia, Latin America and eastern

Europe, and mining in sub-Saharan Africa. In north Africa and the near east, urban pollution is the biggest single source of contamination. "The fundamental step of identifying the party liable for the pollution is still lacking in many states," the report said. "Soil pollution is expected to increase unless there is a rapid shift in production and consumption patterns and a political commitment towards a real sustainable management where nature is fully respected."

Therefore, we should be cautious and should reach for solutions of this problem soil pollution by keeping in mind pollution, factors of pollution and its harmful effects. Pollution means "such harmful changes in physical, chemical and biological properties of soil or land, water and air, which can put or will put harmful effect on life of human and other wanted species". Among the components of environment (soil, water, air, creatures, sound, society, politics, economics etc.) soil, water and air are major components and these also influence our life directly. So often, these are included in defining pollution.

Roughly pollution is defined as unnatural environment changes done by human, which for itself and for other creatures is harmful, is called pollution. The soil contamination is addition of external material to the original and natural soil. Contamination may or may not be harmful, depending on the level of variation and the nature of the contaminant. The agricultural chemicals we use to increase crop yield or to save them from insects and diseases, they can also be included in the class of pollutants in the future. For example, nitrogenous fertilizers increase quantity of nitrate in the underground water by more than 45 ppm, than that will become in the class of pollutants. Similarly, when Thiodan sprayed to kill hoppers in the field of paddy kills also honey bees, then that will also be include in class of pollutants. It means that any substance which instantly benefits us, but in long run it causes harm to us or any living being and material related to us then it will be called pollutant.

Generally, pollution in the soil is produced by biological wastes, agricultural wastes, industrial wastes, residue of various agricultural chemicals and radioactivity. In this section these pollutants are described in short.

Biological wastes: In urban areas biological wastes pollutes very badly land and soils of there. Domestic, municipality and industrial wastes increase gravity of this problem. Polythene bags in the waste material pollute whole environment very much, because they are not decomposed. So, they remain on land as such till sufficient time. Detergents, water softeners, borates, phosphates and other salts coming with wastes pollute soil remarkably. In our country old newspapers and magazines are purchased by street hawkers. These are used in making bags and other such materials. These bags are thrown later. These are decomposed by microorganisms quickly because of being cellulose in paper. So, it does not appear on soil surface for long duration of days.

But polythene, plastics, foils of aluminium and other such materials remain on the soil surface in undecomposed state and pollute the environment. These wastes do not have any importance in agricultural production. So, for the purpose of disposal, these should be sent to factories where other useful goods could be manufactured from these. Although quality of such products might not be of high standards, nevertheless it is good method of recycling of wastes.

Use of wastes left from recycling can be used as landfills. These landfills should be made in waste lands. Also, it should be kept in mind no river or water reservoirs exist in its vicinity. Otherwise, water flowing through these landfills will pollute the river or reservoir. These landfills are compacted by the help of bulldozers.

Agricultural wastes: Wastes produced by animal and plant residues are main two types of agricultural wastes. Wastes produced by animals include cattle dung, litter and forage, left over of animal feeds, dung of horse, sheep and goat etc. droppings of poultry, faeces of pigs and various animals, wastes of meat, poultry, milk production and processing industries etc. Wastes originated by plants are crop residues as straw, husk, residue of plants and nurseries and wastes of food processing industries.

These agricultural wastes and residues should be managed properly by recycling. These should be converted in manure, compost, bio gas, animal feeds and other useful substances. Residues of fruits and vegetables can be used as animal fodder. Thus, dry peel and seed of tomato can be mixed with in poultry feeds. Dungs of horses can be used for the purpose of mushroom production after its proper cleaning. Straw of paddy and wheat, stalks (karabi) of maize, sorghum, bagasse of sugarcane etc. can be used in paper industry to make paper. Furfural can be obtained by hydrolysis of hull of paddy and oat and ear of corn. Synthetic resin and adhesives are forms by furfural. Tartrate is obtained from residues of wine grapes. Monosodium glutamate is obtained from bagasse of sugarcane.

Leather, soap, glue, gelatine, animal food, hormones, vitamin and enzymes are obtained from wastes and residues of meat processing industry. Human waste is considered very bad material. People defecate (discharge waste material from body) in fields of villages. Later, flies sit on this. These flies come in the house, pollute food and water. If defecation is done in small pits in the fields and is immediately covered by soil, the environment also will remain clean and fertility of soil of the field will also increase because human defecation has plant nutrients in sufficient quantities.

Industrial Wastes: There is a tremendous contribution of modern industries in polluting soil. Smoke, soot and various gases emitted in the air by chimneys of these industries pollutes the soil. Here, with heavy metal thrown as debris of industries also pollute the soil. Based on speed and direction of air pollutants released in the air by industries can reach long distances. Ashes and soot released in the air by chimneys of industries fly in suspended condition in the air. When speed of air slows then ashes and soot deposit on soil surface, vegetation and building due to sedimentation. Afterwards remaining ashes comes down with rain water in the rainy season.

Fly ash and general ashes released by of thermal power houses pollutes soil very much. Since fly ash has calcium and iron in sufficient quantities, it can be used in the fields otherwise it is most suitable for cement, bricks, construction materials etc. Ashes are also used as soil amendment and land filling because of its quality. In the soils near aluminium factories excess of fluorides occurs. Due to excess of fluorides teeth become defective and bones becomes curved. Super phosphates, phosphoric acid and ceramic industries also release fluorides in high quantity.

Production of Sulphur dioxide as a waste continues to increase although in small quantity of this gas is produced by natural processes. But it is produced mainly in thermal power plants, metal industries, oil refineries and similar factories. In these factories coal and petroleum products are mostly used. Bituminous coal produces more SO_2 gas than subbituminous coal. Gas arrived in the atmosphere from chimneys of factories is mixed with atmospheric moisture and produce sulphurous acid and sulphuric acid. In rainy season these acids come on the land as acidic rain and spoil vegetation, soil and buildings of concerned area. Aforesaid fact can be understood easily by following equations:

$$SO_2 + H_2O \rightarrow H_2SO_3 \rightarrow H^+ + HSO_3^-$$

$$SO_2 + O_2 + \text{Sunlight} \rightarrow SO_3$$

$$SO_3 + H_2O \rightarrow H_2SO_4$$

$$CaCO_3 + H_2SO_4 \rightarrow CaSO_4 + CO_2 + H_2O$$

Calcium sulphate is highly soluble in water. Also, nitrogenous gases (NO, NO_2, N_2O) emitted out from exhaust pipes of motor vehicles after mixing with atmospheric moisture produce nitrous and nitric acids:

$$2NO + O_2 \rightarrow 2NO_2 + H_2O \rightarrow HNO_3 + HNO_2$$

In addition to these acids, carbonic acid (H_2CO_3) is also found in rain water. Carbon Di-oxide (CO_2) when mixed with water, makes this acid. Besides these various gases dissolved in rain water make water acidic and its pH become very less. All these causes Acid rain. Acid rain is also caused naturally by lightning, volcanic eruptions, and by forest fires.

Acid Rain

In 1974-75 rainfall in Mumbai recorded 4.8 pH and Khemani et.al. (1989), Mahadevan et.al. (1989) and Verma (1989) studied pH of rain water during last 8 to 10 years in cities across the country. According to them mean pH of rain water in Kodai Kanal and Jodhpur was 6.02 and 7.25 respectively. Port Blair and Nagpur had 6.12, Vishakhapatnam had 6.42, Minicoy had 6.50, Pune had 6.60, Allahabad had 6.82 and Srinagar had 7.06 pH readings. The Institute of Chemical Technology, Hyderabad points to a concern about acid rain over oceans. Analysis of rainwater samples from Nagpur, Mohan Bari (Assam), Allahabad, Visakhapatnam and Kodaikanal in the decade 2001-2012 showed a pH level varying from 4.77 to 5.32. This indicated that these places also received acid rain.

The term 'acid rain' was first used in 1852 by Scottish chemist Robert Angus Smith during his examination of rainwater chemistry near industrial cities in England and Scotland. The phenomenon became an important part of his book 'Air and Rain' and the Beginnings of Chemical Climatology (1872). However, it was not until the late 1960s and early 1970s that acid rain was comprehended as a regional environmental issue affecting large areas of Western Europe and eastern North America. Acid rain also occurs in Asia and parts of Africa, South America, and Australia.

Now acid rain and its harmful effects are becoming common in developing countries also. It is a significant problem in the US, Canada, Germany, Scotland, Brazil, India, China etc. Industrialised areas of Indonesia, Malaysia, the Philippines and Thailand will also experienced heavy acid rain deposition in a few years.

SO_2 gas is main contributor of acidic rain. In areas where pH of soil is more, these acids received from acidic rain reclaim land reducing pH of soil. Also, Sulphur coming in soil through acidic becomes a part of sulphur cycle of soil. This sulphur is beneficial for plants. But when pH of soil is neutral or acidic then harmful effect of acidic rain can be seen clearly. Too much lowering of pH exerts negative effect on the growth of plants and balance of nutrients in the soil. Finally, soil fertility and productivity are adversely affected. Although in our country problem of acidic rain and its ill effects are not very grave, but we must think in advance, otherwise raindrops will spoil the land instead of gratification.

Heavy Metals

Various industries, vehicles, sludge and sewage pollute soil by increasing quantity of heavy metals in the soil. Metals having density higher than 5 g /cc are included in heavy metals. Though there are numerous heavy metals, but heavy metals of agricultural importance are Copper (Cu), Iron (Fe), Manganese (Mn), Molybdenum (Mo), Nickel (Ni) and Lead (Pb). All these metals are harmful for human. But relative harmfulness of these are severe. Cadmium is highly poisonous. Mercury, Lead and Nickel are medium poisonous.

Cadmium: Cadmium obtained from metal smelters, incineration of wastes and pyrolytic furnaces of coal and oil. This substance damages kidneys and weakens bones.

Mercury: Mercury comes in the soil from furnaces of coal and oil, smelters, paper industries, chlor-alkali industry and use of processed seeds. Mercury effects central nervous system and kidneys.

Lead: Earlier Lead was a main constituent of petrol. Its compound tetraethyl lead was mixed as an antiknock material. Lead is used in jointing and cementing of pipes, water and cables of electricity and telephone. Smelters, mines and batteries, PVC plastics and paints industries also use lead is remarkably. lead comes in the soil from these sources through air, water or vegetables. As a result of this translocation of calcium, bone marrow, haemoglobin and body parts are adversely affected. Gupta et.al. (1994) studied Pb in six soil series of Pant Nagar. They informed that quantity of this heavy metal is found highest in Ap horizon and lowest in C horizon. Its quantity in Ap horizon was found 9 to 24 ppm while its quantity in C horizon was up to 7 ppm. They also observed that maximum quantity of lead is found in soil organic matter.

Nickel reached in soil mainly as wastes of stainless steel industry, nickel cadmium battery industry, coal and oil furnaces, fertilizers industry and various pigment industries. Nickel mainly influences the possibility of lung cancer.

Copper plays an important role in human, animal and plant nutrition. But its excess also is harmful. This reaches in soil from many domestic goods, chemical industries, copper mines and sprays of cupriferous fungicide, Bordeaux mixture etc.

Zinc reaches in soil mainly from galvanizing units and zinc smelters. Viscose rayon and rat catcher and fungicide industry also pollute soil by zinc.

Manganese reaches in the soil from mines of manganese, dry cell battery, Ferro-manganese industries and burning of fossil fuel. Manganese is not supposed a major soil pollutant.

Excess of iron is found in the soils near iron mines, iron and steel industries and steam power plants.

Besides these heavy metals Arsenic (As) and Chromium also pollute soils. Arsenic reaches in the soil from tanneries, glass and ceramic industry, chemical industry and use of lead arsenate named insecticides. Ignition of petroleum, coal and oil furnaces and effluents of some fertilizers industries also increases quantity of Arsenic up to pollution level. Arsenic can cause skin and lung cancer. Smoking person are most susceptible to Arsenic rising cancer.

Chromium (Cr) is used in the dyeing industries. Wool, leather, textile, glass and ceramic and electroplating industry use chromium in abundance. In dyeing of cotton cloth, specially dyeing of Khaki dress, this metal is used remarkably. This metal reaches in soil by various ways and means from these industries and increases a possibility of cancer in trachea in human.

Sewage and sludge also pollute soil from heavy metals, because they have very high quantity of these heavy metals. Sewage means mainly waste water coming from houses and up to 80% of water supplied to houses comes in this waste water. Remainder 20% is used in drinking, cooking food, cleaning and other house hold works. But now sewage is taken in other sense. Accordingly, sewage includes water from houses, hospitals, laundries, run-off of rain reached in sewer pipes, effluent water of slaughter houses, other industries etc.

Effluent water of industries is full of dissimilarities, in nature, properties and qualities. Effluent water of industries besides having inorganic elements and compounds also contains fats, proteins, cellulose, resin etc. Mixing of wastes of breweries and dairies, tanneries, paper industries, distilleries etc. increases biological demand of oxygen of sewage and pollution level.

Often raw sewage is discharged in rivers, streams or low-lying lands directly or used to irrigate fields. This discharge or use directly for irrigation is a very wrong practice. It results besides soil pollution, increase chances of diseases.

Behaviour of Heavy Metals In Soils and Plants

Heavy metals show very diversified behaviour in the soils and plants according to their own properties, properties of soil, characteristics and species of plants and amount of accumulated heavy metals in plants and soils. Brief discussion about behaviour of major heavy metals in soils and plants is given here,

Cadmium: Cadmium in the soil come by weathering of parent rocks, fertilizers and soil amendments. It is considered that uptake of this element reduces in the presence of organic matter, silicate clays, hydroxides of iron and aluminium and poor aeration. Cadmium is not an essential nutrient for plants. But in some conditions like use of sludge or sewage having excess of cadmium and acidic pH of soil, it accumulates in plants up to very harmful level for human and animals. Major part of Cadmium taken by plants remains in the vegetative branches than seeds of the plants. So, more use of leafy vegetables or fodder can also increase its concentration in human and animals. Besides pH of soil, redox potential also influence uptake of cadmium. In the oxidized state uptake of cadmium is more due to more solubility and in the reduced state cadmium is less soluble.

Mercury: Methylation of mercury is most harmful for soil pollution. This process converts insoluble inorganic compound of mercury to soluble organic compound, Dimethyl mercury (CH_3HgCH_3). Before dimethyl mercury, methyl mercury (CH_3Hg^+) is formed, which is absorbed by plants. Its quantity increases very much in fishes raised in water polluted by methyl mercury. By using such fish in the diet, mercury also reaches in body of men. Methylation of mercury is a microbiological activity. In the mud of lakes methane producing bacteria makes methyl cobalamin compound. This compound reacts with mercury and makes dimethyl mercury. Bacteria perform this reaction in the soil, rivers and skin and guts of fish. Inorganic compound of mercury coming in the soil are absorbed very intensely by organic matter and clays. So inorganic compound becomes insoluble and ultimately unavailable for plants. Inorganic compounds can convert in mercury metal after reduction. In the reduced state, by volatilization these may spread here and there in environment.

Lead: Very minute particles of lead produced by various industrial and transport activities of human reach in the atmosphere. Hence these particles are accumulated in vegetations, leaves of plants, soil surface and open water sources. Particles of lead settled on soil surface can reach in various reservoirs and rivers by runoff water and from there can go in the sea. Lead present on surface of leaves come in animals' body after grazing of such leaves or come on soil surface with rain water. Movement of that towards lower layers happen very less. Lead present in the soil does not available to plants because in common conditions this remains in insoluble state. High acidic conditions increase its solubility. So on amelioration of acidic soils by lime, availability of lead in these soils and uptake by plants decreases.

Arsenic: Besides insecticide Lead arsenate various inorganic and organic compounds of arsenic are also used as herbicides. Behaviour of arsenic is somewhat similar as behaviour of mercury and phosphate ions in the soil. Some bacteria and fungi mix it with methane to make methyl arsenic (CH_3AsCH_3). Negative arsenic tetra oxide ions (AsO_4^-) are adsorbed by hydrous oxides of iron and aluminium as phosphate ions are adsorbed. Anion exchange releases adsorbed arsenic ions for other anions (phosphate) in soil. Excess of arsenic in the soils reduces the availability of iron to plants and general growth of plants. Toxicity of arsenic in the soil can be reduced by use of sulphate, zinc, iron or aluminium, although in normal soils toxicity of arsenic in soils is rarely found.

Copper: Copper is found in the soil as water soluble, exchangeable (Cu^+, Cu^{2+} $CuOH^+$), hydrous oxides (CuO, $CuOH$, $Cu(OH)_2$) and complex (Copper Humate) and minerals. This is available in the soil as monovalent cuprous and bivalent cupric in both forms. But which form and when it will be depends on redox potential. With forming various complexes and chelates of copper in the soils this can be adsorbed by clays. Generally, exchangeable copper is considered in available form in the soils and plants only uptake Cu^{2+}.

Iron: Iron is found in soil as water soluble, exchangeable (Fe^{2+}, Fe^{3+}), hydrous oxides ($Fe(OH)_2$, $Fe(OH)_3$, Fe_2O_3, $3H_2O$) and minerals. In soil, iron remains in relatively immobile state. These can be adsorbed on surface of soil colloids.

Manganese: In the soil, manganese is found in various form of water soluble, exchangeable (Mn^{2+}), hydrous oxides (MnO, $Mn_2O_3.H_2O$) and high manganese oxides (Mn_2O_3, MnO_2). Soil acidity increases due to reduction in pH, availability of manganese increases. Increase in soil pH also increases quantity of insoluble oxides of manganese is the soil. In the soil Mn^{2+} is only such form which plants can use.

Zinc: In the soil zinc is found as water soluble, exchangeable (Zn^{2+}, $ZnOH^{+}$, $ZnCl^{+}$) and oxides and relatively insoluble salts (Zn), Zn $(OH)_2$, $Zn_3(PO_4)_2$). Deficiency of zinc is not found in soil having pH value less than 6.0, because in acidic conditions zinc remains in soluble state. Soluble zinc applied in soil is adsorbed and trapped in lattice structure of colloid complexes. Plants absorb bivalent zinc (Zn^{2+}).

Generally, various responses of copper, iron, manganese and zinc in the soil depends on many factors. In these soil pH, amount of organic matter, submergence of soil, type of soil and redox potential of soil are major factors. On account of more solubility of these elements these are available to plants in more quantity on low pH (acidic state). Similarly, more submergence of soil also increases availability of these.

Agricultural Chemicals

In modern agriculture to increase yield of crops and to save them from various insects, pests, diseases, weeds etc. various chemicals are used. These chemicals may be organic and inorganic. These chemicals used in agriculture, pollute soil and other components of environment. Although problem of these is not remarkable yet. But before these agricultural chemicals let damage extensively to our environment in future, we shall think about harmful effects of these, beneficial and economic management and biological alternatives, well in time.

Agricultural chemical may be divided mainly in two groups,1. Yield increasing chemicals 2. Plant protection chemicals. Plant protection chemicals pollute soil more than fertilizers. Yield increasing chemical are mainly chemical fertilizers. But plant protection chemicals mainly include insecticides, fungicides, weedicides, pesticides and rodenticides mainly.

Pollution of soil by insecticides: Everyone knows the importance of insecticides in agricultural production and human health protection. Insecticides reach in abdomen of insects and destroy sensory and respiratory activities of insects. Insecticides are kept in three major classes of chlorinated hydrocarbons, organophosphates and carbamates.

In chlorinated hydrocarbons DDT and BHC are major insecticides. Insecticides of this class persist in the environment for several years. More persistent of insecticide will pollute greater for longer period. Insecticides of this class are capable to destroy a very large genera of insects. But due to continuous use and long persistent period of 10-11 years of DDT and BHC, insects become tolerant and developed resistant species. Therefore, to solve this problem more toxic chemicals are being discovered and used. These are more adversely affecting the environment.

Organophosphates are less persistent than chlorinated hydrocarbons. These are easily degradable on the surface of plants, under plant and in the soil and water. So, these are less harmful for environment. In this group many insecticides are included namely diazinon, parathion, malathion are major insecticides.

Persistency of insecticides of carbamate group is very less. These are degraded within one to three months. In this group carbaryl, carbofuran, and aldicarb are major insecticides. Environmentalists like very much insecticides of this group, because these are very quickly biodegradable. Therefore, these are not more harmful for mammals.

Insecticides pollute soil by various ways, which are given here

1. Continuous and considerable use of insecticides also destroy other beneficial soil organisms and decrease their population.
2. Decrease of soil flora and fauna enhances population of some harmful organisms in soil.
3. After entering in the body of various soil organisms these insecticides can reach in other soil organisms through food chain.
4. These insecticides may induce mutation in organisms present in soil and can develop resistant species.
5. These can change in genetic, metabolic and reproduction related behavior of microorganisms present in soil.

Degradation of insecticides: Degradation of insecticides used directly in soil or sprayed on the crops can be done by both chemical and biological methods. In chemical method soil particles perform this work, while in biological method both anaerobic and aerobic microorganisms take part. Degradation of insecticides is influenced by soil moisture, quantity of fertilizers broadcasted in the soil, soil pH, soil temperature and cation exchange capacity of soil.

Pollution of soil by Fungicides

Fungicides mainly destroy reproductive organs of fungi present on plant surface. These also completely prevent growth and development of fungi. Continuous use and higher doses are also harmful for soil just like insecticides. In fungicides, there are copper compounds (Bordeaux mixture and Burgundy mixture), Sulphur compounds (Sulphur powder, organo-sulphur, lime Sulphur), organic mercury compounds (agrosan GN, captan etc.), organic tin compounds (dutor), systematic fungicides (plantavax, vitavax), antibiotics (streptomycin, streptomycin sulphate, agrimycin 500 and terramycin) and thiocarbamates group (thiram, ferbam, ziram, zineb, mineb etc.).

Sulphur fungicides do not leave any specific harmful effect in the soil. Sulphur is also a plant nutrient. Fungicides with heavy metals produce largest problem. In these tin and mercury fungicides pollute soils to maximum extent. Copper fungicides have less harmful effect than tin and mercury. Copper is also used in plant nutrition. Persistency of organic fungicides is very less as they are biodegradable. So, their harmful effect is also very less.

Inorganic and organic both types of mercurate have too much persistency. So, these have high polluting effect. After various reactions in soil mercury is released in it. This mercury can be vapourised. This can form new compounds with oxides, chlorides and nitrates present in the soil or take part in methylation process. Among various fungicides mercurial fungicides causes highest pollution of soil.

Pollution of soil by Weedicides

Chemicals destroying various plants are called herbicides. Similarly, chemicals destroying various weeds grown in crops are called weedicides. There are numerous groups of herbicides among them 8-9 are main groups. In these first group is of phenoxy alkyl acid. In this groups 2,4-D, 2,4,5-T and MCPA are major herbicides. Atrazine, simazine, propazine etc. in triazine group. Another group is of various urea consisting diuron, linuron, monuron etc. Dalapon herbicide comes aliphatic acid group. In carbamate and thiocarbamate group butylate, chloro prophane, prophane are included. Amide is a group herbicide in which propenyl, propechlor etc. are included. In benzoic group amiben, di and tricamba and are included.

Herbicides destroy plants by various ways. Some herbicides prevent light reaction in photosynthesis done by plants. Some prevents photo respiration; some chemicals destroy plasma membrane. Some herbicides badly influence nucleic acid and protein synthesis and stop normal translocation of food material in plants. Some other herbicides destroy chlorophyll, irregulate cell division and cell growth, precipitate protein and effect internal metabolism of plants. Collective or single result of all these is death of plant immediately or after some time. When these herbicides can destroy plants' root and shoot then they will certainly exert ill effect on soil and other component of environment.

Based on persistence of various herbicides in soil, these come in categories of pollution. Organic herbicides are best because they are quickly degradable. Although persistence or stability periods of various herbicides of any group is quite different but, carbamate herbicides are considered best. Because these exist in the soil at the most for two months.

2,4-D and 2,4,5-T herbicides exist at the most for six months. Degradation rate of urea and triazine is very less than these. So, these remain in soil for more time. Herbicides with urea exist in the soil for about one and half year. Same situation is of amiben and dicamba. Persistence period of simazine, triazine, atrazine etc. in the soil may be up to 2 years.

However, persistence period of various herbicides depends on amount of organic matter in the soil, amount of clay in the soil, soil moisture, temperature and pH and rate of use of herbicides in the crops or soil. More organic matter and clay, more 2:1 type clay, and alkaline pH increase degradation period of herbicides. So, persistence period of these in soil is more. However, increase in soil moisture and temperature also increase rate of degradation of herbicides which reduces persistence period in the soil.

Pollution of Soil by Rodenticides

Many small mammals significantly harm houses and field crops. In these, mouse, rats, squirrels, bats, rabbits (hares) etc. are major rodents. Because these mammals eat and destroy food stuff by gnawing, so these are called rodents. The chemicals used for killing rodents are called Rodenticides. Rodenticides are mainly two groups- poison and anticoagulant chemicals. The poisonous rodenticide is zinc phosphide (ZnP). This is mainly used to kill mouse. Anti-coagulant chemical is Warfarin. It prevents clotting of blood. So, after its consumption mouse dies due to internal haemorrhage. It is seen that some species of mouse produce resistance against rodenticides.

Behaviour of Plant Protection Chemicals in Soil

Plant protection chemicals sprayed or broadcasted on plants or directly used in the soil does not remain at same place. These reach in atmosphere by air or volatilization. These may reach in various water bodies and soil with rain water and pollute them. These may reach other places by runoff and leaching from soil or can pollute underground water. The use of this water with chemicals in irrigation again bring chemicals in the soil. It means that whichever plant protection chemicals applied on plants or soil surface, much quantity of these remain in the soil after wandering at various places and components of environment. So, the knowledge about behaviour these chemicals in the soil is very necessary.

Fate of Plant Protection Chemicals in Soil

Fate of plant protection chemicals in soil may be in several ways. Adsorption on soil colloids, movement in the soil, evapouration in the atmosphere, runoff and leaching, absorption by plants and soil organisms and decomposition are major steps in the fate of plant protection chemicals (PPC). A brief description of these points is given here.

Adsorption: Generally, adsorption of plant protection chemicals on clays or humus abides the rules of adsorption isotherm. Thus, curves adsorption of these are 'S' or 'L' shaped. Type of soil, nature of plant protection chemical, quantity of moisture in the soil, soil pH and soil temperature specially influence adsorption.

Soils of fine texture adsorb more plant protection chemicals than soils of coarse texture. More quantities of clays and organic matter in the soils also increase adsorption. This is due to the more cation exchange capacity and surface area of clays and organic matter.

Nature of plant protection chemicals or chemical structure of PPC directly influences adsorption, affinity with soil colloids and solubility in soil solution. Generally, it has been seen that adsorption increases if functional groups of R_3N+, $-CONH_2$, -OH, $-NH_2$ and OCOR are present in molecular structure of PPC. Hydrogen bonding and protonation also influence adsorption. PPC having larger molecule are more adsorbed. Also, less solubility of any pesticide in the water results in more.

As a result of less moisture in soil adsorption of pesticides increases and only very trace amount of PPC remain in free state. Water because of its polar nature competes with pesticides for adsorption sites present on the soil particles. Due to less water molecules in the dry soil many adsorption sites remain unoccupied. So, molecules of pesticides are adsorbed on these sites. Increase of amount of moisture in the soil, water molecules are adsorbed on the adsorption sites and these compel molecules of pesticides to remain in the soil liquid. Thus, increase in moisture reduces adsorption.

Soil pH directly influences adsorption of pesticides. Generally, in acidic soils adsorption of pesticides is more. Pesticides making bases are maximum adsorbed on protonation pH. With very small decrease in soil pH from normal pH, negatively charged pesticides begin to become positively charged. These positively charged molecules are adsorbed considerably on the negatively charged soil particles.

With increase in soil temperature pesticides adsorption is very less and solubility of pesticides increases. Because pesticides are desorped from soil and sufficiently reaches in the soil liquid. Adsorption of plant protection chemicals on soil particles is exothermic action. So, with increase in temperature kinetic energy of bound molecules is increased. On account of this electrostatic force attached in attraction of molecules becomes weak. So, separation of pesticides from soil particles begins and adsorption appears to reduce.

Movement: Movement of PPC can be in dissolved state in the soil liquid or in the adsorbed state in soil particles. Dissolved particles in the soil liquid are moved mainly by mass flow and diffusion. In context to pesticides sometimes mass flow is considered as synonym of leaching. Increase in amount of water both mass flow and leaching increases. Transport of soil particles by wind and water erosion also help in the movement of PPC. Very compactly adsorbed plant protection chemicals are also moved with moving soil particles by erosion and go up to the equal distance of soil particles. Slowly degrading pesticides show more possibility of their movement by soil erosion.

Evaporation: In vapour state distribution of pesticides takes place in whole soil profile but their evaporation takes place only from soil surface. In vapour state movement in soil or loss to the atmosphere depends on vapour pressure. Due to high vapour pressure their penetration in soil pores is easy. Besides evaporation many factors like low temperature, less moisture, low organic matter and more quantity of clay decrease evaporation of low volatile pesticides. Flow of air on the soil surface also influences evaporation.

Runoff and leaching: Pesticides fallen on soil surface, their residues and parts present in surface water move with flowing water. At the water stagnation places pesticides also remain stagnant. When leaching begins pesticides also move towards lower horizons of soil. Besides leaching pesticides may move horizontally with horizontal flow of water. Dust storms are capable to take pesticides far away than water.

Adsorption of soil organisms or plants: Residue of pesticides reached in the soil enters in the body of soil organisms by various ways and slowly and slowly accumulate in their bodies. Such accumulation of pesticides in bodies adversely influence normal activities of soil organisms and finally they die. Special effects pesticide residue has been seen on earthworms because earthworms feed the soil. So, residues of pesticides reach in gut of earthworms. It has been seen in the experiments that residue of pesticides in soil organisms are found 6 to 10 times more than soil.

Residues of pesticides reaches in the plants when crops are cultivated in the polluted soils. Besides runners of rhizomes and tubers roots of normal plants also absorb water soluble residue of pesticides with water and nutrients. These residues are distributed in whole plant from roots by xylem vessels. In plants these exert adverse effect on photosynthesis and other metabolic activities, because these produce blockage between enzymes performing these activities and substrates.

Accumulation of residues of these pesticides in plants depends directly on the amount of residue available in the soil. If residues in the soil are more, then absorption and accumulation will be more. Residue present in sandy soil are absorbed quickly by plants than residue present in organic or clayey soils. It's the main reason of adsorption of ions of residues by clays through organic ions.

Decomposition: Residues left from adsorption, movement, evaporation, runoff and leaching and absorption are decomposed. This decomposition may be of two types, Biological and Chemical. Various algae, fungi, actinomycete and bacteria decompose residues of pesticides. Some residues of some pesticides are store of energy for microorganisms. So, organisms depending on organic ions begin to consume these residues and continue to increase their population. They consume these residues till these are finished completely.

Decomposition of residue of pesticides by microorganisms depends on number of chlorine ions present in them. If number of chlorine ions is more in residues, then decomposition of that will take more time. 2,4-D pesticide which has only two chlorine is decomposed quickly in carbon dioxide, water and chlorine by microorganisms. But in 2,4,5-T in which there are three chlorine takes more time for decomposition than 2,4-D. Hexachlorobenzene, which has six chlorine ions, is very much resistant for biological decomposition.

Various soil properties and conditions also govern decomposition of residues of pesticides. Decomposition of residues of pesticides is more and intensely in the irrigated soil than dry or submerge condition and in state of good fertility than low fertility. Adsorption of Atrazine pesticide on soil colloids increase its chemical decomposition. pH of soil helps in chemical decomposition PPC residues. But various pesticides like cytrolene and malathion decompose on alkaline pH. But atrazine and diazinon decompose rapidly on acidic pH.

Pollution of Soil by Fertilizers: Agriculture development simultaneously increasing problem of pollution. Higher productivity of modern agriculture essentially requires use of fertilizers. In ancient times due to less population and absence of fertilizers' industries various organic manures were used in crops according to need and availability. But for sufficient food supply to continuously increasing population charismatic seeds were discovered. These seeds were of very high standard according to productivity and quality. But to get high level productivity of these seeds, uninterrupted supply of nutrients in appropriate quantity with good irrigation facilities and plant protection are too much necessary. Less plant nutrients values of various organic manures does not provide sufficient plant nutrients to achieve full productivity potential of modern seeds. In addition to these organic manures are available in less quantity. Due to less nutritive value and less availability, use of chemical fertilizers is indispensable.

Although continuous discussion is going on about use and merits of manures and fertilizers between supporters of natural farming and modern improved farming. Under middle course of this discussion, it was decided that where supply of manure is available according to productivity of crops, these manures may be used and where to take more production is needed in short time, there fertilizers may be used. Use of fertilizers is immediate as injections of medicines. These show own effect more intensely and remove quickly symptoms of deficiency of nutrients. But improper and unscientific use of fertilizers can also harm instead of benefit. Without soil testing improper use of fertilizers can also pollute underground water with undesirable changes in soil properties.

If crop utilize thoroughly nutrients given by fertilizers, then there is no problem. But problem is when plants cannot utilize nutrients of fertilizers completely. Very reason is that residual effects of fertilizers appear for next two-three years. So, fate of residual fertilizer or nutrients may be mainly of three types:

1. Residual nutrients can remain on soil.
2. These can be leached with runoff.
3. These can enter in the atmosphere in the gaseous state.

In the process of fate residues, nutrients types, physical-chemical nature and biological behaviour of those can decide in the determination of quantity of residues of nutrients. Nutrients used as fertilizers in the soil and residues of those can produce an imbalance among nutrients already existed in the soil. This imbalance may be profitable or harmful for plant nutrition. There is balance between carbon: nitrogen, nitrogen: potassium, carbon: nitrogen: sulphur and zinc: phosphorous etc. If an element is applied in the soil, then that influences availability of any other element. With the application of fertilizers of potassium and ammonium availability of calcium increases, because calcium adsorbed in the soil is displaced by cation exchange. So, utilization of calcareous fertilizers considerably will displace ammonium. This displaced ammonium will reach in the soil solution. If plants cannot utilize ammonium present in the soil, then this after converting in nitrate reaches in the lower horizons and can produce pollution there.

It is a very clear fact that plants do not utilize full doses of fertilizers applied in the crops in the first year. Some parts of remaining quantity of nutrients exist in the soil and some part reach in the lower horizon with soil liquid. From soil liquid some part of nutrient reaches in any river, streams or lakes and there does eutrophication of those. As a result of eutrophication vegetables irrigated with water of such water bodies grow very profusely but they may have excess nutrients up to toxic levels for animals and humans.

Although fertilizers do not pollute the soil directly and apparently but as a result of continuous use of fertilizers various changes in soil properties occur. These changes produce other problems. Type and dose of fertilizers, soil type, local climate, sown crop and crop rotation and facilities of irrigation etc. are some such factors which influence changes to be produced in the soil by fertilizers.

Pollution of soil by Radioactivity

Radioactivity has special contribution in the soil pollution. Common man thoroughly knows about radioactivity, because of two atom bombs fallen on Nagasaki and Hiroshima in August 1945 during Second World War. Besides this production of electricity and radiotherapy in cancer like disease all know. But radioactivity also pollutes soil.

Pollution of radioactive substances in the soil may be of various medium. First and powerful medium is nuclear fallouts. This is produced by experiments of nuclear weapons etc. In this a radioactive dust falls on the soil and plants from atmosphere. Besides radioactive fallouts dumping of radioactive wastes and waste produced from nuclear energy plants are other sources of pollution in the soil.

Radioactive substances behave as normal elements in the soil. Most part of these remain in the soil which may be absorbed by plants. Its some part being swept with runoff can pollute water sources. It is seen in the experiments that 99% of fallouts of strontium came with runoff becomes concentrated up to 10 times in sediments. It is proved that the sediment have more capacity of accumulation of radioactivity the soil. A large part of radioactive fallouts present in the soil and fallen on plants are absorbed by plants. By using these plants polluted by radioactive elements reaches in animals. It is seen in an experiment that concentration coefficient of Sr90 in the soil was 1, which in a pasture existed in the same soil was 21 and in the bones of sheep depending on that pasture was 714.

Uranium, Thorium and Radium are found in rock phosphates naturally. During production of fertilizers by rock phosphate these elements reach in those fertilizers and from there in the soil. Although these fertilizers or rock phosphate do not pollute the soil appreciably.

Soil Compaction

Soil compaction is also an important soil vulnerability. A good structured friable soil is important for the movement of water, air and roots, which are all critical for a healthy soil. Compacted soils lack good soil structure as the pore spaces that are essential in the movement of water, air and roots are compressed and decreased. Some soils are naturally compact which limit water entry, cause waterlogging and restrict plant growth. Soil compaction takes place when soil aggregates or particles are compressed together constantly for a long period of time. However, continuous ploughing at the same depth and continuous heavy machinery movement on soils can compact soils. The threat of soil compaction is greater today than in the past because of dramatic increase in the size and weight of farm equipment and implements. Soil compaction is defined as the artificial and mechanical process of decreasing the volume and increasing the density of the soil by rapidly reducing the air spaces or pore spaces in the soil. Soil compaction is important for agriculture, forestry, horticulture and construction industries.

Effects of soil compactions: Soil compaction shows many negative effects on soil properties and crop yields. Some important effects are summarized here,

Soil compaction reduces soil pore spaces, water infiltration rate into soil, rate of water penetration up to plants' root zone and subsoil, reduces the ability of a soil to hold water and air and reduces crop emergence and declines soil structure and its stability. Soil compaction increases the clodiness, the surface water accumulation, water runoff, soil surface waterlogging and soil erosion. Compacted soils are difficult to till and require more heavy machinery for agricultural operations. The wetter soil due to compaction increases loss of nitrate nitrogen by denitrification. It impedes root growth and volume of soil explored by roots for water and nutrients.

Soil compaction also reduces the ability of crops to take up nutrients and water from soil and their soil inputs use efficiencies. Ultimately soil compaction reduces crop yields up to 50 percent or more.

Factors Affecting Soil Compaction

The soil compaction depends on Water content, Compactive effort, Type of soil and Method of compaction. These are briefly described here,

Water content: Water content has significant role on compaction of soil. At low water content, soil is stiff and soil particles offer more resistance to compaction. As the water content increases, water films are formed around the soil particles and water around the soil particles act as lubricant. Due to this, soil particles come close to each other and make a dense configuration. This helps in the compaction of wet soils.

Compactive effort: As the compactive effort increases maximum dry unit weight increases in a decreasing rate. This enhances the compaction of wetter soils.

Type of soil: Soil type has significant effect on compaction. It is observed that cohesive soil has higher compaction as compared to sands. Clayey and silty soils are most susceptible to compaction because their particles hold more water for longer periods than sands or loams. As a result, clay soils remain in a plastic state, sometimes for the whole year, which means they will compress and shear when a load is applied to them.

Method of compaction: The heavy and more traffic compacts soil to more extant as compared to light and less traffic.

Soil sealing: Soils are also vulnerable to soil surface sealing. Soil sealing is more concerned with urban area as compared to rural area where much of soil and land is used for infrastructure development. Soil sealing is the covering of the soil surface with impervious materials like bricks, concrete, asphalt, plastic, stone etc. Soil sealing reduces or destroys natural soil functions and ecosystem services of the area concerned. Soil sealing can be defined as the destruction or covering of the soil surface or ground by an impermeable material.

Harmful effects of soil sealing are, it is one of the main causes of soil degradation. Soil sealing often affects fertile agricultural land, puts biodiversity at risk, increases the risk of flooding and water scarcity and contributes to global warming. Also, the negative impacts of soil sealing are numerous, from withdrawing fertile soil from biomass production to modifying the microclimate and decreasing biodiversity. Many of the processes are interrelated and propagate further undesirable consequences from local to global levels. These are mostly related with soil organic carbon (SOC), soil water management and the biomass production.

Soil crusting: Crusting is a surface problem and requires treatment at surface only. Soil crusting is defined as a thin layer of dense and impermeable material on the soil surface which is more smooth, compact and packed than the underlying soil material. When soil surface is dispersed by rain or irrigation water then upon drying, a hard crust is formed on soil surface.

Effects of crusting: Soil crusting does not allow younger seedlings or plumules to come out from soil. Because they do not have sufficient energy to break this hard crust. Ultimately, they die. This is true for many crops and particularly for soybean where rains just after sowing of seeds makes a hard crust on soil surface after drying. Consequently, germinating seeds does not come out from soil or their plumules are broken. Therefore, sowing is done again and again to maintain proper plant count.

The effects of soil crusting on soil properties etc. are somewhat similar to soil compaction. Soil crusting reduces water infiltration, increases soil erosion, provides impedance of seedling emergence, reduces aeration of roots, requires more and extra tillage operation needed for seedbed preparation and interculture operations.

Process of soil crusting: The processes and mechanisms of soil crusting or crust formation depend upon disintegration of soil aggregates and their compaction. This disintegration and compaction are caused by the impact of raindrops. The chemical dispersion of the clay particles also takes part in it. The low electrical conductivity of the rainwater as well as the organo chemical bonds between the primary particles of the surface aggregates, decide the rate and degree of dispersion. An interface suction force rearranges suspended clay particles into a continuous dense layer. Finally, an almost impermeable layer is formed right on the surface of the soil.

Types of soil crusts: There are three types of crusts: chemical, biological, and physical crusts. Chemical crusts are the result of encrusted salt on soils in arid or semi-arid regions. Biological crusts are primarily formed by cyanobacteria, algae, mosses, and lichens growing on soils when ponded water stands and retreats on soils with low permeability. Physical crusts are the result of structural degradation of the surface soil and may be classified as structural or depositional crusts. Structural crusts are formed when surface aggregates disintegrate due to the forces of raindrops or irrigation water. They typically consist of two layers: a very thin skin underlain by a thicker (up to one inch thick) washed-in layer. Depositional crusts are formed by transport and deposition of fine particles from eroded surfaces to depositional areas. All these crusts are divided in many types.

Soil Water-logging

Water-logging is an important soil vulnerability because in the world water logged soils are found everywhere; even to the extent that in lower dry region also. Waterlogging has occupied 10 percent area soils are found. According to American system water logged soils are those which on account of ground or capillary fringe water remain saturated in any time of the year. Because of water saturation water logged soils can be kept in the classes of submerged soils, floodwater soils, ground water soils and surface water soils.

Origin of Water logging: In origin of water logging climate and physiographic conditions play major role. In wet climate accumulation of rain water occurs in lower places. So, these areas remain water logged for most of time. Lower flooded plain, delta and lower sea shores help in water logging. Waterlogging may be at upper places also, provided these areas are with heavy texture and impermeable soil materials. Various uneven terrains due to continuous submergence from rain water for sufficient time are also show water logging.

Also, water, biomass, soluble salts and enriched sediments are source of energy. Structural development, translocation of soluble salts and carbonates, accumulation of organic matter, translocation of organic solutes, redox potential, different pedoturbation etc. help for creating waterlogged condition.

Types of waterlogging: Based on physiography, climate, soil type, water table, waterbodies etc. waterlogging may be divided in five types namely,

Riverine Waterlogging, when heavy rain floods rivers, causing water to seep into nearby land.

Oceanic Waterlogging, when extreme weather conditions, bring ocean water to the land which does not move back.

Seasonal Waterlogging, when runoff water of rain accumulates in lowlands and depressions and creates waterlogging.

Perennial Waterlogging, when the water table is high and the soil is unable to drain excess water. This creates perennial waterlogging.

Subsoil Waterlogging, when the soil has shallow impermeable layer which prevents water infiltration. This creates subsoil waterlogging.

Reference:https://www.soilmanagementindia.com/soil/water-logging-types-and...
Water Logging: Types and Effects - Soil Management India

Characteristics of Water Logging

Origin of most waterlogged soils are on very new geological sediments. Therefore, profile development of these is very immature. So, determination of characteristics of these soils can be done with help of characteristics of parental sediments. Besides this, status of weathering of parent material, local diversity, water table, waterlogging status and climatic variations also influence their characteristics up to large extent. Consequently, much diversities are found in the characteristics of the soil. However, general physical-chemical properties of the water-logged soils / waterlogging are described here:

Soil Texture: In most water-logged soils, texture of soil is fine because of more quantity of clay in it. Quantity of clay in these is up to 45 percent. Generally, in water logged soils 33 percent sand, 28 percent silt and 39 percent clay are found. Because of less quantity of organic matter with fine texture, physical properties of these are not good. In rainy season water in huge volumes get filled in these soils turning them into very sticky and plastic. In summer season these become very much hard and with wide cracks. So, in both these seasons cultural practices or operation become very difficult.

SoiI structure: Structure of these soils is coarse prismatic. This structure changes slowly to coarse angular or sub angular fragment shape and finally to platy structure. Because of platy structure in these soils drainage and movement of air is not good.

Soil Reaction: Generally, reaction of waterlogged soils is acidic and pH of soil is about 6.0, pH of waterlogged soil of new Gangetic alluvium is 7.0 or more. Because of parent material pH of water in logged soils of Vertisols is also more.

Carbon: Nitrogen ratio: Like other normal mineral soil, carbon: nitrogen ratio of waterlogged soil remains 10-12:1. Generally, quantity of organic matter of these soils is up to about 3 percent. This quantity varies 0.1 percent to 40 percent. In water logged organic soils (histosols) quantity of organic matter is very high. Due to dry climate in water logged soils of India quantity of organic matter is less.

Nitrogen: In waterlogged soils quantity of total nitrogen varies between 0.05 to 0.45 percent. On an average, quantity of total nitrogen is 0.15 percent which is more than nearby upland soils. Its main reason is fixation of nitrogen by Azolla, blue-green algae and other microorganisms in the soils.

Total and available phosphorous: Total quantity of phosphorous in waterlogged soils is about 400 ppm. This quantity is less for good yield of paddy. In these soils quantity of available phosphorous is very less (about 15 ppm) although level of available phosphorous in water logged soils is more than nearby soils of highland soils because of waterlogged soils release fixed phosphorous. In waterlogged state, with reduction of iron, fixed or occluded phosphorous being released and reaches in soil liquid. Besides waterlogged state of soil tries to bring acidic or alkaline reaction of soil towards neutral pH. As a result of this HPO_4^{2-} ions begins to change towards $H_2PO_4^-$ ions. It is worth knowing that plants uptake $H_2PO_4^-$ ions more than HPO_4^{2-} ions.

Exchangeable potassium and other secondary nutrient: In water logged soils quantity of exchangeable potassium varies from 0.15 to 1.35 meq/100g soil and average is about 0.4 meq / 100g soil. Total potassium (K_2O) is 1.82 percent.

In acidic waterlogged soils naturally have sufficient availability of calcium and magnesium. In the reduced state of waterlogged soils oxidized form of sulphur is changed in hydrogen sulphide (H_2S). Since plants uptake of sulphur as sulphate ion (SO_4^{2-}). In the waterlogged conditions at one side there is deficiency of sulphur simultaneously plants also face the toxicity of hydrogen sulphide.

Cation exchange capacity: Cation exchange capacity of the waterlogged soils is 4.2 to 34.4 milli-equivalent/ 100 g soil. Average cation exchange capacity of these soils is 19.2 milli-equivalent/ 100 g soil. the percent base saturation of waterlogged soils is 85.7 which varies between 59 to 100.

Micronutrients: In the waterlogged soils nearly quantities of iron, manganese, zinc, copper, boron and molybdenum are in the range of 3000 to 100000, 30 to 10000, 3 to 1000, 3 to 500, 0.5 to 300 and 0.2 to 90 ppm, respectively. But average range of these micronutrients is nearly 10000 to 80000, 300 to 1200, 30 to 100, 12 to 50, 9 to 30 and 1 to 10 ppm, respectively. With the increase or decrease in waterlogged state, changes in these quantities are also possible. As a result of these changes deficiencies or toxicities of micronutrients also appear.

Other elements: In some waterlogged soils toxicity of aluminium and hydrogen sulphide also found. In acidic soils having pH below 5, toxicity of aluminium is also found. Acids of lower pH of react with soil minerals and release potassium, magnesium, iron and aluminium present in soil minerals. These released ions reach in soil liquid. In soils these ions are taken up by the crop. Because aluminium is not essential nutrients of plants, so very little amount of aluminium present in the soil liquid also produces Aluminium toxicity. Deficiency of phosphorus is also in linked with toxicity of aluminium. Aluminium present in the soil liquid reacting with phosphorus fixes phosphorus by formation of insoluble aluminium phosphate.

In waterlogged conditions toxicity of hydrogen sulphate is produced mainly by reduction of sulphate. But in soils having excess of iron concentration, H_2S remain low, because there is formation of Fe_2S in these soils. But on account of excess of carbon dioxide, H_2S is released again, because Fe $(HCO_s)_2$ is formed. Also due to excess of concentration of iron if pH less than 5.2, then in such condition S^{2-} is accumulated abundantly in the soil. In soils of less iron oxide, concentration of hydrogen sulphide increases.

Soil Aridity

Soil aridity is also an important soil vulnerability and increasing aridity is a significant hazard for water management, agriculture, forestry, animal husbandry and many more economic activities. The simple definition of aridity is the long-term average dryness of a region which shows limited or very low water content in the soil. It is also related with strong insolation, low air humidity and wind velocity. Aridity enhances soil erosion, salinisation, growth of desertic or xerophytic vegetation etc.

Soil aridity is extremely diverse in terms of their land forms, soils, fauna, flora, water balances and human activities. In soils of arid regions quantity of organic matter is found less, phosphorus less to medium and potassium medium to high. Fertility level of these soils does not obstructive for growth and development of crops. Main obstacle for healthy vegetative production is less quantity of available water in the soil, less water holding capacity and more evaporation.

Land form of arid region: Besides, water and wind erosional landforms, depositional landform is the main feature of arid region. These are Playas, Loess, Sand dunes (Crescent Dunes, Longitudinal dunes and Transverse dunes). Crescent dunes, also known as Barchans, are found in abundance in sandy desert areas. The shape of the crescent-shaped dunes is like the letter "C". The wings of the crescent-shaped dunes come out in the direction of the wind. The longitudinal dunes are formed when the wind direction is constant and there is a lack of sand supply. Generally, longitudinal dunes are more in length and less in height. Transverse dunes are formed perpendicular to the direction of the wind. These are formed when the wind direction is constant and there is a sufficient supply of sand.

Also in arid region sand dunes are of various shape and size. These are parabolic, coalesced parabolic, barchans and shrub coppice type. Parabolic sand dunes are stable, while barchan dunes are instable. In arid region coalesced parabolic dunes are in abundance. These are formed by mutually joining of several parabolic dunes. So, height of these is up to 20 to 25 meters.

Technically aridity is mainly a function of rainfall and temperature parameters of climate and expressed as,

Climatic aridity index = p/ETP

where

P = precipitation

ETP = potential evapotranspiration, calculated by method of Penman, taking into account of atmospheric humidity, solar radiation, and wind.

Arid zones: Three arid zones can be delineated by this Climatic aridity index, namely, hyper-arid, arid and semi-arid. Arid conditions also are found in the sub-humid zone (arid index 0.50-0.75). Out of the total land area of the world, the hyper-arid zone covers 4.2 percent, the arid zone 14.6 percent, and the semiarid zone 12.2 percent. Therefore, almost one-third of the total area of the world is arid land. These zones are briefly described here.

The hyper-arid zone (arid index 0.03) comprises dryland areas without vegetation, apart from a few scattered shrubs. Here livestock raising is frequently practiced by true nomadic. Annual rainfall is low, rarely exceeding 100 mm. The rains are infrequent and irregular, sometimes with no rain during long periods of several years.

The arid zone (arid index 0.03-0.20) is characterized by livestock raising and rarest farming if irrigation is available. Most of the parts have sparse native vegetation comprised of annual and perennial grasses, herbaceous vegetation, shrubs and small trees. There is high rainfall variability with annual amounts ranging between 100 and 300 mm.

The semi-arid zone (arid index 0.20-0.50) can support rain-fed agriculture with more or less sustained levels of production. Sedentary livestock production also occurs. Native vegetation is very diversified including grasses and grass-like plants, semi shrubs, shrubs and trees. Annual precipitation varies from 300-600 to 700-800 mm with summer rains, and from 200-250 to 450-500 mm with winter rains.

Causes of aridity: There are numerous causes of development of aridity. Some important causes or factors are summarized here,

Distance from ocean: Distance of the region from oceanic moisture sources by topography or by distance which provides dry descending air. Therefore, aridity is found mostly in places where anticyclonic conditions are persistent. as is the case in the regions lying under the anticyclones of the subtropics.

Arid zone climate: The arid zone is characterized by excessive heat and inadequate and variable rainfall. Arid zones have three main types of climates, the Mediterranean climate, the tropical climate and the continental climate.

In the Mediterranean climate, the rainy season is during autumn and winter. Summers are hot with no rains; winter temperatures are mild. Wet season starting in October and ending in April or May, followed by 5 months of dry season.

In the tropical climate, rainfall occurs during the summer. The greater the distance from the Equator, the shorter the rainy season. Winters are long and dry.

In the continental climate, the rainfall is distributed evenly throughout the year, although there is a tendency toward greater summer precipitation. Monthly rainfall is less than twice the corresponding mean monthly temperature. Therefore, the dry season extends over the whole year.

Rainfall: Much of the rainfall is finally returned to the atmosphere by evapotranspiration processes from the vegetation or by evaporation from streams and other water bodies. The rainfall distribution in arid zones varies between summer and winter. More annual rainfall is required in summer rainfall areas than in cooler winter rainfall climates to obtain the same amount of water available for the plants.

Temperature: The climatic pattern in the arid zones is indicated by a relatively "cool" dry season followed by a relatively "hot" dry season and ultimately by a "moderate" rainy season. In general, there are significant diurnal temperature fluctuations within these seasons. In many situations, these diurnal temperature fluctuations restrict the growth of plant species. Growth of plants can take place only between certain maximum and minimum temperatures. Extremely high or low temperatures can damage to plants. Plants might survive high temperatures, if they can compensate for the high temperatures by transpiration, but growth will be affected negatively. High temperatures in the surface layer of the soil result in rapid loss of soil moisture due to the high levels of evaporation and transpiration. This condition enhances aridity in soils.

Atmospheric humidity: The moisture in the air has importance for the water balance in the soil. When the moisture content in the soil is higher than in the air, there is a tendency for water to evaporate into the air. When the opposite is the case, water will condense into the soil. Humidity is generally low in arid zones. The presence of dew and mist provide higher humidity in the air. Thus, evapotranspiration is reduced and soil moisture is conserved. This helps in the reduction of soil aridity.

Wind: The arid regions are generally windy. Winds remove the moist air around the plants and soil resulting increase in evapotranspiration. This generates soil aridity.

Landslides

A landslide is the downward mass movement of rock, soil, soil material, debris etc. from a slopy land. Rain, earthquakes, volcanoes, gravity etc. or other factors providing instability to slopes are main reasons of landslide. Mountains specially Himalayans area of our country is a new geographical structure. This has inherently unstable steep slopes, contorted and river stones in layers and fragile ecosystem have made it more unstable. The tremendous enhancement in population required more basic amenities which excessively increased the load on the land.

Now land is being utilized in agricultural and non-agricultural works for the purpose of farming and development. Most of the mountainous soils are not cultivable. Grazing of forest lands by animals, cultivation and cutting of trees for the purpose of fuels and timbers continuously make this area devoid of vegetation. So, problem of landslide is becoming serious day by day. The worst and deadliest landslide occurred in Kedarnath, India in June 2013 in which around 5700 people were died and lot of infrastructures were lossed.

Landslide, generally, occur on steep slopes, where internal geological structures in Hilly region naturally move here and there. When soil and soil material rested on impermeable rock is saturated with excess of rainwater and it is immediately converted in to semifluid mud. Fragments of rocks also lie in this mud. This whole material slides below the slope due to its weight. On account of this sometimes-whole area is submerged or destroyed.

Factors affecting landslide: Although since ancient times landslides continue to occur, but recently their repetition has become more severe. In Himalayan states often large landslides are taking place in every year. Increased repetitions of landslides compel to consider the manmade factors besides the natural reasons. Chiefly deforestation, shifting cultivation, construction of roads, mining and quarrying, irrigation and grazing are included in manmade factors.

Deforestation: Existence of forests is too much necessary for reducing blow of raindrops and runoff and entry of rain water in the soil. Roots of trees enter directly in the soil to provide strength to the hilly slopes. When trees are felled down, then roots begin to rot and their power of soil binding is decreased. Hill slopes become bare due to deforestation. Bare slopes do not hold rain water in full. Therefore, rainwater flows as a torrent, smashing and eroding hill slopes and mingles in flowing water current.

Shifting cultivation: Shifting cultivation is common practice of crop growing in hilly and northeast regions. In this case an area of forest is burned, ground is cleared of vegetation and cultivated for few years and then abandoned until its fertility has been naturally restored. This is practiced for new area also. However, this deteriorates the properties of soil to cause erosion during heavy rainfall. This makes such regions more vulnerable to landslides.

Construction of roads: In hilly areas often, roads are constructed with support of steep rocks. These rocks are cut and exploded to give required width to roads. All these loosen other rocks and near by soil. Also trees etc. are cut in very large number. As a result of this new areas of landslides are developed and old areas again become active. In districts of Tehri Garhwal and Dehradun during roads construction about ten medium size landslides per kilometre were seen. These were produced about 500 cubic meter debris from each landslide. This debris, instead of taking away from the site, was pushed downwards to any water current where it causes sedimentation of water bodies.

Mining and quarrying: Unplanned and uncontrolled mining and quarrying cause much damage in hilly areas. On account of this, balance of soil above rocks and angle of base changes. Also, soil becomes loose and cracks are produced in this. All these cooperate in the landslide.

Irrigation: Maintenance of constructed canals is not good in hilly areas. Surplus water also cannot be removed easily from these. So, water is used to infiltrate continuously and make the super saturated soil. Thus, loose layer of mud is formed inside the soil, which due to its weight slides downwards very easily.

Grazing: Animals have very much contribution in hilly cultivation and economy. Farmers have animals in large number for power for farming, manure for fields and to get milk. Because of large numbers of animals more cattle feeds are required. Often every farmer has 4 to 5 animals. Generally, for grazing an animal 0.25 to 0.3 ha land is required. Thus, four animals require 1.0 to 1.2 ha land which is not easily available in hills at a single place. Conditions of forests and pastures becomes very bad due to excess grazing by animals in less space, more lopping of trees for the purpose of their fodder and eating of bushes and sprouts of plants by goats etc. The marginal lands of steep slopes are also ploughed to grow more fodder. As a result of this erosion and later landslides begin to happen.

Effects of Landslides

Landslides result in destruction of property. A significant landslide can remarkably decline the economy of that area, region or country. After a landslide, the affected area needs urgent and immediate rehabilitation. This rehabilitation requires huge money to invest. However, some effects of landslide are summarised here,

Destruction of infrastructure: The force flow of mud, debris, rocks etc. due to landslide can cause serious damage to property. Infrastructure such as roads, railways, leisure destinations, buildings and communication systems are destroyed by a single landslide.

Loss of life: People living at the foot of hills and mountains are at a greater risk of death by landslides. A substantial landslide carries along huge rocks, heavy debris and heavy soil with it. This kind of landslide has the capacity to kill innumerable people and animals.

Shape of landscapes: The afterward seen of landslide area is quite devasting because a beautiful scene is changed into a rugged landscape. The pile of soil, rock, trees, debris etc. cover downhill land utilized by the local people for agriculture and other purposes.

Impact on river etc.: The impact on dams and streams is one of the negative effects of landslides. The soil, rock, trees, debris etc. sliding downhill can find way into rivers and block their natural flow. Many river habitats like fish etc. can die due to interference of natural flow of water. People depending on the river water for household activities and irrigation will suffer if flow of water is blocked. Dams, on the other hand, have the potential to flood the other side. Outburst floods can suffocate streams with a massive volume of fresh sediment. Alternatively, they can submerge the flooded area and kill the vegetation etc.

Effect on agriculture: Landslides also show negative effect on agriculture and agricultural lands. Debris scattered across farms, fields, pastures, orchards etc. destroy valuable agricultural areas. Seeds, plants, food stocks and grazing areas are all destroyed. These lands can be unavailable for years, putting farmers' livelihoods at risk. Furthermore, the landslides also destroy enormous areas of forest, wildlife sanctuaries and fertile soils of slopes.

Abandoned mines

Mining also is a very old and important industry just like agriculture. Raw materials are made available to the various industries from mines. But in most mines mining work is done by unscientific manner. Digging for the purpose of mining, explosion, road construction and cutting of trees very badly influence agricultural lands, soil, water and worsen the economic condition of habitants.

Effects of abandoned mines: Globally it is not possible to count the number of abandoned or orphaned mines, but these are certainly several million. Similarly, the environmental, social and economic problems associated with abandoned mine sites are serious and global. Abandoned mines may pose unique and complex challenges, often leaving many negative effects. Some effects of abandoned mines are briefly given here,

Contamination of environment: Abandoned mines contaminate the surrounding soil, water tables and waterbodies with heavy metals and other toxicants such as mercury, arsenic, fluorine, sulfuric acid and selenium. Mining, particularly coal mining releases air pollutants gases like sulphur dioxide, carbon monoxide etc. which affect human health. Also, substantial amounts of methane gas which accelerate global warming.

Drainage: Acid and metalliferous drainage (AMD), also known as acid mine drainage, is one of the negative environmental impacts on water quality that can result in a subsequent loss of biodiversity. AMD impacts can occur from underground workings, open pit mine faces, waste rock landforms, and tailings storage areas.

Social and economic issues: Social and economic impacts on countries and individual communities can be due to issues such as, a loss of the productive land, loss or degradation of groundwater, pollution of surface water by sediments or salts, changes in river regimes, air pollution from dust or toxic gases or risk of falls into shafts or open pits.

Marc Hufty Professor, Interdisciplinary Programmes The Graduate Institute, Geneva. The consequences of abandoned mines - Global Challenges https://globalchallenges.ch/issue/6/abandoned-mines-the-scars-of-the-pas

5

Restoration and Management of Soil Vulnerabilities

In the earlier chapter the major soil vulnerability namely, water and wind erosion, depletion of soil organic matter, loss of plant nutrients, salinity, alkalinity, acidity, pollution and contamination, compaction, sealing, crusting, water-logging, aridity, landslide and abandoned mines were described. In this chapter two more parts, Importance of remote sensing (RS) and geographical information system (GIS) in soils and Watershed management are also included after seeing their importance in agriculture and soils. Restoration means that by various actions or processes to return something to its original owner, site, place, or condition. In this context our aim is to restore our mother soil from various vulnerabilities by different actions and processes to its natural and original condition up to the maximum possible extent. Management is defined as the scientific methods to attain targeted production through planning, organising, training and fixing targets for employees. Therefore, after the restoration, soils should be managed according to scientific method of action by scientifically selected and trained staff to perform the assigned responsibilities, effectively, efficiently and rapidly within stipulated period. On these guidelines restoration and management of vulnerabled soils is described here,

Importance of Remote Sensing (RS) and Geographical Information System (GIS) in Soils

1. Importance of remote sensing in soils: The lack of data of soil properties is a limiting factor for adaptation of new techniques in agriculture in day to day changing environment. Therefore, instantaneous availability of data on soil properties is very essential and remote sensing is a tool to gather such data. The spatial and temporal variations in soil properties, soil capability, soil vulnerability, soil restoration and management are most important to know about changes in it to assign its best suitable use. The remote sensing techniques can instantly provide detailed characteristics of soil across space and time without disturbing soil and going at place of study. However, ground truth is also necessary to validate remotely sensed data. Remote Sensing includes mechanisms, methods for identifying specific soil properties, data acquisition, interpretation, validation etc.

Our body has five sensory organs namely eyes, ears, nose, tongue and skin. Among these eyes, ears, nose can remotely see objects, hear sounds and smell, respectively without coming in contact and discriminate them. Therefore, Remote Sensing (RS) means acquiring information about a phenomenon, object or surface while at a distance from them without coming in any actual contact with them. This name is attributed to recent technology in which satellites and spacecraft are used for collecting information about the surface of earth. This was an outcome of developments in various technical fields from 1960 onward. Dr. Farouk El-Baz, an American geologist and pioneer of space photography is known as the "Father of Remote Sensing" and Pisharoth Rama Pisharoty (10 February 1909 – 24 September 2002) was an Indian physicist and meteorologist. Pisharoth Rama Pisharoty is the father of remote sensing in India

Remote sensing system: The remote sensing is basically a multi-disciplinary science which includes a combination of various disciplines such as optics, spectroscopy, photography, computer, electronics and telecommunication, satellite launching etc. All these technologies are integrated to act as one complete system itself. This system is known as Remote Sensing System. There are number of stages in a Remote System, working as links in a complete, and each of them is important for successful operation.

Principle of Remote sensing: Detection and discrimination of objects or surface features means detecting and recording of radiant energy reflected and emitted by objects or surface materials. Different objects return different amount and kind of energy in different bands of the electromagnetic spectrum, incident upon it. This unique property depends on the property of material (structural, chemical and physical), surface roughness, angle of incidence, intensity and wavelength of radiant energy.

Stages of action in Remote sensing: There are about 8-10 stages of action in the collection of data by remote sensing technique. These are depicted in Fig. 5.1.

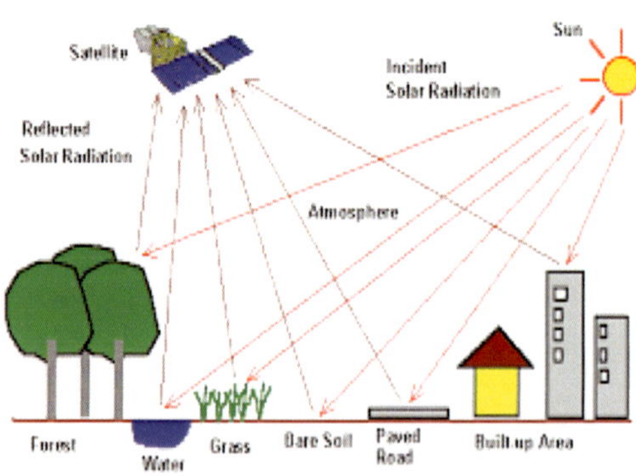

Fig. 5.1: Stages of Remote Sensing

The figure 5.1 will help to understand the various stages involved in remote sensing technique.

1. **Source of electromagnetic radiation (EMR):** The sun is an important source of energy for remote sensing. Sometime object under provides self-emitted EMR.

2. **Transmission of energy from source to the surface of the earth and its interaction with intervening atmosphere:** When the energy is emitted by the source then during its transmission, absorption and scattering of the electromagnetic radiation take place. Basically, the energy from the source propagates with the speed of light.

3. **Interaction of energy with the object, target, surface of earth or self-emission:** The energy emitted from the source, incident up on numerous objects and interacts with target object. This incident energy may be reflected, absorbed, transmitted or self-emitted. The diversified objects behave dissimilarly for incident radiation. Therefore, it is essential to critically record the characteristic and behaviour of the EMR.

4. **Transmission of energy from object to sensor:** The energy from the source after reflection or emission is transmitted from the target object to remote sensor placed on suitable platform through intervening atmosphere.

5. **Detection of energy by the sensor:** The energy detected by sensor is converted into photographic image or electrical output. The photographic, non-photographic plates or electrical output measuring devices are used by the sensors.

6. **Transmission and recording of sensor output:** The data received by the sensor is sent towards the earth station. The sensor electronically sends the received data to either single or various earth receiving stations on the ground.

7. **Preprocessing of sensor output and generation of data products:** Once the earth station receives the data. They undergo for preprocessing for geometric and radiometric corrections. After this data products are generated depending upon data processing techniques.

8. **Collection of ground truth and other ancillary information:** The ground truth data and other collateral information are collected for the validation of processed data. Accuracy check with base map is done.

9. Data analyses, interpretation are done, thematic report, final report of targeted object are prepared with the help of GIS and other data base and beneficiaries are informed about the outcome.

Basic concepts of remote sensing: Everything in nature has its own unique distribution of reflected, emitted and absorbed radiation. These spectral characteristics, if inventively exploited, can be used to distinguish one thing from another or to obtain information about shape, size and other physical and chemical properties. At temperature above absolute zero, all objects radiate electromagnetic energy by virtue of their atomic and molecular oscillations. The total amount of emitted radiation increase with the body's absolute temperature and peaks at progressively shorter wavelengths. The sun being a major source of energy, radiation and illumination, having a power peak around 0.5 um, allows to capture reflected light with conventional and nonconventional) cameras and films.

Characteristics of Solar Radiant Energy: The sun is the strongest and most important source of radiant energy for remote sensing. The energy in solar irradiation comes in the form of electromagnetic waves of wide spectrum. Longer wavelengths (infrared) have less energy and shorter wavelength (visible and ultraviolet) have more energy. The solar spectrum extends approximately from 0.15 um to 4.0 um near ultraviolet, visible and near infrared regions. The maximum irradiance occurs at 0.47 um. The visible band from 0.40 um to 0.74 um receives about 40 percent of the total solar energy.

The solar constant includes radiation over the entire electromagnetic spectrum. It is measured by satellite which is about1.361 kilo watts per square meter (kW/m^2). The flux density or intensity of the solar beam received per unit area at the top of atmosphere when sun is at its mean distance from the earth is called the solar constant.

Electromagnetic radiation: Electromagnetic radiation in the field of classical physics is defined as the waves of the electromagnetic field. These waves propagate through space and carry electromagnetic radiant energy. The different types of these waves are microwaves, radio waves, visible light, X-rays, ultraviolet rays and gamma rays. The electromagnetic waves are synchronized oscillations of electric and magnetic fields. These fields are perpendicular to each other like a transverse wave. The velocity of the electromagnetic waves is equal to the speed of light.

Thermal Emission of Radiation: Thermal emission is a process of emitting radiation above zero temperature. All Objects at all temperature emit electromagnetic radiation at all wavelengths. The thermal emission of radiation is due to the conversion of heat energy, which is the kinetic energy of the random motion of the particles of the matter, into electromagnetic energy. Thermal emission of radiation depends upon two parameters, namely, Absolute Temperature (T) and Emissivity. The total thermal radiation from a body increases with four powers of T. The absolute temperature is given by T=273 + temperature in degrees centigrade in units of degree Kelvin (^0K).

The emissivity of a substance is related to its absorptive ability. Good absorbers are good radiators, whereas poor absorbers are poor radiators. An ideal thermal emitter is called a blackbody, where $\in=1$.

Interaction of EMR with the Earth's Surface: Radiation from the sun, when incident upon the earth's surface, is reflected by the surface, transmitted into the surface or absorbed and emitted by the surface. The EMR, on interaction, experiences several changes in magnitude, direction, wavelength, polarization and phase. These changes are detected by the remote sensor and enable the interpreter to obtain useful information about the object of interest. The remotely sensed data contain both spatial information (size, shape and orientation) and spectral information (tone, colour and spectral signature).

Types of remote sensing: There are two types of remote sensing, Passive remote sensing and Active remote sensing.

Passive remote sensing: It is the study of the structural aspects of an object or the land surface by measuring the energy reflected or emitted by the object itself. In passive remote sensing solar radiation reflected from the surface is usually measured. Generally, passive remote sensing is applied for the study of Agriculture, Biodiversity, Disaster management, Forest, Land use etc.

Active remote sensing: Active remote sensing does not depend upon the solar energy that is reflected or emitted back by the objects or the land surface. Active remote sensors emit their own energy source which is directed towards the object or area of study. The amount of radiation that is reflected back is detected and measured by the active remote sensors.

The active remote sensing is used for Groundwater and surface water measurements, Study of water bodies, Soil moisture, Vegetation, Snow, Topographic mapping, Weather forecasting etc. There are three main types of Active remote sensing namely, Active optical Remote Sensing: This type of remote sensing emits radiation in the visible range which hits the target and the back scattered light is analysed. Active Microwave Remote Sensing: This is the microwave emitting remote sensing and Active thermal Remote Sensing: This form of remote sensing is used to detect the thermal infrared radiations.

Two types other of remote sensing are, Imaging (Image forming) and non-Imaging (non- image forming). Image forming systems are again of two types – framing type and scanning type. In the framing type, entire frame of image is acquired instantaneously in the basic image unit e.g., in a frame camera used in photography. In scanning type, the information is acquired sequentially from the surface in bits of picture elements or pixels, point by point and line by line, which may be arranged after acquisition into a frame format.

Non imaging types of sensors are used to record a spectral quantity or a parameter as a function of time or distance (such as Gamma radiation, magnetic field, temperature measurement etc.) They are mostly used for ground observation and study of atmosphere and meteorology. These sensors do not form image and as such and are not used in operational remote sensing but give detailed information on spectral characteristics of the target.

Sensors used in remote sensing: Sensor is an electronic circuit which can record the electromagnetic radiation incident upon it. A sensor is a device comprising of optical component or system and a detector with electronic circuit. It senses a variation in input energy to produce a variation in another or same form of energy. All sensors employed on earth observation platforms use electromagnetic radiation to observe terrain features. Various types of sensors are used because of requirements and purpose.

Types of remote sensors: Remote sensors can be broadly classified as passive sensors and active sensors. Sensors which sense natural radiation, either emitted or reflected from the earth are called passive sensors. Such sensors, which produce their own electromagnetic radiation, are called active sensors. The sensors may be broadly be divided and classified based on their working principles and recording mechanisms in to Photographic camera. The vidicon using detecting media as television camera, The optical scanner, Microwave radiometer and Microwave radar are other sensors.

Remote Sensing & GIS Applications: Lesson 4 Sensors …

Spectral Signature: Various materials of the earth's surface have different spectral reflectance characteristics. Spectral reflectance is responsible for the colour or tone in a photographic image of an object. Trees appear green because they reflect more of the green wavelength. And ripe tomatoes look red because they reflect more of red radiation. The values of the spectral reflectance of objects averaged over different and well-defined wavelength intervals. These wavelength intervals comprise the spectral signature of the objects or features by which they can be distinguished. The spectral characteristics of various natural objects are extensively measured and recorded to obtain the necessary ground truth for the interpretation of multispectral imagery. A Figure 5.2 of spectral signature of soil, vegetation and water is given here,

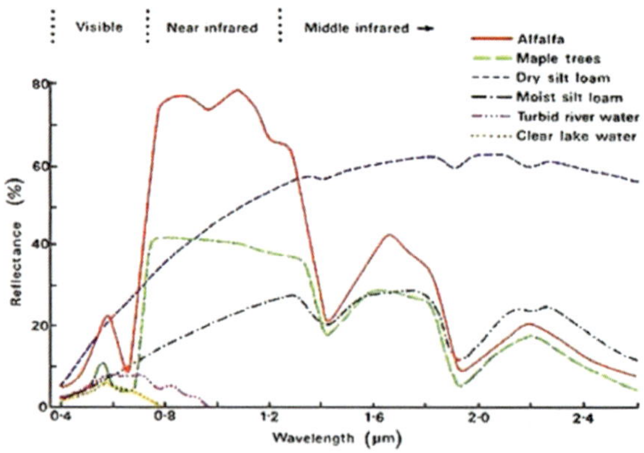

Fig. 5.2: Spectral signature of soil, vegetation and water.

Platforms for remote sensing: Sensors should be mounted on suitable stable platforms for remote sensing activities. These platforms can be ground based air borne or space borne based. As the platform height increases the spatial resolution and observational area increases. Thus, higher the sensor is mounted; larger the spatial resolution and synoptic view is obtained. The types or characteristics of platform depend on the type of sensor to be attached and its application. Depending on task, platform can vary from ladder to satellite. For some task sensors are also placed on ground platforms. Though aircrafts and satellites are commonly used platforms, balloons and rockets are also used. Three types of platforms are used to mount the remote sensors, Ground Observation Platform, Airborne Observation Platform, and Space-Borne Observation Platform.

Atmospheric Effects: The sun is the source of radiation, and electromagnetic radiation (EMR) from the sun that is reflected by the earth and detected by the satellite or aircraft – borne sensor must pass through the atmosphere twice, once on its journey from the sun to the earth and once after being reflected by the surface of the earth back to the sensor. Interactions of the direct solar radiations and reflected radiation from the target with the atmospheric constituents interfere with the process of remote sensing and is called as Atmospheric Effects.

Atmospheric Windows: The atmosphere selectively transmits energy of certain wavelengths. The spectral bands for which the atmosphere is relatively transparent are known as atmospheric windows. Atmospheric windows are present in visible and infrared regions of the EM spectrum. In the visible part transmission is mainly affected by ozone absorption and by molecular scattering. The atmosphere is transparent again beyond about = 1mm, the region used for microwave remote sensing.

Orbits: The satellites always move in a plane called the orbital plane. This passes through the centre of the earth. The orbits could be circular or elliptical depending on the eccentricity. Therefore, many orbits are possible due to the inclination of the orbital plane. If the orbital inclination is zero or 180^0, the orbital plane lies in the equatorial plane and orbit plane is called as equatorial orbit.

When the inclination is 90^0, the satellite moves over the poles and the centre of the earth as well as north and south poles. This lies in the orbital plane and called as polar orbit. An orbit with inclination near to 90^0 is called near polar orbit. There are also various orbits having orbital inclination in between 0^0 to 180^0 and generally called as inclined orbits.

When the satellite height increases, the time taken for one revolution or period also increases. At about 35786 km. height the period is equal to one siderial day (23 hours, 56 minutes and 4.091 seconds). This orbit is called as geosynchronous orbit as the orbital period is synchronized with the rotational period of the earth.

A satellite in a circular orbit as geosynchronous with zero inclination, the satellite will appear stationary with respect to the earth over the equator. This is called the geostationary orbit. This is an ideal orbit for communication satellites and also used for weather satellites as the satellite continuously sees the one area.

The precession rate of the orbit is a function of satellite altitude and orbital inclination. The precession rate can be made so exactly to compensate the mean rate of earth about the sun. In this way the satellite always sees the point on earth when the sun is shining on earth from same angle and same sun time. Such orbit is called as sunsynchronous orbit. This orbit is always near polar.

2. Importance of Geographical Information System (GIS) in soils: The role of GIS in soil is many faced. GIS is often used in conjunction with remote sensing and Global Positioning System (GPS) technology to benefit soil studies. The use of GIS in soil mapping is called Digital Soil Mapping (DSM). This involves creation of geographically referenced soil databases from spatially explicit environmental features and field surveys. The data that goes into creating these DSM use remote sensing technologies. Good quality soil maps are the baseline for many areas to decide land use planning and management.

GIS can be used to make estimates on soil erosion, conservation and land degradation status. These are worldwide important in agriculture, the environment and in urban areas. In parts of the world where soil loss is affecting people, nature and food production, information like rainfall data, soil type, land use, vegetation cover and the digital elevation model (DEM) can be combined to make an estimate of the erosion risk of an area.

Soil contamination and remediation is becoming a big topic around urban areas, waste or mine sites. Contamination from metals and other substances can make the soil toxic for years to come unless the contamination can be identified and removed. Use of powerful GIS software it's possible to more precisely measure contaminants, reduce the labour required in the field and target treatment methods more efficiently.

The soil material and soil moisture content is an important part of understanding flood risk. Climate change is putting additional pressure on water resources globally. Different soil types hold water in different ways and it is important to understand soil moisture content for various reasons. Most obviously, it benefits farmers to grow crops, also helps to understand groundwater recharge and ultimately better management of water resources. Soil moisture data can be collected either at a small scale for specific farms or regional areas or large scale. This data has the potential to be used to monitor droughts, predict the weather and many other possibilities.

Applications of GIS in Soil Science - July 21, 2023 (mapscaping.com)

A geographic information system (GIS) is a computer system for capturing, storing, checking, and displaying data related to positions on Earth's surface. GIS can show many kinds of data on one map, such as streets, buildings, and vegetation. This enables people to see, analyse, and understand patterns and relationships more easily. (https://education.nationalgeographic.org/resource/geographic-information-system-gis/)

The recent digital and analogue electronic devices facilitate the inventory of resources and the rapid execution of arithmetic or logical operations. These Information Systems are undergoing much improvement and they can create, manipulate, store and use spatial data much faster rate as compared to conventional methods. An Information System is a collection of data and tools for working with data in analogue form or digital form about the phenomena in the real world.

Spatial information is always related to geographic space. i.e., large-scale space. Geographic space is the space of topographic, land use, land cover, climatic, cadastral, and other feature of the geographic world. Geographic information system technology is used to manipulate objects in geographic space, and to acquire knowledge from spatial facts.

Geoinformatics: Spatial data handling involves many disciplines. The discipline that deals with all aspects of spatial data handling is called geoinformatics. This is combination of geo-sciences and information technology.

Objectives of GIS: Objectives of GIS are, Maximises the efficiency of planning and decision making, provides efficient means for data distribution and handling, eliminates redundant data base, minimises duplication, Capacity to integrate information from many sources and Complex analysis/query involving geographical referenced data to generate new information.

GIS data models: Conversion of real-world geographical variation into discrete objects is done through data models. It represents the linkage between the real-world domain of geographic data and computer representation of these features. Data models are of two types: Raster and Vector.

In raster type of representation of the geographical data, a set of cells located by coordinate is used; each cell is independently addressed with the value of an attribute. Each cell contains a single value and every location corresponds to a cell. One set of cell and associated value is a LAYER. Raster models are simple with which spatial analysis is easier and faster. Raster data models require a huge volume of data to be stored, fitness of data be limited by cell size and output is less beautiful. The raster model has evolved to model such continuous features. A raster image comprises a collection of grid cells rather like a scanned map or picture.

Vector data model uses line segments or points represented by their explicit x, y coordinates to identify locations. Discrete objects are formed by connecting line segments which area is defined by set of line segments. Vector data models require less storage space, outputs are appreciable, Estimation of area/perimeter is accurate and editing is faster and convenient. The vector model is extremely useful for describing discrete features, but less useful for describing continuously varying features such as soil type.

Besides these two, there is one data model is also available. This is Triangulated Irregular Networks (TIN) data model. TIN data model of GIS represents terrain or other surface features as a set of irregularly spaced points. These points are then connected by lines, which form a network of triangles. TIN data models are useful for applications involving terrain analysis, such as hydrological modelling, land-use planning, and environmental impact assessment. Some examples of TIN data models are Esri TIN and GRASS TIN. Figure 5.3 of raster and vector models is given here,

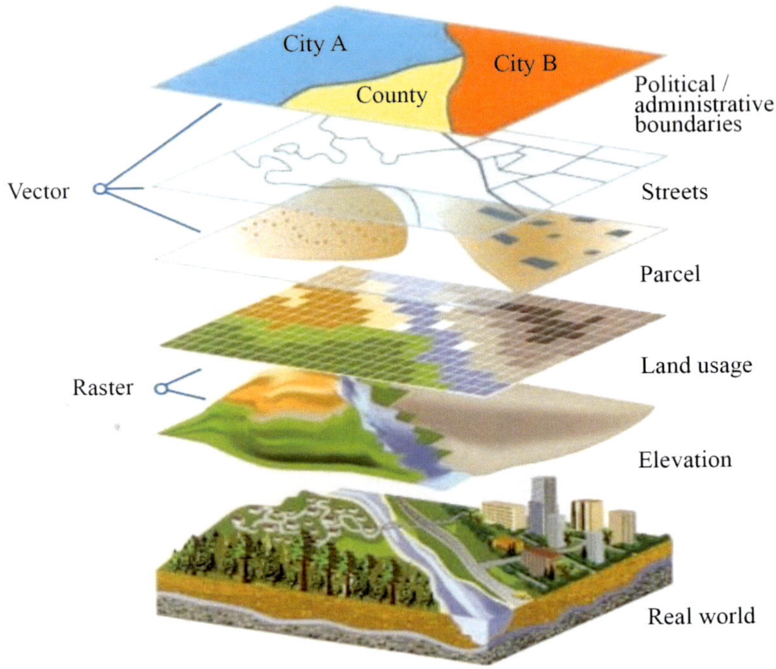

Fig. 5.3: Vector and Raster models. *Source*: 68 GIS Data Models:- Discovering New Insights From GIS Data Models - (geolearn.in)

Layers and Coverages: All data of a particular level of classification like buildings, roads, rivers or vegetation type are grouped into layers or coverages.

The concept of layers is the part of both vector and raster models. The layers can be combined with each other in various ways to create new layers that are a function of the individual ones. The characteristic of each layer within a layer-based GIS is that all locations with each layer may be said to belong to a single arial region or cell, whether it be a polygon bounded by lines in vector system, or a grid cell in a raster system. But it is possible for each region to have multiple attributes.

Data structures in GIS: There are number of different ways to organize the data inside the information system. The choice of data structure affects data storage volume and processing efficiency. Many GIS software's have specialized capabilities for storing and manipulating attribute data in addition to spatial information. Relational, Hierarchic and Network are three basic data structures.

Relational data structure organizes the data in terms of two-dimensional tables where each table is a separate file. Each row in the table is a record and each record has a set of attributes. Each column in the table is an attribute.

Hierarchical data structure stores the data in a way that a hierarchy is maintained among the data items. Each node can be divided into one or more additional node. Stored data gets more and more detailed as one branches further out on the tree.

Network data structure is like to hierarchy structure with the exception that in this structure a node may have more than one parent. Each node can be divided into one or more additional nodes. Nodes can have many parents. The network data structure has the limitation that the pointers must be updated every time a change is made to database causing considerable overhead.

Restoration and management of soil from water and wind erosion: Restoration of eroded soils can be done by numerous engineering, agronomical, agroforestry, biological etc. methods. These methods are different for water eroded and wind eroded soils. These methods are described here.

Soil Erosion Control

Irrelevant development in many sectors significantly increases soil erosion which make slopes and embankments unstable. Ultimately this destroys infrastructures, ecosystems, water supplies and increases flood hazards. Soil erosion is done by water and wind. Hence methods of control of erosion by water and wind are different. In erosion first there is separation of soil from its earlier place and then its transport to other places. So, it should be our first attempt that separation of soil could be at lowest level and if separation takes place then transport of it should be minimum. It means that such conditions and situations should be developed in the soil and near about that transport of soil could be minimum and negligible.

In context of water erosion, the erosion control method should minimize the impact of raindrops, make soil resistant for separation and reduce the shearing power of runoff. Soil can be protected from impact of raindrops by various vegetative or artificial covers. Soil can be made separation resistant by improving soil structure and increasing stability of soil structure. Engineering methods can help in the reduction of quantity and velocity of runoff and its power of soil shearing.

Two principles are useful in prevention of wind erosion. Firstly, velocity of wind near soil surface could be minimized and secondly soil surface could be made resistant to wind drag. Most simple method to minimize wind velocity is to prepare vegetative cover on soil surface. Vegetative cover will reduce the blowing power of higher wind velocity. Hence soil separation will be minimum. Moisture conservation in soil, soil structure improvement and stability and various cultural operations make soils wind drag resistant. Methods of water and wind erosion control are discussed here:

Water erosion control: According to earlier mentioned principles, water erosion control methods can be kept in three groups:

1. Minimization of effect of raindrops,
2. Making soils separation resistant and
3. Minimization of soil shearing power of runoff.

1. Minimization of effect of raindrops: Most easy method to minimize the striking or blow power of raindrops is to cover soil surface. This cover may be of any type, natural or artificial and biological or chemical. Factors affecting efficiency of cover are many in which nature of cover, continuity, structure, density and height important.

In vegetative cover soil erosion is controlled by the help of various vegetations and crops. More erosion occurs in such areas where vegetation is scarce. Fifty cm high vegetative cover from soil surface remarkably reduces hitting power of raindrops. But excessively high vegetative cover is not much effective. There is more possibility of soil erosion below sparsely grown tall trees.

Often it is found that erosion losses are negligible in lands covered by grasses as compared to fallow lands. Also, soil erosion is more in the fields of cereals like sorghum, bajra and maize than fields of pulse crops, green gram and cowpea. Its main reason is that sorghum, bajra and maize are grown in wide lines, height of these crops is more and these crops do not cover soil completely. While cowpea and green gram are grown in narrow lines, height of these crops is less and these after spreading completely cover soil. So, these crops save the soil from blow of raindrops.

Mulching is second method to save the soil from blow of raindrops. Besides the importance of mulch in control of soil temperature. Its importance in erosion control is well established and recommended. When vegetative cover is unable to control soil erosion, then mulch easily reduces soil loss. Crop residue broadcasted on the soil surface as mulch just halve the soil erosion. Mulches spreaded on soil surface stop scouring and blowing of soil. Also, mulches reduce evaporation, enhances infiltration, control weeds, improves soil structure and ultimately enhance yield of crops.

It is necessary to pay attention on quantity of crop residues or other waste materials to be used as mulch. Excess quantity of mulch adversely affects the proper growth of plants. Use of 1-7 tons per hectare straw is the range for mulching. However, in a paddy field of with 2% slope, placing of spear grass mulch at the rate of one ton/ha recorded very less soil loss i.e., 0.9 ton/ha only. Generally, application of 1.5-2.0 tons, husk or cow dung manure as mulch is sufficient. Besides of husk and cow dung manure, crop residues, as leaves, stubs, parts of stem, saw dust, scrapings of wood or planks, different types of canvasses, nariyal and jute mats as well as strips or shredding of plastic and fine gravel are used as mulch. Organic mulches reduce more runoff and erosion than inorganic mulches. Any mulch which improves more physical properties of soil, reduces more soil erosion. If plastic mulch is used continuously for several years, then it will adversely affect the rate of water infiltration in the soil. Although mulch of fine gravels prevent soil from impacts of raindrops but its use hinders in smooth cultural operations. Mulch of plants residues is best mulch to prevent soil from erosion. But availability of plant residues in huge quantities is a problem, because mostly plant residues are burnt in field itself or used in non-agricultural purposes.

2. Making soils separation resistant: After protecting soil from blows or hits of raindrops by vegetative cover or by mulching then it is essential to make soil separation resistant by interchanging its physical properties. Soil particles should be mutually linked with such a force that these cannot be easily separated or dislodged from their previous place. Soil can be made separation resistant by lowering its bulk density, improving soil structure and its stability, increasing particles having more than 0.5 mm mean weigh diameter, increasing saturated hydrolic conductivity and increasing water holding capacity. All these properties can be obtained by mixing organic matter in soil or application of organic mulches and among organic mulches or organic matters, residues obtained from pulses are most important.

Besides mulches and organic matters, some chemicals are also used to improve soil structure and to bring stability in it. These chemicals increase stability of soil aggregates as a result of this splash erosion is reduced. Materials improving physical characteristics of soil are called soil conditioners. In these besides sawdust, peat and compost many inert and synthetic materials are included.

Synthetic polymers help clay particles to remain mutually interlinked by changing inter particles strength of clays. The interchange of inter particle strength of clays is obtained either by the adsorption of polymer on clay particles or adherence of many soil particles on a single particle of polymer. In both conditions flocculation of clay particles increases which brings stability in inter particle arrangement.

Application of chemicals such as polyvinyl alcohol, polyvinyl acetate, waterproof asphalt emulsion, natural rubber, polysaccharide and hydrated lime provide stability to soil structure. Also, some biopolymers like chitosan, starch xanthate, cellulose xanthate, acid-hydrolysed cellulose, microfibrils and polyacrylamide reduce shearing of soil by erosion.

In many countries these chemicals are used in abundance. Use of asphalt emulsion prevents water and wind erosion. Its application increases infiltration rate of soil. Application of natural rubber emulsion provides hydrophobic property to soil. So, soil particles do not disperse easily in the water and soil aggregates do not scatter. Use of hydrated lime is very much beneficial in the easily dispersible soil. It increases flocculation of soil which prevents soil from internal erosion.

3. Minimization of soil shearing power of runoff: Minimization of quantity and speed of runoff drastically reduces its soil cutting power. Therefore, runoff does not excessively erode soil and erosion begins to be under control. Mechanical methods are most important to minimize soil shearin power of runoff. These methods help to agronomic and other methods applied for erosion control. Main purpose of mechanical methods is to increase time for accumulation of runoff and to reduce runoff speed. Mechanical methods include drainage channels, bunding, ditches, terraces, contour bunds and contour farming, basins, sub soiling, diversion channels and grassed water ways. Brief description of these methods is given here:

1. Drainage channels: Construction of drainage channels is first step to control soil erosion. These channels carry water of stormy rain, flood water or runoff formed by these from affected area to another. If runoff remain persists in affected area, then erosion will be much more terrible. These channels control soil erosion of affected area. These channels are constructed in such a way that these channels will carry superfluous water to the other side by controlled manner. So that there will be no any loss of agricultural land as well as other infrastructures present beneath these channels. A figure of lined and unlined drainage channels is shown here:

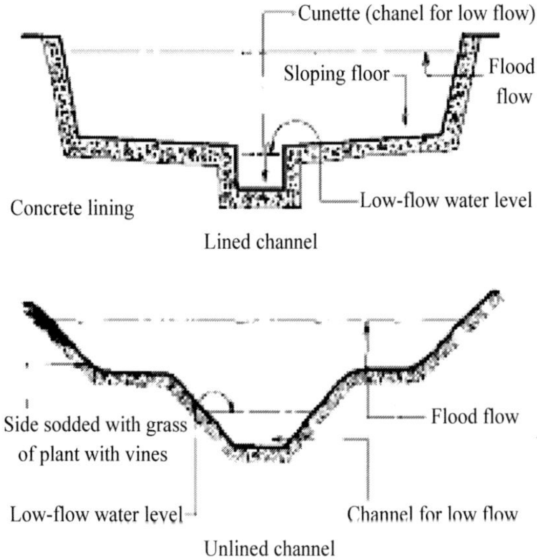

Fig. 5.4. Lined and unlined drainage channels

2. Bunding: Bunding is another method to prevent runoff. In this method rain water is checked or stopped in such a manner that all the water is infiltrated in the soil of affected area. When all water is infiltrated, then there will be no runoff. Bunding can very well be done on levelled or less slopy lands. In this method all the area is divided in to small plots and bunds are made in two angles and two directions. In this way small hexagonal beds are formed in the whole area. In sloping lands more stronger bunds are made than levelled lands, because in sloping land if any bund breaks, then water will flow speedily breaking all downward bunds. This results a flood like situation and water is accumulated in lower part of slope. Therefore, strength of bunds should be monitored time to time and if any breakage is seen that should be repaired immediately.

Bunding reduces monthly runoff and peak rate up to 62% and 40%, respectively in small watershed. In a study comparing effects of bunding and without bunding, it was found that bunding reduces runoff from 22 to 14 percent and soil loss from 33 to 1.0 ton/ ha.

There are different types of bunds according to topography of related area. These may be normal, graded or contour. Shallow, medium and slight deep soils are suitable for bunding. In hilly areas having 2 to 6 percent slope graded bunding, in areas having small gullies and ravines contour bunding and in areas of shallow black soils, both graded and contour bunding are found appropriate.

3. Ditches: Bunding is useless for erosion control in deep black soils. In summer season cracks are formed in these soils and infiltrated water begins to flow from these cracks. These result widening of cracks. In rainy season water swiftly flows through these cracks and break bunds at many places. This problem arises due to presence of more amount of 2:1 type clay in soil which has swelling and shrinking property. So, in deep black soils bunding is not successful for the purpose of soil erosion control.

Central Soil and Water Conservation Research and Training Institute has found that ditches are most suitable for erosion control in deep black soils. Since in this method bunds are not made, so there is no fear of their breakage and other related losses. Water filled in ditches maintains soil moist up to 4-5 meters. So, it is very much helpful in profile development and crop production.

Preparation of ditches: In this method about 5.56 m wide and 76 cm deep ditches are dug. Width of the bottom of ditch may be up to 61 cm. In these ditches less slope (4:1 to 5:1) is kept at runoff side i.e., in which direction runoff is coming and more slope (1.5: 1) at other side i.e., in which direction runoff is going. Thus, water accumulates towards more slope side. Distance between two ditches is determined by contour lines and length may be kept up to 400 meters on the basis of availability of terrain. In the ditches after each100 meters, 4–5-meter land is left as such without making ditches for the use of farmers, animals, agricultural implement etc. These ditches are called conservation ditches.

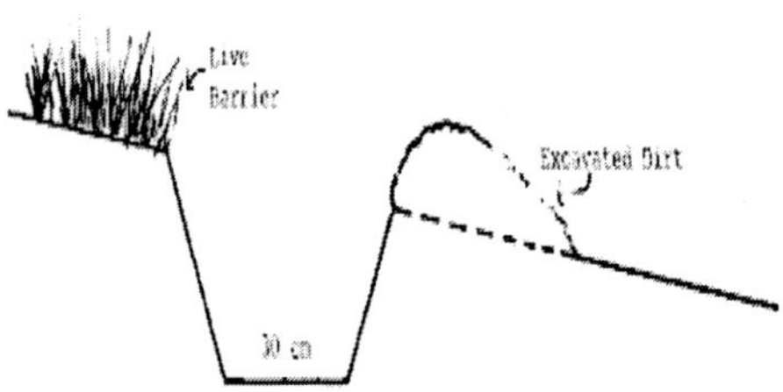

Conservation ditch

Contour ditch

Fig. 5.5: Conservation ditch and contour ditch

In the area of deep black soils these ditches control 70 to 90 percent of annual runoff and 90 to 100 percent of soil loss. If a little bit soil loss occurs, is deposited in the bottom of the ditch. Accumulated or collected water in these ditches can be used for agricultural purposes by lift pump or other local methods. In construction of these ditches about 7 percent land of whole area is used. But this area can be utilized for planting of various vegetations, grasses, trees etc. useful for fodder, forage, fuel and agriculture on both sides of ditches. By this way these ditches control soil erosion and develop healthy environment.

4. Terraces: Terraces control soil erosion by reducing the speed of runoff by disrupting intensity of water. Scientifically constructed terraces conserve soil and water properly in all types of soil. In terraces good yield of crops can be obtained by using crop rotation, ploughing, improved seeds, plant, manure or fertilizers and plant protection chemical according to need.

Terrace is an earthen structure in hilly areas where steep and large slopes are converted into many small levelled or horizontal strips that intercept runoff, increase infiltration and allow soil particles to settle down. Terracing involves excavation or digging of soil from the upper part of the slope and using it at lower part of slope to make flat land. Walls of terraces are supported by rocks or other similar barriers to protect from collapsation.

Types of terraces: There are two main types of terraces,1. Broad-base terraces and 2. Bench terraces. In a simplified manner the types of terraces are bench terrace, irrigation terrace, channel terrace, intermittent terrace, convertible terrace and natural terrace. The classification, brief description and figures of some main terraces are given here:

Broad-base terraces: These terraces are suitable for cultivation and constructed to remove or retain water on sloping lands. These are again divided in two namely graded (drainage) and levelled (absorption or ridge). Graded terraces are further divided in, with proper channel and without proper channel.

Bench terraces: Bench terraces are generally used to reduce the land slope. Types of bench terraces are levelled, sloping outward, sloping inward and Puerto Rican.

1. Bench terrace: Such terraces are constructed across slope as steps in the areas of 20-30% slope. In this land surface is cut to make approximately levelled surface. If there is little slope, that can be kept outwards or inwards. There is a standing wall between two steps. This wall is supported by rocky stones, bricks or wood. This wall is covered by vegetation in very stable soil. The outer part of terrace is little bit raised. Length of bench terraces may be as desired, but less than 100 meters long terraces are best. The levelled land surface of terraces reduces speed of runoff and raised outer sides prevent water to go out. So, water is gradually infiltrated in the soil. Bench terraces can reduce runoff up to 50 percent and soil loss up to 98 percent. In paddy growing area of Himalaya and area having 16-30 percent slope of Nilgiris bench terraces are found beneficial for the purpose of erosion control.

1. **Irrigation terrace:** Irrigation terraces are flat. Their outer side is sufficiently raised. There is a channel or drain type structure for retaining irrigation water. Such terraces are constructed mostly in the field of paddy and their land also can be used for growing of fruit trees and other cash crops.

2. **Channel terrace:** Open channels are constructed across land for proper water drainage. When this channel is used as terraces, then it is called

channel terrace. Similar terraces are constructed in slight sloping areas. Runoff moves outside gradually through these terraces and due to slow speed of runoff water soil erosion does not occur. Channel terraces may be on contour lines. Contour channel terraces are found beneficial to control soil erosion in the area of black soils of India.

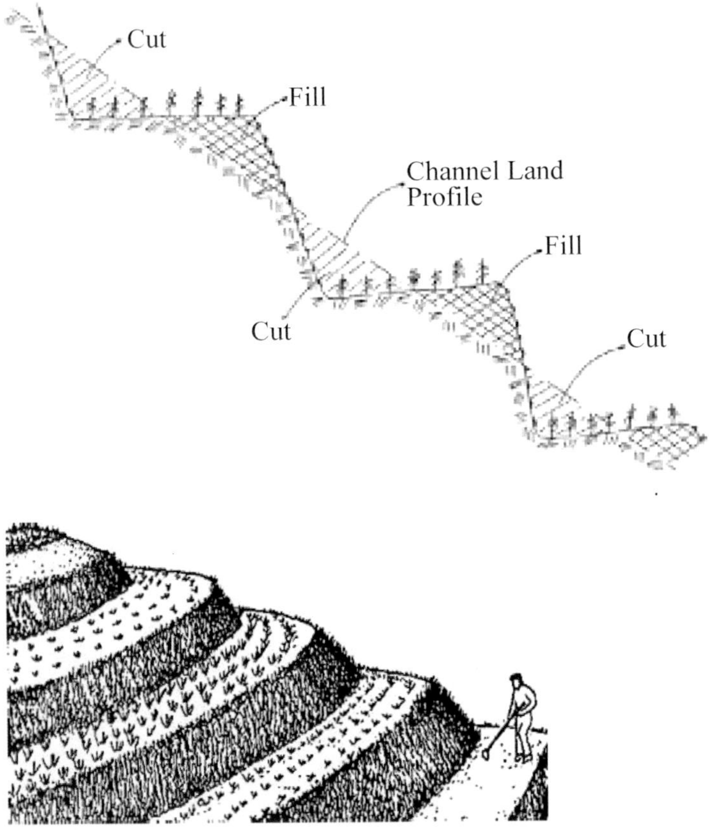

Fig. 5.6: A simple technique to construct terrace

4. Puertorican terrace: Puerto Rican terraces are economical and eco-friendly soil erosion control method developed at Puerto Rico and adopted by many countries including India. Bench terracing is generally recommended as soil erosion control measure for the hilly regions. Many farmers hesitate to adopt bench terrace technology as it initially requires heavy investment. In Puerto Rican method of terrace formation, the expenditure incurred is only one sixth of the cost as compared to mechanical terracing.

Puerto Rican terraces are formed naturally without disturbing the topsoil by using grasses or other local vegetation as vegetative barrier. On a sloping land selected for terrace formation, suitable species are planted in single or double row on contour at pre-determined spacing. The inter space between the two hedge lines of barrier is cultivated to grow crops. As a natural process, the tilled soil slowly moves towards the vegetative barrier and gets deposited against it leading to the formation of terraces in 3 to 4 years time.

This technology developed by this centre Goa was successfully adopted in the farmer's field. Shri C. Ragu of Melkavatti village in the Nilgiris had adopted this technology in 0.4 ha of sloping land under the Farmers Participatory Action Research Programme implemented by this Research Centre. It was observed that within a period of three years, terraces have been naturally formed in his field with minimum runoff, soil loss and nutrient loss as the chief benefit.

Apart from fodder production of 8 t from the grass raised, Shri Ragu had also obtained 22 % additional crop yield in comparison to the yield obtained before adoption of this technology. Two more farmers also got the same benefit following this technology. A photograph of Puerto Rican terraces with standing crop is given here:

Fig. 5.7: Puertorican terrace An economical and eco-friendly conservation measure getting attention in Nilgiris (2012) https://agrigoaexpert.res.in-images-success story https://www.google.com/search?q=puerto+rican+terrace

Relation between length and slope of terraces: Terraces should be designed and constructed: based on topography and climate of area concerned. If terraces are constructed scientifically then runoff water flows slowly and slowly towards specified direction. Therefore, erosion also does not occur. In terraces runoff water flows to downward from upper part of slope. Therefore, quantity of runoff water tremendously increases at final drainage site i.e., at the bottom of slope. If at this site slope will be less then sediment will be deposited more. So, terraces should be more gradually sloping towards final drainage site. Therefore, terraces with variable slopes should be constructed here.

In areas having excess rainfall, slope should be kept 2.5 cm per 33 meters during first 135 meters length, 5.0 cm per 33 meters during second 135 meters length, 7.5

cm per 33 meters during third 135 meters length and 10 cm per 33 meters during fourth 135 meters length.

In areas having less rainfall, it is necessary that soil should absorb maximum quantity of rainfall. So, for such areas terraces with lesser slope should be constructed. At the places of insufficient rainfall, first 170 meters length of terraces is kept nearly level without slope, for the second 170-meter length slope should be kept 2.5 cm per 33 meters and for third 170 meters length slope should be kept 5.0 cm per 33 meters.

5. Contour dam: Lines of equal height made on right angles towards common flow of runoff water are called contour lines. Dams constructed on contour lines are called contour dams and cultivation done on land existed between contour lines is called contour cultivation or contour farming. Contour dams are most suitable for areas having less than 600 mm annual rainfall, permeable soil and less than 6% slope. In such areas control of soil erosion and absorption of rain water in the soil is essential for crop cultivation.

Fig. 5.8: Contour dam

Contour dams absorb the runoff water stored at the upstream side of the dam. Appropriate height of the dams is necessary to prevent overflow during floods and monsoon as the whole runoff water cannot be absorbed in soil. This may create water logging also. The surplus water causing over flow is checked by providing weirs on the dams. These weirs pour excess water into the next bund.

Formation of contour lines in the field: Contour lines in the field are formed after determining the lines of equal height with the help of proper equipments like Engineer's Transit Farm level, Abney level or common level of carpenters. After determination of contour lines these can be drawn on the land with the help of plough, stake or thick nail. On the basis of topography of land, if actual contour lines are drawn very accurately. Then it can be seen that at some places lines are very nearer and at places they are very far instead of being parallel. Therefore, the land left among these lines is very wide at some places and very narrow at other place. Sometimes it is very difficult to operate agricultural implements easily in these narrow lands and perform other cultural practices. However, this problem can be solved by making somewhat parallel contour dams at equal height after altering some predetermined points. Generally, for the ease of implements' movement and cultural practices only 3 percent alteration is allowed.

Construction of contour dams: After the formation of contour lines small contour dams are constructed across of slope on these contours. Each dam reduces erosion by stopping intense flow of water. Therefore, soil can absorb most of the water due to slow speed of flow. Remainder water from absorption by soil flows downwards with slow speed and benefits crops grown there. Land between two contour dams facing each other has its own catchment area, reservoirs and command area.

Maintenance of contour dams: Time to time supervision of contour dams is very much necessary. Negligence in maintenance and repair of any one dam can harm whole area. Breaking of any dam possibility break other dams. So, all breakage, cracks, burrows etc. of dams should repaired immediately. Dams situated at lower part of the area become weak by movement of animals etc. and upper part of dams breaks by their hoof. These activities reduce height of dams. Appropriate repairing of such dams should be done and constructed again up to the required height. Burrowing animals also damage dams by making burrows in them. Proper measures should be adopted to control burrowing animals and burrows should be closed properly by suitable materials. Drainage channels of the dams should be repaired immediately. All repairing and maintenance works should be completed before rainy season. However, damages arose during rainy season should be eliminated immediately

6. Basin listing: This method is good for slightly sloping areas. Basin listing provides maximum time to infiltrate rain water in soil. Therefore, small basins are listed to stop runoff alongside contour lines or across the slope. First of all, drainage water falls in these basins. An agricultural implement named basin lister is used to make basins. If basin lister is not available then ploughing is done parallel to contour lines and furrows are prepared. These furrows are closed in between and a series of shallow basins is formed.

Fig. 5.9: Field with a Series of Shallow Basins

1. Sub soiling: Sub soiling is suitable for the soils having hard layers in their lower horizons. These hard layers decrease soil permeability. So, water does not easily enter in to soil. In such situations most of the rainfall water accumulates on soil surface as runoff and erodes soils. A specific implement known as sub soiler is used to break these hard layers. Breaking of hard layers provides passage to move water. So, water begins to go fastly towards lower horizons and does not accumulate as runoff on the soil surface.

The subsoiler operates up to 45 to 60 cm. or more depth. This helps to break hard layers of soil formed by continuous ploughing, clay layer or other such hard layer. Besides soil erosion control, breaking of hard layers has other benefits also. The roots of plants can reach up to more depth and uptake more nutrient and water from there. So, growth and development of plants will be better. Soil aeration increases due increase in volume of micropores. Moisture holding capacity of soil increases. Conservation of moisture increases in concerned area and soil erosion stops.

8. Diversion channel: Diversions are constructed at the point where the soil is susceptible for erosion due to the attainment erosive velocity by surface runoff. The diversions are constructed across the prevailing slope and divert it across the slope of grassed waterways. Both diversion and ditches are in form of channel or trench with gentle slope. This is a ditch, which to divert run off before cultivable land carry it on the other side. As a result of this cultivable land is saved by erosion. This method is used to stop runoff and soil loss in cultivable land having 10 to 16 percent slopes.

9. Grassed water ways: Grassed waterways are also called as vegetated drainage channels and vegetated waterways. Grassed waterways are broad, shallow and some what trapezoidal, parabolic or tringular channels. The triangular waterways are changed to parabolic or trapezoidal shapes due to deposition of sediments in due course of time. These channels move runoff water from land without causing soil erosion.

Grassed waterways are natural or constructed channels. These transport concentrated flow of water at safe velocities from the catchment or cropped area to any river, reservoir, stream etc. without causing soil erosion. Constructed grassed waterways are generally used at terraced and bunded cropped areas, diversion channels, spillways, contour furrows, etc. for disposal of the runoff. The grassed waterways are located in such area so that runoff from all sides should come in it.

The grassed waterways outlets are constructed prior to the construction of terraces, bunds etc. because grasses take time to get established on the channel bed. Generally, it is recommended that there should be a gap of one year so that the grasses can be established during the rainy season.

The flexible vegetative cover in the waterway slows the water flow and protects the channel surface from the eroding forces of runoff water. The soil and climate are two major factors in selection of vegetations to be grown in grassed waterways. Normally local vegetation should be selected for grassed waterways. However, it was observed that cultivation of canna, vetiver, torpedo grass, para grass, doob grass, crown grass etc. are good enough for grassed waterways. Sometimes, a light seeding of small grain is also used to develop a quick cover before the grasses are fully established in the waterway. The figures of grassed waterways are given here:

Fig. 5.10: Grassed water ways

Methods of Wind Erosion Control

In wind erosion control primarily two tasks are considered. Firstly, to provide vegetative cover so that wind velocity on soil surface would be reduced. Secondly to make soil wind drags resistance. Plantation, pasture development, sand dune stabilization, field strip cropping, stubble mulching and proper tillage practices are necessary to solve the purpose of both tasks. In these except tillage all other operations minimize wind velocity on the soil surface. Vegetative cover and mulch besides decreasing wind velocity make soil wind erosion resistant and improve physical properties of soil.

1. Plantation: In areas of wind erosion plantation is very beneficial. Appropriate trees for wind erosion areas are neem (*Azadirachta indica*), babul (*Acacia nilotica*), vilayati babul (*Prosopis juliflora, Acacia tartalis*), siras (*Albizzia lebbeck*) etc. But these trees should be grown as wind break and shelter belt instead of growing as such.

Wind barrier: Wind barrier is prepared with the help of various trees and bushes in the wind break and shelter belt. Wind barrier protects agricultural area and other related structures. Wind break and shelter belt of trees and bushes remarkably minimize wind in the leeward direction. This changes microclimate of that area. Minimization of wind velocity and change in microclimate by wind break depend on wind speed, direction and turbulence and length, height, thickness and porosity of wind break.

Shelter belt is defined as a wide belt of trees, shrubs and grasses, planted in rows which goes right across the land at right-angle to the direction of the prevailing winds to deflect air current, to reduce wind velocity and to give general protection to cultivated areas and other infrastructure against wind erosion and desiccating effect of the hot winds in lee-ward side.

Wind break can be made of single row or many rows. Single row wind break is sufficiently open. So, wind easily passes through it. In wind break having multiple rows a solid wall is formed by planting trees and bushes of different heights of various species. In this wind can not enter easily. It has been observed that wind break having three rows of vilayti babul (*Prosopis juliflora*), neem and black siris (*Albizzia lebbek*) present on the sides of highways of western Rajsthan was sufficiently effective. A report of Central Arid Zone Research Institute, Jodhpur indicated the importance of shelter belt in soil erosion control as total soil loss was 3 ton/ha in the area of shelter belt while in unsheltered area soil loss was 72 ton/ha. The figures of single row and multiple rows shelter belts are given here:

Fig. Single row wind break

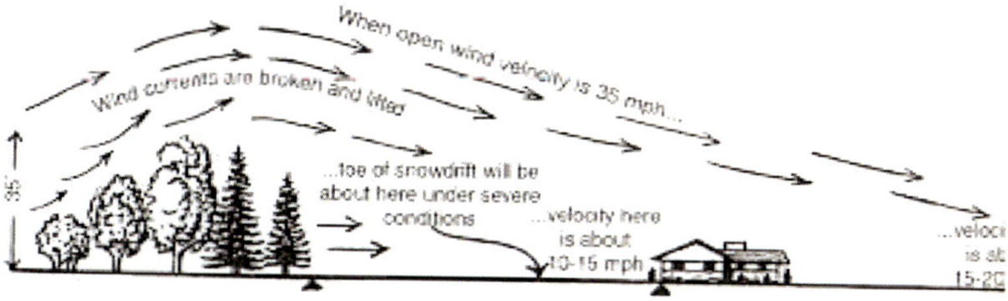

Fig. 5.11: Multiple row wind break

A study of Gupta et al. (1983) at Rajasthan reported soil and nutrients loss from three shelter belts of *Acacia tartalis*, *Casia siamea* and vilayati babul (*Prosopis juliflora*) during April, 20 to June. 26 in relation to fallow land. Mean results of their study are given in Table 5.1.

Table 5.1: Effect of various trees on losses of soil and nutrients

Trees	Soil loss (kg / ha)	Nutrient loss (kg / ha)		
		N	P	K
Acacia tartalis	351.2	123.0	65.0	667.4
Cassia siamea	184.3	64.5	34.1	350.1
Prosopis juliflora	308.0	105.0	55.5	670.0
Fallow	546.8	191.4	101.2	1039.0

It is clear from table 5.1 that in comparison of fallow land loss of soil and nutrients in the conserved area from shelter belts was remarkably less. Shelter belt of *Cassia siamea* was best in the control of soil and nutrients loss.

Plants for shelter belts and wind breaks: Numerous plants are suitable for growing as shelter belts and wind breaks. However, the choice of species to be raised in shelterbelt is governed by the climate, soil and topography of the area.

Locally available plants species should be selected because they can be easily established. Exotic plants may also be used to improve the efficiency of the shelterbelts. The plants can be selected either by plants species or by rows. Plants should be selected from these lists according local utility. The examples for both are given here:

1. Plants species for shelterbelts

Grasses: *Cenchrus barbatus, Saccharum spontaneum, Saccharum munja, Panicum turgidum, Panicum antidotale.*

Shrubs: *Calotropis procera, Crotolaria burhia, Calligonum polygonoides, Clerodendron phlomoides, Cassia auriculata, Dodonaea viscosa, Jatropha curcas, Leptadenia spartivm, Agave spp., Sesbania aculeata.*

Small trees: *Acacia jacquemontii, Acacia leucophloea, Balanites aegyptiaca, Capparis aphylla, Salvadora oleoides.*

Large trees: *Acacia arabica, Acacia senegal, Acacia cyanophylla, Albizzia lebbek, Azaadirachita indica, Dalbergia sissoo, Lannea coromendelica, Parkinsonia aculeata, Prosopis cineraria, Prosopis juliflora, Pongamia pinnata, Tecoma undulata, Tamarix articulatat, Eucalyptus* spp., *Acacia tortilis.*

2. Plants species for rows

In the first row should be grow following trees or vegetation:

Euphorbia roylena, Eyphorbia nerifolia, Saccharum munja, Arundo donax, Sesbania aegyptiaca, Lawsonia alba, Agave americana, Agave sisalena, Ipomea carnea, Vitex negundo, Calotropis gigintea, Tamarix articulata, Zizyphus nummularia, Taphrosia candida, Cajurina equisetifolia.

In the second row should be grow following trees and vegetation

Zizyphus jujuba, Parkinsonia aculeata, Prosopis juliflora, Acacia arabica, Pongamia glabra, Cassia siamea, Acacia catechu, Boswellia serrata, Acacia auriculiformis, Casuarina species., Coconut palm, Borassus flabellifer.

In the third row should grow following trees and vegetation

Acacia arabica, Albizia lebbeck, Delbergia sisso, Eugenia jambolana, Cedrela toona, Toona ciliate, Tectona grandis, Mangifera indica, Delbergia latifolia, Tamarindus indica, Terminalia species, Eucalyptus species, Hardwicia binata, Gravillea robusta, Casuarina species, Cashew nut, Artocarpus oxidentalis, Palm, Bamboo,

2. Pasture development: Wind erosion also significantly occurs in pastures and grazing lands. Such terrain where cultivable land is not available, development of pasture is a good option for wind erosion control. However, controlled grazing should be adopted. Controlled grazing is a simple method for soil erosion control or soil conservation in which grazing of animals on grasslands and within forests is restricted. When animals continuously graze or over graze on pastures, they even uproot grass and smaller plants. This makes soil loose and susceptible for erosion. Therefore, pasture can be divided in some parts and fenced or hedged to make protected enclosures. This allows animals to be confined to graze in one part while other parts remain ungrazed. Then animals are allowed to graze in next part and simultaneously in other parts. In this way land, can be grazed for 1 to 2 weeks and then ungrazed for several weeks to allow grass and plants to regrow. This is also called as pasture rotation.

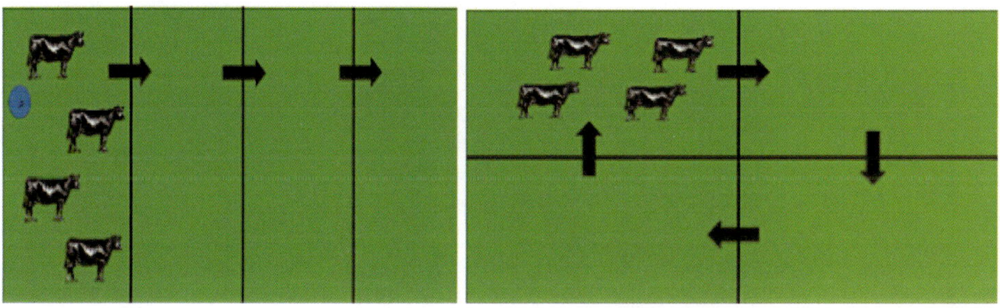

Fig. 5.12: Controlled Grazing and Pasture Rotation

However, for the purpose of appropriate pasture development, management and erosion control some measures indicated here should be adopted:

1. Number of animals in grazing areas should be limited. 2. After dividing pasture into parts put a part of pasture in grazing at a time so that vegetative growth may be continued in other parts of pasture. 3. Make necessary arrangement of irrigation in pasture so that vegetation of grazed part of pasture may grow and develop immediately. 4. Necessary artificial or natural fencing should be provided to stop entry of animals in ungrazed or excessively eroded parts. 5. Grow wind break belts for the protection of sources of water of pastures. 6. Contour trenching, contour furrowing and staggered pitting should be adopted for the conservation of range lands.

Experimental evidence indicated that in a pasture near Bikaner wind erosion was nearly negligible. But drift of soil in the fallow sandy land next to this pasture was 1449 to 5560 ton/ha during a period of 75 days (April to June). Growing grasses in pastures also increase soil fertility with providing stability to soil. In the period of twelve years cultivation of *Cenchrus ciliaris* and *Lasiurus jindicus* grasses enhanced 16 to 43 percent in soil organic matter and 20 percent in soil nitrogen.

3. Stabilization of sand dunes: Sand dunes can be stabilized by using indigenous grasses and other plants which help in the reduction of the velocity of air waves and absorb their energy. Soil displaced and transported from sandy terrains also affect near by good terrains. The control of displacement and movement of sands in these areas can be done by enriching whole area with vegetation besides wind break etc. These vegetations should be able to tolerate blows of sandy particles with reducing the velocity of wind. These vegetations should have deep roots and strong stems so that this could tolerate blasts of violent wind without falling on ground.

Vegetation enrichment of sand dunes affected area can be done even in May and June because in these months moisture also remains up to 2.5 percent in lower parts of sand dunes. This will help in easy establishment of vegetation. In the process of sand dunes stabilization first of all, small or micro wind breaks should be planted or constructed from dried brush woods. *Zizyphus nummularia* and *Crotolaria burhia* are suitable plants for small windbreaks. Dried brush woods of *Zizyphus nummularia, Crotolaria burhia, Avena persica, Prosopis cineraria* and *Leptodentica pyrodentica* are used for micro windbreaks.

Fig. 5.13: Micro windbreak

Now after establishing small or micro windbreak various trees, bushes, creepers and grasses should be planted in the leeward side of micro windbreak of these areas. The appropriate plants may be selected from the list given here:

1. **Trees:** *Acacia senegal, A. tortilis, A. nubica, A. jaquimonti, A. bavenosa, Prosopis juliflora, Albizia lebbeck, Cordia rothii, Dalbergia sissoo, Zizyphus jujuba*
2. **Bushes:** *Calligonum polygonoides, Cassia auriculata, Ricinus communis, Zizyphus nummularia*
3. **Creepers:** *Citrullus colosyntbes, Cucumis Spp.*
4. **Grasses:** *Lasiurus scindicus, Panicum turgidum, Erianthus munia*

4. Strip cropping: Strip cropping is more effective for erosion control in the area where terraces are not practically feasible. In the larger terrains displacement of soil is more due to more violent winds. Particles of sand flying with wind are more erosive. Strip cropping reduces soil erosion from wind and protects growing crops from damage by wind-borne soil particles. By adopting strip cropping soil particles cannot fly up to greater distances, because velocity of wind is reduced.

Strip cropping is the practice of growing strips of crops having poor potential for erosion like cereals, cotton, tobacco, sugar beet, potato etc., alternated with strips of crops having good potential for erosion control like fodder crops, grasses, small grains etc. Also strip cropping in ratio of 6:1 of grasses and pulses (mung and moth) is considered very good for erosion control and productivity. Strip cropping of selected crops should be done on right angle of direction of wind.

Fig. 5.14: Strip Cropping

5. Mulching of stubbles: In soil erosion control and moisture conservation mulching is very much important. Stubble mulching is especially useful in wind erosion control. In this method stubbles of previous crop are left in the field and sowing of next crop is done with minimum cultural operation. Stubbles are crop residues namely, paddy straw, wheat straw, cotton stalk, maize stalk etc. Although in our country this method is not in use, because of shortage of fodder in areas having wind erosion so all residues are taken off from the field for animal fodder as well as burning of rice and wheat straw in field itself.

Also, carbon: nitrogen ratio of stubble of cereal crops is high. Therefore, immobilization of nitrogen takes place which reduces nitrogen supply to next crop. So, next crop does not grow well. But in view of loss of soil by erosion it should be adopted. It is seen in the experiments that height of stubbles on soil surface directly influences erosion. In Rajasthan stubble mulch of millet (bajara) up to height of 10-15 cm showed soil loss only 22.5 ton/ha, while soil loss was 1449 ton/ha in unmulched area. This very clearly indicates the importance of stubble mulching in soil erosion control. The quantity stubbles required for erosion control depends upon speed of air, soil movement rate and other conservative measure adopted. More quantity of stubbles will be needed if wind speed and soil movement are more and any conservative method is not adopted.

6. Cultural practices: Proper tillage is very important in controlling wind erosion. More tillage enhances wind erosion. Minimum tillage along with mulch of crop residues stop wind erosion. At places of wind erosion prepare fields in such a way that unevenness on soil surface is maintained. Uneven soil surface stops erosion, while fine and smooth soil surface increase it. Tillage should be done across the direction of wind. Cultural practices, which leave crop residues on the soil surface are helpful in wind erosion control. Furrow sowing of crops save plants from blows of wind and increase number of alive plants.

In areas of having intense wind erosion spring tooth harrow is better than spike tooth harrow. Spring tooth harrow brings more clods on the surface by piercing soil to the greater depth. More clods on soil surface reduces wind erosion. Field cultivators are good for making mulch of stubbles of crop and deep or shallow ploughing of the soil can be done. Minimum tillage can be done by field cultivators to maintain clods of useful size on the soil surface.

If there is extreme loss of vegetative cover of any terrain then instantaneous ploughing is very useful for wind erosion control. Ploughing done on right angle with direction of wind to produce unevenness, clods and rip in the field. All these restrict forces of wind. Normally 7.5 to 10 cm depth of immediate ploughing produces sufficient unevenness on the soil surface. However, this method is temporary, because clods break immediately. In areas of wind erosion deep ploughing is more beneficial. Deep ploughing up to depth of 40 to 120 cm is done by Mold board or disc plough. This forms more clods on soil surface if subsoil contains more clay.

7. Other measures of wind erosion control: Use of various adhesive substances and moisture conservation are included in other measures of wind erosion control. Many types of petrochemicals and adhesive substances of biological origin control wind erosion temporarily. In these anionic asphalt emulsion, latex emulsion, liquid protein colloid and fermented corn extracts are major substances. There is less in wind erosion areas. So, in such areas conservation of moisture is necessary. In addition to mulching, cultural practices, wind break and vegetative cover by increasing irrigation facilities moisture can be maintained in the soil. It is seen in the experiments that in the soil up to depth of 20 cm application of 40 tons/ha exfoliated vermiculite and bentonite increase accumulation of moisture in the soil. Likewise, application of pond sediment @ 76 tons/ha enhances moisture accumulation capacity of soil up to 8.5 to 11.5 percent. Compacting the soil by roller i.e. by increasing bulk density, infiltration rate of water is reduced. Therefore, water being remained in upper horizons of soil to benefit next crop.

Erosion control in gully or ravine

Ravines are locally called as Bihad. Rugged soil is terrible form of erosion. This destroys land completely. Ravine and gully erosion affects 1% of India's land area. Zones of severe ravine trenching are found along the margins of the Gangetic Basin and in the semi-arid northwest. Every year about 8000 hectares cultivable good land of the country is converted in ravines. Ravines are spread over an estimated area of 3.67 million hectares along the rivers and their tributaries. Madhya Pradesh, Utter Pradesh, Rajasthan and Gujrat are mostly ravine affected states. In addition to these Punjab, W. Bengal, Bihar and Tamil Nādu also are partly ravine affected states.

When cultivable lands of any area are eroded and become ravine then numerous problems arise including shortage of food grains, fodders, fuel, timber etc. Water level goes down. This creates scarcity of drinking water for the animals and human beings as well as water for irrigation. In these circumstances social problems also arise. Ultimately these ravines serve as shelter for dacoits and criminals and surrounding areas become their working place.

Ravine reclamation programme: Successful ravine reclamation programme requires the support and involvement of the local cultivator and local community, reform of local land tenure and social arrangements. Ravine reclamation is currently rated as a high national priority and India has the Third World's leading soil conservation movement. Ravine reclamation programme is not so easy to handle as compared to normal soil conservation work. After seeing all the problems arose due to ravines, Government of India established three centres Agra (U.P.), Kota (Rajasthan) and Basad (Gujrat) for conservation of ravine lands. These centres are under ICAR, Indian Institute of Soil and Water Conservation, Dehradun.

A stepwise ravine reclamation programme is outlined here: 1. Survey, Planning and Investigation of Land 2. Treatment of Level Land 3. Marginal Land Protection for Intensive Agriculture 4. Main Ravine Land Reclamation 5. Ravine Stabilisation through Biological Measures.

1. Survey, Planning and Investigation of Land: Before starting any ravine reclamation programme, it is necessary to know the proper knowledge of land and water resources of that particular area. Also about soils, vegetation, socio-economic status of beneficiaries etc. constituted the important aspects. A complete watershed of any ravine system should be considered as one planning unit. It should be well identified on a small- scale map (1: 6000). These units may be further classified based on detailed contour maps.

The preliminary studies conducted at Agra, Kota and Basad indicated that useless annual grasses are grown in ravines or gullies and they are also over grazed. The quality of these grasses is progressively deteriorating. So, in areas having up to 6 percent slope first of all fencing and ridging should be done and after that grazing should be completely banned. This will enhance the growth of grasses and improve their quality. Ultimately soil loss will also decrease.

2.Treatment of Level Land: The levelling of less slopy ravines is necessary for systematic handling of runoff from top of watershed. This will benefit soil and water conservation work and check the further spread of ravines. The levelled lands should be used for cultivation of locally available non-erosive crops and their crop sequence.

3.Marginal Land Protection for Intensive Agriculture: Based on the rainfall, soil, topography and cropping system contour bunding should be done. Under rainfed conditions contour bund farming, contour ridge and furrow help to hold and conserve large portion of rainfall. Small gullies should be slightly levelled after clearing them. Afterwards in them a series of contour bunds should be constructed. Cross section and vertical intervals of these bunds should be 0.9 to 1.3 m^2 and 0.9 to 1.9 m, respectively. Stability of these bunds, appropriate use of rain water and production of fodder can be attained by growing various grasses namely, *Dicanthium annulatum, Cenchrus ciliaris, Andropogon spp etc*. Production of these grasses will solve the problem of fodder also.

4. Main Ravine Land Reclamation: The soil conservation engineering structures like drop spillways, drop inlets, chutes, gully plugs, check dams etc. should be established at the selected points to prevent further extension of ravines. Also, a series of check dams of soil and bricks should be constructed after proper levelling and clearing of medium sized gullies for their improvement. Vertical intervals of these dams and outer slope between two dams should be kept about 1.20 meters and about 0.75 percent, respectively. Bench terraces should be made with 8-15 percent side slope. Earthen dams should be stabilized by planting *Dicanthium annulatum* or *Cynodon dactylon* grass. After sometime small and medium gullies can be utilized for the purpose of agricultural production.

Best land utilization in deep gullies or ravines is to plug them. In the condition of abundant runoff gully plug of earth having 2.2 m^2 cross section is constructed. An outlet pipe is attached in those plugs. Size of pipe depends on discharge of the gully. Furthermore, if the runoff is not from the head, then gully plug without pipe is also constructed. Cross section of these plugs is 1.1 m^2. Here grassy ramps are also made on gap of 45 to 60 meters below 22.5 cm from upper head of these.

Fig. 5.15: Gully Plug and Check Dam

5. Ravine Stabilisation through Biological Measures: The most economic use of ravine land is the growth of permanent vegetation. The deep ravines may be stabilised by completed ban on grazing. Regular afforestation and engineering measures are also necessary along with farm forestry, grass land and pasture development. The thorny species are natural vegetation of ravine areas. However,

Dalbergia sissoo, Acacia catechu, Acacia arabica and *Dendroclamus strictus* will also good in the ravines. Simultaneously grasses should also be planted in ravines to solve fodder problem. *Cenchrus ciliaris, Dicanthium annulatum, Cynodon dactylon, Cymbopogon nardus, Heteropogon contortus, Chrysopogon montanous, Saccharum spontaneum* are good for ravines.

After plugging deep gullies or ravines plantation programmes should also be run in specific part as *Prosopis juliflora* on hump, *Acacia nilotica* on side slopes and *Dendroclamus strictus* like trees are planted in the bottom of ravines. Grasses as *Dicanthium annulatum* and *Sanchrus siliaris* should also be planted. These trees and grasses will solve problem of fuel and fodder.

Bhabhar grass (*Eulaliopsis binata*) is also found more promising in the rugged ground of Mahi River in Gujrat. This grass is fibrous. This is used to make rope and paper. Furthermore, orchard of jujube (Ber) also can be planted in degraded and denuded ravines of Agra and Kota. Banarasi, Ponda, Safeda and Chhuhara varieties of jujube give profit very early. Papaya, guava, lemon and banana can be grown successfully in Basad area. Thus, rugged or ravine areas also have ample capacity of production of fruits, timber, bamboo, fuel and fiber and give employment also. Therefore, ravines should be made useful for development of nation by proper management.

Restoration and Management of Soil from Depletion of Organic Matter

We have come across that soil organic carbon content fell from 1% to 0.3% in 70 years in India and such a drastic fall in the SOC content affects the productivity of soil as micro-organisms do not survive, a key factor to provide nutrients for plants. Intensive cultivation of crops without providing proper compost to the soil is the reason behind the fall in SOC content. Farmers should reduce their over-dependence on pesticides and fertilisers. Biofertilizers and compost can increase the SOC level of the soil. Presently about 63 Pg soil organic carbon has been reduced to 22.72 ± 0.93 Pg. Soil organic carbon is a component of soil organic matter and SOM is the store house of SOC. When soil organic matter is restored, ultimately soil organic carbon is restored. The restoration of SOC or SOM is very much essential for long continuous sustainability of agricultural production.

Now all are concerned about the increase in organic carbon or organic matter of soil. There are many methods to increase SOC. These methods include composting, conservation agriculture, reduced or zero tillage, cover cropping, green manure crops, crop residue mulching, mixed cropping with livestock trees and pastures, biochar application etc. Some of these methods are already described in last part of section 3.6 of this book. Remaining methods are described in nutshell.

1. Composting: We've all heard of composting at some point in our lives, and it is an age-old practice. Soils can be kept fertile and rich in SOC by composting specially by the application of organic compost. Organic compost is decomposed organic matter added to soil in order to improve its quality. Different microorganisms can be added externally to the pile of compost to increase the process of decay more efficiently. Organic compost helps improve the texture of the soil, reduces compaction, and increases the amount of water and air that can be held by the soil. Application of organic compost in any agronomic and horticultural operation, it becomes a more sustainable, chemical-free, and health and environment-conscious effort.

The ideal organic compost is crumbly, smooth in texture and dark brownish in colours. Mature organic compost smells like a forest on a rainy day. If there are any sour and ammoniacal smells, it will need some more time for maturity. After maturity it becomes about 33 % of the original size of the pile.

Benefits of Organic Compost: There are many benefits by using organic compost in the soil.

It helps in increasing organic matter, organic carbon and plant nutrients in soil. Organic composting is an excellent way to sequester carbon in the soil. By doing so, organic composting can help to improve soil quality and fertility and reduce greenhouse gas emissions. It helps to balance the soil pH and contains natural soil organisms like worms which helps to maintain good health in the garden. These organisms help to break down organic matter and release nutrients that plants can use and can reduce the need for pesticides and fertilizers. Organic compost significantly increases aeration, porosity, water holding capacity etc.

Organic waste to organic compost: There are many day-to-day waste materials which can be used as compost material and make good quality organic compost. This is just like making gold from waste. Fruit and vegetable waste, leftovers of cooked plain rice, Stale bread, Crushed Eggshell, used tea leaves, Coffee grounds, non-glossy shredded papers, Paper towels, toilet paper, napkins, Chips of untreated wood, Wooden popsicle sticks, wooden toothpicks, Grass clippings, Dried fallen leaves etc. are few organic wastes used to make organic compost.

How to Use Organic Compost: There are so many ways to use compost in your yard, farm or garden. Here are few of them.

Use as Mulch: Compost can be used as mulch effortlessly and efficiently. Compost is spread at the top of the soil as a thick layer and this can protect the soil around the plants. The mulch retains moisture in the soil, suppresses weeds, and keeps plants' roots cool in summer. Also make sure that organic compost should spread around the base of plants without piling it against the stems. Spread mulch generously around trees and shrubs, especially young plants. If it is hot outside, wet the compost before spreading it to avoid it drying out plants. After some time, compost breaks down and disappears, so it must be replenished regularly.

Use as Compost Tea: The nutrients of compost can directly be applied to the plant roots. Compost can be brewed as Compost tea. The composition can be steeped into the Emulsion of liquids and this process helps to extract nutrients in a more concentrated form. Plants also can easily absorb it. It usually takes 24 to 36 hours to prepare compost tea. Compost tea can help plants to grow faster. It plays the role of 'Magic Elixir,' which contains many bacteria, protozoa, fungi, and maybe some nematodes that are much absorbable.

Spread on the Lawns: Organic compost can be spread over the lawn. This will add valuable nutrients and organic matter that will help grow thick, healthy grass. This is also used on lawns before planting as well as later for maintenance purposes. A 3-5 cm layer of compost can be added to the lawn before the seed is sown to enrich the soil with nutrients.

Use For Growing Fruits and Vegetables: The appropriate time to compost the fruit tree is early spring if the expected harvesting time is in the fall. Use compost as mulch for plants. Compost is nitrogen rich and full of macro and micro-nutrient nutrients. Therefore, it is ideal for fruit trees. Melons, tomatoes, brinjals, cucumbers and squash need plenty of nitrogen for their growth. So, adding compost to the planting holes is recommended at the time of transplanting seedlings. During the growing season, top-dress compost on the soil several times to achieve the most effective production.

Use In Fall Perennials and Spring Bulbs: It is ideal to add 2-4 cups of organic compost to the planting hole of perennials. It will help the plant to grow vigorously and healthy. It will make them bloom longer. Also, it benefits soil organisms, which help the ecosystem. Furthermore, bulbs that have been recently divided should have compost added to the planting hole to help them thrive once they emerge from their winter dormancy.

2. Conservation agriculture: Conservation Agriculture (CA) as described by FAO of UNO is a farming system that can prevent losses of arable land while regenerating degraded lands. It promotes maintenance of a permanent soil cover, minimum soil disturbance, and diversification of plant species. It enhances biodiversity and natural biological processes above and below the ground surface, which contribute to increased water and nutrient use efficiency and to improved and sustained crop production. Principles of conservation agriculture are universally applicable to all agricultural landscapes and land uses with locally adapted practices. Soil interventions such as mechanical soil disturbance are reduced to an absolute minimum or avoided, and external inputs such as agrochemicals and plant nutrients of mineral or organic origin are applied optimally and in ways and quantities that do not interfere with, or disrupt, the biological processes.

Conservation agriculture facilitates good agronomy, such as timely operations, and improves overall land husbandry for rainfed and irrigated production. Complemented by other known good practices, including the use of quality seeds, and integrated pest, nutrient, weed, water management etc. Conservation agriculture is a base for sustainable agricultural production and intensification. It opens increased options for integration of production sectors, such as crop-livestock integration and the integration of trees and pastures into agricultural landscapes.

3. Reduced or zero tillage: The influence of reduced or zero tillage on the restoration of soil organic carbon over conventional tillage is very well documented in numerous findings. Organic farming improves soil quality and soil carbon storage. Reduced tillage under organic farming conditions may further enhance this potential. The soil organic carbon (SOC) stocks of reduced tillage compared with mouldboard ploughing in nine organic farming field trials of 8 to 21 years old in France, Germany, the Netherlands and Switzerland with the same soil core sampling up to 100 cm depth and analytical protocol.

SOC stocks increased with increasing clay-to-silt ratio, precipitation and organic fertiliser input. Across sites, reduced tillage in comparison with ploughing increased SOC stocks in the surface layer (0–10/15 cm) by 20.8% or 3.8 Mg ha^{-1}, depleted SOC stocks in the intermediate soil layers to 50 cm soil depth with a maximum depletion of 6.6% or 1.6 Mg ha^{-1} in 15/20–30 cm and increased SOC stocks in the deepest (70–100 cm) soil layer by 14.4% or 2.5 Mg ha^{-1}.

The subsoil SOC stock increase may be linked to the inherent soil heterogeneity. Cumulative SOC stocks increased by 1.7% or 1.5 Mg ha^{-1} (0–50 cm, n = 9) and 3.6% or 4.0 Mg ha^{-1} (0–100 cm, n = 7) by reduced tillage compared with ploughing with estimated mean C sequestration rates of 0.09 and 0.27 Mg ha^{-1} yr^{-1}, respectively. Under reduced tillage, biomass production was 8% lower resulting in a decrease of crop C input by 6%. However, this reduction may have been outbalanced by increased C inputs from weed biomass resulting from a higher weed incidence in reduced tillage. Thus, reduced tillage in organic farming has the potential to increase total SOC stocks, while crop management has to be improved to increase productivity.

Furthermore, soil tillage disruption is a major cause of loss of organic matter and a decline in the number and stability of soil aggregates as natural habitats are transformed into agriculture. Compared to conventional tillage, zero tillage cropping systems typically demonstrate increased aggregation and soil organic matter. The most significant pillar of conservation farming is zero tillage. The need for an hour is conservation farming. It is a win-win operation for farmers as well as for the environment. The goal of Tillage was to establish a soil environment conducive to plant growth, but to have negative effects on soil resources, structure and eventually on the environment in the long run. Zero tillage could enhance the chemical properties and environment of the soil in the long run. zero tillage, which eliminates the adverse effects of tillage and retains soil resources and eventually contributes to sustainable agriculture.

4. Cover cropping: Importance of cover crops in agricultural management from soil erosion control to restore soil organic carbon (SOC) is an open secret. Numerous earlier and ongoing findings very clearly indicated influence of cover crops on soil carbon restoration. Cover cropping simultaneously increased yields and SOC in 59.7% of 434 paired observations. Increases in SOC directly increased crop yields in soils with initial SOC concentrations below 11.6 $g kg^{-1}$. A change in soil organic carbon from 5 $g kg^{-1}$ to 6 $g kg^{-1}$ increased yields by +2.4%. These yield benefits of SOC did not decline as nitrogen inputs increased or when legume cover crops were used. It suggested that fertility inputs cannot substitute for SOC effects. Regardless of direct effects of SOC increases on yields, integrating legume cover crops into systems with simplified rotations or with nitrogen inputs < 157 $kg ha^{-1} season^{-1}$ N led to the largest yield increases (up to +24.3%. Legumes also increasing SOC more than non-legumes (up to +1.5 $g kg^{-1}$). Cover cropping benefits both food security and climate. It simultaneously increases yields and SOC. Also includes direct yield benefits from SOC increases on low carbon soils.

Also cover crops (CCs) can increase soil organic carbon (SOC) sequestration by providing additional OC residues, recruiting beneficial soil microbiota and improving soil aggregation and structure. The various CC species that belong to distinct plant functional types (PFTs) may differentially impact on SOC formation and stabilization. Biogeochemical theory suggests that selection of PFTs with distinct C:N ratio of litter should influence the pathways and magnitude of SOC sequestration. Soil samples collected under monocultures of three CC PFTs (legume [crimson clover]; grass [triticale]; and brassica [canola]) and a mixture of these three species, from a long-term CC experiment in Pennsylvania, USA. This study showed that bulk SOC content was higher in all CC treatments compared to the fallow.

5. Mixed farming with trees and livestock: The impact of Shifting Cultivation (SC} and fallow length on soil-Carbon depends on the farming system. Main driver of deforestation and degradation of miombo woodland is shifting cultivation (SC). SC is the main farm system and livelihood of the people in miombo woodlands. The studies reporting all the stages of agriculture-induced ecological succession (woodland, shifting cultivation, fallow vegetation, and reconversion to shifting cultivation or to woodland) are lacking. Therefore, response of soil-C to (1) dry miombo woodland (DM) conversion to SC, (2) SC conversion to fallow vegetation, (3) reconversion of fallow vegetation to SC, (4) the length of the cropping and fallowing periods, and (5) land-use intensification, was studied.

The study was carried out in a landscape mosaic consisting of treeless SC, tree-based SC, grass fallow (treeless SC-fallow), and woody fallow (tree-based SC-fallow). DM (dry miombo woodland) conversion to treeless SC resulted in depleted soil-C by up to 72% and depletion increased with the increasing length of the cultivation period. The fallow period (grass fallow) recovered part of the soil-C lost during the cultivation period. Grass fallow lands were approximately threefold larger in soil-C stocks than treeless SC lands. Soil-C increased with increasing grass fallow age. No significant changes in soil-C were observed when DM was converted to tree-based SC, or when tree-based SC lands were left under fallow (woody fallow); consequently, soil-C did not vary with the increasing lengths of cultivation and fallow periods.

Woody fallows were up to 45% higher in soil-C stocks than grass fallows. Reconversion of grass fallow to treeless SC resulted in soil-C loss; no changes were observed when woody fallows were reconverted to tree-based SC. Land-use intensification by increasing the frequency of SC cycles did not alter soil-C in either treeless or tree-based SC. It was concluded that, under the study area conditions, the response of soil-C to agriculture-induced land-use changes depends mainly on the presence or absence of the arboreal component. Maintaining or introducing the arboreal component is recommended to avoid or mitigate soil-C loss following DM conversion to SC.

It was also observed that ruminants including domestic livestock, have been accused for damaging impacts on the global environment and human well-being. However, with appropriate management, ruminant livestock can play a significant role in efforts to reverse environmental damages caused by human mismanagement and neglect. Worldwide, at least one billion people live in grazing ecosystems depend on them for their livelihoods, usually through livestock production, and for other ecosystem services. Long-term rangeland sustainability and ecological resilience, agricultural production policies are urgently needed globally to transform current damaging industrial inorganic input and agricultural practices to resource conservation practices that enhance ecosystem function.

The farmers and ranchers who apply regenerative management practices to restore ecosystem functionality create sustainable, resilient agroecosystems cost-effectively. The enhanced management of grazing resources, domesticated ruminants can be used to produce higher permanent soil cover of litter and plants, which are effective in reducing soil erosion and increasing net biophysical carbon accumulation. Incorporating forages and ruminants into regeneratively managed cropping systems can also elevate soil organic carbon and improve soil ecological function and reduce production costs by eliminating the use of annual tillage, inorganic fertilizers and biocides.

Ecosystem services that are enhanced using regenerative land management include soil stabilization and formation, water infiltration, carbon sequestration, nutrient cycling and availability, biodiversity, and wildlife habitat, which cumulatively result in increased ecosystem and economic stability and resilience. Many of these producers have used Adaptive Multi-Paddock (AMP) grazing management as a highly effective approach for managing their grazing lands sustainably. This approach uses short-duration grazing periods, long adaptively varied post-grazing plant recovery periods requiring multiple paddocks per herd to ensure adequate residual biomass, and adjustment of animal numbers as environmental and economic conditions change. Using this approach, farmers and ranchers have achieved superior ecosystem and profitability outcomes.

Restoration and Management of Soil from Loss of Plant Nutrients

Nutrients depleted soils have significant effects on health of human beings and livestock depending on such soils. Without fertile, nutrient-rich soil, plants are unable to get what they need to grow nutritious food. Crops grown in depleted soils simply do not have enough nutrients to produce nutrient-rich food to maintain us health. In fact, nutrient depleted soils are a crucial to retard the nutritional quality cereals, fruits, vegetables, fodder etc. for the last many years. The depleted nutrition of our food begins with depleted nutrients in the soil.

Generally, the methods to restore the soil fertility and soil organic matter or soil organic carbon can also be used to restore lossed plant nutrients. This is possible because plant nutrients are also a component of soil fertility and soil organic matter. The methods to restore depleted plant nutrients from soil are, 1. Tillage, 2. Crop cultivation with legumes, 3. Broadcasting of compost and manures, 4. Application of organic fertilisers and biochar 5. Use of inorganic fertilizers, 6. Use of concentrated organic manures 7. Crop residue management and 8. Bokashi. Except Bokashi all methods are already described. Therefore, Bokashi is described here,

Bokashi: Bokashi is the technique of inoculation to create compost. Nearly odourless and effective within days. Bokashi is especially popular in making compost from kitchen scraps and wastes. It relies on effective microorganisms to decompose the material. These microorganisms are introduced as a living culture, usually with wheat bran, and work to break down waste through fermentation. Developed in Japan, bokashi is gaining in popularity and for nutrients depleted soils, bokashi composts can speed up the process of restoration of lossed plants nutrients.

Restoration and Management of Soil From Salinity, Sodicity and Acidity

These soils are also come under Salt-affected and other problematic soils. These can be divided into saline, saline-sodic, sodic and acidic soils depending on salt contents, type of salts, amount of sodium, alkalinity and acidity present in soil. Each type of salt-affected soil have different characteristics. Therefore, restoration and management techniques are quite different for each other.

1. Soil salinity: After the estimation of electrical conductivity, pH and quantity of exchangeable sodium percentage, the restoration and management programme of soil salinity is carried out. The reduction of salts to the desire extent in saline soils is a major activity to restore these soils. Soil is leached by clean water. Proper drainage is provided in soil for leaching of salts. If work of leaching and drainage does not accomplish in a right way, then salts will not be forced out from soils or move to lower layers of soils. This will increase possibility of salinity.

At the time of restoration first identify reasons of formation of salinity and arrange to remove of those reasons, otherwise restored soil will be spoiled again. Suppose soil has been spoiled by saline irrigation water, at first it will be arranged for clean water for irrigation. Then work of leaching and putting out of salts from soil will be accomplished correctly.

Economic aspects and time taken in improvement of saline soils should also be kept in mind. According to these methods of restoration should be adopted. However, for the purpose of permanent restoration several methods should be operated together. But it will take more time and expenditure in the restoration works.

Sometimes farmers scrap salts deposited on soil surface by spade etc. and throw them away from field at any barren land. But this is a temporary method because dissolved salts will again come on soil surface due to vertically moving water. After the evaporation the salts will again deposit on soil surface after the evaporation of water. Therefore, leaching and application of amendments are the ways to restore soils from salinity. Here method for removal of salts by leaching is described:

Suitable time for leaching: Capacity of leaching depends on weather. Sumer season is right time for leaching in Punjab, Haryana and Bihar. In summer water table is also very much deep and water absorption capacity also is more. If leaching is not be accomplished in summer, then this work should positively be done in the beginning of rainy season.

Preparation of field for leaching: Soil surface should be loose for more effective leaching. Therefore, deep ploughing up to 15-20 cm of field should be done without turning top soil. So, mouldboard plough should not be used. After the levelling the field 6 meters long and 5 meters wide beds should be made in it. According to need size of beds may be increased or decreased. Ridges of beds should make very strong to beat and hammer these so that water filled for purpose of leaching cannot come out of beds. These beds should be left after filling water at the rate of about 1500-2000 cubic meters per hectare in them.

After two or three days when water is absorbed by soil, then again water should be filled in them at the rate of 1500-2000 cubic meter per hectare. After a week unabsorbed water should drain out from beds and again fill fresh water. Before draining out water every time electrical conductivity of soil should be tested so that speed of restoration and reduction in quantity of salts can be known.

Generally, it is assumed that for leaching of salts up to depth of 1 meter in saline soils water should be filled up to height of 1 meter on soil surface. So, requirement of total quantity of water shall be 1x100 x100 cubic meter per hectare. But it is seen by experiments that by half quantity of water this work can be accomplished. Based on concentration of salts and type of salt present in saline water, requirement of water for the leaching can be known.

The concept of 'leaching requirement' was first developed by the US Salinity Laboratory in Riverside, California. It has been defined as the fraction of the irrigation water that must be percolated out of the bottom of the root zone or up to any specific depth in order to prevent average soil salinity from rising above some specifiable level. Generally, leaching requirement is expressed in ratio of electrical conductivity of irrigation water and drainage water.

If in plant root region electrical conductivity of drainage water is to be kept 5 and electrical conductivity of irrigation on water is 1, then leaching requirement will be kept 1/5 = 0.2 or 20 percent. It is implied that 20 percent irrigation water should pass through plant root region. But required quantity of water for purpose of leaching cannot calculated by this formula. To know quantity of water for leaching will be adopted following formula:

$$\text{Leaching water requirement (LWR)} = D \log \frac{(S_1)^T}{(S_2)}$$

LWR = Water requirement for purpose of leaching (cubic meter)

D = deviation coefficient

S_1 = Quantity of salts in soil (ton / hectare)

S_2 = Quantity of required salts in soil (ton /hectare)

T = Salinity type coefficient, chlorides 0.90, sulphate and chloride 0.95. sulphate and carbonate 1.0.

Necessarily, ample rain water should be used for leaching of salts from soils, because quantity of salts in it is zero or equal to nil. In Rajasthan, salts can be expelled up to 40 cm depth of soil by rain water. Efficiency of leaching also depends on soil texture. Leaching happens very well in soils of coarse texture than fine texture. In soils having fine texture method of salt flushing also can be used. Soils having fine texture due to less infiltration rate water enters in the soil slowly and slowly. So, it will take more time in leaching. Therefore, flushing with leaching is necessary. In flushing method after filling with water field is left and when after three- or four-days salts in that are dissolved, then water with salts is drained out of field. In method of flushing, it should to keep in mind that water with salts do not reach in another field.

Drainage: Drainage of water left on soil surface after leaching or flushing is to drain out from soil. Drainage can be of surface and subsurface both types. These both types of drainage have its own characteristic features and limitations. In surface drainage water filled at soil surface is removed whereas, in subsurface drainage water present in sub soil and plant root region is removed. Besides vertical drainage also occurs where in useless or leftover water is drained out by tube wells.

However, before adoption of any type of drainage, it should be also kept in mind fluctuation of water table. Soils in which about uniform water table remains whole year such stable water indicates that drainage system of soil is not good. But soils where fluctuation in water table is sufficient according to season, drainage system of those soils is good.

Drainage system should be so that in addition to drain out useless water at a definite time this should be manageable easily. Drainage system should not obstruct in cultural practices in fields and soil should come easily and immediately in the culturable condition after rain. It should not hurdle timely preparation of field for rabi crop.

In the field surface drainage, open channels and for sub surface drainage, underground channels are used. Drainage channels in the field are constructed parallel to irrigation channels at depth of 60 to 90 cm from soil surface. This channel is 90 cm deep and wide 113cm at top and 50 cm at bottom. Water of irrigation accumulates in it. This channel is connected to a contact channel with 1:1 slope having 120 cm depth, 300 cm width at top, 60 cm width at bottom. This contact channel is connected with main channel. Main channel is constructed 3 meter deep and 1 meter wide at bottom with 1:2 slopes.

Management of restored saline soils: Management of saline soils or soil salinity, there is extremely necessary to select salt tolerant crops and their salt tolerant varieties, crop rotations, water management practices, and proper quantity of fertilizers. After restoration of saline soils appropriate salt tolerance crops and their varieties are grown according to recommended package of practices which help in further reducing soil salinity and produce more yield. Salt tolerance of few crops based on electrical conductivity for medium texture soils are given in Table 5.2.

Table 5.2: Salt tolerance of crops

Tolerant (5-10 mmhos/cm)	**Medium tolerant (3-5mmhos/ cm)**	**Intolerant (1.5-3 m mhos/cm)**
Common crops		
Barley, sugar beet, cotton, wheat, safflower, Sorghum, pearl millet, mustard	Paddy, sugarcane, maize groundnut, sunflower, Sorghum, tobacco, castor, Soybean, oat.	Linseed, sesame, gram, black gram, Broad been, green gram, Pigeon pea

Fruit		
Date palm	Fig, pomegranate, grape, Guava, jujube	Orange, apple, peach, Apricot, plum.
Vegetables		
Sugar beet, amaranth, spinach Turnip	Tomato, eggplant, onion, Cucumber, cauliflower, pea, cabbage, kohlrabi, Potato, carrot, pumpkin, Sweet potato, bitter gourd Ridged gourd, round gourd, okra, green chili	Radish, cowpea, bean
Grass and forages		
Cynodon dactylon, Sudan grass Barren grass, para grass, lucerne	Dhaincha, anjan grass (*Cenchrum setigers*), Karnal grass, Janeshwar grass	Berseem

After the selection of crops, salt tolerant varieties are selected. Salt tolerant varieties of various crops are shown in table 5.3

Table 5.3: Salt tolerant varieties of crops

Crop	Salt tolerance (mmhos/ cm)	Varieties
Paddy	5 5-10	Jaya, IR8, Ratna, Cabery, Masuri, Pankaj, CSR4 Damodar, Getu, Pasal,
Wheat	5 5-10	WG357, C306, K65, HD1553, HD2009, HD2009, UP 215 Kharchia
Barley	5	Azad, Jyoti, Amber, DL300, BG25, DL48, K125
Bajra	5	PHB12, PHB14, PHB50, HS1
Mustard	5	Varuna, PR10, Prakash
Linseed	5	Neelam, Mukta
Safflower	5	K65
Sugarcane	5	Co1148, COSHA767, BOO91
Cotton	-	H777, Shyamli, Lohit
Oat	-	Cant
Jujube	-	Karaka

New life for saline soil: Raylene Nickel (2020) indicated stepwise summary for providing management of saline soils by growing different crops successfully. As indicated, that those dead zones in fields where crops fizzle out and give way to kochia and foxtail barley are places where perennial grasses might lend a hand. Research suggests these may reclaim soil from excess salinity. At North Dakota State University's Carrington Research Extension Center, research specialist Ezra Aberle has for six years evaluated perennial grasses, legumes, and cover crops for their performance in soils of varying salinity.

Plantings were evaluated under both conventional-till and no-till management. Some species of cover crops were grown under no-till stabilized soil salinity. Testing for electrical conductance (EC) indicates the salinity level – or salt content – in soil. "Soil that has an EC rating of 4 or higher is considered saline, even though such soil may not show the telltale whiteness typically associated with soil salinity. Those white areas in fields may actually have EC readings of 6 to 30." Therefore, a three-tier approach was suggested for managing soil salinity on matching crop types to EC levels. Aberle suggests a three-pronged approach. These approaches are given here,

1. **Plant cash crops that tolerate salt:** With an EC level of about 4 or lower, soil salinity can be stabilized by growing cash crops adapted to moderate soil salinity. The most salt-tolerant crops are barley, camelina, rye, safflower, sunflower, and sugar beets. Replace lot of these crops with the least-tolerant crops like dry beans, soybeans, corn, and field peas.

2. **Grow cover crops that are salt tolerant:** After evaluating 35 species of cover crops grown in areas with ECs of 6.5 or lower. Some selected crops were seeded as the full-season cover crops in June and July as high soil moisture receded enough to allow equipment into saline areas. Cover crop species with reasonable seed costs that performed best were rape, barley, rye, Siberian millet, and sorghum-Sudan grass. Planting cover crops was able to maintain soil EC, but salinity did not improve

3. **Plant perennial grasses in the worst saline areas:** When EC is above 6.5 or 7, consider growing alfalfa and perennial grasses in these areas. These plants tolerate higher levels of salinity. Twelve varieties of alfalfa were included in this research; 11 of these were salt-tolerant. The salt-tolerant varieties do have better germination in saline conditions, but seed costs are significantly higher. When trying to reclaim high-saline soils growing saline-tolerant alfalfa with more moderate seed cost above the rim of the white area in a 40 to 60 feet wide strip is quite acceptable. The growth of wild foxtail barley and kochia in areas with a whitened soil surface suggests that certain perennial grasses may survive there, as well.

Livestock producers with field areas that are highly saline, planting AC Salt lander green wheatgrass or NewHy hybrid wheatgrass varieties produce well in saline soil and yield good-quality forage. Plant any of the tall wheatgrasses, like slender wheatgrass, western wheatgrass, or Canada wild rye grass to use moisture and produce good ground cover. These produce lower quality forage, but they also have lower seed costs.

The perennial grasses are sod forming and, thus, tend to creep into the deepest areas of saline seeps. As grasses get started, their use of soil water lowers the EC in adjacent areas, where the perennials are then able to take hold. Over time, the wheatgrasses can choke out foxtail barley. While Garrison creeping foxtail, a domestic grass species adapted to wet soils, has previously been billed as a good fit for saline areas, but it is less tolerant of salt than some other grass species.

After six years of growing salt-tolerant perennials, saline areas showed improvements in soil ECs, with levels declining from 7 to 10 to 4 to 8. Total reclamation of saline soil is a long-term process. It took decades for some soils to become excessively saline and it will take many years to restore soil to its more natural state.

Give Grass a Good Start: To start salt-tolerant perennial grasses in saline areas, seed the grasses into saline areas at the end of June, after soil dried enough to permit passage of equipment. Before seeding roll the tilled fields to create a firm seedbed. The seeding depth is about ½ inch, and the seeding rate was 8 to 12 pounds per acre. Seed costs came to about $70 per acre. Try to plant just prior to a rain, so the seed would have water to germinate. In the first year, cut the growth with a rotary mower three or four times to control weeds. In each of the next five years, cut the grass for hay.

2. Soil sodicity: In sodic soils excess exchangeable sodium percent is main problem. So, its removal is main activity for restoration of sodic soils. In restoration of sodic soils, it is necessary to displace sodium by calcium and then leaching of displaced sodium. In improvement of sodic soils importance of leaching is as essential as in the improvement of saline soils. However, application of appropriate amendment is also essential for improvement of sodic soils along with leaching of alkalis.

Soil amendments make soil neutral by decreasing excess pH of soils. These react with free sodium carbonate present in soil and displace sodium by calcium. Thus, lessen harmful effect of sodium. Selection of suitable amendment for restoration of sodic soils depends on the percentage of exchangeable sodium present, amount of calcium present and presence or absence of pebbles in soil. The three major groups of amendments used for reclamation of sodic soils are 1. Soluble calcareous substances- Gypsum and phospho gypsum, 2. Acidic substances- Pyrite, sulphur, sulphuric acid, spent wash of dairy, sugar industry waste and 3. Miscellaneous substances- Press mud, molasses, compost, cow dung, green manure, hay of paddy, husk, sadabahaar, water hyacinth, Satya nashi and blue green algae.

In our country mainly gypsum and pyrite are applied to removing soil sodicity. Sulphuric acid is also a successful chemical ameliorant. But its application is limited because of more carefulness in its use, more expensive and not available easily due its misuses. Descriptions of some major ameliorants are given here:

Gypsum: Chemical formula of gypsum is $CaSO_4.2H_2O$. This is white coloured. This is found under soil up to depth of 2 to 20 meters in arid climatic zones. It is found mainly in Nagpur, Pali, Bikaner, Jaisalmer and Ganga Nager of Rajasthan,

Porbandar, Kuchh and Bhavnagar of Gujrat. It is also found here and there in Tamil Nadu, Utter Pradesh and Himachal Pradesh. Its total estimated quantity is nearly 10 lakh tons, in which 9 lakh tons are found only in Rajasthan. This is taken out from soil by mining from aforesaid places. Besides mines gypsum is a byproduct of phosphoric acid producing factories. Such like gypsum is called phosphogypsum. In northern India gypsum is marketed by Fertilizer Corporation of India, Rajasthan Mine and Mineral Ltd. and Rajasthan State Industrial Corporation Ltd.

Purity of gypsum: Fertilizers Corporation of India has fixed three grades of gypsum: grade 1- 80 to 90 percent gypsum, grade 2- 80 to 85 percent gypsum and grade 3- 70 to 80 percent gypsum. Low grade gypsum with at least 70% purity and 16 percent sulphurised, is used as soil amendment which should also not contain more than1.0% Sodium oxide. High grades of gypsum are used in factories of fertilizers, cement etc.

Fineness of gypsum: Gypsum to be used for the purpose of amelioration of sodic soils should be fine, thoroughly ground and passed through 2 mm sieve to increase its effectiveness.

Solubility of gypsum: Generally, solubility of gypsum is only 0.25 percent. But its solubility in sodic soils increases up to 10 times. Thus, it requires more water for its solubilisation. Similarly, use of compost, manure of cow dung and green manures increases solubility of gypsum. In normal conditions to dissolve 10 to 12 tons gypsum, 90 to 120 cm water per hectare is needed. Gypsum application increases water absorption capacity of soil.

Quantity of gypsum required for reclamation of sodic soils: Quantity of gypsum required for sodic soil improvement depends up on texture of sodic soils, quantity of exchangeable sodium in soil and type of the crop to be sown. In amelioration of sodic soil, soil having fine texture will require more gypsum than soils having coarse texture. If soil has more exchangeable sodium and to grow sodium intolerant then more gypsum will be required.

The laboratory method proposed by Schoonover is widely used for assessing the gypsum requirement of sodic soils. The formula is given here:

$$\text{Gypsum requirement (GR) meq / 100 gram soil} = \frac{\text{CEC x}\left(\text{IESP} - \text{FESP}\right)}{100}$$

Where

CEC = Cation exchange capacity.

IESP = Initial exchangeable sodium percent.

FESP = Final exchangeable sodium percent.

Definition of ESP is that up to which extent (percent) Sodium occupies adsorption complex of any soil is called as exchangeable sodium percent (ESP). It is calculated by the formula:

$$ESP = \frac{\text{Exchangeable sodium}\left(\text{centimol / kg soil}\right)}{\text{Cation exchangeable capacity}\left(\text{centimol / kg soil}\right)} \text{ x } 100$$

Both formulae are used to calculate amount of gypsum required to ameliorate any sodic soil. Suppose, cation exchange capacity of any soil is 30 and in this exchangeable sodium is 20 and to bring this quantity of exchangeable sodium up to 3 then how much gypsum will be needed?

Solution

$$\text{Initial exchangeable sodium percent} = \frac{20 \text{ x} 100}{30}$$

$$= \frac{2000}{30} = 66.67 \text{ percent}$$

$$\text{Final exchangeable sodium percent} = \frac{3x100}{30}$$

$$= \frac{300}{30} = 10 \text{ percent}$$

$$\text{Gypsum requirement} = \frac{30 \text{ x}\left(66.67 - 10\right)}{100}$$

$$= \frac{30x56.67}{100}$$

$$= \frac{1700}{100}$$

$$= 17 \text{ meq gypsum}$$

It is clear from this calculation that for amelioration of each 100 gram of sodic soil 17 meq of gypsum will be required. This 17 meq can be converted in to milligram gypsum by multiplying it with molecular weight of gypsum (172) and dividing by valency of Calcium (2) and 1000. This can be written as:

Milligram Gypsum / 100 gram soil

$$= \frac{\text{meq gypsum x molecular weight of gypsum}}{1000\,x\,valency\,of\;calcium}$$

$$= \frac{17 \text{ x}172}{1000\,x\,2} = 1.462 \text{ mg gypsum /100 gm so}$$

We know that weight of upper 15 cm deep soil is 2.2 million kilo gram and this fact will be used to calculate amount of gypsum required for one hectare soil as,

1.462 mg gypsum /100 gm soil

14.62 x 2200 = 32164 kg gypsum / ha = 32.164 Ton gypsum / ha. Thus, 32.164

Ton pure gypsum will be required to restore, earlier supposed sodic soil. But purity of gypsum in hand is only 75%. Therefore, more amount of gypsum will be required as,

$$\frac{32.164 \text{ x}100}{75} = 42.89 \text{ Ton gypsum / ha.}$$

Generally, it is seen that good yields of paddy, wheat and barley can be obtained only by applying half dose of gypsum as indicated by laboratory testing of gypsum requirement of sodic soil. It is not necessary that whole soil profile should be made sodium free. Sodium can be removed only from upper 15-20 cm layer of soil by applying gypsum. So, that crops like paddy, wheat, sugarcane, rye and barley can easily be grown. Therefore, gypsum should be broadcasted on soil surface and thoroughly mixed in upper 8-10 cm soil by desi or mouldboard plough.

Chemical reaction of gypsum in soil: Calcium of gypsum displaces sodium present on clay of sodic soil and adsorb there. By this way displaced sodium ion after reacting sulphate ions of gypsum makes sodium sulphate. Sodium sulphate is water soluble. Therefore, it drains out from soil by leaching and flushing or goes in lower layers of the soil by percolating water. The reaction of displacement of sodium by gypsum is given here in the equation,

$$Na[Clay]Na + CaSO_4 \rightarrow [Clay]Ca + 2Na_2SO_4$$

Pyrite: Chemical name of pyrite is iron sulphide (FeS_2). This is a major mineral of iron and sulphur. In addition to these silica, aluminium, carbon, magnesium, calcium, zinc, copper and manganese are also found in this mineral.

In India pyrite is found sufficiently in Bihar and Rajasthan. Mines of pyrite are present in Bihar at Amjhor and in Rajasthan at Saladipura. It is divided into two classes according to quantity of sulphur: Pyrite with 40 % sulphur and pyrite with 10 % sulphur. Application of pyrite with 40 % sulphur is done chiefly in manufacturing of sulphuric acid while pyrite with 10 % sulphur is used as soil amendments.

At time of mining pyrite is obtained as large sized clods. These are crushed in fine powder and small fragments. Powdered Pyrite used in agriculture should pass cent percent through 5mm sieve and 50 percent through 0.85 mm sieve. The effectiveness of pyrite increases with its fineness. More fineness more effective. In our country mining and marketing of pyrite is done by Pyrite, Phosphate and Chemicals Ltd.

Quantity of pyrite required: First pyrite requirement is determined to know the quantity of pyrite for amelioration of sodic soils. Because pyrite has more quantity of sulphur than gypsum, so it is required in lesser quantity than gypsum. Generally, gypsum requirement of sodic soil is determined and said quantity is multiplied by 0.73 to know the pyrite requirement for the improvement of soil sodicity.

Chemical reaction of pyrite in the soil: When pyrite is applied in the soil, then this is converted in sulphuric acid and ferrous sulphate by chemical and microbiological activities. Both chemicals help in improvement of sodic soils. Sulphuric acid lowers pH of the soil. After lowering of soil pH availability of phosphorus, zinc, iron, boron etc. is increased.

$$2FeS_2 + 7O_2 + 2H_2O \rightarrow 2FeSO_4 + 2H_2SO_4$$

Pyrite Oxygen Water Ferrous Sulphate Sulphuric acid

Sulphuric acid dissolves pebbles and changes calcium carbonate present in pebbles to calcium sulphate.

$$CaCO_3 + H_2SO_4 \rightarrow CaSO_4 + CO_2 + H_2O$$

Calcium carbonate Sulphuric acid Calcium sulphate Carbon dioxide Water

Calcium of calcium sulphate displaces exchangeable sodium of clay of sodic soils. Now sodium by mixing with sulphate makes sodium sulphate. Sodium sulphate is water soluble. Therefore, it drains out from soil by leaching and flushing or goes in lower layers of the soil by percolating water. The reaction of displacement of sodium by calcium sulphate is given here in the equation,

$$Na[Clay]Na + CaSO_4 \rightarrow [Clay]Ca + 2Na_2SO_4$$

But in such sodic soils in which calcium carbonate is in less quantity, hydrogen ions of sulphuric acid produced from pyrite directly displaces sodium adsorbed on clay.

$$2FeS_2 + 7H_2O \rightarrow 2FeSO_4 + 2H_2SO_4$$

$$Na[Clay]Na + H_2SO_4 \rightarrow [Clay]Ca + 2Na_2SO_4$$

Also, this acid reacts with bicarbonates and carbonates of sodium and neutralises them as given here:

$$2NaHCO_3 + H_2SO_4 \rightarrow NaSO_4 + 2H_2O + 2CO_2$$

$$Na_2CO_3 + H_2SO_4 \rightarrow Na_2SO_4 + H_2O + CO_2$$

Various reaction of pyrite with sodium and its salts present in soil go on simultaneously. Its result is to ameliorate soil sodicity by placing calcium or hydrogen at the place of displaced of sodium.

Just as it is clear from equations that water and air both are essential to change of pyrite in to sulphuric acid. So, simply broadcasting of pyrite on the surface of water saturated soil, water as well as air will be available simultaneously. Consequently, sulphuric acid will rapidly be formed. This easily and swiftly ameliorates soil sodicity.

Sulphuric acid: Sulphuric acid is mostly used in advanced and prosperous countries for amelioration of sodic soils in very less time. Sulphuric acid is very much costly than pyrite. But its more effectiveness, need in small quantity and suitable for sensitive crops cultivation in acid treated soils. Application of sulphuric acid is more profitable over its cost. In our country in Kanpur and Indore it is seen that soil amelioration by sulphuric acid than gypsum is more profitable. Although in our country due to uneasy availability of sulphuric acid and necessary implements, this method is not in use.

Sulphuric acid produced by pyrite, ameliorates the soil, so direct application of required quantity of sulphuric acid also ameliorates soils. It is expressed by equations:

$$CaCO_3 + H_2SO_4 \rightarrow CaSO_4 + CO_2 + H_2O$$

$$Na[Clay]Na + CaSO_4 \rightarrow [Clay]Ca + 2Na_2SO_4$$

Programme of restoration of soil from sodicity: Amelioration of sodic soils should be a well-managed stepwise programme. Any change in steps, no adoption of any step or doing it some other time will not completely restore soil from sodicity. Amelioration programme of sodic soils can be implemented in the steps given here:

Area selection and survey → irrigation water management → bunding → ploughing and levelling → leaching and flushing of salts → application of soil amendments and leaching → cultivation of tolerant crops.

Area selection and survey: For removal of sodicity such soil should be selected, where amendment application or amelioration can be done easily. Higher water table, thick layer of pebbles and excess of salts with alkalis make difficult to improve soil from sodicity. So for selection of soil, soil survey and soil testing of such areas should be given priority whose water table is low and in which pebbles are not present should or pebbles are in very small quantity. So selected area should be demarcated.

Irrigation water management: Sufficient supply of water is necessary for reclamation of soil sodicity. When more saline and alkaline water spoils soil, good water is useful in reclamation of soils. If there is no river or canal near the selected area, then for the availability of water construction of tube well is necessary. Water of tube well also should be tested for its salt and sodium content so its utility can be known for removing sodium from sodic soils.

Bunding: Around the selected area high ridging or bunding should be done. Under the ridging of area road for transportation, channels of irrigation and drainage etc. should be constructed. After this whole area is divided in small beds. According to slope of the field area size of beds may be 200 square meters to 500 square meters.

At more slopy land smaller beds are needed. But it should be kept in mind that size of beds should be sufficient for easy movement of tractors and agricultural implements. Around these beds strong ridges or bunds strong should be made. Size of ridges should be 120 x 45 x 30 cm. It means that at land surface breadth is 120 cm, height is 45 cm and at top breadth is 30 cm. According to need size of ridges can be reduced or increased.

Ploughing and levelling: After ridging beds should be ploughed up to 15- 20 cm depth. After ploughing uneven or undulated places should be levelled by the help of spade or small mechanical leveller. In levelling soil in more quantity should not be lifted from one place and do not put to another place. If more quantity of soil is put to another place, then there will be unequal distribution of salts in bed. This will create difficulty in soil amelioration. In those areas where hard layer of pebbles is present, it should be broken by pan breaking implement or subsoiler.

Leaching and flushing of salts: In the prepared soil, water should be filled up to 8-10 cm on the soil surface. After 3 to 4 days this water should be drained out by the help of drainage channels. So, by leaching and flushing of soils quantity of salts will be reduced. By adopting this method soil amendments is required in small quantity.

Application of soil amendments and leaching: Gypsum requirement is assessed by soil testing. Therefore, gypsum is broadcasted homogenously in the field. Broadcasted gypsum is mixed in the soil up to 8- 10 cm depth by indigenous plough or cultivator. Now after filling up to 8 to 10 cm water in the field, it is left for leaching for 7 days. The leaching is done two or three times. After leaching remaining water in the beds should be drained out by drainage channels.

If pyrite is to be used, then moist soil is essential. On the water saturated surface of soil after homogenous broadcasting of pyrite, it is left for 15 to 20 days. When surface of soil attains radish colour, then it should be mixed in the soil in the depth of 8 – 10 cm by the help of indigenous plough or cultivator. After filling up to 8- 10 cm water in the field it is left for 7- 8 days. When water present on soil surface is dried up, then water up to 8 – 10 cm us filled again and leached. After two - three leaching, remaining water should be drained out by drainage channels. If any reason pyrite is moist, then pyrites should be broadcasted to put on gloves of rubber instead of naked hands. Otherwise, acid of pyrite will develop blisters on the hands.

After the performing leaching, flushing and application of soil amendments, at first as sodium is also an alkali. Therefore, alkali- tolerant crops and alkali tolerant variety should be grown in the ameliorated soil. Alkali – tolerant crops are given in Table 5.4.

Table 5.4: Alkali tolerant crops based on pH (1:2 soil: water suspension)

Most tolerant (pH 9.8 and more)	**Tolerant (pH 9.3 – 9.8)**	**Medium tolerant (pH 8.8 – 9.3)**	**Sensitive (less than pH 8.8)**
Bermuda (doob) grass	Paddy, sugar beet	Wheat, barley	Mostly all fruits and pluses
Karnal grass	Cotton, sun hemp	Safflower	
Rhodes grass	Para grass, Sudan grass, Napier grass	Sorghum, bajra, rai, mustard, maize, guava sunflower, berseem, jujube, Lucerne	

Management of sodic soils: There is acute deficiency of organic matter in sodic soils. Therefore, FYM, compost and or green manures should be essentially applied in these soils. Generally, alkaline or sodic patches are seen even in the reclaimed soil. These patches can be removed by the cultivation of green manures. Decomposition of these organic manures produces organic acids which reduce the soil pH and dissolve pebbles.

Dhaincha is a appropriate crop for green manuring. It contains more amount of calcium and sap of leaves shows pH less than 4. Therefore, use of green manure of Dhaincha very rapidly reduces the pH of the sodic soil. Simultaneously, its calcium removes exchangeable sodium from the clays.

In the reclaimed sodic soils after cultivating rice and wheat crops, Dhaincha should be grown in summer and it should be turned in soil in the month of June. After this paddy should be transplanted in the month of July. If Dhaincha is not cultivated in summer then, it should be essentially grown during in rainy season. It is observed that in reclaimed sodic soils cultivation of Dhaincha and turn that in soil in alternate years at least for ten years is remarkably beneficial. The appropriate crop rotations for reclaimed sodic soils is given in Table 5.5.

Table 5.5: Appropriate crop rotations for reclaimed sodic soils

Year	**Rainy season**	**Winter season**	**Summer season**
First	Paddy/Dhaincha	Wheat	Dhaincha
Second	Paddy/ Pearl millet	Rye/ wheat/ barley	Dhaincha
Third	Paddy/ Dhaincha (Sesbania)	Berseem/ Rye/ Potato/ Coriander/Linseed/ Toria/ sugarcane	Sugarcane/ Ladies finger
Fourth	Paddy/Maize/ Barley	Rye/ Potato/ Wheat/ Linseed	Sugarcane/ Moong bean

3. Soil acidity: Active acidity plus reserved acidity is called total acidity or titratable acidity. Total acidity is considered for reclamation of soil acidity. Generally, agricultural lime is used as soil amendment to restore acidic soils. The description of agricultural lime is given here:

Agricultural lime: Such soil amendment as calcium carbonate, magnesium carbonate or dolomite or as such other substance which are used to neutralize of soil acidity and provide calcium and or magnesium to plants are included in the class of agricultural lime.

Calcium oxide (CaO), calcium carbonate ($CaCO_3$), calcium bicarbonate ($Ca(HCO_3)_2$), calcium magnesium carbonate ($CaCO_3.MgCO_3$) and calcium silicate ($CaSiO_3$) are used. The application of these substances displaces hydrogen ions present on soil complex by calcium ions. Thus, clay complex becomes calcareous and displacement of hydrogen ions increases soil pH. Some reactions of displacement of Hydrogen ions are given here:

$$H[Clay]H + CaO \rightarrow [Clay]Ca + H_2O$$

$$H[Clay]H + Ca\,(OH)_2 \rightarrow [Clay]Ca + 2H_2O$$

$$H[Clay]H + CaCO_3 \rightarrow [Clay]Ca + CO_2 + H_2O$$

$$H[Clay]H + Ca\,(HCO_3)_2 \rightarrow [Clay]Ca + 2H_2O + 2CO_2$$

$$H[Clay]H + CaCO_3.MgCO_3 \rightarrow Mg\,[Clay]Ca + 2H_2O + 2CO_2$$

Calcium silicate byproduct of steel industry is also used in reducing acidity of soil. But in it quantity of iron should be less, otherwise availability of phosphorus will be reduced.

$$H[Clay]H + CaSiO_3 \rightarrow [Clay]Ca + CO_2 + H_2SiO_3$$

Liming of acidic soils increases soil pH which increases availability of various nutrients and exchangeable calcium and magnesium in the soil. Toxicity of manganese, iron and aluminium is terminated. Microbial activities begin appropriately and soil structure is improved so the roots are properly distributed in the soil.

Lime requirement to reclaim soil acidity: Lime requirement of the soil means that quantity of lime which changes acidic soil to any specific state with the reference of pH or soluble aluminium. It can be said in common words the lime requirement is that quantity of lime which is required to bring pH of soil in between 6 and 7.

There are several methods to find out the lime requirement of soil. But method of Shoemaker etal. (1961) is used in laboratories. In this method a buffer liquid is used. This buffer liquid is prepared by mixing 1.8 gram nitro phenol, 2.5 ml tri ethanol, 3.0 gram potassium chloride, 2.0 gram calcium acetate and 53.2 gram calcium chloride are dissolved in one litre distilled water. Dilute sodium hydroxide solution is added in it till pH becomes 7.5.

No lime is required if soil pH is more than 5.4. If pH is less than 5.4 then in 5.0 gram soil, 5.0 ml distilled water and 10.0 ml buffer liquid are mixed and it is intermittently shaken for 20 minutes. After this pH of suspension is measured by the help of pH meter.

So far, the lime requirement is found out to remove soil acidity with the help of pH of soil: buffer suspension and table 5.6 given here,

Tabl 5.6: Lime requirement to obtain desirable pH

pH of soil buffer suspension	Requirement of pure lime ($CaCO_3$, ton/ hectare)		
	Desirable pH		
	6.0	6.4	6.8
6.7	2.5	3.0	3.5
6.6	3.5	4.3	4.8
6.5	4.5	5.6	7.3
6.4	5.8	6.8	7.8
6.3	6.8	8.0	9.3
6.2	7.8	9.3	10.5
6.1	8.8	10.5	12.0
6.0	9.8	11.8	13.6
5.9	11.0	13.1	15.1
5.8	12.0	14.3	16.3
5.7	13.1	15.6	17.8
5.6	14.1	16.8	19.3
5.5	15.1	18.1	20.8
5.4	16.3	19.3	22.3
5.3	17.3	20.6	23.6
5.2	19.6	22.8	26.6
5.0	20.6	24.1	28.1
4.9	21.6	25.4	29.6
4.8	22.8	26.6	32.1

These quantities are for cent percent pure lime, but agricultural lime is not cent percent pure. So, based on its purity quantity of lime shall be increased by calculating separately. As soon as purity will decrease, quantity of lime will increase.

In the soil when exchangeable aluminium saturates more than 60 percent of cation exchange capacity, then quantity of aluminium in soil liquid increases significantly and lime requirement is calculated as,

Lime requirement = 15 x meq Aluminium / 100 gram soil

(meq $CaCO_3$ / 100 gram soil)

Application of lime: More fine lime will be much effective and fineness of lime should be at least 60 mesh. Rate of activity of lime depends on the condition and characteristics of the soil and properties of lime. Generally, lime should be applied in the field one month before sowing of crop. In kharif season, application of lime before commencement of rain in the field provides sufficient time for the preparation of field and timely for sowing of kharif crop. If there is sufficient moisture in the field, lime can also be used in Rabi (winter) season. After broadcasting lime on the surface of soil, mixing of it in the field by indigenous plough, disc harrow or cultivator is more beneficial. After proper improvement first acid tolerant crops should be grown. A list of acid tolerant crops is given here,

Sweet corn, cucumbers, beans, broccoli, turnips, squash, onions, cranberries, blueberries, huckleberries, beech, willow, oak, dogwood, azaleas, rhododendrons, hydrangeas, camellias, daffodil, rice, potato, cassava, mango, cashew, citrus, pineapple, cowpeas and certain evergreen perennial grasses.

Restoration and management of polluted and contaminated soils

FAO reviewed technologies for remediation of polluted and contaminated soils for both point source and diffuse pollution. The technologies are categorized in two levels: first as either biological or physical/chemical treatments, and second as either *in situ* or as *ex situ*. Nature-based (biological) solutions are grouped separately because of their likelihood to provide a more sustainable solution in terms of both environmental impact (maintaining soil biodiversity and soil functions) and economic feasibility.

In situ technologies are those where the soil remains in the ground during the treatment. *In situ* techniques may involve some manipulation of the soil to introduce substances to stimulate the remediation. Some of the technologies separate or concentrate the contaminants such that the polluted fraction can be extracted, and treated on-site or disposed of elsewhere. The main advantage of *in situ* treatment is that it allows soil to be treated without being excavated and transported, resulting in potentially significant cost savings and fewer environmental changes. *In situ* treatment can also retain some of the soil's characteristics, such as structure, organic matter, and biodiversity that are often difficult to restore after more invasive treatments. However, *in situ* treatment can require longer time periods as variability in the environmental conditions of the site and soils can affect its efficiency. There is also less certainty about the uniformity of treatment and its efficacy is more difficult to verify.

Ex situ technologies require the soil to be excavated. The excavated soils maybe treated on-site (and the cleaned soil replaced) or transported for treatment elsewhere. *Ex situ* technologies tend to be more controllable and require less time, however they have the additional expense of excavation and transportation. As *ex situ* treatments tend to degrade the structure and organic matter in the soil, its rehabilitation can be more complex, costly and time-consuming.

1. Agricultural wastes: Our soils are farmland soils and mostly used for agricultural purposes. Therefore, more emphasis is given on remediation of pollution and contamination caused due to agricultural production. Agricultural wastes and residues are first in this category. Thus, description is being started from these.

The agricultural wastes and residues should be managed properly by recycling. These should be converted in manure, compost, bio gas, animal feeds and other useful substances. Residues of fruits and vegetables can be used as animal fodder. Thus, dry peel and seed of tomato can be mixed with in poultry feeds. Dungs of horses can be used for the purpose of mushroom production after its proper cleaning. Straw of paddy and wheat, stalks (karabi) of maize, sorghum, bagasse of sugarcane etc. can be used in paper industry to make paper. Furfural can be obtained by hydrolysis of hull of paddy and oat and ear of corn. Synthetic resin and adhesives are forms by furfural. Tartrate is obtained from residues of wine grapes. Monosodium glutamate is obtained from bagasse of sugarcane.

Leather, soap, glue, gelatine, animal food, hormones, vitamin and enzymes are obtained from wastes and residues of meat processing industry. Human waste is considered very bad material. People defecate (discharge waste material from body) in fields of villages. Later on, flies sit on this. These flies come in the house, pollute food and water. If defecation is done in small pits in the fields and is immediately covered by soil, the environment also will remain clean and fertility of soil of the field will also increase because human defecation has plant nutrients in sufficient quantities. Here we shall discuss only manure of cow dung, compost and cow dung gas for the purpose of proper management of agricultural residue and wastes.

Manure of Cattle dung: The manure of cattle or cow dung is called farm yard manure (FYM). It will be more proper because besides cow dung it contains animal urine, litter, left over fodder etc. Also dungs of horses, droppings of poultry and litter are also mixed in it. But quantity of cow dung is more so it is called manure of cow dung or FYM.

Preparation of Farm Yard Manure: Biological wastages by which manure of cow dung or FYM is prepared are cow dung, urine, soil or litter, leftover fodder, wastage of kitchen and house, washing etc. Everyday putting all these in pit or accumulating all these in piles are existing method of preparing manure. This is not good for prepare better manure. So, a very much effective method to prepare manure of cow dung is method of segmental filling of pits. By segmental filling of biological material in pits manure of high quality is prepared. In preparing manure of cow dung by this method those all facilities and benefits are obtained which are obtained as a result of disintegration of material collected in pits. This method is favourable of circumstances available in our country. According to number of animal measurement of pits is given in Table 5.7.

Table 5.7: Measurement of pits to prepare manure of cow dung

Number of animals	Length (Meter)	Width (Meter)	Depth (Meter)
2 to 5	6.5	1.0	1.0
6 to 10	8.0	1.2	1.0
11 to 20	10.0	1.4	1.0
More than 20	10.0	1.6	1.0

Filling of mixed biological waste material obtained from animal yard is started in one segment from one end of pit. Length of this segment of pit is 1 meter. To fill the segment up to top towards length in the pits, 1.1-meter-high bamboos are buried in both sides. Mobile screen made of stems of cotton or any other material is set up between bamboos. Everyday collected material pouring in this segment is covered by soil. Such filling should be continued till that time unless this segment of pit is filled up to 50 cm. above of surface of this land.

After filling first segment material should be covered by soil and screen of stems of cotton stalk should be shifted to second segment 1 meter ahead. So, that for filling fresh empty segment become available. If material does not available in sufficient quantity from animal yards, then mixture of leaves, straw of paddy, shells of ground nuts, straw husk etc. should be filled. This will enhance the manure. This method prepares manure within three months. About 10 to 12 tons manure is prepared in a year from a pair of bullocks.

***Gobar* Gas or biogas:** Cattle dung can be used in gobar gas or bio gas plant for the purpose of fuel and manure. Problem of energy is solved up to some extent by gobar gas. Higher temperature of tropical zone is good for fermentation of cattle dung.

Method of Construction of Gobar Gas Plant

A well-shaped round pit of 120 cm depth and 210 cm diameter is made. All sides of this pit are left 23 cm space again 240 cm deep pit is made. Layering of bricks and plastering of cement are done in the surface of this pit. Wall of the well is raised up to 180 cm height from bottom. On this place now 20 cm wide channels which has slope towards inner pit from surface of land, is constructed. Second end of the 20 cm wide channel remain in reservoirs constructed on the surface of land. This reservoir is 90 cm far from the well and 45 cm high from land surface. This reservoir is constructed 60 cm long and 60 cm deep.

Everyday slurry of cow dung is poured from this reservoir. After joining the channel laying of remaining well is done so that layering of bricks reaches up to 15 cm from land surface. Only in the opposite direction of the reservoir a brick is left for a place to construct a channel. After formation of gas comes out from pit through this channel. This decomposed cow dung is collected in the drying bed or in cow dung collecting pit.

This decomposed and rotten cow dung is used to make manure for fields. A small pit is constructed outside the well. This pit is 23 cm long, 23 cm wide and 30 cm deep. In it vapour of gas come out from the pipe which is converted in water. In this pit a round nozzle is fitted for time-to-time release of water formed.

In the curvature of well 180 cm long three angles or strips of iron are affixed at equal distance. Pulleys are fitted above these angles or strips of iron one-by-one. One 360 cm long wire or chain of iron is fitted at one end of this and connected with drum of iron and second end being above pulley relates to a wooden box. Eight bricks are kept in a wooden box so that drum can move up and down.

On the formation of gas drum comes above and gas as is being used in like manner comes down. This drum of iron has 150 cm diameter and 120 cm higher which is kept opposite to the wall. This separates outer air so that rotting of cow dung can take place very well and obtain gas sufficiently. A nipple is affixed in the upper part of the drum with which a 180 cm hose pipe is attached. Second end of hose pipe is connected to line so that gas can reach in line.

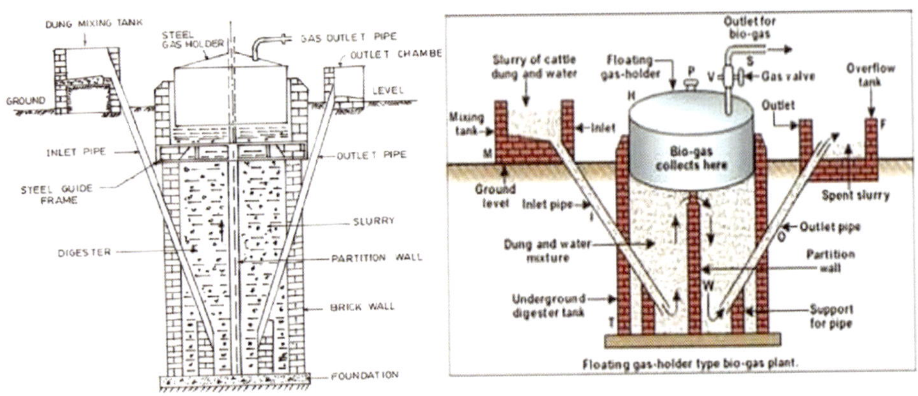

Fig. 5.16: Gobar gas plant

Method to Use of gobar gas plant: First, outlet of gas pipe is closed. Now about 50 kg cow dung mixed with water in the form of slurry is poured in the reservoir daily. This goes in the lower part of the well through pipe and decompose there. Decomposition of cow dung produces gas and rotten cow dung comes out by channel above the pit or is collected in another pit from where it is used to make manure. At the time of gas use weight should be lessened by taking out the bricks from the box. Thus, gas comes out because of weight of drum. This gas can be used for cooking food or lighting the lamps.

Daily about 30 cubic meter gas and 4 to 8 Tons (about 16 Bullock Carts) cow dung manure per year can be obtained from two pairs of animals. Nitrogen up to 2% is found in this manure while normal cow dung manure contains only 0.5% nitrogen.

This gas has 50-60% methane, 30-40% CO_2 and 10% H_2. Energy power of this gas is 650 BTU per cubic foot. While energy power of coal gas is 400-500 BTU per cubic foot. Therefore, manufacture and use bio or cow dung or gobar gas is beneficial in all respects.

Compost: Composting is another method of pollution control. Compost is defined as the decomposed organic material obtained after the controlled aerobic, anaerobic, mesophilic and thermophilic biological decomposition of biodegradable materials. Composting is done to prepare compost manure. Generally, carbon and nitrogen containing wastes are used in composting. These wastes come from farm refuse as straw, chopped fodder, leaves, straw of paddy etc. Also shells of ground nut and eggs, sugarcane tops, bagasse and other agricultural and industrial wastes, rubbish and dust obtained from cities and villages, human and animal faeces, debris received from slaughter houses, ash, sawdust, seaweed, newspapers, mud, dirty water and even any useless organic material can be used for composting. If these materials are collected from rural area, then it is called as rural compost and collected from urban area, then it is called as urban compost. The composting may be on small scale and large scale and from home to farm or factory level.

Basics of production of manure is that after initial fast integration and becoming loss of excess of carbon from carbonaceous material. Conservation of nitrogen from nitrogenous materials including human faeces and decomposition at proper level so the final product can be used as manure. Thus, when prepared manure is taken out from ditches and pits after digging, then often clearly seen different layers of composing material used. This material then is thoroughly mixed manually and can be used as manure in fields. It further decomposes in the soil to make plant nutrients available present in it.

Composting is done worldwide. Various countries like Bangladesh, Colombia, India, Sri Lanka etc. making compost and using as organic fertilizer in fields. In India, first of all such work was done by Dr. Haward in Indore and Dr. Fowler in Bangalore. They mixed the cow dung and human faeces i.e., nightsoil in leaves, straw and debris received from cities to increase the activity of bacteria. In India four methods namely Indore method, Bangalore method, Coimbatore method and NADEP method are used for composting.These methods are described here:

Indore Method to Make Compost: In the Indore methods, after spreading waste material, debris, rubbish in to cattle shed as bedding it is dampened by animal urine, cow dung and human faeces matter and made 4 to 6 meter long, 1 meter wide and 1-meter-high heap of it. Layering process continued for about a fortnight. A thin layer of well decomposed compost is spread over top and the heap given turning. Old compost acts as inoculum for decomposing the material. The heap is left undisturbed for about a month. Under these circumstances disintegration process begins fastly and temperature increases in short time span. When this material is overturned then in aforesaid process further more fastly occurs. Because by such overturning the material receives air abundantly. If necessary, by spraying water or urine, material can be damped further more so that material would be decomposed completely and within three months period uniform manure of brown colour is formed. Under this aerial method organic matters are damaged heavily and up to 40-50% nutrients are lost. In manure prepared by Indore method quantities of nutrients are as nitrogen 0.8%, phosphoric acid 0.3% and potash 1.5%.

Banglore Method to Make Compost: In the Banglore method compost is made from garbage and wastes of colonies. Its guidance to prepare manure in pits in India was given by Dr. Acharya. Dr. Acharya developed this method for the utilization of town residues and night soil. In this method, mainly night soil and other waste material collected from cities and colonies are used. This hot fermentation method or Bangalore method is adopted in many cities and colonies of India. In this method foul or unpleasant smelling material can be converted in bio-composed manure of high quality.

Practically in this method compost making centres are established at the outskirts of cities or near boundaries of population so night soil, rubbish or waste material collected from roads can conveniently be carried to centres. In large cities such two or three centres can be established so that night soil, debris or waste etc. can be collected easily from various areas. Now nightsoil is not available because of ban on open defecation under Swachh Bharat Mission. However, wastes of Sulabh Shauchalaya can be used instead of nightsoil.

In this method pits or ditches of appropriate size based on the population of the city are dug in line at the distance of 1.5 meters. Sufficiently wide space is left among the ditches for the movement of vehicles to carry composting materials from roads to centres and produced compost from centres to roads. First, refuse is filled in the trench to about 15 cm height. Afterward a layer of human faeces i.e., night soil is spread over this to a layer of 5 cm above it. After filling the pit with refuse and night soil in alternate layers, the pit is filled to 15 cm above the ground level with a final 15 cm layer of refuse on the top.

This may be made dome-shaped and covered with thin layer of soil or mud to prevent moisture loss and breeding of flies. Generally, except the upper layer decomposition of material present in these pits or ditches occur in absence of sufficient air. Thus, anaerobic disintegration or decomposition happens slowly but its nutrient loss is not more. The materials in ditches or pits are allowed to remain as such without any turning and moistening for about three months. The compost obtained by this method contains 1.5% N, 1.0% P_2O_5 and 1.5 % K_2O.

Coimbatore Method to Make Compost: Coimbatore method is anaerobic decomposition in starting followed by aerobic fermentation. The Coimbatore method of composting was introduced by Manickam in 1967. Composting occurs mainly in trenches/pits under shade. In the initial stage of composting, anaerobic decomposition takes place followed by aerobic decomposition. Weeds, straw, leaves, vegetable wastes, animal excreta, bone meal, urine, water etc. are used for composting.

In this method composting is done in pits of different sizes depending on the waste material available. Generally, size of pits is kept about 3.6 m long, 1.8 m wide and 1.0 m deep. A layer of waste materials is first laid in the pit. It is moistened with a suspension of 5-10 kg cow dung in 2.5 to 5.0 litre of water and 0.5 to 1.0 kg fine bone meal sprinkled over it uniformly. Similar layers are laid one over the other till the material rises 75 cm above the ground level. It is finally plastered with wet mud and left undisturbed for 8 to 10 weeks. Plaster is then removed, material moistened with water, given a turning and made into a rectangular heap under a shade. It is left undisturbed till its use.

NADEP method to Make Compost: The NADEP method of compost making was first developed by a farmer named Narayan Deorao Pandharipande of Pusad, Maharashtra (India). Therefore, name of this method was given as the abbreviation of his name. In this method all the good projections of earlier methods are included. Hence, nutrient losses are very less. If chemical fertilizers are added in composting materials during preparation of manure, then it will produce nutrient enriched organic fertilizer which is beneficial for crops as chemical fertilizers without any adverse effect on soil.

NADEP method compost making requires the construction of simple, rectangular tanks made up of bricks and mud paste or cement mortar on the land surface. The size of each tank is 3 metre long, 1.8 m wide and 0.9 m high and thickness of walls is about 0.23 m. In the long walls of tanks holes of size 0.3m x 0.23m is made to provide necessary air to compost by removing one brick from each third line after the height of 0.3 m or three lines of bricks from bottom as given in Figure 5.17 Inner walls of tanks are plastered with dung and soil paste.

Fig. 5.17: NADEP compost tank

Each tank should be filled in one day and not beyond two days to get good quality compost. Each tank requires agricultural waste (dry and green) 350 kg, cattle dung dry 11 kg or biogas slurry 100 kg, fine sieved soil 350 kg and water 100 litre. Tanks are filled in definite layers. First of all, 15 cm layer of agricultural waste is spread and moistened with water. Above this slurry of dung or biogas slurry is sprinkled. After this a layer of 45 kg of fine sieved soil is made. By this way whole tank is filled. Finally, a 5 cm thick layer of soil is spread on the top of tank. No cracks are allowed to develop. If there is any crack, the same is immediately closed by dung paste or slurry. Sixty percent moisture is maintained in tanks by sprinkling water on the top and holes. The tanks are covered with a thatched roof to prevent excessive evaporation of moisture.

After four months the compost is taken out from tanks and well mixed. Now the compost is used in fields or kept at any safe place. NADEP compost contains 1.2-1.7 percent nitrogen, 0.6-0.9 percent phosphorus and 0.6-0.8 percent potash besides other plant nutrients.

Compost making on land surface: Compost making on land surface is the simplest method to make compost. Compost may also be prepared as piles or heap of refuse and debris of cities on the surface of land, if facilities of digging ditches or pits are not available. These piles may be of 1 meter wide, 1 meter high and of any length. The piles are made just like a rectangular open box by fencing and staking with twigs, leaves, grasses etc. The ideal location for these piles is dry, shady and well drained dump area.

In these piles layers of composting materials are made in same manner as indicated in ditch or pit method. However, there is no need to overturn of material in piles.

So, this method takes place rapidly and within three- four months compost becomes ready for use.

Fig. 5.18: Open box and heap methods of composting

2. Industrial Wastes: There is a tremendous contribution of modern industries in polluting soil. Smoke, soot and various gases emitted in the air by chimneys of these industries pollutes the soil. Here, with heavy metal thrown as debris of industries also pollute the soil. Based on speed and direction of air pollutants released in the air by industries can reach long distances. Ashes and soot released in the air by chimneys of industries fly in suspended condition in the air. When speed of air slows then ashes and soot deposit on soil surface, vegetation and building due to sedimentation. Afterwards remaining ashes comes down with rain water in the rainy season.

Prevention of acidity of rain water: Several measures can be adopted to prevent acidity of rain water and to control its ill effects on environment. Some basic measures are, 1. Use of alternative mode of transport and car-pooling 2. Use of clean fuel 3. Alternate source of energy and energy conservation 4. Management of flue gas and 5. Liming. These measures are briefly described here:

1. Use of alternative mode of transport and car-pooling: Automobiles are generating lots of acid rain producing gases. Therefore, alternate mode transport like buses, electric buses, hybrid buses, automobiles with catalytic converters, car sharing, pooling, bikes, bicycles, walking etc. should adopted for reducing gaseous emissions. To reduce quantity of gasses in smoke of vehicles efficiency of these should be increased. Efficiency of now days used vehicles is very much less (about 10%). Catalytic converter should also be installed in automobiles.

2. Use of clean fuel: Use cleaner fuels like natural gas, hydrogen gas, electricity etc. and low oxides of sulphur and nitrogen containing coal. Use the coal with low sulphur content and wash the coal to remove sulphur of coal to be used in coal fired power plants. Also, the chimneys of factories should be such managed that the gases should not come out of them. The columns of active carbon dust can be filled in chimneys. Active carbon has more surface area and electric charge. By virtue of these it can absorb various ions and gasses.

If absorbed materials are useful, then these are separated from carbon. Active carbon can separate up to 90% SO_2. Separated SO_2 can be used to make sulphuric acid. With dry absorption by carbon wet absorption method is also used to separate SO_2. In this method dolomite or oxides of manganese are used. Oxides of nitrogen can be separated partly by using sprays of water. Scrubbers are also used to remove sulphur dioxide from gasses coming out from the chimneys of power plant. Coal and other fossil fuels burning processes can also be altered for lesser emission of oxides of nitrogen from power plants.

3. Alternate source of energy and energy conservation: Use of non-renewable source of energy should be reduced and use more alternate or renewable source of energy available from water, wind, and sun. Energy conservation is as important as energy production. Therefore, energy should be conserved to prevent acid rain. All the electrical and electronic gadgets should be shut off while leaving the room or working place. The heating and cooling systems significantly consume energy. These can also be used only when in need.

4. Management of flue gas: Flue gas desulphurization (FGD) is a set of technologies used to remove sulphur dioxide (SO_2) from flue gases produced from industrial combustion at petrol refineries, chemical manufacturing industries, mineral ore processing plants, power stations etc. The removal of sulphur dioxide is essential for a safe and clean environment where toxic emissions are kept to a safe low. Most FGD systems have two stages: one for removing fly ash and another for removing SO_2.

Sulphur dioxide is an acidic gas. Therefore, it is essential to neutralise it before disposal. An alkaline-based sorbent is used to bring the pH level of the sulphur dioxide gas near to neutral. Scrubber systems are one of the most efficient ways of reducing sulphur dioxide emissions caused by industrial combustion, and a classic example of an acid-base chemical reaction performed on an industrial scale. Flue gas desulfurization systems (FGD) or scrubbers are devices capable of sulphur removal efficiencies between 50% to 98%. These systems greatly help in reducing the SO_2 emissions. Typically, the highest removal is achieved by wet scrubbers and the lowest by dry scrubber.

5. Liming: Liming reverses the effect of acid rain on waterbodies as well as soils. The liming materials used to improve the acidic soils can also be used here. In waterbodies the direct application of fine enables immediate acid neutralization. This increases pH of water. The adverse effect of acidic rain on the soil can also be ameliorated by application of lime. In this method firstly pH of rain water is tested. If pH of rain water is acidic then pH of affected soil is tested. If necessary for increasing pH, the lime requirement is determined and lime is added in soil accordingly.

3. Heavy metals: Pollution from heavy metals is one of the significant environmental concerns facing the world today. Heavy metals polluting soil can harm crops, change the food chain, and endanger human health. Thus, the overarching goal for humans and the environment should be the avoidance of soil contamination by heavy metals. The removal of heavy metals from contaminated soil can be accomplished using various physical, synthetic, and natural remediation techniques (both *in situ* and *ex situ*).

The most controllable (affordable and eco-friendly) method among these is phytoremediation. The removal of heavy metals can be accomplished using phytoremediation techniques, including phytoextraction, phytovolatilization, phytostabilization, and phytofiltration. The bioavailability of heavy metals in soil and the biomass of plants are the two main factors affecting how effectively phytoremediation works.

Restoration of soils polluted by heavy metals should be started from point source itself. It means effluent water of any factory containing heavy metals should be removed at factory and clean water should be drain to soil or reservoirs. Similarly, instead of using raw sewage for the purpose of irrigation, processed sewage should be used for irrigation. Many impurities are separated from processed sewage. During proper processing of raw sewage toxic matters are converted into nontoxic organic matters, which can be used in plant nutrition or others. After treatment of sewage, remaining clean water can be used in agricultural operations.

Treatment is sewage is done in three levels. In first level sewage water is passed through various lattices, grit channels and traps. By passing through lattices soil, small flora and fauna and residues present in sewage are separated. Grit channels check it settling down minerals. From grit channels sewage goes to another level, where oily layer floating on water surface is separated. Finally, this water goes in centrifugal chamber, where centrifugal forces accumulate separated particles towards periphery and clean water at the centre. This water is stored in several tanks for sufficient time for slow sedimentation and hence treatment of second level begins.

In second level various microorganisms are used. In sewage, organic matter is present in sufficient quantity. Carbohydrate, fat, protein and other compounds present in organic matter are converted in normal compounds after decomposing by microorganisms. So decomposed sewage can be used as manure. In decomposition of sewage mainly those microorganisms are used which decompose organic matter. Sewage is passed through trickling filters for decomposition and for easiness of decomposition in these filters supply of microorganisms is done. As at first sugars are decompose, so filters will be will be supplied sugar decomposing microorganisms. Sometimes instead of passing sewage through trickling filters it is passed through sludge plants.

Activate sludge means sludge with organic matter, various bacteria and ciliated protozoa. Processed sewage of first level is sent in the activated sludge plants. In these plants aerators are fitted which supply oxygen to bacteria and protozoa. Supply of oxygen and organic matter increase bacterial population swiftly and they decompose more quantity of organic matter present in sewage. Excess population of bacteria is controlled by bacteriophagous protozoa, Colpidium and Glaucoma. Sewage treated by bacteria is useless because it has excess of microorganisms and organic matter at various stages of decomposition. So, processing at third level is necessary.

Sewage received from trickling filters or activated sludge plants is sent to the plant of third level. In this plant many activities are done according to nature and quality of pollutant. In these (i) removal of nutrients by autograph plants and microorganisms, (ii) flocculation and adsorption of colloidal size particles, (iii) oxidation by the help of potassium permanganate, (iv) reverse osmosis by artificial pressure through semi permeable membrane, (v) chlorination to remove orthophosphate and precipitation with the help of calcium, (vi) change in nitrogen of nitrates by the help of bacteria are major activities.

As a result of these and other such activities sufficiently clean water is obtained which is very useful for irrigation and industries. After separation of clean water from sewage remaining sludge is used in various works. After proper management of polluted soils by heavy metals, tolerant crops should be grown in these soils. But such tolerant crops should be grown which are not used for human and animal food, otherwise heavy metals will come again in food chain.

4. Agricultural chemicals: Use of agrochemicals is indispensable for fulfilling foods demands of ever-increasing population. Therefore, use of these cannot stop suddenly. Residues of agrochemicals show harmful effect on soil ecosystem. So, protection of soil from these is very essential. An appropriate method of restoration followed by proper precautions, harmful effects of agrochemicals on environment can be reduced. Bioremediation is best suitable method to restore soils polluted by agrochemicals. Numerous plants and soil organisms are helpful to perform this task of restoration.

However, to stop use of agrochemicals a cordial programme should be followed in which several methods can be used elegantly. Main purpose of this programme is to prevent abrupt or gorgeous reproduction of pests. This will reduce growth and development of pests. Therefore, these will always remain behind from growth and development of crops and vegetables. Adopting coordinated programme on perennial crops than annual crops and on insects than plants diseases is easy.

Under this programme various physical or mechanical means to kill and catch pests by various methods; selection of crops, their proper species, cropping pattern and soil, use of clean fertilizers, and interval in time of sowing; biological means, as conservation of useful organism, use and conservation host and weed eater insects and microorganisms and biotechnological means as use of sterility producing chemicals, growth regulators, attractants and repellents should be done. Use of aforesaid methods control of pests will be successful. Also, if pests remain alive then finally chemicals should be used in very low dose. So, there will be lesser residues to cause very low soil pollution.

5. Radioactive wastes: Radioactive wastes are mainly in solid state. Radioactive wastes are of less, medium and high activity types. In the management of these wastes at first volume of these are reduced by incineration of these on 1000° - 1100°C. Afterwards ashes having little volume is compacted on high pressure (about 50Kg/cm). This compacted material after keeping in proper vessels is buried in pits or immersed in the deep sea. Vessels are madc so strong that they cannot be destroyed by weight of soil or water. According to radioactivity, depth may be up to 400 meters. After dumping, sites are guarded continuously. Also, it is kept in mind that there should no soil erosion and no any type of digging at those sites.

Restoration of soil polluted by radioactive elements is done by scrapping of first upper 10-15 cm soil surface and dumping that at appropriate place, Deep ploughing, leaching, intensive farming with use of fertilizers and use of soil amendments are necessary. Soil amendments are used according to soil pH. Besides it various chelates help in lowering radioactivity of the soil. In this use of DTPA with algae and fungi is very much useful.

Restoration and Management of Soil from Compaction, Sealing, Crusting

The rapidly increasing world's population is expected to reach 8.5 billion by 2030 and more than 10 billion by the end of this century. This requires increased food production to meet the demand without overwhelming the available resources.The mechanised farming has significantly increased agricultural production. But also caused many problems of compaction, sealing, crusting etc. Here restoration and management of soil from compaction, sealing and crusting are described,

1.Soil Compaction: The soil compaction mostly decreases soil porosity and increases soil bulk density. The management of soil compaction, especially in arid and semi-arid regions may be achieved through suitable application of some or all the described techniques i.e., addition of organic matter, controlled traffic, mechanical loosening such as deep ripping, selecting a rotation which includes crops and pasture plants with strong tap roots able to penetrate and break down compacted soils.

Addition of organic matter into the soils: The organic matter recollects soil water and helps soil to rebound beside compaction. The maintaining an adequate amount of organic matter in the soil stabilizes soil structure, makes it more resistant and reduces bulk density and soil strength. Since organic materials possess lesser bulk density and higher porosity than that of mineral soils mixing them with soil would improve soil bulk density and porosity. Though plant residues are a common source of organic matter in the soil and manure is also used extensively by farmers to decline soil compaction and improve soil fertility.

Managing traffic and tillage: The traffic management is a system that maintains a zone more proper for plant growth by restricting soil compaction to the traffic lanes and a loose rooting zone. The wheel traffic in a well-ordered agriculture traffic system, combined with conservation tillage enhances the sustainability of cropping, improved infiltration, increases plant-available water, and reduces soil erosion caused by runoff. The energy requirements for different types of tillage are related to tillage depth, soil type, degree of compaction, soil moisture at the time of tilling and to the type of implements to use.

Deep cultivation and deep ripping: The deep cultivation or deep ripping is a significant preparation for eliminating soil compaction, destroying hard pans and ameliorating hard setting of soils. It has become a common management practice used to shatter dense subsurface soil horizons that limit percolation of water and penetration of roots. The increasing soil compaction related with mechanized agriculture and the availability of more powerful tractors and better sub-soilers, ripping can become a favourable soil management exercise.

Deep ripping of compacted soil may also improve soil health and ability of plants to resist disease. The growth-stunting disease was eradicated completely by deep cultivation as with compacted soil layers. Though, choosing the correct soil moisture and the right combination of ripping depth and tine spacing can lessen the cost of ripping. Ripped soil must be allowed to settle down before seeding, otherwise seeding depth is difficult to regulator and seeds may be placed below the preset seeding depth.

Plants with strong tap roots: The hydrostatic pressure (Turgor) inside the elongating area of the root affords the force required to push the root cap and meristematic region through the resisting soil. If the hydrostatic pressure is not sufficient to overcome wall resistance and soil impedance, elongation of that particular root tip ceases. The result of roots on soil structure depends on the types grown, soil constitution and environmental aspects.

Plants grown in compacted soil have a smaller number of lateral roots with less dry matter as compared to plants developed under controlled situations at both low and high soil water contents. Soil compaction could be due to adverse effects upon plants growing in the soil by increasing the mechanical impedance to the growth of roots, altering the extent and configuration of the pore space. Plant species that have the ability to penetrate soils with high strength usually possess a deep tap root system.

2. Soil sealing: Soil sealing caused by rainfall etc. can be removed by mulching and light ploughing. However, soil sealing caused by cement, concrete, stone, rock fragments, bricks, asphalt etc. can be desealed by mechanical followed by biological methods. First all the pieces and fragments of cement, concrete, stone, rock, bricks, asphalt etc. should be collected by hands or by a mechanical collector and deposited at appropriate sites. Then field should be heavily irrigated to loosen and swell the soil. When soil achieved the moisture at about field capacity then it should plough again. After ploughing if buried pieces of artefact seem on soil surface, these should collect again.

Now a ten cm thick mulch of partially decomposed organic materials along with sufficient number of earthworms should be evenly spread on soil surface. This will increase organic matter status soil, make soil more perforated and enhance the easy entry of water and nutrients in soil. Afterward few bushy shrubs like cranberry etc. should be grown till soil attains fertility to cultivate normal crops

3. Soil crusting: There are several methods to restore soil from crusting. First, protect the soil from the impact of raindrops by keeping it covered. This means no-tillage and no removal of crop residue from the row. If a crop leaves little crop residue, it may be necessary to plant a cover crop to avoid crust formation.

Second, the surface soil structure needs to be improved. This can be done by increasing the organic matter content in the very top of the soil, as is achieved by using no-tillage continuously, and by adding organic matter in the form of manure, compost, cover crops, and crop residues. Dense root systems help make soil less sensitive to crusting. If any tillage is done, limit it as much as possible and leave the soil rough with as much residue cover as possible to limit crust formation. Simultaneously soil biological activity is another way to make soil resist crusting. Earthworms will bury through crusts when they are moist, deposit casts on the soil surface and these casts exhibit high resistance against disintegration.

If field suffers from crust, then crust should break up by rotary hoe. Another tactic is to plant seeds close together because they will be able to push through a crust by joint force. This tactic could be applied by planting soybeans with a drill with about 20 cm row spacing. Increasing seeding rates may also help for the same reason. Deep planting should be avoided in a crust-sensitive soil to allow the seedling to emerge before the crust re-forms. Avoid very early planting, so crusts do not have time to form above the seedlings prior to emergence.

Restoration and Management Soils from Water-logging, Aridity

Restoration and management soils from water-logging and aridity is very much different than normal soils because these two situations are extreme ends of each other. Restoration of these require specific techniques which are very much time taking, laborious and highly expansive. Even though restored soils again develop water logging and aridity due to prevailing climatic conditions. Therefore, it is best suitable that these soils and conditions should be used as such with necessary changes for crop cultivation etc.

Waterlogging: Management of waterlogged soils is very different from normal soils. Waterlogging can be efficiently regulated by modulating land shapes, mechanical as well as bio-drainage, and controlled irrigation measures. Tolerant or resistant varieties and proper nutrient management would be much more effective for better survival of agricultural crops in waterlogged soils. Here brief description of major facts related to management of waterlogged soils are given.

Fig. 5.19: A waterlogged soil (Photo Dr Venkatesh Bharadwaj)

Preparation of waterlogged submerged soils for crop: Waterlogged soils are generally used for paddy cultivation. Primarily puddling of soils is done for preparing the field. Puddling means to make impervious mixture of soil with water. In this method the water filled soil is ploughed again and again, so that large aggregates of soil are broken in small particles. In an experiments it has been observed that about 40 percent soil aggregate was divided in small particles than 0.005 mm and mean weight diameter of particles is reduced to 0.36 mm from 1.7 mm.

During puddling reorientation of soil particles takes place and large particles of soils settle at first. Particles of medium size settle on these large particles. Finally smallest particles of soils are settle at the top. This setting makes very much dense arrangement destroying non-capillary pore space and reducing void ratio. So, puddled soil is that whose structure has been destroyed, whose aggregate cannot be detected and which is a homogenous structural mass of fine aggregates and fractions. Although in puddling sufficient energy is required but it has many benefits which are given here.

Benefits of puddling: Puddling is necessary for transplanting paddy. Puddling destroys weeds, provides good physical, chemical and biological conditions and proper management of water in the soil for paddy crop.

Both primary and secondary tillage operations are done for puddling. In primary tillage land is ploughed deep as a result of which weeds and plant residues of previous crop turned in the lower surface of the soil. In secondary tillage harrowing is done two to three times in water filled soil. By secondary tillage remaining weeds, their seeds and other plant residues are reached to lower areas of puddle soil. This area is of reduced state. Here due to lack of oxygen seeds of weeds do not germinate.

Second major benefit of puddling is to conserve water in the soil. Puddled soil can retain any level of water on soil surface than natural unpuddled soil. Its main reason is to reduce non- capillary pore space in the puddled soil. A compact state of soil particles is formed. These both states reduce hydraulic conductivity and water percolation of the soil. The drainage of water from the puddled soil also is less due less hydraulic conductivity. Puddling also conserves plant nutrients dissolved in the water.

Water management in the waterlogged soils: In the waterlogging excess water essentially requires proper management. Generally, crop of paddy needs water for the purpose of preparation of land, seepage, infiltration, percolation and evapotranspiration. About 60 cm water is required for preparation of land. This includes evaporation of water from puddled soil before transplanting, presence of water on the soil surface, absorption water by the soil and infiltration and percolation of water below plough layer. Generally, quantity of infiltration (seepage) and percolation is 0 to 6 mm per day which may become 20 mm per day. Variation in infiltration and percolation is very localized and this depends on proximity of drain, extension and mutual attachment of waterlogged area, soil properties, water table and maintenance of water level. Transpiration from the waterlogged soils is 4 to 9 mm per day and it is affected by solar energy, moisture and wind speed.

Fertilizer management in the waterlogged soil: Development of new fertilizer responsive varieties of crops increased the importance and demand of fertilizers.

Yield potential of these varieties without fertilizer application is slightly more than old varieties. But by use of fertilizers yield of new varieties is three times more than old varieties. So, fertilizers have special importance in the crop production. Fertilizer is very much costly means for farming so its judicious use is extremely necessary at proper time in the appropriate quantity.

In the waterlogged soil because of presence of excess of water and reduced conditions proper management of fertilizers becomes further more essential. These conditions influence availability of nutrients present in the fertilizers by different ways than normal soils. Farmers do not have sufficient money to buy fertilizers in desired quantity. Therefore, proper fertilizers management practices should be communicated to them. So that they can obtain more yield by using less fertilizer. This will provide more profit for farmers, saving of energy and decrease in pollution cause by fertilizers with reduction in cost of cultivation. Here about management of some nutrients in the waterlogged soils is described.

Nitrogen: Development of such methods of fertilizer application and crop management becomes necessary where by maximum yield can take by minimum use of fertilizers. In the waterlogged soils in the reference of yield of paddy, it has been observed that ammonical and ammonium producing, both types of fertilizers are equally efficient. Currently in cultivation of crops usage of urea is increasing continuously than other fertilizers of nitrogen. Its main reason is to be its low price. Now urea fertilizers with various coating are produced. Release of nitrogen from coated urea is very slow which is helpful for clean environment also. Nitrogen is lost easily in paddy fields. Therefore, it should apply in three split doses. Fifty percent of nitrogen should be applied just before transplanting and remaining two split doses of 25% each should apply at tillering and flowering stages.

It is essential to add the nitrogenous fertilizers in the field on time. Little quantity of fertilizer should be used at the time of highest need. It has been observed in submerged paddy that less quantity (20 kg. per hectare) of nitrogen added in the maximum tillering and flowering stages helps most in increasing yield of grain. Similarly high dose of nitrogen applied after panicle initiation stage does not carry any benefit.

Method of fertilizer application also influences the efficiency of nitrogen fertilizers. It is also seen that in water logged soils use of nitrogen fertilizers in reduced layer of the soil is more beneficial. Urea or ammonium sulphate put in the soil at 10 cm depth gave highest yield. Band placement of mud balls or briquettes of urea is also good for higher production. But in waterlogged condition to put nitrogenous fertilizers at more depth, use of mud balls or briquettes is labour consuming and expensive. So, among common farmers it is not in use. Use of nitrogenous fertilizers at time of preparation of land, thoroughly mixing them in the soil and afterwards presence of water on the soil surface is very beneficial for of production.

Phosphorus: Soil reaction, weathering state, type of clay minerals, quantity of water, cropping intensity, crop rotation etc. are factors influence management of phosphorus. In water logged soils use of single super phosphate is the source of phosphorus just like normal soils. Use of rock phosphate in the acidic soils also is beneficial. Cheap rock phosphate reduces cost of crop cultivation, but for direct use in the soil it should be fine ground. Fine granules (60- 140 mesh) of rock phosphate can be used directly in the soil and result of these are very good.

Primarily phosphorus is used at the time of transplanting, but on deficiency it may be used before active stage of tillering. In submerged condition because of more availability with time, less leaching losses and rapid translocation from old leaves to new leaves, use of this in divided doses is not beneficial. On account of immovability of phosphorus deep placement also is not beneficial. In waterlogged soils despite more availability of phosphorus its recovery in the paddy is less, so to grow phosphorus deficiency tolerant varieties of paddy as IR 28, 29, 30 and 34.

Potassium: In submerged soils response of potassium in the soils is less than nitrogen and phosphorus, although management of potassic fertilizers is somewhat like to nitrogen. In waterlogged soils use of muriate of potash (potassium chloride), fertilizer mixture and potassium sulphate all three are used. But use of potassium sulphate is more beneficial in high pH soils. Potassic fertilizers are used with final preparation of the field. But at the time of various growth stages of paddy more availability of potassium in the soil is good for yield of paddy. More absorption of potassium at the stage of maximum tillering increases number of panicles and grains. In the stage of panicle initiation absorbed potassium increases weight of grains. If potassium is to be used in less quantity, then its used remains good at active stage of tillering.

Zinc: In waterlogged soils deficiency of zinc already exists and more uptake of zinc by new improved varieties of paddy, use of urea in place of ammonium sulphate , use of phosphatic fertilizers, use of concentrated fertilizers and multiple cropping make more grave this problem. Zinc deficiency can be treated by zinc oxide or zinc sulphate application.

Silicon: Silicon produces lodging resistance and protection from insects (stem borer) and disease (blast) in the paddy. Crop of paddy needs sufficient silicon and uptake of this by paddy may be up to 1000 kg Silicon oxide /hectare. Silica slag is main fertilizer of silicon. This has Calcium silicate. in the waterlogged soils 1.5 to 2.0 tons / ha silica slag is sufficient for the paddy crop. Its boasal dressing on the soil surface or in the initial stage of growth of plants top dressing are equally beneficial.

Bio-drainage for restoration of water logging: Bio-drainage may be defined as "pumping of excess soil water by deep-rooted plants using their bioenergy." The bio-drainage system consists of (i) fast growing tree species, (ii) larger canopy area, and (iii) plants with luxurious water consumption and higher transpiration rate. As for example, *Syzygium cumini, Pongamia pinnata, Terminalia arjuna, Casurina glauca, Eucalyptus tereticornis* are some most effective and used bio-drainage plants. As curative (for waterlogged areas) and preventive (for potentially waterlogged areas) measures bio-drainage is most useful technique.

Now lotus is also grown in water logged soils nearby cities for makhana (foxnut), lotus stem and flowers.

Fig 5.20: A water-logged soil with lotus crop (Photo by Dr Venkatesh Bharadwaj)

Soil Aridity: Successful restoration strategies of soil aridity is to increase water availability and reduce evapotranspiration. Water availability can be increased by water harvesting, increased infiltration and increased water retention. These will increase plants and their growth in arid soils. However, foremost requirement for proper management of soil aridity is to provide stability to unstable dunes. This aim primarily may be achieved by three activities. First activity is to prevent conserved dunes from all types of biotic interferences. Second activity is to prevent collapsing sand from dunes by growing plants of local bushes in parallel strips. Third activity is to made sand dunes thoroughly stable by growing local trees, bushes and grasses.

Pasture development: An appropriate management of sandy plains of arid region is development of pastures and cultivation in strips. In marginal and sub marginal lands of such areas instead of cultivation of any type vegetation, pasture should be developed. In developed pastures time bound and systematic programme should be conducted grazing. Grasses grown in pastures increase fertility of soil as well as stability of sand dunes and sandy plains. In the soils of grassy plains there were 16- 43 percent increase in organic matter and 20 percent increase in nitrogen. In sandy plains for soil and water conservation contour furrows, staggered pits and contour ditches can be made.

Crop cultivation: Pearl millet (Bajra) is main crops of arid region of Rajasthan which is up to 50-70 percent area. Afterwards pulse crops and sesame come. In the arid region of Gujrat according to quantity of rain groundnut, sorghum and cotton crops are grown. Strip cropping of these has many benefits. Strips cultivation of green gram and vetch with bajra or grass in ratio of 6:1 has been found very beneficial. In arid regions to prepare field shallow ploughing should be done according to distance of rows of plants. As a result of this lesser area of soil will expose or open. Hence, there will be less soil erosion. After harvesting, crops residues should be left on the soil surface. This will check the evaporation and maintain soil stablity.

Moisture conservation: In the soils of arid regions there is essential need of moisture conservation. In these soils 40-60 percent water of rainfall is destroyed by deep percolation. Afterwards excess of evaporation also sand away very much quantity of water in the atmosphere. This problem can be solved by increasing water holding and water retaining capacity of soil and preventing evaporation and percolation losses of the soils. Use of exfoliated vermiculite and bentonite clays increase water accumulation up to 12-17 percent at 20 cm depth of the soil.

Similarly, use of sediment of ponds also increases water supply capacity of soils of arid region. Because particles of aforesaid substances are very much fine, these after mixing with coarse particles of soils make little fine of their texture. As a result of this micro pore space increases. The increase in pore space, water holding and water retention capacity also increases. All these increase yield and water use efficiency of crops.

In arid region evaporation of water can be prevented by using various mulches spread on the surface of soil. Mulches of dry grass and husk of bajra have been proved more useful. Deep percolation of water is prevented by making sub surface barriers in the soil. These barriers are made up of bentonite, asphalt and or such other substances. These barriers prevent leaching losses of nutrients mainly nitrogen from the soil and stop deep percolation of water.

Irrigation: Excess water percolation capacity of soils and high transpiration more water is needed for the purpose of irrigation of crops in arid region. Because of less availability of water it is essential to take special consideration for time and method of irrigation. It is observed in the experiments that in such area light irrigation on an interval of 8-10 days is very effective than deep irrigation on more days interval. In irrigation methods, use of sprinkling method, drip method and pitcher method are quite good. These methods besides saving of water also enhance water use efficiency and fertilizer use efficiency. Ultimately crop's yield increase remarkably.

Water harvesting: In arid regions water harvesting is also necessary. In arid regions water is harvested by different methods like runoff farming, water spreading, micro catchments, inter row water harvesting systems and traditional water harvesting systems like tanka, Nadi, khadin.

Runoff farming is like Water Harvesting but for Irrigation Purposes. When the harvested runoff water from un-cropped areas is directed to a cropped area, this technique is called runoff farming. Soil profile acts as a water storage container, but storage in ponds or cisterns is also feasible. In water spreading method flood waters are deliberately diverted from their natural courses and spread over adjacent plains. In micro catchment method rainwater catchment basin is built around the plant. In inter-row water harvesting systems water collected from ridge is stored in furrow and the crop is benefited by high moisture. Ridge and furrow system is better for getting higher yield of pearl millet than flatbed method. Ridge provides partial shading of the furrow for 6 to 7 hours a day and reduces evaporation.

Tanka, nadi, khadin are the important traditional water harvesting systems of Rajasthan. Tanka is an underground tank or cistern constructed for collection and storage of runoff water from natural catchment or artificially prepared catchment or from a roof top. Nadi or village pond is constructed for storing water from natural catchments. Khadin is unique land use system wherein runoff water from rocky catchments are collected in valley plains during rainy season. In Khadin method crops are grown in the winter season after water is receded in shallow pond on the residual moisture.

Restoration and Management of Soils of Landslides

A landslide is the downward mass movement of rock, soil, soil material, debris etc. from a slopy land. Rain, earthquakes, volcanoes, gravity etc. or other factors providing instability to slopes are main reasons of landslide. Restoration of landslide areas requires knowledge of geotechnology of area, nature of subsurface conditions, evaluation of the soils and hydrology, and development of a site-specific restoration method. Preventive and remedial both types of measure should be adopted to control landslides. Preventive measures include search of unprotected areas, land capability classification, afforestation, raising of less animals of improved breed, planning of developmental schemes, consolidation of holdings, population control and cooperation and training of local public and institutions etc. Brief description of some measures out of these is presented here:

Search of unprotected areas: Extensive surveys should be done to search unprotected sites and underground faults. After extensive analysis of results obtained by these surveys priority of unprotected sites should be fixed. The people of unprotected sites should be shifted to protected area and at unprotected places remedial measures should be adopted.

Land capability classification: In hilly areas utilization of land should be according to land capability classification. In hilly and Himalayan areas of our country, American land capability classification system cannot be adopted as such. Because according to this system whole Himalayan area is not suitable for cultivation. But seeing the social-economic requirements of here cultivation cannot be stopped outright. Therefore, new land capability classification should be developed for hilly area. In this classification clear demarcation should be between agricultural and non-agricultural lands. Then non-agricultural lands should not be utilized for agriculture. Generally, utilization of sites having more than 15 percent slope should be utilized for orchard, forage production and forestry instead of crop cultivation. Generally, in hilly areas three steps system as timber, fuel and silvipastoral trees at top of slope, fruit trees in middle part and farming in the lower part of slope should be followed.

Afforestation: Landslide and erosion can be stopped easily by forestry. Rehabilitation, protection and upgradation of old forests should be done. At the time of selection of trees for the establishment of new forests, requirements of local public should be fully considered. In hilly areas such trees should be planted that could provide food, forage, manure, fuel and forest product in abundance. Yonchok, walnut, chestnut, pecannut, wild apricot, bird cherry, emblic myrobalan etc., fruits tree; oak, gravia, bauhinia, for forage; hata and other bushes, for fuel, rhododendron, oak and vasaka for leaf manure and tree cotton, mulberry, bamboo, ringal etc. for fibre should be grown.

Raising of less animals of improved breed: Excess number of animals in hilly area should be reduced. Too much animals put pressure on pastures and forests. So, in hilly areas such breeds of cows and buffaloes should be raised that together with milch animals can also work in fields. So, less animals would be kept there and forage should be given to them at their dwelling places. Grazing in forests and pastures should be reduced. Also, recurrent and strip grazing should be adopted. In hilly area goat husbandry should be reduced as goats eat newly grown saplings and even climb on trees to eat up leaves and twigs of them.

Planning of developmental schemes: In hilly areas such developmental schemes should be conducted that really can develop that area. Before construction of roads, building, canals etc. places of these should be finalized after taking advice with geological and soil survey departments. Necessary trees and bushed should be planted previously in plan period for the purpose of soil conservation. So that before completion of the plan, well grown vegetation can provide stability to the developmental works. Pakke (metalled) drains should be constructed at required places to prevent stagnation and seepage of water. However, herring-bone drainage system good for such places. A generalized figure of herring-bone drainage system is given here:

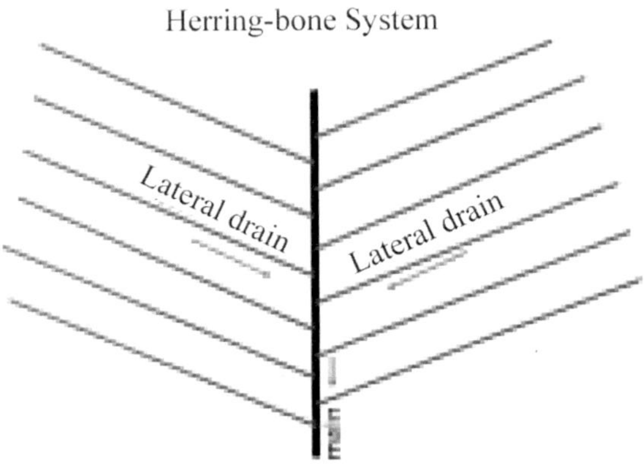

Fig. 5.21: Herring-bone Drainage System

Consolidation of holdings: About 97 percent holdings are less than one hectare which are scattered in 5 to 6 distant places. So, proper care of them is not possible and they quickly affected by landslide. Therefore, small holdings should be consolidated for proper care and management.

Population control: Now population control is utmost necessary for welfare of all. All the negative aspects of higher population are also responsible of landslides.

Cooperation and training of local public and institutions Gram Sabha, Van panchayat and Yuvak Mangal Dal and other such institutions should jointly take part for afforestation and pasture development and management. Rural specially women should be taught about importance of soil conservation. They should be trained for improved farming, gardening, animal husbandry and social forestry.

Remedial Measures: Based on local problem and situations remedial measures should be adopted. The whole area should be encircled to restrict all types of biological activities. Then work of contour wattle, mulch, check dam, retaining walls, diversion drains, gabion structure and terracing should be done. The stability of all such structures and slope should be ascertained by planting trees. Plants *Lannea coromendelica*, *Vitex negundo, Erythrina suberosa* and *Salix tertrasperma* for contour wattling, *Chrysopogon fulvus* or mulch and *Saccharum spontaneum* for grass can be planted.

The designing of landslide remedial measure is done after studying field investigation data and causes of landslide. The remedial measures are broadly come into three main categories viz. Reinforcement measures, Retaining structures and Surface- sub surface drainage measures. The restraining structures are designed to retain rock/soil and are generally economicaly cheaper as compared to other categories of remedial measures. But these have numerous drawbacks such as water drainage problem in sausage walls, base sliding failure in concrete gravity walls, need of bearing surface in case of anchored walls etc.

Second major type of remedial measures is Earth Reinforcing structures which involve insertion of tensile resisting materials such as steels rods, metal strips, geosynthetics of cloths and steel angles into the soil to improve stability. Reinforcement remedial measures are widely used because of flexibility, ease of construction and are inexpensive as compared to large retaining walls. But these methods have limitations such as soil nailing technique must occur above groundwater level and away from seepage location.

Moreover, the frozen water can damage the shotcrete and place additional loads on the anchor plates due to soil expansion that results in overall reduced effectiveness. To avoid this surface runoff water should be intercepted before reaching a soil nailed wall, and any collected water should be disposed of rapidly. Third major category is Drainage measures which is most effective in the geological condition which allow interference with natural water regime.

Restoration and Management from Abandoned Mines

Mining also is a very old and important industry just like agriculture. Raw materials are made available to the various industries from mines. But in most mines, mining work is done by unscientific manner. Digging for the purpose of mining, explosion, road construction and cutting of trees very badly influence agricultural lands, soil, water and worsen the economic condition of habitants. Therefore, reclamation and restoration of damaged mined area is very necessary and urgently required to save the natural resources.

Reclamation is the process of getting something useful from the waste. Similarly, reclamation of mines are activities which produce land, water and others conditions which meet social needs. Restoration is the activity to bring something to former conditions. Restoration of mining areas means to restore the degraded and abandoned mining areas by adopting suitable land utilizing methods. Mine is an excavation made in earth to extract minerals, ores, coal, precious stones etc. Mining is the activity, occupation or industry related with extraction of minerals, ores, coal, precious stones etc.

Although there are many methods of mining. But four main mining methods are: 1. open surface (pit), 2. underground, 3. placer and 4. in-situ mining. Open Surface (pit) mining is typically used for more shallow and less valuable deposits. Underground mining is more expensive and often used to extract deeper deposits.

Placer mining is used to sift out valuable metals from sediments in river channels, beach sands, or other environments. In-situ mining is primarily used in mining of uranium. The method involves dissolving the mineral resource in place then processing it at the surface without moving rock from the ground.

There are mainly six stages in a sequence during the life of a mine. The sequence of stages is prospecting, exploration, development, exploitation, closure and reclamation. Reclamation of the mine site has become an essential part of the mine life cycle. It is necessary due to laws for abandoned mines and the demands of society for a cleaner environment. Indian Bureau of Mines (IBM, Nagpur) had identified abandoned / orphaned mines which had been left un-reclaimed prior to the promulgation of rules about Mine Closure Plan in April, 2003.

Through a special study at national level, 297 abandoned mine sites were identified. Out of the 297 abandoned mine sites, IBM identified 106 abandoned mine sites belonging to Public Sector Undertakings, major and other private sector companies requiring reclamation / rehabilitation. Out of the above 106 sites, 24 mine sites become operational again, Thus, requiring reclamation and rehabilitation in respect of 82 abandoned sites only. These 82 sites are as Andhra Pradesh-12, Goa-1, Gujrat-1, Jharkhand-8, Karnataka-2, Madhya Pradesh-36, Orissa-4, Rajasthan-2, Tamilnadu-12, Uttar Pradesh-1 and Uttarakhand-3.

Reclamation technology for abandoned mines: Generally, abandoned mines have bare lands, loose soil mounds, minerals and rock wastes, piles of condemn machinery etc. Rock wastes create more problem for reclamation and restoration. However, all theoretical, practical, economic and environmental aspects should be considered before starting any reclamation programme. A general reclamation programme is multidisciplinary in methods and techniques. It consists levelling and contouring of land, topsoil substitution on the land if possible, sowing and planting of local crops, vegetation and trees, long term careful monitoring and evaluation of programme. In case of open pit mining the reclamation involves replacement of sizeable amount of rock waste, backfilling of mined pits, levelling and covering of waste dumps with topsoil, planting of local vegetation to consolidate the material and then erection of fencing to prevent livestock.

Reclamation of abandoned coal mines: Eco-friendly use of closed/abandoned/ discontinued coal mines in India was suggested by ministry of coal (2022). The various coal mines have been closed/abandoned/discontinued due to depletion of reserve, spontaneous heating, safety reasons, inundation, adverse techno-economics, financial losses, adverse geo-mining conditions, surface constraints, conversion to opencast mine, etc. The abandoned quarries are used for dumping / filling of fly ash in to the worked out area for suitable reclamation, development of eco and mine tourism parks, afforestation, pisciculture, source to supply of drinking water and other domestic use and generation of sand from overburden of opencast mines. The mine closure and reclamation activities also consist extensive plantation on overburden dumps and vacant lands as part of post closure reclamation works, developing agricultural fields, eco-parks, fruit bearing and vegetable crops for the benefit of surrounding community, creating water bodies in the final voids which augment ground water recharge and also serve as source of water for agriculture.

Reclamation of abandoned bauxite mines: Reclamation of back filled bauxite mine areas through afforestation activities at Lohardaga, Jharkhand during 2010-2013 was conducted by TERI, New Delhi. They reported that the aluminium industry in India is highly concentrated with relatively few players dominating the market. Hindustan Aluminium Company Ltd. (HINDALCO) is the largest among them, controlling about 40% of market share. The industry is one among the flagship companies of Aditya Birla Group of Industries. HINDALCO strategically established its processing units closer to the mine areas in different parts of the country to achieve maximum benefits. Thus, in order to fulfil the raw material requirement, the company took many bauxite mines on leases in Lohardaga, Gumla and Latehar districts of Jharkhand State. The Pakhar and Bagru plateaus are situated just 30 to 35 kms away from the district head quarter Lohardaga and to develop green cover in at least 25-hectare area of backfilled and over burden dump of Pakhar bauxite mines and Bagru hills bauxite mines in Lohardaga. Technology adopted for reclamation of abandoned mines in Pakhar and Bagru plateaus was that after the completion of mining leases at Pakhar and Bagru plateaus, company started back filling process with laterite, morrum and normal soil layer at the top. After backfilling the mine areas, rehabilitation and restoration through afforestation activities was carried out and finally handed over to the concern owners.

TERI, New Delhi provided technical assistance to vegetate the reclaimed back filled mine areas in more systematic and scientific methods for various vegetative measures which were in accordance to provide short, medium and long-term benefits to the marginal communities of surrounding villages. The mine backfilled area and over burden dump of Pakhar and Bagru plateaus were afforested initially for three years with tree species such as Harar, Bahera, Amla, Arjun, Jamun, Bael, Neem, Ritha or any other medicinal tree species suitable for local microclimate. This yielded fruit, fuel, timber, fodder etc. and provided other benefits like employment to the local community, carbon sequestration, oxygen, biomass availability etc.

Later on, the seeds of the grass species were sown to stabilize the soil through their roots. Grasses have fibrous roots, which decrease the soil erosion by increasing porosity of soil and their sod forming tendencies finally produce an organic matter rich layer of soil. Growth of grasses is very useful in restoring the slopes of mine areas since it stabilizes soil, conserves soil moisture and may compete with weeds. This initial grass cover allowed the development of diverse, self-sustaining plant communities. Thus, an inhospitable bauxite mine backfilled area was developed into a productive land through systematic and scientific afforestation techniques.

Reclamation of back filled bauxite mine areas through https://www.teriin.org/project/reclamation-back-filled-bauxite-mine...

Reclamation of abandoned iron ore mines: In the year 2016, ICAR-Central Coastal Agricultural Research Institute, Goa reported that India is an important producer of iron ore in the world contributing more than 7 per cent of the production and ranking fourth in terms of quantity produced following China, Brazil and Australia.

Iron ores are generally, classified as haematite, magnetite, limonite and siderite. In India, iron ore mostly occurs as haematite or magnetite and the former is about 75% of the total. The waste piles created are often unstable at steep slopes. Since, Goa experiences high rainfall with higher intensity, the combination of unstable piles and heavy rains leads to the increased possibility of runoff.

Many mines and their waste piles are located close to residences and agricultural fields. The silt transported from the waste piles are being accumulated layer by layer in the adjacent agricultural fields and the land becomes unsuitable for cultivation. Another major problem faced by the Goa iron ore mines is that the waste material that emerge from the excavation of ore contains a type of material that does not allow vegetation to grow on the waste piles.

The institute indicated the methodology for the rehabilitation of iron ore mine dumps by various horticultural crops for biological reclamation of iron ore overburdens and control measures for reduction of soil erosion. The field experimental study was conducted at one of the mine dumps of M/s. Sesa Goa Limited at Sonshi in South Goa.The mine dumps selected was having maximum of magniferrous clay with an area of 0.9 ha. The selected area of 0.9 ha was equally divided into four blocks. The treatments of the amendments were, T1-Lateritic soil + Mine reject soil (1:1), T2-Farm Yard Manure + Mine reject soil (1:1), T3-Pressmud + Mine reject soil (1:1) and T4-Mine reject soil alone (control).

In each treatment planting of Cashew grafts (Vengurla 4), Mango (Ratnagiri Alphonso) grafts, Jamun (Gokak-1), Kokum grafts (Konkan Amurta), Jackfruit (Firm type), Tamarind (SMG - 4) and local seedlings of Aonla and Neem was taken up with onset of monsoon in a spacing of 6 m X 6 m. Contour staggered trenches with the dimensions of 2 m x 0.5 m X 0.5 m at 4 m vertical interval were also made. Also, *Stylosanthes hamata* grass and Mucuna seeds were sown as cover crop in between the tree plants uniformly in all the treatments.

The survival percentage and crop growth parameters of Cashew, Mango, Jamun, Kokum, Jackfruit, Tamarind, Aonla and Neem were monitored for the period of 8 years from 2002 to 2009. Silt reduction through silt trapping units was observed for the period of three years from 2002 to 2004. Effect of various soil amendments and cover crop on nutrient status was periodically monitored at every 3 years interval for 9 years period from 2002 to 2011.

The collected long-term data were analysed and conclusions and recommendations were drawn as, 1. Biological reclamation through establishment of horticultural crops in iron ore overburden is one of the environmentally sustainable options for rehabilitation of iron ore overburdens. 2. Soil amendment and excavation of staggered contour trenches are essentially required while planting horticultural crops in magniferrous clay iron ore over burdens. 3. Cashew, aonla, mango and jamun out performed in growth and canopy establishment compared to jack fruit, kokum, neem and tamarind. 4. Aonla outperformed and attained maximum growth parameters with the laterite soil amendment with mine overburden soils. Cashew, Mango and Jamun attained better growth parameters with farm yard manure amendment with mine over burdens. 5. *Stylosanthes hamato* was proved to be best cover crop for rehabilitation of iron ore overburdens as compared to mucuna. 6. Staggered contour trenches and *Stylosanthes hamato* as cover crop reduced the soil erosion to the extent of 86 per cent and *Stylosanthes hamato* and litter generated from horticultural plants increased the nutrient contents marginally in the iron ore overburdens in long term.

Finally, it was recommended that Cashew, Mango, Aonla and Jamun are possible horticultural crops for biological reclamation of iron ore overburdens. Either laterite soil or farm yard manure in the ratio of 1:1 is recommended as soil amendment for planting horticultural crops in iron ore overburdens. Staggered contour trenches with the dimensions 2 m X 0.5 m X 0.5 m at 4 m vertical interval are good for controlling soil erosion.

Reclamation of abandoned limestone mines: In hilly areas mining has very much adverse effect. After mining waste materials are thrown downwards from steep slopes of mountain. This waste material begins to flow downwards ruining fields, trees, public construction etc. coming in route and reaches in any stream or reservoir. There it causes sedimentation of these water bodies. So, for the purpose of reclamation of such areas various works related with engineering, soil and plantation are necessary as done in areas of Mussoorie and Sahastradhara lime quarries of Uttarakhand by soil and water conservation management.

Mussoorie is famous for her natural prettiness and tourism. On account of heavy mining work for limestones, dangers of explosion, road construction, deforestations etc. were noticed to mountains of Mussoorie. The mountains were converted into a barren area devoid of vegetation, full of gullies after erosion and a natural imbalance in whole Doon valley. The late PM Smt. Indira Gandhi was also sought court order for immediate stop of quarrying of lime stone. After considering seriousness of problem, the Supreme Court had stopped mining from 55 mines of the area since June 1985.

Abandoned mines area situated in Sahastradhara (Mussoorie) is extended in 64 hectares. Its height is 840 to 1310 meters from sea level. This area is divided in two watersheds. Soils and sediments discharged from these watersheds fall in Baldi river by two main water courses of Bhandunala and Dhandola streams. Baldi river is tributary of Ganga. A study of Central Soil and Water Conservation Research and Training Institute, Dehradun reported that before reclamation about 32000 tons per year or 550 ton/hectare/ year debris come down on roads and destroy them creating a havoc for transportation, tourism and others. Central Soil and Water Conservation Research and Training Institute, Dehradun provided the methodology for the reclamation of Khariva-Dhandola mining area since 1983. Its results are being presented here.

At first topographical survey of the area was done for slope, height, quantity of runoff etc. Thereafter to lessen the slope and control of soil, sediment and runoff survey of external form was done by construction of various structures on proper places and its profile was prepared.

After doing so Gabion structures were constructed on predetermined sites. Width of these structures begins to increase with enhancement of slope. Gabion structures were constructed 2 m wide on less than 15 percent slope, 3 m wide on 15-25 percent slope and 4 m wide on 25-40 percent slope. At extreme slopes check dams of bristle or log woods were constructed.

Fig. 5.22: Check Dams

On unstable slopes contour wattling was constructed and planting of suitable locally available grasses and trees were done. At the areas of less slope interceptor were constructed with filling stones in networks of GI wire in the upper area of watershed to control the debris etc.

Fig. 5.23: Recharging structure at Vaniputhurmedu in Vaniputhur Panchayat in T.N. Palyam Union

The stability of constructed mechanical structures and area is attained by planting of fast-growing local vegetation such as *Salix tertrasperma, Arundo donax, Ipomoea cornea, Vitex negundo, Leucaena leucocephala, Lannea grandis, Erythrina suberosa, Bauhinia retusa*, joint Napier, Bhabhar grass, *Chrysopogon fulvus* etc.

So, by adopting various measures debris stops above and does not reach up to roads. Reticular Gabion structures stop soil loss. Thus, filtered water come outside. This water was very much clear, pure, clean and potable. Mean quantity of calcium, magnesium, sulphate and bicarbonate was found 70, 31,151 and 152 ppm, respectively in the water.

Watershed management: The land of countries since ancient times are producing cereals, fodders, fibres, fuels etc. by cultivation to persist since then. The construction of water reservoirs, ponds etc. were outstanding examples of water conservation science and skills of that time. These waterbodies were used for irrigation and other activities. It has been mentioned that when sediments were deposited in reservoirs and ponds, the king himself along with his subjects used to throw sediments after filling very first scuttle with spade. At that time whole soil and water resources were very efficiently used and conserved.

But unplanned use of land and water were started with the advent of 20th century due to population explosion and other related factors. Illegal and unscientific cutting of forests also began. All these were very remarkably eroding soils which ultimately decreased the area of watersheds. Again, many troubles like interruption of water current, sedimentation in canals, rivers and reservoirs, floods, decrease in water table, drought, decrease in productivity, loss of properties and communication system, loss of animals, people etc. have been arisen. Therefore, proper management of watershed is necessary to resolve all these problems.

Watershed, also called as drainage basin and catchment area, besides a hydrological unit, has an ecological, economic, social and political existence. Watershed plays critical role in resolving food, social, economic and life support services necessary for all. The watershed is the area of land that drains or sheds water into a specific receiving waterbody i.e., rivers, lakes, ponds, streams, estuaries, oceans etc. through a single outlet or coming out of drainage area is called watershed. One watershed is separated from another by a natural boundary known as the water divide or the ridge line. When rainwater or melted snow move, they collect, transports sediments and other materials and deposits them into the receiving waterbodies.

Watershed management is defined by various ways like any human action aimed to ensure sustainable use of natural resources in a watershed and provide solutions to threats. But well accepted definition given by Tejwani (1982) is "Optimum and sustained production and reasonable utilization of land and water resources without conveying minimum loss or trouble to natural resources is called watershed management." This definition is necessarily related with soil and water conservation. Accordingly, watershed management includes proper land use, its protection from all forms of land degradation and vulnerabilities, formation and maintenance of soil fertility, water conservation for crop cultivation and others, proper management of water for drainage, protection from floods, reduction in sediments, enhancement in productivity of all land uses etc. Thus, area of watershed management is very wide and extensive.

Types of watersheds: Watersheds are classified according to their size, drainage, shape and land use pattern. However, size (ha) classification of watersheds and one generalized figure are given here:

1. Macro watershed (> 50,000 ha)
2. Sub-watershed (10,000 to 50,000 ha)
3. Milli-watershed (1000 to10000 ha)
4. Micro watershed (100 to 1000 ha)
5. Mini watershed (1-100 ha)

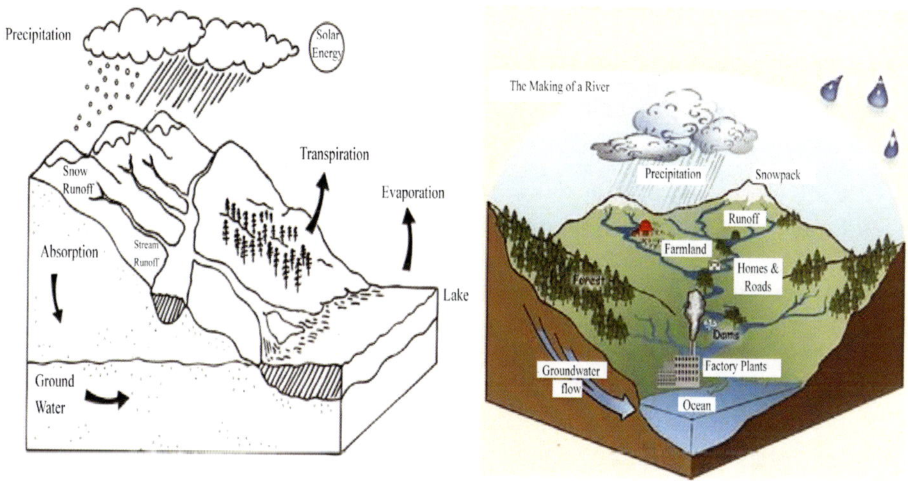

Fig. 5.24: Watersheds

Watershed Management Programme: Watershed management is a coordinated programme. This programme is conducted conjointly by several departments. Hydrology, geology, engineering, soil science, horticulturc, forestry, agronomy, economics etc. are some important departments which are specially required in watershed management. Although there is no border line between their areas of responsibility as each department is independent but inter-related or interconnected with each other.

Integrated Watershed Management Programme (IWMP) was implemented by Department of Land Resources of Ministry of Rural Development. The main objective of IWMP is to restore ecological balance by harnessing, conserving and developing degraded natural resources such as soil, vegetative cover and water after integrating three area development programmes namely; Desert development programme, Drought prone areas programme, Integrated watershed management programme and Integrated wasteland development programme. Later, Pradhan Mantri Krishi Sinchayee Yojana was formulated by combining by ongoing projects namely, accelerated irrigation benefit programme of the Ministry water resources, river development & Ganga rejuvenation, integrated watershed management programme of department of land resources and on farm water management of department of agriculture and cooperation.

Pradhan Mantri Krishi Sinchayee Yojana

https://www.pmksy.gov.in

Steps in Watershed management: Generally, Watershed management programme is under following five points,

1. **Collection of preliminary information and policy determination:** Preparation for any work requires the input of considerable effort by various technicians, scientists and economists. First, based on problems of area concerned, past experiences and national policies watershed is selected. After wards preliminary data of watershed, as nature, size and shape of the watershed, topography of area concerned, soils and their properties of the area, land use pattern, land capability classification, condition of crops, rainfall and water drainage, soil erosion, socio economic condition of inhabitants etc. are collected. These informations are collected by various means and ways like surveys, base map of patwari, aerial photographs, satellite imageries etc.
2. **Necessary organization for the implementation of programme:** Numerous organisations like National Rainfed Area Authority (NRAA), State Level Nodal Agencies (SLNA), Local voluntary organisations and Self-Help Groups, National Level Data Centre and National Portal, District Watershed Development Unit (DWDU) etc. are the part of watershed management programme. In districts where the watershed area is around 25,000 hectares, a DWDU is established which has one Project Manager and 3-4 subject matter specialists to assist in formulating district level strategies in coordination with SLNA.
3. **Research, demonstration and preparation of recommendations:** With the help of preliminary data, an outline of management programme of watershed is prepared. Researches are carried out according to need of scientific area. Within it first development of water resources is done. Subsequently development of community land is done and after adopting erosion control methods improved methods of agriculture, horticulture and forestry are demonstrated to beneficiaries and inhabitants of area concerned. Lastly salient results are recommended for other peoples.
4. **Training of inhabitants of area with the help of trained employees:** The watershed management programme must have well-trained employees to very well perform the responsibilities assigned to them. They should well verse to train the inhabitants about various activities and requirements for the management of their watershed.
5. **Co-operation of local people:** Cooperation of local people in management of watersheds is very much necessary and important. When earnings and

by-products of forests etc. are linked with income of local people, then they will emotionally cooperate in watershed management programme. Therefore, management and conservation of resources of ecosystem should be in such lines that beneficiaries may begin to get economic benefits.

References

गर्ग.ए.सी. अदनानी, एम. ए. एवं अब्राहम, टी पी ;1997. जैविक खाद , भारतीय कृषि अनुसंधान परिषद, नई दिल्ली।

गौड,कृपा शंकर ;1965 भारत की भौगोलिक समीक्षा, हिन्दी प्रचारक पुस्तकालय वाराणसी।

घिल्डियाल, ब्रजनंदन प्रसाद एवं त्रिपाठी, रमेश प्रसाद ;1985 मृदा भौतिकी, प्रकाशन निदेशालय, गो. ब. पंत कृषि एवं प्रोद्योगिक विश्वविद्यालय, पंतनगर ।

जौहरी, भवदीश नारायण एवं मेलकानिया, निरंजन प्रसाद; 1988, ग्रामीण पर्यावरण प्रदूषण एवं रोकथाम, प्रकाशन निदेशालय, गो. ब. पंत कृषि एवं प्रोद्योगिक विश्वविद्यालय, पंतनगर ।

तमहणे, आर पी मोतीरमानी,डी पी एवं बाली, वाई पी ;1970द्ध मिटिटयां उष्ण कटिबन्धीय एशिया में उनका रसायन एवं उर्वरता, प्रंटिस हाल ऑफ इंडिया, प्रा, लि, नई दिल्ली।

त्रिपाठी, रमेश प्रसाद, राठौर तुलसीराम एवं सिंह, हरिहर प्रसाद ;1988 मृदा अपरदन एवं भूमि संरक्षण, प्रकाशन निदेशालय, गो. ब. पंत कृषि एवं प्रोद्योगिक विश्वविद्यालय, पंतनगर ।

परीक, राम प्रयाग; 1985 कृषि में सूक्ष्मजीव निवेशन, प्रकाशन निदेशालय, गो. ब. पंत कृषि एवं प्रोद्योगिक विश्वविद्यालय, पंतनगर ।

भटनागर, ओ के ;1991 भारत की ऊसर मृदाए और उनका सुधार, विस्तार निदेशालय कृषि मंत्रालय भारत सरकार।

भारद्वाज वेंकटेश ;1997 मृदा विज्ञान के मौलिक सिद्धांत, प्रकाशन निदेशालय, गो. ब. पंत कृषि एवं प्रोद्योगिक विश्वविद्यालय, पंतनगर ।

यादव जें एस पी एवं गुप्ता आई.सी. ;1984 ऊसर भूमि का सूधार भारतीय कृषि अनुसंधान परिषद नई दिल्ली ।

यादव जय सिंह पाल शर्मा, अवधेश कुमार एवं राठौर, तुलसीराम ;1987 पर्वतीय एवं वन मृदा प्रबंध प्रकाशन निदेशालय, गो. ब. पंत कृषि एवं प्रोद्योगिक विश्वविद्यालय, पंतनगर ।

रामनाथ वेदालंकार 1991 आर्ष ज्योति, समर्पण संस्थान साहिबाबाद।

रामनाथ वेदालंकार ;सं02007 वैदिक सूक्तियां, प्रकाशन मंदिर गुरूकुल कागडी विश्वविद्यालय, हरिद्वार।

राय माधुरी मोहन ;1986 पौधो के पोषक तत्व एवं उर्वरक उपयोग, प्रकाशन प्रकाशन निदेशालय, गो. ब. पंत कृषि एवं प्रोद्योगिक विश्वविद्यालय, पंतनगर ।

शर्मा, अवधेश कुमार ;1979 प्रायोगिक मृदा एवं भूमि उपयोग सर्वेक्षण, प्रकाशन निदेशालय, गो. ब. पंत कृषि एवं प्रोद्योगिक विश्वविद्यालय, पंतनगर ।

शर्मा, अवधेश कुमार एवं सुमन कुमार ;1990 मृदा सर्वेक्षण प्रकाशन निदेशालय, गो. ब. पंत कृषि एवं प्रोद्योगिक विश्वविद्यालय, पंतनगर ।

12th International Congress of Soil Science (1982) Nonsymbiotic nitrogen fixation and organic matter in tropics, Symposia Papers I, New Delhi.

12th International Congress of Soil Science (1982) Review of Soil Research in India Part I, New Delhi.

12th International Congress of Soil Science (1982) Review of Soil Research in India Part II, New Delhi.

12th International Congress of Soil Science (1982) Souvenir, New Delhi.

A.J. (Tom) van Loon Geological Institute, Adam Mickiewicz University, Maków Polnych 16, 61-606 Poznan, Poland; tvanloon@amu.edu.pl, tom.van.loon@wxs.nl 2.9: Nomenclature of sedimentary rocks

Abrol I.P. (1982) Reclamation and management of salt affected soils, Review of Soil Research in India, part II, New Delhi

Adhikari, S.; Gupta. S.K. and Banerjee, S.K. (1993) Heavy metal content in city sewage and sludge, J. Indian Soc. Sci.., 41:170-172

Agarwal R.R. (1967) Soil Fertility in India, Asia Publishing House, Bombay.

Agnihotrudu, V. and Mithayantha, M.S. (1978) Pesticide Residue –A Review of Indian Work, Rallis India Ltd. Bangalore.

Akihiko Ito and Rota Wagai (2017) Global distribution of clay-size minerals on land surface for biogeochemical and climatological studies. Scientific Data volume 4, Article number: 170103 (2017) Published: 22 August 2017.

AK Priya · 2023, Clean-Up of Heavy Metals from Contaminated Soil by MDPI https://www.mdpi.com › ...

Alexander, M. (1961-1977) Introduction to Soil Microbiology, John Wiley and Sons Inc. New York.

Alexander, T.M. (1972) Soils of India. Fertilizers Association of India, New Delhi.

All India Coordinated Research Project on Long Term Fertilizers Experiments, Annual Report 1983-84 and 1984-85, New Delhi.

Alsaedi S. A., Ibrahim Bakri Razaq and Norrya A Ali. Effect of Soil Textural Classes on the Biological Nitrogen Fixation by Bradyrhizobium Measured by 15N Dilution Analysis. January 2016 DOI: 10.21123/bsj.2016.13.4.0734

Ambasht, R.S. and Ambasht P.K. (1992) Environment and Pollution (an ecological approach), Student's friends & Co. Lanka, Varanasi.

Applications Of GIS In Soil Science - July 21, 2023 (mapscaping.com)

Barrante, James, R. (1977). Physical Chemistry for the life sciences. Prentice Hall, Inc. Englewood Cliffs, New Jersy.

Baver, L.D; Gardner, W.H. and Bardnev, W.R. (1976) Soil Physics, Wiley Eastern Ltd. New Delhi.

Bear, F.E (1976) Chemistry of the Soil, Oxford & IBH Publishing Co. New Delhi.

Beckett, P. H. T. (1964) Studies on soil potassium, I and II. J. Soil Sci.15, 1–8 and 9–23.

Bennett, K. (2016) Abandoned mines- environmental, social and economic challenges. Proceedings of 11 th international conference on Mine Closure, Australian Centre for Geomechanics, Perth, pp. 241- 252.

Bharadwaj, Venkatesh (1999) Fundamentals of soil science Laboratory manual, College of Agriculture, GBPUA&T, Pant Nagar.

Bharadwaj, Venkatesh (2002) Manual for practicals Salt affected and other problem soils, College of Agriculture, GBPUA&T, Pant Nagar.

Bhargava, G.P. and Bhattacharjee, J.C. (1982) Morphology, genesis and classification of salt affected soils, Review of soil Research in India Part II, 508-528.

Bhatia, K.S. and Varna, R.D. (1982) Physico-chemical and erosional behaviour of red and black soils of Bundelkhand Region of Uttar Pradesh. J. Indian. Soc. Soil Sci. 30:523-527.

Bhattacharyya, T., Pal, D.K., Mandal, C., Chandran, P., Ray, S.K., Sarkar, Dipak, Velmourougane, K., Srivastava, A., Sidhu, G.S., Singh, R.S., Sahoo, A.K., Dutta, D., Nair, K.M., Srivastava, R., Tiwary, P., Nagar, A.P. and Nimkhedkar, S.S. (2013) Soils of India: Historical perspective, classification and recent advances in knowledge; A review. Current Science 104, 1308-1323.

Biological Reclamation of Iron Ore Overburdens, Technical Bulletin No. 57 (2016) ICAR-Central Coastal Agricultural Research Institute, Goa.

Biswas, T.D. (1982) Management of soil physical conditions for soil productivity, J. Indian Soc. Soil Sci.

Biswas, T.D. and Mukherjee, S.K. (1987) Text book of soil science. Tata McGraw Hill Publishing Co. Ltd. New Delhi

Bitton, Gabriel and Marshall, Kevenc (1980) Adsorption of micro-organism to surface. John Wiley & Sons, Inc., New York.

Black, C. A. (1973) Soil Plant Relationship, Wiley Eastern Pvt. Ltd. New Delhi.

Brady, N.C. (1988) The Nature and Properties of Soils, Eurasia Publishing House (P) Ltd, New Delhi.

Brunauer, S., Emmett, P.H. & Teller, E. (1938) Adsorption of gases in multimolecular layers. J. American Chemical Society. 60:309

Buol, S.W.; Hole and Mc Cracken, R.J. (1975) Soil Genesis and Classification, Oxford and IBH Publishing Co. New Delhi

Carole Santi, Didier Bogusz and Claudine Franche. March 2013 Biological nitrogen

Central Arid Zone Research Institute, (1988-89) Annual Report, Jodhpur.

Central Board of Irrigation and Power; (1979) Water Research in India, Publication No. 78 (Revised) New Delhi.

Central Soil and Water Conservation Research and Training Institute (1987) Sahastradhara Mine rehabilitation Research Project, Dehradun.

Central Soil and Water Conservation Research and Training Institute (1989) CSWCRTI, An Introduction, Dehradun.

Central Soil and Water Conservation Research and Training Institute, (1987) Conservation ditching for soil and water conservation in low rainfall Vertisol Region, Bull No. I-19/D-15, Dehradun.

Central Soil and Water Conservation Research and Training Institute, (1988) Annual Report, Dehradun.

Chapter I. The arid environments - Food and Agriculture Organizati…

Chaudhary, T.N. and Sandhu, B.S. (1982) Soil temperature and plant growth, Review of Soil Research in India, part I. New Delhi.

Chen, W. (2010) Chlorine nutrition of higher plants: progress and perspectives. fao.org https://agris.fao.org › agris-search

Chhibber, R.K., Ghosh, P.C. and Satyanarayan, K.V.S. (1961) Studies on the physical properties of some of the Himachal Pradesh soils formed on different parent materials in relation to their erodibility. J. Indian Soc. Soil Sci. 9:187-192.

Chopra, S.L. and Shree Ramulu U.S. (1982) Soil pesticide interaction, Review of soil research in India, part I, New Delhi.

Christenson. C.B. (1988) Soil fertility and fertilizer management in semi-arid tropical India, International Fertilizer Development Centre, Alabama.

Cline, M.G. (1949) Basic principles of soil classification, Soil Sci.67, 81-91.

Coffey, George Nelson (1912) Soil Map of the USA. http://ui.adsabs.harvard.edu.

Critical Review of Different Types of Remedial Measures for Landslides (trb.org) trid.trb.org/view/1698426

D.K. Pal, C. Mandal, P. Chandran, S.K. Ray, Dipak Sarkar, K. Velmourougane, A. Srivastava, G.S. Sidhu, R.S. Singh, A. K. Sahoo, D. Dutta, K. M. Nair, R. Srivastava, P. Tiwary, A. P. Nagar and S.S. Nimkhedkar and T. Bhattacharyya. Soils of India: historical perspective, classification and recent advances. Current Science, Vol. 104, NO. 10, 25 May 2013 130

Dalal, R.C. (1977) Soil organic Phosphorous, Advances in Agronomy, 29:83-113.

Dargan, K.S. (1979) Agronomic and cultural practices for crop production in salt affected soils. Curr. Agric.3:1-20.

Das, D.K. (2017) Introductory Soil Science. Kalyani Publishers, New Delhi.

Das, D.K. and Das S.C. (1970) Study of silt and sand mineralogy in relation to the genesis of clay minerals in some black and brown soils, J. Indian Soc. Soil Sci.18- 473-478.

Das, S.C. and Chatterjee. R.K. (1982) Chemical composition of soils of India, Review of Soil Research in India, Part I, New Delhi.

David Dent and Edward Cocking (2017) Establishing symbiotic nitrogen fixation in cereals and other non-legume crops: The Greener Nitrogen Revolution Agriculture & Food Security, volume 6, Article number: 7 (2017)

Deckers, J; Spaagaren, O. and Dondeyne, S. Soil Survey as A Basis Land Evaluation. http://www.eolss.net

Definitional dictionary of Soil science. Commission for scientific and technical terminology, New Delhi, 1977.

de Gruijter, J. J. (1977). Numerical classification of soils and its application in survey. [external PhD, WU]. Pudoc. https://edepot.wur.nl/202256

Deshpande, S.B.; Farenbacher, J.B. and Beavers, A.H. (1971a) Mollisols of Tarai Region of Uttar Pradesh, Northern India, Part I, Morphology and mineralogy, Geoderma 6:179-193.

Deshpande, S.B.; Farenbacher, J.B. and Roy, B.W. (1971b) Mollisols of Tarai Region of Uttar Pradesh, Northern India, Part II, Genesis & classification, Geoderma 6:195-201.

Dhir, R.V. and Jain, S.V. (1982) Morphology, genesis and classification of desert soils, Review of soil Research in India Part II, New Delhi, 474-484.

Digar, S and Barde N.K. (1982) Morphology, genesis and classification of red and laterite soils, Review of Soil Research in India Part II New Delhi.468-507.

Directorate of Extension (1982) Soil Survey and their utilities, Ministry of Agriculture, New Delhi.

Donahue, R.L, Miller, R.W. and Shickluna, J.C. (1987) Soils: An introduction to soils and plant growth. Prentice hall of India Pvt. Ltd. New Delhi.

D P Kelly, I R McDonald, A P Wood . Proposal for the reclassification of Thiobacillus novellus as Starkeya novella gen. nov., comb. nov., in the alpha-subclass of the Proteobacteria, Int J Syst Evol Microbiol. 2000 Sep. PMID: 11034489 DOI: 10.1099/00207713-50-5-1797

e-Krishi Shiksha ecoursesonline.iasri.res.in/mod/page/view.php?id=124941 Remote Sensing & GIS Applications: Lesson 4 Sensors ...

e-Krishi Shiksha ecoursesonline.iasri.res.in/mod/page/view.php?id=125149 Soil Mechanics: LESSON 15. Factors Affecting ...The Causes and Effects of Soil Compaction – Soil Erosion.com

Eco-friendly Use of Abandoned Coal Mineshttps://www.pib.gov.in › Press Release Page 07-Feb-2022

ecoursesonline.iasri.res.in/mod/page/view.php?id=149590 (2011) Organic Farming NADEP Method.

ecoursesonline.iasri.res.in/mod/page/view.php?id=2103 S&WCE: Lesson 7 Gully Erosion

Ecoursesonline.iasri.res.in/mod/page/view.php?id=2149 Lesson 27 Grassed Waterways. Grassed waterways

Erick K. Towettab Keith D. Shepherdb Jerome E. Tondohc Leigh A. Winowieckid Tamene Lulsegede Mercy Nyamburab Andrew Silab Tor-G.Vågenb Georg Cadischa 2015 Check https://www.sciencedirect.com/science/article/pii/S2352009415300018

FAO.org https://www.fao.org/3/T1696E/t1696e06.htm Factors and Mechanism of Crust Formation Chapter 5

Fertilizer Association of India (1988-89) Fertilizer Statistics, Part II -8.

Foth, H.D. (1984). Fundamentals of Soil Science. John Wiley & Sons, New York.

Freundlich, H. (1926) Colloid and capillary chemistry. Methaen, London.

Frédéric Zakhia and Philippe De Lajudie (2001). Taxonomy of rhizobia

GAEC 6: Soil and Organic Matter – Maintenance January 2021 For further information please contact: Welsh Government Natural Resources Wales.

Gangwar, M.S.; Sachan, R.S. and Rathore, T.R. (1982) Morphology, genesis and classification of Tarai soils. Review if soil Research in India, part II, 535-551.

Gapon, E.N. (1933) J. General Chemistry, U.S.S.R., 3:144

Garrison, Sposito (1981) The thermodynamics of soil solutions. Oxford Clarendon press, New York.

Gaur, Y.D. and Sen, A.N. (1979) Cross inoculation group specificity in Ciser-Rhizobium symbiosis. https://doi.org/10.1111/j.1469-8137.1979.tb02305.x

George Joseph (2009) Fundamentals of remote sensing. Second edition, Universities Press (India) Pvt. Ltd. Hyderabad, India.

Ghildiyal, B.P. (1969) Indian J. Agric. Sci. 37:757.

Ghosh, S.K.; Raychaudhuri, S.P. 1974. Distribution of clay minerals in Indian soils. Bulletin, Indian Society of Soil Science 9: 143-164.

Ghosh, S.K. and Kapoor, B.S. (1982) Clay minerals in Indian soils, review of Soil Research in India Part II New Delhi.

Girmay Kalayu 2019 Phosphate Solubilizing Microorganisms: Promising Approach biofertilizers…https://www.hindawi.com/journals/ija/2019/4917256

GIS Data Models:- Discovering New Insights From GIS Data Models - (geolearn.in)

Glasstone, Samuel (1961). Text Book of Physical Chemistry. Second edition, 14th printing, D. Van Nostrand Co. Inc. Princeton, New Jersy.

Glasstone, Samuel (1975). An Introduction to Electrochemistry. Affiliated East West Press Pvt. Ltd. Madras.

Govindrajan, S. V. (1965) World Soil Resources, Report No.26.FAO/UNESCO, Rome.

Govindrajan, S. V. and Gopala Rao, H.G. (1978) Studies of Soils of India, Vikas Publishing House, New Delhi.

Gupta, S.K.; Bharadwaj, Venkatesh and Sharma, A.K. (1994) Lead (Pb) Status of Mollisols as influenced by physico- chemical properties, drainage condition and parent material of soil. 15th World Congress of soil Science, July 10-16, 1994 Acapulco, Mexico, 174-175.

Gupta J.P.; Rao, G.G.S.N.; Gupta, G. N. and Raman Rao. B.V. (1983) Soil drying and wind erosion as affected by different types of shelter belts planted in the desert region of western Rajasthan, India. J. Arid Environ.6:53-58.

H.L.S. Tandon Soil Nutrient Balance Sheets in India: Importance, Status, Issues, and Concerns www.ipni.net/.../$FILE/bc-india_Nov 2007_p15.pdf

Hansford T. Shacklette and Josephine G. Boerngen through U.S. Geological Survey Professional Paper 1270 (1984).

Haridasan, M. and Chhibber, R.K. (1971) Effect of physical and chemical properties on the erodibility of some soils of the Malwa Plateau. J. Indian Soc. Soil. Sci. 19:293-298.

Hartemink, A.E.; Krasilnikov, P and Bockheim, J.G. (2013) Soil maps of the world, Geoderma, 201-208:256-267.

Hassani Amirhossein, Adisa Azapagic and Nima Shokri Nature Communications Global predictions of primary soil salinization under changing climate in the 21st century volume 12, Article number: 6663 (2021)

http://commons.wikimedia.org/wiki/File:Saltation-mechanics.gif)

http://www.lbk.ars.usda.gov/wewc/rweq/readme.htm

https://agrigoaexpert.res.in-images-success story

https://education.nationalgeographic.org/resource/geographic-information-system-gis/

https://electronicsdesk.com/remote-sensing.htm What is Remote Sensing? Principle, Stages of Operation, …

https://environmentgo.com/effects-of-landslide 10 Positive and Negative Effects of Landslide - Environment Go!

https://iiss.icar.gov.in/eMagazine/v2i1/11.pdf· PDF file Management Strategies for Waterlogged Soils …

https://microbenotes.com/microbial-degradation-of-hemicellulose Microbial degradation of hemicellulose (Enzymes, Steps, …

https://onlinelibrary.wiley.com/doi/full/10.1111/gcb.16296

https://www.agric.wa.gov.au/measuring-and-assessing-soils/what-soil-organic-carbon December 2020

https://www.agriculture.com/crops/cover-crops/new-life-for-saline-soil#:~:text=Plant%20cash%20crops%20th

https://www.climate-policy-watcher.org/nitrogen-removal/the influence...09/03/2021 The Influence of Temperature on the Nitrification Rate - Nitrogen …

https://www.earthreminder.com/organic-compost-benefits-examples-how-to-use-it

https://www.fao.org/3/AD082E/AD082e01.htm 185. PRINCIPLES OF GULLY CONTROL 236. (Table 4: State/UT wise Coastal length in India, NHO 2011, Status Report on Coastal Protection and Development in India, Page 7, Central Water Commission, New Delhi December, 2016).

https://www.fao.org/3/T1696E/t1696e10.htm Chapter 9

https://www.fao.org/land-water/land/land-governance/land-resources...

https://www.feedipedia.org/node/565. Feedipedia. Azolla Animal feed resources information system Downloaded: 1522.

https://www.google.com/search?q=puerto+rican+terrace

https://www.howden.com/en-gb/articles/flue-gas-desulfurization-acid-rain (2020)

https://www.mdpi.com/2073-445X/11/6/840 Land | Free Full-Text | Impact of Soil Sealing on Soil …Web June 3, 2022

https://www.nature.com/articles/s41893-023-01131-7

https://www.phytojournal.com/archives/2020/vol9issue5/PartA… · PDF file Effect of long term zero tillage on soil organic carbon and

https://www.researchgate.net/figure/Reflection-by-vegetation-water-and-soil-Spectral-signatures_fig2_322095499

https://www.slideshare.net/ravikumar5191/stem-nodulating-74301492 Ravi Kumar Hoogar and Pal B. 04/04/2017 Stem Nodulating Bacteria and their Application in Agriculture

https://www.soilmanagementindia.com/soil/water-logging-types-and...

https://www.thegef.org/topics/land-degradation Land Degradation | Global Environment Facility

https://www.theguardian.com/environment/2021/jun/04/soils-great...WebJun 4, 2021

ICAR-IASRI 2013 http://ecoursesonline.iasri.res.in › mod › page › view AE: Lesson 8. Charges on Soil Colloids

ICFRE, 2020. A Handbook on Mine Reclamation. Indian Council of Forestry Research and Edu[ca]tion, Dehradun, India.

Indian Council of Agricultural Research (2011) Hand Book of Agriculture. New Delhi.

Indian Society of Soil Science (1974) Mineralogy of soil clays and clay minerals, Tech. Bull. No.9, New Delhi.

Indian society of soil science (1976) Acid soils of India, their genesis, characteristics and management, Bull.No.11, New Delhi.

Indian society of soil science (1984) Soil science in India. Tech Bull. No.14., New Delhi.

International Rice Research Institute (1978) Soils and Rice, Philippines.

International Rice Research Institute (1985) Wetland soils: Characterization, classification and utilization, Philippines.

IUSS Working Group WRB. 2022. World Reference Base for Soil Resources. International soil classification system for naming soils and creating legends for soil maps. 4th edition. International Union of Soil Sciences (IUSS), Vienna, Austria. Citation: IUSS Working Group WRB. 2022. World Reference Base for Soil Resources. International soil classification system for naming soils and creating legends for soil maps. 4th edition. International Union of Soil Sciences (IUSS), Vienna, Austria.

Jan Środoń, Institute of Geological Sciences PAN, Senacka 1, Krakow, 31–002 Poland; e-mail: ndsrodon@cyf-kr.edu.pl

Jenny, H. (1941) Factors of Soils Formation. McGraw – Hill Book Company. New York

jnkvv.org/PDF/13042020102823Compacted Soil- RK Thakur.pdf · PDF file Compacted soils: Characteristics, causes, management and ...

Joffe, J.S. (1949) The A B C of soils. Pedology Publication, New Brunswick, N.J.

Journal of the Indian Society of Soil Science, Vol. 64 (Supplement) pp S41-S52 (2016) The 27th Dr. S.P. Raychaudhury Memorial Lecture* Soil Diversity in India Tapas Bhattacharyya Dr. B.S. Konkan Krishi Vidyapeeth, Dapoli, Ratnagiri, 415 712, Maharashtra.

Kandrika Sreenivas, V.K. Dadhwal, Suresh Kumar, G. Sri Harsha, Tarik Mitran, G. Sujatha, G. Janaki Rama Suresh, M.A. Fyzee, and T. Ravisankar. Digital mapping of soil organic and inorganic carbon status in India https://doi.org/10.1016/j.geoderma.2016.02.002 Get rights and content

Kanwar, J.S. (1978) Soil Fertility, theory and practice. ICAR, New Delhi.

Kanwar, J. S. (Ed.) (1976) Soil fertility: Theory and Practice, Indian Council of Agriculture Research, New Delhi.

Khemani, L.T.: Momon, G.A.; Prakash, Rao, P.S. Safai, P.D., Singh, G and Kapoor R.K. (1989) Spread of acid rain over India. Atoms. Environ.23:757-762.

Khemani, L.T.: Momon, G.A.; Prakash, Rao, P.S. Safai, P.D. Singh, G.; Chatterjee, R.N. and Prakash, P. (1989) Long term effects of pollutants on pH of rainwater in north India. Atmos, Environ. 23:753-756.

Kirsch. Helmut (1968) Applied Mineralogy. Chapman and Hall Ltd. New Fetterhane.

Klingelbiel, A.A. and Montgomery, P.H. 1961. Land capability classification, Soil Conservation Service, USDA Hand Book, Washington, DC.

Kolay, A.K. (1993) Basic Concepts of soil science. Wiley Eastern Ltd. New Delhi.

Konhke, H. (1968) Soil Physics. Tata McGraw – Hill Publishing Co. Ltd. Bombay.

Kovacs, M. (1985) Pollution Control and Conversion, Ellis Horward, Ltd. England.

Krishnamoorthy, C., Davis, L.E. and Overstreet, R. (1948) Science, 108:439.

Kuhad, M.S.; Malik, R.S.; Singh, R. and Singh Anoop. (1989) studies on mobility and accumulation of heavy metals in agricultural soils receiving sewer water irrigation. J. Indian Soc. Soil Sci.37:290-294.

Kuhad, M.S.; Malik, R.S.; Singh Anoop. And Dahiya, I.S. (1989) Background levels of heavy metals in agricultural soils of Indo –Gangetic plains of Haryana. J. Indian Soc. Soil Sci. 37:700-705.

Ladha, J.K.; Pareek, R.P. and Baker, M. (1990) Stem nodulating legume Rhizobium Symbiosis and its agronomic use IRRI report.

Lagerwerff, J.V. and Bolt, G.H. (1959) Soil Science, 87:217

Landey, R.J.; Hirekerun, L.R. and Krishnamoorthy, P. (1982) Morphology, genesis and classification of black soils. Review of Soils Research in India. Part II New Delhi 484-497.

Langmuir, I. (1918). The adsorption of gases on plane surfaces of glass, mica and platinum, J. Am. Chem. Soc. 40: 1361-1382.

Lascar, S. and Govindrajan, S.V. (1980) Erodibility of the soils of Tripura, Indian J. Agric. Sci. 50:161-167.

Li Thuo Xin (1982) Nitrogen fixation by Azolla in rice fields and its utilization in China. Symposium paper I, 12th International Congress of Soil Science, New Delhi, 83-95.

Loehr, R.C. (1977) Pollution Control for Agriculture, Academic Press Inc. New York.

M. L. Jackson, Y. Hseung, R. B. Corey, E. J. Evans and R. C. Vanden Heuvel Weathering Sequence of Clay-size Minerals in Soils and Sediments II. Chemical Weathering of Layer Silicates. First published: 01 January 1952

Mahadevan, T.N.; Negi, P. and Meenakshi, V. (1989). Measurement of elemental composition of aerosol matter and precipitation from a remote background site in India. Atmos. Environ. 23:869-874.

Mandal, S.C.; Sinha, H. (1975). Acid soils of India and Liming, ICAR, Tech. Bull .(Agric.) No. 51

Mann, H.S. and Dhir, R.P. (1982) land use management and practices in arid zone, Review of Soil Research in India, Part II, New Delhi. 655-665.

Marbut, C.F. (1928) A scheme of soil classification, Proceedings paper, Ist International Congress of Soil Science, Commission V, 1927, 31-33.

Marbut, C.F. (1935) Soils of the United States, part III, In Atlas of America agriculture, USDA, US Govt. Print office, Washington DC,1-98.

Marc Hufty Professor, Interdisciplinary Programmes The Graduate Institute, Geneva. The consequences of abandoned mines - Global Challenges https://globalchallenges.ch/issue/6/abandoned-mines-the-scars-of-the-pas

MarioCorrochano-Monsalve,aCarmen González-Murua,aAdriánBozal-Leorri,aLuisLezama andbBeñatArtetxeb Mechanism of action of nitrification inhibitors based on dimethylpyrazole: A matter of chelation. Science of the Total Environment, Volume 752, 15 January 2021, 141885 https://doi.org/10.1016/j.scitotenv.2020.141885Get rights and content

Maron, S.H. and Prutton, C.F. (1972) Principles of Physical chemistry. Oxford & IBH Publishing Co. Pvt. Ltd. New Delhi.

Mavi, H.S. (1994) Introduction to Agrometeorology, Oxford & IBH Publishing Co. Pvt. Ltd New Delhi.

Mehra R.K. (2006) Text book of soil science, Published by: Directorate of Information and Publications of Agriculture, ICAR, New Delhi.

Michael A. Ruggiero, Dennis P. Gordon, Thomas M. Orrell, Nicolas Bailly, Thierry Bourgoin, Richard C. Brusca Thomas Cavalier-Smith, Michael D. Guiry, Paul M. Kirk Published: April 29, 2015. A Higher-Level Classification of All Living Organisms https://doi.org/10.1371/journal.pone.0119248

Michael Golden, Erika Micheli, Craig Ditzler, Hari Eswaran, Phillip Owens, Ganlin Zhang, Alex McBratney, Jon Hempel, Luca Montanarella and Peter Sehed (2010) The 19th World Congress of Soil science, Soil Solutions for a Changing World, 1-6 August, 2010, Brisbane, Australia.

Miller, R.W. and Donahue, R.L. (1992). Soils: An introduction to Soils and Plant Growth. Prentice –Hall of India Pvt. Ltd New Delhi.

Milovsky A.V. and Kononov, O.V. (1985) Mineralogy, Mir Publisher, Moscow.

Minister of Supply and Services Canada (1976) Glossary of Terms in Soil Science, Agriculture Canada Publication 1459/E

Mishra, A.S.; Bhatt, P.N. and Venkataraman, C. (1988) soil conservation, Prospects and Challenges, CSWCRTI, Dehradun.

Mishra, Gopal, (2012). Unified Soil Classification System. https;//theconstructor.org.

Mitra, R.P. (1959) some recent developments in field of clay research. J. Indian Soc. Soil Sci. 7:207-211.

Mohr, E.C.J. and Van Baren, F.H. (1954) Tropical Soils: A critical study of soil genesis as related to climate, rock and vegetation. The Royal Tropical Institute, Amsterdam., Inter science, New York.

Murthy, R.S.; Hirekerur, L.R.; Deshpandey, S.B. and B.V. Venkata Rao (1982) Benchmark Soils of India, NBSS & LUP, Nagpur.

Murthy R.S. (1982) Souvenir, 12th International Congress of Soil Science, New Delhi.19-23.

Narayana, N. and Shah, C.C. (1966) Physical Properties of Soils. Manaktalas Bombay.

Nature of mixed layer clays and mechanisms of their formation and alteration. Annual review of earth and planetary sciences 27:19-53. May 1999. https://doi.org/10.1146.

Niir Board (2004). The Complete Technology Book on Bio-Fertilizer And Organic Farming.

Noller, B.N. and Chadha, M.S. (1990) Chemistry and the environment. Common Wealth Science Council, London.

Om Prakash Pandey (29 April 2020) Geodynamic Evolution of the Indian Shield: Geophysical Aspects pp 331-349, Indian Crust 1. CSIR-National Geophysical Research Institute, Hyderabad India. Part of the Society of Earth Scientists Series book series (SESS)

Organic Farming: Composting - Tamil Nadu Agricultural University www.agritech.tnau.ac.in/org_farm/orgfarm_composting.

Panda, H. (2011). Manufacture of Biofertilizer and Organic Farming.

Peiman Zandi,a JianjunYang, abKatarzyna, MożdżeńcBeata and Barabasz-Krasnyc 2020, Chapter Five - A review of copper speciation and transformation in plant and soil/wetland systems ,Advances in Agronomy Volume 160, Issue 1, Pages 249-293.

Penn State Extension https://extension.psu.edu › soil-crusting 20-Apr-2022

Pereira, Ana L. (2017) The Unique Symbiotic System between a Fern and a Cyanobacterium, Azolla-Anabaena azollae: Their Potential as Biofertilizer, Feed, and Remediation. Submitted: March 29th 2017 Reviewed: July 25th 2017 Published: December 20th 2017 DOI: 10.5772/intechopen.70466

Planning Tank https://planningtank.com/.../types-of-remote-sensing Types of Remote Sensing | Planning Tank

Poonia, S.R. (1998) Sorption/exchange of some nutrients and non-nutrient cations in soils. J. Indian Soc. Sci., 46:505-528.

Poonia, Shefali (2011) RHIZOBIUM: A Natural Biofertilizer. International Journal of Engineering and Management Research Vol. 1, Issue-1, November-2011.

Pradhan Mantri Krishi Sinchayee Yojana https://www.pmksy.gov.in

Rajit Sengupta published: Friday 27 August 2021 Land degradation in India hurts farmers and forest dwellers the most (downtoearth.org.in)

Rajpal (1992) Some aspects of cation exchange equilibrist in soils. J. Indian Soc. Soil Sci., 40:628-647.

Rao, M.V.R (1963) Identification of clay minerals in some Indian soil clays by X-ray diffraction. J. Indian Soc. Soil Sci. 11:321-323.

Rathinasamy, A. and Saliha, B.B. (2018) Fundamentals of Soil Science. Scieentific Publishers, Jodhpur.

Rattan, Lal (1990) soil erosion in the tropics: Principles and Management, Mc Graw – Hill Inc. New York.

Reclamation of back filled bauxite mine areas through ...https://www.teriin.org/project/reclamation-back-filled-bauxite-mine...

Reduced tillage in organic farming affects soil organic carbon stocks in temperate Europe – www.ScienceDirect.com

Roonwal, G.S. and Garalapuri, V.N. (1982) Mineralogy of coarse fraction of soils, Review of Soil Research in India, Part II, New Delhi, 711-717.

Ropera, M. M. and Guptab, V. V. S. R. (2016) Enhancing Non-symbiotic N2 Fixation in Agriculture a CSIRO Agriculture, Private Bag No. 5, Wembley WA 6913 Australia CSIRO Agriculture, PMB No. 2, Glen Osmond SA 5064 Australia Received: November 6, 2015 Revised: March 24, 2016 Accepted: April 2, 2016. The Open Agriculture Journal Content list available at: www.benthamopen.com/TOASJ/ DOI: 10.2174/1874331501610010007

Ruggiero, Michael A.; Gordon, Dennis P.; Orrell, Thomas M.; Bailly, Nicolas; Bourgoin, Thierry; Brusca, Richard C.; Cavalier-Smith, Thomas; Guiry, Michael D.; Kirk, Paul M.; Thuesen, Erik V. (2015). "A higher level classification of all living organisms". PLOSONE. 10 (4): e0119248. Bibcode:2015PLoSO..1019248R. doi:10.1371/journal.pone.0119248. PMC 4418965. PMID 25923521

Russell's Soil Conditions and Plant Growth (Eleventh Edition). Edited by Alan Wild. Harlow: Longman Scientific and Technical (1988)

Sandeep Kr Tiwari Updated 20 July, 2022 12:46 PM IST Published on 19 July, 2022 5:52 PM IST https://krishijagran.com/agripedia/how-to-prepare-nadep-compost/

Sanoria, C.L.; Singh, K.L.; Ramamurthy, K and Maurya, B.R. (1982). Field trials with Azospirillum brasilense in an Indo-Gangetic alluvium. J. Indian soc. Soil Sci. 30:208-209.

Sapkota, Anupma, November 13, 2020 Microbial degradation of cellulose (Enzymes, Steps, Mechanisms) https://microbenotes.com ›

Sapkota Anupama November 17, 2020 Microbial degradation of lignin (Enzymes, Steps, Mechanisms) https://microbenotes.com ›

Sarma, U.A.K. and Sidhu, P.S. (1982) Genesis and transformation of clay minerals, Review of Soil Research in India, Part II, New Delhi, 713-724

Schofield, R.K. (1955). Can a precise meaning be given to 'Available Soil Phosphorus'? Soils Fertil.18:373-375.

Schofield, R.K. and Taylor, A.W. (1955). The measurement of soil pH. Soil Science Society America Proceedings.19:164-167.

Sehgal, J. (1996) Pedology: Concepts and Applications, Kalyani Publishers, New Delhi.

Sehgal J, Sharma, P.K. and Vadivelu, S. (1987) Field Manual, Soil Resource Mapping of different states in India. Tech. Bull. NBSS Publ. 13, NBSS & LUP, Nagpur, 73 p.

Sethunathan, N. (1982) Biodegradation of agrochemicals in soil environment, Review of soil Research in India, Part I, New Delhi269-277

Sharma, K.N.; Singh, Bijay and Rana, D.S. (1980) Fertility status of a Tolewal loamy sand after six years of its use under nine multiple cropping systems. J. Indian Soc. Soil Sci. 28: 173-177.

Sharma, R.R. and Datta Biswas, N.K. (1972) Erodibility of hill soils of Sutlej catchment area in Himachal Pradesh. Indian J. Agric. Sci. 42:161-169.

Shweta Nandanwar, Yogesh Yele, Anil Dixit, Dennis Goss-Souza, Ritesh Singh, Arti Shanware and Lalit Kharbikar (April 8th 2020). Effects of Pesticides, Temperature, Light, and Chemical Constituents of Soil on Nitrogen Fixation, Nitrogen Fixation, Everlon Cid Rigobelo and Ademar Pereira Serra, Intech Open, DOI: 10.5772/intechopen.86128.

Simonson Roy, W. (1967) Outline of a generalized theory of soil genesis, in selected papers in soil formation and classification. SSSA special publication series No.1 Madison, Wisconsin, USA.

Singh, A. (1967). Long-te'm effects of green man"res in subtropics, Indian J Agric Sci. 37:226-2³3.

Singh, H.P. (1982) Management of desertic soils, Review of soil research in India Part II, New Delhi,676-688.

Singh, Vijay (2018) https://www.researchgate.net/publication/331983832_Integrated_Plant_Nutrient_Management

Singh N.T. (1989) Salt affected soils: Some reflections on the state of knowledge J. Indian Soc. Soil Sci. 37:620-644.

Singh P.K. (1982) Azolla as an organic nitrogen fertilizer for medium and low land rice, Review of Soil Research in India, Part I, New Delhi,236-242.

Singh Satya Shila, Upadhyay Ram Sanmukh and Mishra Arun Kumar (2008) Physiological interactions in Azolla-Anabaena system adapting to the salt stress. Original articles Journal of Plant Interactions Volume 3, 2008 - Issue 3 Pages 145-155 | Received 10 Apr 2008, Published online: 17 Sep 2008

Sinha, A.K. and Ghildiyal, B.P. (1982) Soil Aeration, Review of Soil Research in India, Part I, New Delhi 77-80.

Siva Kumar, M.V. K. et al. (2004) Satellite remote sensing and GIS applications in Agricultural meteorology. WMO. Switzerland.

Smart P., Maisonneuve H. and Polderman A. (eds) Science Editors' Handbook European Association of Science Editors. www.ease.org.uk Written 2003, revised 2013© The authors, 2013 Andrew Alden Updated January 23, 2020 Densities of Common Rocks and Minerals.

Soil Orders | Soil Genesis and Development, Lesson 6 - Global Soil Resources and Distribution - passel (unl.edu)

Soils: Their Chemistry and Fertility in Tropical Asia R. V. Tamhane, D. P. Motiramani and Y. P. Bali in Collaboration with R. L. Donahue, New Delhi: Prentice–Hall of India (1965) pp. 475.

Sokal and Sneath (1963) Principles of Numerical Taxonomy, and Principles of Numerical Taxonomy, 1973, W.H. Freeman, San Francisco.

Sood, M.C. and Chaudhary, T.N. (1980) Soil erosion and runoff from a sandy loam soil in relation to initial clod size, tillage time, moisture and residue mulching under simulated rainfall. J. Indian Soc. Soil Sci. 28: 24- 27.

Spaaragaren, O.C. (1994) world reference Base for soil Resources. Food and Agriculture Organisation, Wageningen/Rome.

Sparks, Donald, L. (1986) Soil Physical Chemistry, CRC Press Inc. Boca Raton, Florida.

Stevenson, F.J. (1986) Cycles of Soils. John Wiley, New York.

Subha Rao, N.S. (1982) Advances in Agricultural Microbiology. Oxford and IBH Publishing Co. Pvt. Ltd. New Delhi.

Subha Rao, N.S.; Sen, A.N. and Dadarwal, K.R. (1982) Rhizobium research in India, Review of Soil Research in India. Part I. New Delhi.211-224.

Tan, Kim H/ (1993) Principles of Soil Chemistry 2nd Ed. Marcel Dekker, Inc. New York.

Tapas Bhattacharyya, Dilip Kumar Pal, Champa Mandal and Mariappan Velayutham Organic carbon stock in Indian soils and their geographical distribution September 2000 Current Science 79(5)

Tarquinio Mateus Magalhães. January 2023, Trees in agricultural landscapes maintain soil organic carbon following miombo woodland conversion to shifting cultivation. Geoderma Volume 429, 1 116241.

Tejwani, K.G. and Bharadwaj, S.P. (1982) Soil and water conservation research, Review of Soil Research in India Part II New Delhi, 608-621.

The Grid News https://www.offthegridnews.com/survival-gar…6 Effective Methods To Replenish Nutrients In Your Soil WebJun 12, 2013 ·

The Statesman Tuesday, 31 January, 2023 Acid Rain: Causes, consequences and control of the disaster: SNS | New Delhi | July 25, 2022 7:20 pm

Tisdale, S.L. and Nelson, W.L. (1970) Soil Fertility and Fertilizers. Collier MacMillan Ltd. London.

Trends in Plant Science Volume 19, Issue 12, December 2014, Pages 757-763 Parasponia: a novel system for studying mutualism stability Author links open overlay panel Jocelyn E.Behm1ReneGeurts2E. TobyKiers1https://doi.org/10.1016/j.tplants.2014.08.007Get rights and content Sarthaks.com - https://www.sarthaks.com/892258/describe-the-modes-of-biological-nitrogen-fixation

Tripathi, B.R.; Prasad, R.N. and Bishnoi, S.K. (1982) Acid soils and their management, Review of Soil research in India, Part II, New Delhi. 597- 607.

Twenty Fifth Plus Acid-Loving Plants For Your Garden • Insteading Source: https://insteading.com/blog/acid-loving-plants/ 270. https://www.fao.org/3/cb4894en/online/src/html/chapter-13-4.html Technologies for remediating polluted soils - Food and Agriculture ...

Umit Baris Kutman (2023) Mineral nutrition and crop quality - https://www.sciencedirect.com/science/article/pii/B9780128197738000204

USDA (1978) Soil Taxonomy, Agriculture Handbook No. 436 Soil Conservation Service, US Department of Agriculture and Soil Taxonomy, Twelfth Edition 2014)

V.A. Kovda and E.V. Lobova, editors, Geography and classification of soils of Asia /. (Geografiya i klassifikatsiya pochv Azii); translated from Russian [by A. Gourevitch] Published by Jerusalem: Israel Program for Scientific Translations, 1968.

Vanselon, A.P. (1932) Soil Science: 33:95158.

Varshney, M.C.; Pillai, P Balakrishna (2004) Text Book of Agricultural Meteorology published by ICAR New Delhi.

Venkataraman, G.S. (1982) Nitrogen fixation by blue green algae and its economic importance. Symposia Papers I, 12th International Congress of Soil Science, New Delhi. 69-82.

Venkatesh Bharadwaj (1998) Krishijanit kachara evam uska prabandhan,Varshik Sanghosthi, Gramin evam Krishi vikas samiti Haldwani. 19-21.

Verma G.S. (1989) Background trends of pH of precipitation over India. Atmos. Environ. 23:747-751.

Wadia, D.N. (1953) Geology of India, MacMillan & Co. London.

Water harvesting Methods of arid and semiarid regions - Agri learner www.agrilearner.com/water-harvesting-methods

White, R.E. (1987) Introduction to the principles and practices of soil science. Oxford Blackwell Scientific.

Wild, Alan (1988) Russell's Soil Condition and Plant Growth. ELBS/Longman, U.K.

World Bank (1988) Vetiver Grass: A Vegetative Method of Soil and Moisture Conservation, IInd Edition.

World's soils 'under great pressure', says UN pollution report | Soil ...

www.cs.mcgill.ca › ~rwest › Wikipedia › wpcd › Nitrogen fixation. HTML. The azolla foundation. org/azolla/the-azolla-superorganism

www.fao.org. What is Conservation Agriculture? | Conservation Agriculture | Food and Agriculture Organization of the United Nations (fao.org)

www.fao.org/3/a0100e/a0100e04.htm

www.fao.org/3/t0122e/t0122e03.htm

www.fao.org/soils-portal/soil-degradation-restoration

www.frontiersin.org/articles/10.3389/fsufs.2020.534187/full Managing Grazing to Restore Soil Health, Ecosystem Function, and Ecosystem Services

www.isro.gov.in/earth-observation/land-degradation Land degradation map of India (generated using LISS-III data of 2015-16) by ISRO Land Degradation - ISRO

Index

A

Abandoned mines 15, 371, 374, 452, 453, 431, 532, 533

Absorption of nutrients by plants 176, 291, 292

Accessory minerals present in coarse fraction of soils 242, 248

Acid rain 256, 423, 424, 428, 429, 416, 549

Active absorption 291, 293

Adsorption by soil 158

Adsorption isotherms 159

Adsorption of ions 176, 272, 438

Agricultural chemicals 220, 324, 397, 426, 433, 518

Agricultural lime 504, 505

Agricultural wastes 426, 427, 507, 508

Air exchange between soil and atmosphere 114

Algae 8, 12, 13, 23, 189190, 191, 193, 195, 198, 203, 208. 209, 210, 211, 222, 223, 224, 255, 438, 443, 445, 497, 519,

Alluvial 4, 19, 50, 73, 77, 81, 82, 83, 89, 94, 95, 150, 151, 180, 280, 311, 379, 386, 388, 389, 400, 415, 423

Alluvial soils 50, 73, 81, 82, 83, 180, 280, 388, 389, 415, 423

Aluminium octahedron 262, 303

Ammonification 334, 335

Ammonium fixation 300, 301

Amount of nutrients in plants 288

Anabaena azollae 224, 225, 226

Analysis of soil samples 39, 295, 298

Anion exchange in soils 184

Approaches of soil loss measurement 397

Areas affected by gully erosion 388

Arid zones 447, 448, 449

Asymbiotic nitrogen fixation 207, 209, 211

Atmospheric effects 459

Atmospheric windows 459

Autotrophic cyanobacteria 207, 208

Azolla 223, 224, 225, 26, 227, 445

Azolla pinnata 223, 224, 225, 227

B

Base saturation of soil 185, 306

Basic concepts of remote sensing 456

Behaviour of heavy metals in soils and plants 431

Behaviour of plant protection chemicals in soil 436

Besin listing 473

Biochemical process of asymbiotic nitrogen fixation 211

Biochemical process of nitrification 337

Biochemistry of inorganic sulphur reduction 354

Biochemistry of nitrogen fixation by *Azolla-anabaena* symbiosis 223, 225

Biochemistry of nitrogen fixation by non-legume crops 223

Biochemistry of oxidation of inorganic sulphur 552

Biochemistry of phosphorus mineralization 347

Biochemistry of sulphur mineralization 349

Biochemistry of symbiotic nitrogen fixation 220

Biological and biochemical properties of soil 186

Biological methods of soil fertility evaluation 298

Biological nitrogen fixation 186, 187, 204, 205, 206, 207

Biological weathering 1, 17, 228, 254

Biologically active chemicals in soil 201

Black soils 81, 83, 85, 86, 87, 151, 180, 280, 383, 414, 467, 468, 470

Bokashi 491

Boron nutrition of plants 310

Branches of soil science 13, 14

Buffer curves 175

Buffering agents 174

Buffering in soils 173
Bunding 25, 131, 466, 467 468, 483, 502

C

Calcium nutrition of plants 306
Calculation of percent pore space 121
Calculation of quantities of fertilizers for fertilizer mixture 325
Calculation of soil weight 118
Carbon cycle 331
Cation exchange capacity of soils 180
Causes of aridity 448
CEC determination 181
Characteristics of clay minerals 227, 271
Characteristics of solar radiant energy 456
Characteristics of water logging 444
Chelates 271, 309, 323, 358, 432, 519
Chemical composition of soils 150
Chemical denitrification 340, 342
Chemical potentials in soils 152
Chemical properties of minerals 239
Chemical properties of soil 37, 50, 58, 150
Chemical rocks 232
Chemical weathering 247, 252, 253, 254, 255
Chlorine nutrition of plants 311
Chlorite 6, 31, 180, 234, 235, 241, 242, 250, 263, 267, 268, 269, 270, 272, 279, 280, 304, 307
Class of soil structure 104, 107
Classification of gully 386
Classification of parent material 17, 18
Classification of soil 49, 66, 100, 188, 379
Classification of soil according to WRB 65
Classification of soil erosion 379
Classification of soil organisms 188
Clay minerals in different soil orders 279
Clay minerals in Indian soils 280
Climate 1, 3 4, 13, 14, 21, 22, 27, 28, 31, 32, 33, 35, 39, 50, 52, 59, 61, 63, 67, 69, 71, 73, 75, 76, 77, 78, 80, 82, 84, 85, 87, 88, 89, 91, 93, 94, 100, 111, 126, 128, 130, 143, 144, 150, 204, 222, 250, 255, 270, 279, 280, 282, 286, 311, 312, 327, 329, 337, 339, 365, 369, 379, 386, 391, 392, 397, 408, 409, 411, 415, 418, 425, 440, 444, 445, 447, 448, 471, 475, 477, 489
Coastal erosion 280, 391, 392
Collection of soil samples 295, 296
Components and phases of soil 5
Components of hydrological cycle related to soil erosion 393
Composition of soil air 112
Composition of soil organic matter 359
Compost 111, 203, 293, 300, 312, 313, 328, 357, 367, 368, 416, 427, 465, 485, 486, 487, 491, 497, 498, 504, 508, 511
Compost making on land surface 514
Compost tea 487
Composting 485, 486, 511, 512, 513, 514
Conservation agriculture 485, 487, 488
Constants and units to denote energy state of water 142
Contour dam 472
Control and management of soil temperature 130
Copper nutrition of plants 309
Cover cropping 485, 489
Cross inoculation crops 212
Cross inoculation group and bacteria legume association 213
Crustal segments of India 79
Crystal system 236
Cultural practices 339, 401, 413, 445, 473, 481, 482, 494

D

Data structures in GIS 463
Decomposition of individual organic compounds 361
Decomposition of soil organic matter 360, 369
Deep cultivation 64, 520
Deep ripping 519, 520
Deficiency symptoms of nutrients on plants 286, 294
Denitrification 203, 300, 319, 328, 334, 338, 339, 340
Depletion of organic matter in soil 416
Desert and arid soils 81, 87
Determination of pH 172
Determination of total quantity of nutrients 295
Development of soil salinity 418, 419
Diagnostic horizons 52, 54, 55, 59, 63, 64, 65
Diffuse double layer 155, 156, 276, 277
Ditches 223, 224, 407, 466, 467468, 474, 511, 512, 514, 527
Diversion channel 474

Donan or membrane potential 155
Drainage 6, 7, 10, 111, 12, 19, 25, 27, 28, 32, 33, 34, 36, 50, 61, 67, 73, 83, 86, 89, 92, 93, 102, 130, 131, 141, 143, 144, 179, 196, 202, 267, 306, 307, 333, 338, 339, 340, 341, 376, 385, 386, 387, 396, 399, 418, 421, 445, 452, 466, 467, 469, 471, 473, 474, 492, 493, 494, 502, 503, 523, 525, 526, 530, 531, 538, 540
Drainage channels 466467, 473, 474, 494, 503

E
Earth 1, 2, 3, 4, 13, 14, 30, 43, 45, 59, 65, 72, 79, 80, 127, 128, 135, 203, 227, 228, 230, 231, 233, 239, 240, 241, 242, 246, 250, 256, 279, 330, 331, 348, 373, 374, 379, 393, 424, 425, 454, 455, 456, 457m, 458, 459, 460, 461, 484, 531
Ectomycorrhiza 192, 193
Effect of soil erosion 375, 376, 378,
Effects of abandoned mines 452
Effects of landslides 451
Electromagnetic radiation 455, 456, 458, 459
Element of World Reference Base (WRB) 63
End products of soil organic matter decomposition 364
Endopedons 54, 55
Energy concept of soil water 132
Entry of nutrients ions in roots 291
Epipedons 54, 55
Equipments and instruments used in soil survey 37
Erosion control in gully or ravine 482
Essentiality of plant nutrients 283
Estimation of soil loss by wind 408
Exposed surfaces or broken edges 274, 275

F
Factors affecting ammonia volatilization 342
Factors affecting amount of nutrients in plants 288
Factors affecting Anion exchange capacity 185
Factors affecting asymbiotic nitrogen fixation 209
Factors affecting bulk density of soil 118
Factors affecting CEC 188
Factors affecting denitrification 339
Factors affecting growth and nitrogen fixation by azolla 226
Factors affecting landslides 450
Factors affecting nitrification 336
Factors affecting nutrients absorption by plants from soil 292
Factors affecting organic matter decomposition 365
Factors affecting phosphorus mineralization 346
Factors affecting soil compaction 442
Factors affecting soil organisms 201
Factors affecting sulphur mineralization 350
Factors affecting symbiotic nitrogen fixation 217
Factors affecting temperature of soil 126
Factors affecting the rate of weathering 225
Factors affecting thermal conductivity of soil 125
Factors influencing fertilizer placement 318
Factors influencing wind erosion 411
Factors of soil formation 17, 20, 21, 79
Family 52, 53, 61, 63, 67, 192, 211, 212, 221, 222, 287288, 289, 335, 353, 420
Fertilizer recommendation 298
Fingers feel method to determine soil texture 103
Flow of nutrients from soil to plant roots 290
Flow of water and soil in rills 384
Foliated rocks 234
Forest and hill/mountain soils 90
Formation and development of soil structure 108
Formation of sodic soils 421
Fragmental rocks 232
Functions of plant nutrients in plants 285
Fungi 8, 12, 13, 23, 188, 189, 190, 191192, 193, 195, 196, 197, 198, 202, 203, 208, 222, 299, 309, 335, 346, 347, 348, 353, 360, 361, 362, 363, 364, 365, 367, 432, 434, 438, 487, 519

G
Genesis of soil 17, 110
Geography and cartography of soil 71
Geoinformatics 461
Geological and mineralogical properties of

soil 227
GIS data models 461, 462
Glacier erosion 392
Gobar gas or biogas 509
Gobar gas plant 509
Grade of soil structure 105, 107, 108
Grassed waterways 474, 475
Gravitational water 140, 141, 143
Great group 36, 52, 53, 54, 59, 60, 61
Ground water 2, 9, 32, 42, 56, 136, 141, 142, 143, 176, 338, 418, 41944, 533
Groups of silicate clays 265
Growth of plants in saline soils 420
Growth of plants in sodic soils or soil sodicity 422
Gully erosion 380, 383, 385, 388, 389, 414, 482
Gypsum 33, 44, 46, 57, 87, 111, 112, 179, 219, 232, 233
Gypsum requirement 498, 499, 500, 503

H

Heat balance of earth and soil 126
Heavy metals 162, 176, 429, 430, 431, 435, 452, 517, 518
Hill / mountain soil 92
History of WRB 62

I

Identification of minerals 236, 237
Identification of soil colour 144
Igneous rocks 3, 8895, 230, 231, 241, 243, 244, 245, 246, 250, 311
Importance and specialties of WRB 63
Importance of cation exchange 179
Importance of Geographical Information System (GIS) in soils 460
Importance of remote sensing in soils 453
Importance of soil colloids 259
Importance of soil texture 101
Influence of organic matter on soil properties 357
Inorganic colloids of soil 260
Integrated plant nutrient management system (IPNMS) 327
Integrated watershed management programme (IWMP) 540
Interaction of EMR with the earth's surface 457
Ion exchange equations 176
Iron nutrition of plants 309
Isomorphous substitution 274, 269, 272, 273, 274, 276,

K

Kaolinite group 263, 265

L

Lacustrine 19
Land capability classification 34, 49, 67, 68, 69, 528, 529, 540
Land degradation 372, 373, 460, 538
Land suitability classification 49, 69, 70
Landscape 3, 4, 37, 47, 73
Landslides 387, 393, 449, 450, 451, 528, 530
Laterite and lateritic soils 76, 81, 88
Lime potential 157
Lime requirement to reclaim soil acidity 505
Loss of plants nutrients 417

M

Magnesium nutrition of plants 307
Main characteristics of master horizons 42, 44
Maintenance of organic matter in soil 368, 369
Management of sodic soils 504
Management of soil structure 111
Manganese nutrition of plants 309
Manure of cattle dung 508
Marine 19, 232, 233, 311
Measurement of wind erosion 407
Measuring of water erosion 396
Mechanism of buffering 174
Mechanism of wind erosion 404
Metamorphic rocks 91, 229, 231, 234, 235, 241, 243, 244, 245, 246, 248, 249
Methods of fertilizer application 313, 524
Methods of pH determination 172
Methods of wind erosion control 475
Mineral matter 5, 6, 7, 8, 17, 18, 23, 29, 118, 119, 120, 124, 143, 144, 152, 187, 190, 228, 250, 288, 357, 359, 360
Mineral nutrition of crop plants 300
Minerals 1, 3, 5, 6, 13, 17, 18, 23, 26, 27, 29, 31, 44, 46, 57, 76, 84, 85, 86, 87, 88, 93, 94, 109, 120, 124, 125, 133, 144, 150, 151, 163, 164, 176, 179, 180, 185, 188, 189,

203, 227, 228, 229, 231, 233, 234, 235, 236, 236-242, 245, 246, 247, 248, 249, 250, 251, 252, 253, 254, 255, 256, 257, 266, 267, 268, 269, 270, 271, 272, 273, 274, 279, 280, 282, 300, 303, 304, 306, 307, 309, 348, 349, 356, 358, 386, 397, 414, 417, 418,. 421, 425, 432, 446, 417, 425, 531, 532
Minerals present in fine fraction of soils 242, 249
Minerals present in soils 242, , 414
Mixed and interstratified minerals 270
Mixed farming with trees and livestock 489
Mixed layered minerals 242, 250
Molarity 182
Molybdenum nutrition of plants 310
Morphology of nodules 216
Morphology of soil 42, 47
Movement or flow of water in soil 135
Movement or flow of water vapour in soil 139
Mycorrhiza 192, 203, 347

N

NADEP method of compost 513
Necessities or requirements of biological Nitrogen fixation 204
Nematode 195, 196
New life for saline soil 495
Nitrification inhibition 338
Nitrogen cycle 25, 332, 333, 334, 335, 339
Nitrogen fixation in soil 203
Nodulation on stems of legume crops 221
Normal ranges of soil 165
Numerical classification of soil 66
Numerical on pH 167

O

Objectives of GIS 461
Objectives of soil survey 34, 35
Occurrence of minerals 227, 241
Optical properties of minerals 228, 239
Orbits 459
Organic colloids of soils 277, 288
Organic matter 5, 6, 7, 8, 11, 12, 13, 15, 18, 22, 23, 24, 27, 28, 29, 30, 31, 39, 41, 42, 44, 45, 53, 54, 55, 59, 64, 83, 85, 86, 87, 89, 90, 91, 92, 94, 103, 104, 106, 108, 109, 110, 111, 112, 113, 117, 118, 119, 120, 124, 125, 143, 144, 149, 150, 151, 152, 162, 163, 174, 176, 179, 180, 181, 187, 189, 190, 192, 193, 193, 194, 195, 196, 198, 199, 200, 201, 203, 208, 209, 210, 219, 221, 226, 232, 250, 254, 255, 271, 281, 288, 289, 300, 301, 303, 304, 305, 308, 309, 309, 310, 311, 327, 333, 332, 339, 340, 342, 345, 348, 350, 351, 354, 356, 357, 358, 359, 360, 361, 364, 365, 366, 367, 368, 369, 370, 371, 374, 376, 377, 378, 395, 399, 400, 412, 414, 415, 416, 424, 425, 430, 431, 432, 433, 436, 437, 438, 444, 445, 447, 453, 465, 479, 485, 486, 487, 488, 491, 504, 507, 517, 518, 519, 520, 521, 527, 533
Organic parent material 18
Organisms 1, 3, 4, 5, 6, 8, 10, 12, 13, 14, 21, 22, 23, 27, 29, 106, 112, 123, 163, 174, 186, 187, 188, 189, 191, 194, 195, 196, 98, 199, 201, 202, 203, 204, 207, 211, 223, 224, 227, 231, 232, 250, 254, 255, 299, 331, 332, 334, 344, 356, 357, 359, 361, 362, 363, 365, 416, 434, 436,485, 486, 487, 518
Organisms in soil 10, 186, 187, 254, , 365, 434
Organisms involved in nitrogen fixation 207
Organogenic rocks 232
Origin or source of negative charge 274
Other measures of wind erosion control 482
Other properties of clays 278
Oxidation - reduction potential 153
Oxidation - Reduction Potential (Redox potential) 116
Oxidation of inorganic sulphur 351, 352, 353
Oxygen diffusion on rate (ODR) 115

P

Parent material 1, 4, 5, 13, , 14, 17, 18, 19, 20, 21, 22, 25, 27, 28, 29, 31, 32, 33, 42, 43, 44, 46, 47, 50, 52, 64, 73, 75, 80, 81, 82, 83, 84, 85, 87, 88, 89, 91, 92, 93, 94, 95, 144, 150, 189, 229, 231, 250, 267, 279, 280, 385, 418, 423, 444, 445
Parent material transported by gravity 20
Parent material transported by snow 20
Parent material transported by water 19
Parent material transported by wind 19
Particle density 11, 117, 119, 120, 121, 122,

391
Passive absorption 291
Pasture development 475, 478, 479, 484, 527, 530
Peaty, marshy and skeletal soils 94
Phases of soil 5, 8, 152
Phosphate potential 157
Phosphate retention 302
Phosphorus cycle in soil 344
Phosphorus immobilization 344
Phosphorus mineralization 344, 346, 347
Phosphorus nutrition of plants 301
Physical properties of minerals 237
Physical properties of soil 10, 99, 465, 476
Physical weathering 251, 255, 256
Physiographical regions or divisions of India 79
Physiographical zones or regions of India 79
Plant analysis 294, 295
Plant and algae symbiosis 223
Plant tissue analysis 295
Plantation 475, 476, 485, 533, 535
Plants for shelterbelts and wind breaks 477
Platforms for remote sensing 459
Pollution of soil by fungicides 434
Pollution of soil by rodenticides 436
Pollution of soil by weedicides 436
Potassium cycle in soil 348
Potassium fixation in soil 305
Potassium nutrition of plants 304
Potassium potential 157
Pradhan Mantri Krishi Sinchayee Yojana 540
Prediction of soil loss by water 398
Preparation of fertilizer mixture with two nutrients in one fertilizer 326
Preparation of field for leaching 492
Preparation of multinutrient fertilizer mixture 324
Preparation of soil maps 72
Prevention of acidity of rain water 515
Procedures of IPNMS 329
Process of nitrogen fixation by symbiotic bacteria in legume 213
Process of soil formation 29, 30, 31
Processes and cycles of molecules and elements in soil 330
Processes of soil organic matter decomposition 360
Programme of restoration of soil from sodicity 502
Properties and processes of soils 99
Protozoa 12, 23, 189, 195, 203, 219, 334, 487, 518,
Puddling 145, 522, 523
Puertorican terrace 470
Purpose of soil classification 49
Pyrite 9, 233, 241, 242, 257, 352, 497, 500, 502, 503

R

Raindrop erosion 384
Ravine reclamation programme 483
Reaction of phosphorus in acidic soil 302
Reclamation technology for abandoned mines 532
Red soils 18, 81, 84, 86, 87, 143, 144, 280
Reduced or zero tillage 485, 488
Reduction of inorganic sulphur 354
Relief and topography 25, 255
Remote sensing system 454
Renewal of air in soil 114
Restoration and management of polluted and contaminated soils 507
Restoration and management of soil from compaction, sealing and crusting 519
Restoration and management of soil from depletion of organic matter 485
Restoration and management of soil from salinity, sodicity and acidity 491
Restoration and management of soil from water and wind erosion 463
Retention of water in soil 132, 140
Revised universal soil loss equation (RUSLE) 402
Revised wind erosion equation (RWEQ) 410
Right time of fertilizer application 312
Rill erosion 380383, 384, 397, 402,412, 413
Rock 18, 22, 28, 29, 31, 39, 41, 42, 44, 45, 46, 76, 80, 86, 92, 99, 150, 193, 194, 228, 229, 230, 231, 232, 233, 235, 240, 241, 243, 246, 250, 251, 252, 256, 257, 267, 314, 325, 376, 393, 403, 407, 417, 441, 449, 451, 452521, 525, 528, 531
Runoff farming 528

S
Saline and alkali/ salt affected soils 81, 93
Sedentary parent material 18
Sedimentary rocks 3, 18, 88, 229, 231, , 232, 233, 241, 247, 248, 249, 267
Sensors used in remote sensing 458
Sesquioxide or hydrous oxide clays 270
Sheet erosion 15, 69, 380, , 383, 384, 392, 399, 402
Shelter belt 476, 477
Silicate clays 44, 46, 108, 161, 162, 163, 260, 265, 273, 274, 431
Silicon tetrahedron 260, 261, 263
Smectite group 263, 265
Soil - plant relationship 10
Soil acidity 179, 422, 423, 424, 425, 432, 504, 505, 506
Soil aeration 11, 13, 112, 113, 116, 141, 202, 336, 339, 340, 346, 474,
Soil aeration and plant growth 113
Soil air 7, 8, 99, 112, 113, 14, 115, 120, 187, 332
Soil alkalinity and soil sodicity 420
Soil analysis 294, 295, 296
Soil as a natural body 1, 14
Soil as a natural medium for plant growth 8
Soil colloids 22, 108, 148, 174, 178, 179, 180, 181, 185, 194, 227, 257, 259, 260, 290, 309, 310, 313, 335, 338, 343, 421, 432, 436, 437, 439
Soil colour 12, 42, 87, 99, 127, 143, 144, 145
Soil compaction 12, 118, 196, 414, 441, 442, 519, 520
Soil consistence 11, 99, 145, 279,
Soil crusting 90, 443, 521
Soil degradation 15, 371, 372, 417, 442
Soil density 99, 117
Soil erosion by water and wind 374
Soil erosion control 463, 470, 474, 476, 479, 481, 489
Soil fertility and integrated plant nutrition 282
Soil fertility evaluation 294, 298
Soil macrofauna 199
Soil magna fauna 195, 200
Soil map preparation 38, 40, 41
Soil mapping 37, 41, 69, 71, 73, 460
Soil mesofauna 196
Soil microfauna 196
Soil micromorphology 42, 47
Soil monolith 42, 48
Soil organic matter 6, 7, 119, 120, 193, 199, 201, 217, 219, 221, 301, 327, 332, 357, 358, 359, 360, 361, 364, 365, 369, 370, 371, 374, 416, 430, 453, 479, 485, 491
Soil organism 6, 8, 12, 14, 29, 115, 187, 188, 189, 194, 201, 202, 203, 357, 361, 365, 434, 436, 438, 487, 518
Soil phase 37, 41, 62, 71
Soil pollution and soil contamination 425
Soil pore space and porosity 120
Soil profile, soil horizons and layer 42, 43
Soil reaction 93, 164, 209, 210, 217, 219, 336, 337, 350, 445, 525
Soil salinity 418, 419, 420, 492, 493, 494, 496
Soil salinity, sodicity, acidity 418
Soil sealing 442, 521
Soil series 36, 37, 40, 41, 46, 61, 71, 430
Soil sodicity 420, 422, 497, 501, 502
Soil structure 10, 11, 29, 50, 99, 104, 105, 106, 107, 108, 109, 110, 111, 118, 145, 188, 192, 194, 196, 278, 357, 380, 400, 414, 441, 464, 465, 466, 505, 520, 521
Soil survey 13, 33, 34, 35, 36, 37, 38, 39, 40, 41, 42, 47, 51, 54, 67, 72, 74, 75, 78, 95, 103, 502, 529
Soil survey interpretation 33, 40, 42
Soil survey methodology 33, 38
Soil survey report presentation 42
Soil survey techniques 37, 42
Soil taxonomy 13, 42, 49, 51, 52, 54, 61, 62, 68, 74, 75, 81, 84, 85, 88, 89, 92, 93, 279, 280
Soil temperature 12, 52, 59, 61, 99, 114, 123, 125, 126, 127, 128, 129, 130, 131, 202, 209, 211, 300, 301, 336, 358, 434, 436, 437, 465
Soil temperature fluctuations 129
Soil texture 10, 27, 32, 50, 99, 100, 101, 102, 103, 104, 107, 118, 145, 180, 181, 217, 220, 260, 278, 300, 301, 306, 321, 342, 375, 377, 389, 399, 445, 493
Soil texture and plant growth 104
Soil vulnerability 15, 371, 374, 441, 443, 446, 453
Soil vulnerability, soil degradation, soil vulnerability index 371

Soil water 7, 11, 99, 112, 120, 127, 131, 132, 133, 134, 135, 140, 143, 153, 157, 173, 187, 210, 220, 290, 309, 395, 422, 442, 443, 444, 496, 520, 525
Soil water potential 133, 134
Soils of india 18, 76, 80, 81, 85, 91, 95, 96, 106, 150, 400, 418, 445, 470
Soils of indian states 95, 96
Specialties and drawback of soil taxonomy 51
Spectral signature 457, 458
Stabilization of sand dunes 479
Stages of action in remote sensing 454
Steps in watershed management 540
Stream erosion 380, 389
Strip cropping 475, 480, 481, 527
Sub group 52, 53, 61, 82, 84, 86
Sub order 52, 58, 59, 61
Subordinate distinction of master horizons 45
Suitable time for leaching 492
Sulphur cycle in soil 349
Sulphur immobilization 351
Sulphur nutrition of plants 308
Sulphuric acid 94, 257, 352, 353, 424, 428, 497, 500, 501, 502
Survey of soil 33
Symbiotic nitrogen fixation 207, 209, 211, 217, 219, 220, 222, 300
Symbiotic nitrogen fixation by legume crops 211
Symbiotic nitrogen fixation in non-legume plants 222
System of soil classification 13, 50, 51, 54, 74

T

Tarai and sub-montane soils 27, 81, 89, 90, 151, 280
Terraces 25, 394, 402, 466, 468, 469, 470, 471, 472, 475, 480, 484
Thermal capacity 123
Thermal diffusivity 126
Thermal emission of radiation 456
Thermal transmission in soil 124
Time – Extensive references across manypages, including: 1, 4, 8, 9, ..., 537
Time of fertilizers and manure application 311
Transport of eroded soil in water current 389
Transported parent material 18, 27, 150
Types of adsorption 158
Types of biological nitrogen fixation 207
Types of ion exchange 178
Types of nodules 216
Types of remote sensing 457
Types of remote sensors 458
Types of soil colloids 260
Types of soil erosion 379, 383
Types of soil forming processes 17, 30
Types of soil structure 104, 105
Types of soil survey 33, 35, 41
Types of soils of Asia 75
Types of soils of World 73
Types of soils of World, Asia and India 73
Types of terraces 469
Types of watersheds 538
Types of weathering 251

U

Unfoliated rocks 234
Units of cation exchange capacity 181
Universal soil classification system and unified soil classification system 67
Utility of soil survey 33, 34

V

Vermiculite 180, 242, 249, 255, 257, 267, 268, 269, 270, 272, 279, 300, 301, 304, 305, 482, 527
Vertical sub divisions and discontinuities 42
Vesicular arbuscular mycorrhizae 192
Virus 189, 194, 371
Volatilization of ammonia 342, 343
Volumetric heat capacity of soil 123

W

Water erosion 358, 375, 379, 388, 396, 403, 407, 411, 437, 464
Water erosion control 464
Water harvesting 35, 526, 528
Water logging 11, 113, 114, 202, 220, 444, 472, 522, 525
Water movement or flow under stratified soils 139
Water movement or flow under unsaturated condition 138
Water movement under saturated condition 136

Watershed management 34, 453, 537, 538, 540, 541
Watershed management programme 540, 541
Weathering of rocks and minerals 1, 23, 29, 179, 188, 227, 250
Weathering products 6, 31, 245, 257, 421
Weathering sequence 227, 256, 257, 265
Weathering sequence and products 256
Wind barrier 408, 476
Wind breaks 477, 480Wind erosion 371374375376403404404, 405, 407, 408, 409, 410, 411, 414, 453, 463, 464, 466, 476, 478, 479, 481, 482
Wind erosion equation (WEQ) 408
Wind velocity profile 403
World Reference Base for Soil Resources (WRB) 62

Z

Zeta potential 155, 156
Zinc nutrition of plants 308